Plastisphere

The plastisphere provides a unique man-made habitat for organisms and microbial communities. Investigating the interactions and functions of these organisms helps us understand how fouling communities can settle on these novel platforms and be moved beyond their natural ranges as well as their contribution to nutrient cycling and biogeochemical processes. This book examines the plastisphere and reveals a secret world of microbes that have adapted to live on plastic surfaces. It explains the ecological effects of plastic pollution, human health concerns, microbial ecology, and mitigation techniques and advocates sustainable solutions. Readers will dive into this enthralling ecosystem rich with microscopic life and explore its mysteries in this thought-provoking book.

Features:

- Presents recent insights into the fundamentals of the plastisphere, with a specific focus on environmental interactions.
- Introduces the work of global experts and their ongoing studies, creative approaches, and new discoveries.
- Highlights the macro and molecular interactions of environmental domains with the plastisphere.
- Explores the notion of bioinvasion that can threaten the health and productivity of our ecosystems.
- Includes detailed multidisciplinary data on the behavior of plastic materials and environmental microbiome.
- Explains the main factors causing ecosystems to evolve and to live in human-made plastic environments.
- Helps readers from various backgrounds to gain a thorough understanding of the topic of the plastisphere.

This book is an insightful reference for researchers, academics, students, and professionals in Environmental Science, Microbiology, Environmental Engineering, Oceanography, and Material Science.

Plastisphere

The Ecosystem of Plastics

Edited by

Meththika Vithanage, Adil Bakir, and Nicole R. Posth

CRC Press
Taylor & Francis Group
Boca Raton London New York

CRC Press is an imprint of the
Taylor & Francis Group, an **informa** business

Designed cover image: 1. Photo taken by Nicole R. Posth, 2. Photo taken by Peter Barry and image courtesy of Cefas © Crown copyright, 3. Image taken by Daniel Jolly and image courtesy of Cefas © Crown copyright, 4. Photo taken by Umberto Binetti and image courtesy of Cefas © Crown copyright.

First edition published 2025
by CRC Press
2385 NW Executive Center Drive, Suite 320, Boca Raton FL 33431

and by CRC Press
4 Park Square, Milton Park, Abingdon, Oxon, OX14 4RN

CRC Press is an imprint of Taylor & Francis Group, LLC

ISBN: 978-1-032-71780-7 (hbk)
ISBN: 978-1-032-71782-1 (pbk)
ISBN: 978-1-032-71781-4 (ebk)

DOI: 10.1201/9781032717814

Typeset in Times
by codeMantra

Contents

Preface

Plastic offers a new niche for organisms, the plastisphere. The term "plastisphere" was first coined by Tracey Mincer, Linda Amaral-Zettler, and Erik Zettler in 2013 to describe the diverse microbial life found on marine plastic debris. This plastic-bound microbial community was diverse, and distinct from the surrounding water and sediment, building an understanding that the plastisphere represents a novel ecological niche formed by the introduction of plastic debris of various sizes from micro to macro into the environment.

In the last decade, continuing research has broadened our understanding of the plastisphere as an ecosystem, with themes of research focused on community composition, functional potential, substrate specificity, environmental control, and role in food web dynamics, among others. The application of the term in the literature is evolving, in line with our understanding of the plastisphere's complexity, but also of our advancements in tracking plastics as a pollutant across temporal and spatial scales.

With this evolution of knowledge and perspective in mind, we have invited contributions representing global input from researchers exploring the plastisphere from an ecological perspective. The chapters presented here explore new research, methodology, and review of the aquatic, atmospheric, and terrestrial plastisphere, with a view to the cycling, transport, deposition, and degradation of plastics in these ecosystems. In addition, the chapters herein highlight the implications of this new ecological niche for human and environmental health, food security, contaminant remediation, and sustainable ecosystem management.

We sincerely thank the authors for their contributions. It is our hope that this compilation enhances understanding of the state of the field and inspires ongoing work to define the impact of plastic pollution and develop sustainable future practices.

Nicole R. Posth

Meththika Vithanage

Adil Bakir

About the Editors

Prof. Meththika Vithanage is a professor in the Deological Survey of Denmark and Greenland and the founding director of the Ecosphere Resilience Research Centre at the University of Sri Jayewardenepura, Sri Lanka. Prof. Vithanage also serves as an Adjunct Professor at the Institute of Agriculture, the University of Western Australia, and the University of Petroleum and Energy Studies, India, with previous affiliations at the National Institute of Fundamental Studies, Sri Lanka. With an interdisciplinary focus spanning water chemistry, environmental remediation, microplastics, and waste biomass conversion, she has been recognized as a highly cited researcher by Clarivate. Her accolades include the prestigious 2026 TWAS Award in Earth, Climate and Environmental Sciences for her groundbreaking research on biochar, underscoring her commitment to advancing sustainable solutions. Prof. Vithanage's publication record includes over 275 SCI-indexed journal articles, 60 book chapters, and 10 co-edited books, making her a prolific contributor to the global scientific community, and her work has had a profound impact on the fields of environmental sustainability and resilience. Beyond academia, she has actively engaged with the public through more than 100 newspaper articles, furthering scientific literacy and advocacy.

Adil Bakir is a principal scientist working on microplastics at the Centre for Environment, Fisheries and Aquaculture Science (Cefas) in the UK. Adil has been involved in the field of microplastic research for about 10 years starting as a postdoctoral researcher for a Department for Environment, Food & Rural Affairs (Defra) funded project at Plymouth University in the UK investigating whether microplastics pose a physical and/or chemical threat in the marine environment. In 2014, he joined the School of Earth & Environmental Sciences at the University of Portsmouth (UK) as a senior scientific officer and was working closely with industry and environmental groups on projects ranging from high-frequency nutrient monitoring to the development of new applications in passive sampling. Adil joined the marine litter team at Cefas in 2018 and is currently leading the Cefas microplastic laboratory aiming at developing faster and cheaper techniques for the monitoring of microplastics in environmental matrices.

Nicole R. Posth is a geomicrobiologist and sediment biogeochemist studying microbial element cycling, biomineralization, and microbe-surface interactions. Bridging field and laboratory experiments, her projects range from interpreting the role of microorganisms in the evolution of life on Earth to the impact of microbial life in driving biogeochemical cycles in the Anthropocene.

Nicole is an Associate Professor of Geomicrobiology in the Sedimentary Systems Research Group at the Department of Geosciences & Natural Resource Management (IGN), University of Copenhagen. She is the Head of the Geomicrobiology & Sediment Biogeochemistry Laboratory focused on the interactions between microorganisms, minerals, and plastics mediated within biogeochemical cycles in water and sediment at the macro-, micro-, and nanoscale.

List of Contributors

Praveen Abhishek
Ecosphere Resilience Research Center
Faculty of Applied Sciences
University of Sri Jayewardenepura
Nugegoda, Sri Lanka

Frederick O. Ahmadu
Afe Babalola University
Ado-Ekiti, Nigeria

Rana Al-Jaibachi
Department of Biosciences
University of Sheffield
Sheffield, United Kingdom

Ahamed Ashiq
Beaty Water Research Centre
Department of Civil Engineering
Queen's University
Kingston, Canada

Thilakshani Atugoda
Faculty of Environment
Science and Economy
University of Exeter
Exeter, England

Craig Baker-Austin
Centre for Environment, Fisheries and Aquaculture Science (Cefas)
The Nothe
Weymouth, United Kingdom

Tharindi Bandara
Molecular Microbiology and Human Diseases Research Programme
National Institute of Fundamental Studies
Kandy, Sri Lanka

Peter J. Barry
Centre for Environment, Fisheries and Aquaculture Science (Cefas)
Lowestoft, United Kingdom

Nilendu Basak
Department of Microbiology
University of Kalyani
Kalyani, Nadia, West Bengal, India

Bijay Kumar Behera
Biotechnology Laboratory
ICAR-Central Inland Fisheries Research Institute
Kolkata, West Bengal, India
and
Rani Lakshmi Bai Central Agricultural University
Jhansi, Uttar Pradesh, India

Idil Can-Tuncelli
Faculty of Aquatic Sciences
Department of Seafood Processing Technology
Istanbul University
Istanbul, Türkiye

Ankita Chatterjee
Department of Biotechnology
School of Applied Sciences
REVA University
Bangalore, Karnataka, India

Manish Chaudhary
School of Environment & Natural Resources
Doon University
Dehradun, Uttarakhand, India

Atif Aziz Chowdhury
Faculty of Agricultural, Environmental and Food Sciences
Bolzano, Italy

Amelia Curd
Ifremer
DYNECO
Plouzané, France

Heinrich T.J. Dahms
Department of Zoology
Kingsway Campus, University of Johannesburg
Auckland Park, South Africa
and
Institute for the Alpine Environment
Eurac Research
Bolzano, Italy

Simone Dahms-Verster
School of Geography, Archaeology and Environmental Studies
University of the Witwatersrand
Wits, South Africa

Tatenda Dalu
Aquatic Systems Research Group
School of Biology and Environmental Sciences
University of Mpumalanga
Nelspruit, South Africa

Sujata Dey
Biotechnology Laboratory
ICAR-Central Inland Fisheries Research Institute
Kolkata, West Bengal, India

Maya Shakunt Dodhia
Geology-Sedimentary Systems
Department of Geosciences & Natural Resource Management (IGN)
University of Copenhagen
Copenhagen, Denmark

Gabin Droual
Ifremer
DYNECO
Plouzané, France
and
DECOD (Ecosystem Dynamics and Sustainability)
IFREMER, INRAE, Institut Agro—Agrocampus Ouest
Nantes, France

Randima Edussuriya
Ecosphere Resilience Research Center
Faculty of Applied Sciences
University of Sri Jayewardenepura
Nugegoda, Sri Lanka

Nuray Erkan
Faculty of Aquatic Sciences
Department of Food Safety
Istanbul University
Istanbul, Türkiye

Yves Filion
Beaty Water Research Centre
Department of Civil Engineering
Queen's University
Union Street, Kingston, Canada

Omowunmi H. Fred-Ahmadu
Afe Babalola University
Ado-Ekiti, Nigeria

Koushik Ghosh
Aquaculture Laboratory
Department of Zoology
The University of Burdwan
Golapbag, West Bengal, India

Elena Gorokhova
Department of Environmental Science
Stockholm University
Stockholm, Sweden

Richard Greenfield
Department of Zoology
Kingsway Campus, University of Johannesburg
Auckland Park, South Africa

Matthew Elmo Gruwell
Aquatic Systems Research Group
School of Biology and Environmental Sciences
University of Mpumalanga
Nelspruit, South Africa

Ishani Gunasekara
Ecosphere Resilience Research Center
Faculty of Applied Sciences
University of Sri Jayewardenepura
Nugegoda, Sri Lanka

Shashini Gunathilaka
Molecular Microbiology and Human Diseases Research Programme
National Institute of Fundamental Studies
Kandy, Sri Lanka

Saghar Hendiani
Globe Institute
University of Copenhagen
Copenhagen, Denmark

Irushi Hettige
Ecosphere Resilience Research Center
Faculty of Applied Sciences
University of Sri Jayewardenepura
Nugegoda, Sri Lanka

Jasintha Jayasanka
Faculty of Technology
Department of Biosystems Technology
University of Sri Jayewardenepura
Pitipana, Homagama, Sri Lanka

Keshani Jayathilaka
Ecosphere Resilience Research Center
Faculty of Applied Sciences
University of Sri Jayewardenepura
Nugegoda, Sri Lanka

Dinesh Karthiban
Department of Biotechnology
School of Applied Sciences
REVA University
Bangalore, Karnataka, India

Neelakanta Sarvashiva Kiran
Department of Biotechnology
School of Applied Sciences
REVA University
Bangalore, Karnataka, India

Ayushi Kothari
School of Environment & Natural Resources
Doon University
Dehradun, Uttarakhand, India

Hoi Shing Lo
Department of Environmental Science
Stockholm University
Stockholm, Sweden

Sumedha Madawala
Department of Botany
University of Peradeniya
Peradeniya, Sri Lanka

Dhammika N. Magana-Arachchi
Molecular Microbiology and Human Diseases Research Programme
National Institute of Fundamental Studies,
Kandy, Sri Lanka

Devrim Memis
Faculty of Aquatic Sciences
Department of Aquaculture
Istanbul University
Istanbul, Türkiye

Anjali Mishra
School of Environment & Natural Resources
Doon University
Dehradun, Uttarakhand, India

Morgan le Moigne
Ifremer
COAST
Concarneau, France

Pule Mpopetsi
Aquatic Systems Research Group
School of Biology and Environmental Sciences
University of Mpumalanga
Nelspruit, South Africa

Thomas Neu
Department of River Ecology
Helmholtz Centre for Environmental Research (UFZ)
Magdeburg, Germany

Kulathi Nishshankage
Ecosphere Resilience Research Center
Faculty of Applied Sciences
University of Sri Jayewardenepura
Nugegoda, Sri Lanka

Collins Oduro
Department of Geography and Spatial Information Techniques
Ningbo University
Ningbo, China

Oluwafemi T. Ogunmodede
Afe Babalola University
Ado-Ekiti, Nigeria

W.M.W. Perera
School of Applied Sciences
University of West England (UWE)
Bristol, United Kingdom

Asal Peydaei
Geology-Sedimentary Systems
Department of Geosciences & Natural Resource Management (IGN)
University of Copenhagen
Copenhagen, Denmark

Nicole R. Posth
Geology-Sedimentary Systems
Department of Geosciences & Natural Resource Management (IGN)
University of Copenhagen
Copenhagen, Denmark

Kumudi Premarathna
Department of Botany
University of Peradeniya
Peradeniya, Sri Lanka

Ajaya Kumar Rout
Department of Aquaculture
Rani Lakshmi Bai Central Agricultural University
Jhansi, Uttar Pradesh, India

Karina Krarup Sand
Globe Institute
University of Copenhagen
Copenhagen, Denmark

Edmond Sanganyado
Department of Applied Sciences
Northumbria University
Newcastle Upon Tyne, United Kingdom

Debby Schmidt
Geology-Sedimentary Systems
Department of Geosciences & Natural Resource Management (IGN)
University of Copenhagen
Copenhagen, Denmark

Uvin Eksith Senadheera
Faculty of Technology
Department of Biosystems Technology
University of Sri Jayewardenepura
Pitipana, Homagama, Sri Lanka

Shako Senyolo
Department of Zoology
Kingsway Campus, University of Johannesburg
Auckland Park, South Africa

Madushika Sewwandi
Ecosphere Resilience Research Center
Faculty of Applied Sciences
University of Sri Jayewardenepura
Nugegoda, Sri Lanka

Conrad Sparks
Faculty of Applied Sciences, Centre for Sustainable Oceans
Cape Peninsula University of Technology
Cape Town, South Africa

Surindra Suthar
School of Environment & Natural Resources
Doon University
Dehradun, Uttarakhand, India

Franco Teixeira de Mello
Departamento de Ecología y Gestión Ambiental
Centro Universitario Regional Este, Universidad de la República–Uruguay
Maldonado CP, Uruguay

Gokhan Tuncelli
Faculty of Aquatic Sciences
Department of Aquaculture
Istanbul University
Istanbul, Türkiye

Meththika Vithanage
Ecosphere Resilience Research Center
Faculty of Applied Sciences
University of Sri Jayewardenepura
Nugegoda, Sri Lanka

David I. Walker
Centre for Environment, Fisheries and Aquaculture Science (Cefas)
The Nothe
Weymouth, United Kingdom

Manura Weerasinghe
Ecosphere Resilience Research Center
Faculty of Applied Sciences
University of Sri Jayewardenepura
Nugegoda, Sri Lanka

Naicheng Wu
Department of Geography and Spatial Information Techniques
Ningbo University
Ningbo, China

Chandrashekar Yashaswini
Department of Biotechnology
School of Applied Sciences, REVA University
Bangalore, Karnataka, India

Pinar Arslan Yüce
Faculty of Science
Çankırı Karatekin University
Department of Biology
Çankırı, Türkiye

1 The Global Microplastic Cycle in the Ecosphere

Pınar Arslan Yüce

1.1 INTRODUCTION

During and after the Industrial Age, urbanization, mechanization, industrialization and the human population increased rapidly. All of these have caused anthropogenic activities on the environment, thus causing stress effects on the biota (Kabir et al. 2023; Muruganandam et al. 2023).

In the 1960s, one of the important works drawing attention to these anthropogenic activities was published: *Silent Spring* by Rachel Carson. This work, which draws attention to the effects of chemicals resulting from industrialization on American bald eagles and their future generations, laid the foundations for the intensification of studies on the effects of chemicals on biota and abiotic environment compartments (Carson 1962). These chemicals were later called persistent organic pollutants (POPs). These compounds, which are resistant to chemical, photolytic and biological degradation in the environment, are being tried to be controlled by national and international regulations. The main characteristics of these compounds are that they remain in the environment for a long time and that most of them are lipophilic. The polycyclic aromatic hydrocarbons (PAHs), polychlorinated biphenyls (PCBs), polybrominated diphenyl ethers (PBDEs), and organochlorine compounds (OCs) are examples of POPs (Alharbi et al. 2018; Sheriff et al. 2022). They have the ability to be found and accumulate in abiotic compartments of environment such as soil (Reid et al. 2000; Wang et al. 2007; Klánová et al. 2008; Ren et al. 2018), water (Bizzotto et al. 2009; Bao et al. 2012; Duttagupta et al. 2020; Arslan and Ozeren 2022a; Şimşek and Bilgili 2022) and sediment (Reid et al. 2000; Yuan et al. 2023; Gómez-Gutiérrez et al. 2007; Klánová et al. 2008; Bizzotto et al. 2009; Sakan et al. 2017; Duttagupta et al. 2020; Arslan and Ozeren 2022a; Şimşek and Bilgili 2022). Apart from intensive studies on the abiotic part of the environment, there are also many studies examining the bioaccumulation characteristics of these chemicals and their effects on organisms, either collected from the natural environment or with experimental setups in the laboratory environment (Geyer et al. 2000; Goerke et al. 2004; Bizzotto et al. 2009; Herceg-Romanić et al. 2014; Sun et al. 2015; Chiesa et al. 2018; Arslan and Ozeren 2022b; Arslan et al. 2023; Boitsov et al. 2024; Chen et al. 2024; Hartmann et al. 2024; Pereira et al. 2024). The use, production and distribution of these substances, which have been intensively researched scientifically, are limited or prohibited by national and international regulations or conferences (UNEP 2009).

Approximately 10 years after the *Silent Spring*, studies were published on the presence of plastics carrying PCBs on the surface of the western Saragossa Sea (Carpenter and Smith 1972) and the southern coastal waters of New England (Carpenter et al. 1972). Carpenter and Smith (1972) detected many lump-shaped plastic particles and showed that these particles formed adhesion surfaces for the attachment of diatoms and hydroids. In this study, the authors predicted that we will face significant plastic pollution in water and soil as a result of increasing plastic production in the coming years and inadequate implementation of current waste disposal. In another study, Carpenter et al. (1972) showed that plastic beads have an average size of 0.5 mm, had different pigmentations, carried biofilm-forming bacteria on their surfaces, and absorbed PCBs in the seawater they were found in. They also predicted that vertebrates in this aquatic ecosystem consumed these plastics as food, which could lead to intestinal obstruction in small fish. With these studies, the existence of substances that have an important pollutant potential for the environment has been revealed.

DOI: 10.1201/9781032717814-1

Plastics emerged as substances resulting from the polymerization of monomers obtained mainly from petroleum or its derivatives. The term plastic is derived from the word "pliable" meaning "easily shaped" (Tan et al. 2024). The first plastic called *Bakelite* was introduced in 1907 (Cole et al. 2011), and with the development of technological equipment over the years, it has become possible to make plastics lighter, more inert and more corrosion resistant (Cole et al. 2011; Tan et al. 2024).

While the stagnation in global plastic production due to the COVID-19 epidemic reached 390.7 million tons in 2021 (Plastics Europe 2022), it reached 400.3 million tons in 2023 in the postpandemic period (Plastics Europe 2024). Plastics, which were initially produced only from fossil sources, are produced from fossil sources, recyclable sources and bio-based sources, according to 2021 data (Plastics Europe 2022). According to 2023 data, it is seen that it is produced from fossil sources, mechanically recyclable sources, chemically recyclable sources, bio-based sources, and carbon captured sources (Plastics Europe 2024). Plastics with such a high production have countless uses. The packaging industry uses plastic materials extensively. It is also used in many areas such as the construction industry, automotive, electrical products, sports products, textile and agricultural applications (Plastics Europe 2022).

Industrially produced plastics are used in many different areas for consumption purposes and have an important place in exports and imports in the global economy. Plastics used for short or long periods can be reused or generating energy or recycling (Plastics Europe 2022) (Figure 1.1).

Fossil-derived plastics consist of a wide variety of chemical substances. These plastics are examined in seven categories by the Society of Plastics Industrialists Association (SPI) according to their recyclable properties (Tan et al. 2024). Polyethylene terephthalate (PETE or PET) is in the first category and is recyclable. It is generally used in areas such as bottles, jars, plastic bags, carpets, curtains and clothes (Askar et al. 2023; Tan et al. 2024). High-density polyethylene (HDPE) is in the second category and is recyclable. It is generally used in bags, engine oil cans, boxes of personal care products and toys (Bataineh 2020; Tan et al. 2024). Polyvinyl chloride (PVC) is in the third category and is recyclable. It is generally used in the fields of bags, shoes, hoses and construction materials (Lieberzeit et al. 2022; Tan et al. 2024). Low-density polyethylene (LDPE) is in the fourth category and is recyclable. It is generally used in areas such as grocery bags, refrigerator storage bags and films (Menzel

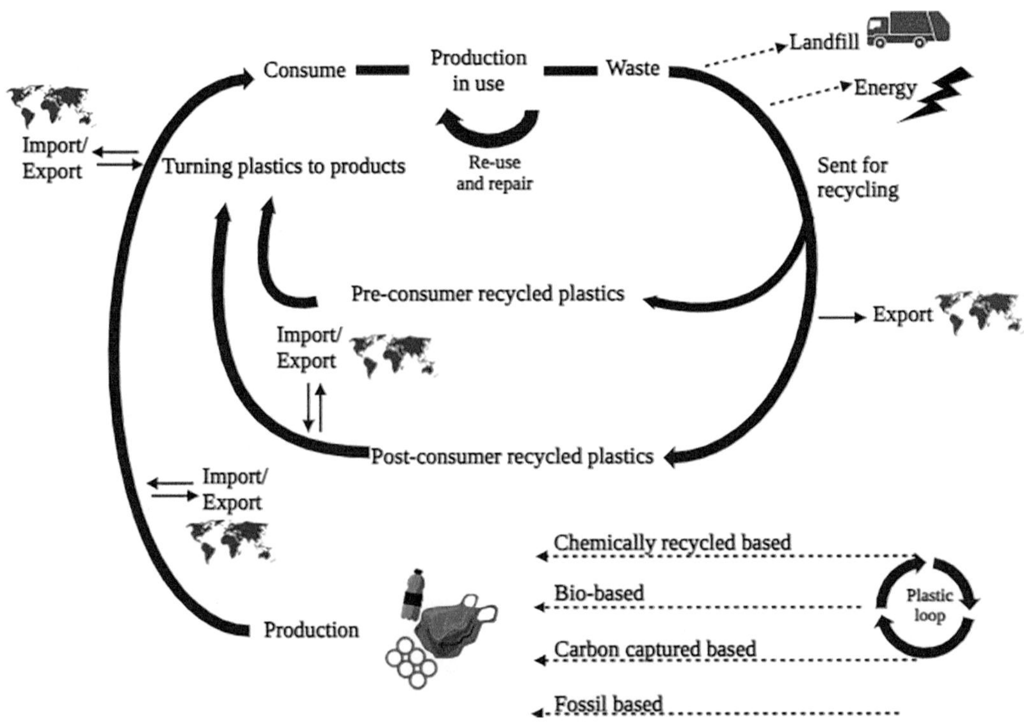

FIGURE 1.1 Plastic production, distribution, consumption and reusability. (Adapted from Plastics Europe 2022.)

et al. 2022; Tan et al. 2024). Polypropylene (PP) is in the fifth category and is nonrecyclable. It is generally used in products such as sanitary cloths, fibers, food and beverage storage containers, bottles, disposable plates, forks and spoons (Latifi et al. 2022; Tan et al. 2024). Polystyrene (PS) or Styrofoam is in the sixth category and is a nonrecyclable material. It is used in areas such as disposable beverage cups, food storage containers and packaging products (Jeon et al. 2023; Tan et al. 2024). Those in the seventh category that cannot be recycled are products such as polycarbonates, acrylics, fiberglass and nylon, which are referred to as various plastics. It is used in different areas such as glasses, plastic technological devices (such as CDs), feeding bottles and medical products (Tan et al. 2024). According to 2023 data, the most commonly used plastics are PP, LDPE, PVC and HDPE (Plastics Europe 2024).

Plastic materials are very durable in nature and require centuries to decompose. Biodegradation of these materials is a very slow process. Various abiotic factors (such as sunlight and wind) break down large pieces of plastic into smaller sizes (Bajt 2021). Plastic sizes are divided into three classes according to their dimensions. Macroplastics are plastics larger than 5 mm. While microplastics (MPs) are plastics with sizes between 1 µm and 5 mm, nanoplastics are plastics smaller than 1 µm (Dissanayake et al. 2022). These plastics particles are found in biological materials such as human blood (Mintoo and Kousar 2024), milk (Basaran et al. 2023) and fish (Steer et al. 2017; Gündoğdu et al. 2020; Munno et al. 2022), as well as in abiotic compartments such as soil (Chai et al. 2020; Bai et al. 2024), water (Egessa et al. 2020; Tavşanoğlu et al. 2020; Ferguson et al. 2024) and sediment (Cho et al. 2024; Ferguson et al. 2024; Suteja et al. 2024; Wang et al. 2024).

1.2 MICROPLASTICS

The term microplastic was first used by Thompson et al. (2004). Thus, the scientific world has encountered MPs and their effects on the environment, and an increasing number of studies have been published every year (Yu et al. 2019). In studies on MPs, they are classified according to their sources, sizes, morphological shapes and colors, and chemical structure of plastics (Figure 1.2).

In terms of occurrence, the plastics are examined in two parts: primary and secondary. Primary MPs are microplastics produced in microscopic size. It is especially used in personal care and medical application products. Also included in this class are acrylic or melamine MPs used in machines that use air blasting to remove rust or paint. On the other hand, secondary MPs are defined as those derived from large plastic pieces in the terrestrial or aquatic ecosystem. It consists of plastics whose integrity deteriorates physically, chemically or biologically over time (Cole et al. 2011; Yu et al. 2019; Dissanayake et al. 2022).

Plastic sources	Plastic sizes	Plastic morphology	Plastic color	Degree of light transmittance of plastic	Chemical structure of plastics
• Primary • Secondary	• Nanoplastics • Microplastics • Mesoplastics • Macroplastics • Megaplastics	• Filaments • Fragments • Films • Foams • Pellets, granules or beads	• Black • Gray • White • Red • Orange • Yellow • Green • Blue • Purple • Pink • Brown • Multicolored • Colorless	• Opaque • Transparent	• PET • PVC • PP • PS • Polyethylene • Polyester • Polyamide • Polyurethane

FIGURE 1.2 The features of plastic particles.

TABLE 1.1
The Size Categories of Plastics

Plastics Name	Relative Size	Common Size Divisions	Measurement Units
Mega	Very large	>1 m	Meters
Macro	Large	25 mm and 1 m	Meters Millimeters
Meso	Medium	5 and 25 mm	Millimeters
Micro	Small	1 μm and 5 mm	Millimeters Microns
Nano	Extremely small	<1 μm	Nanometers

Source: Adapted from GESAMP 2019.

Plastic particles are named according to their size. Nanoplastics are particles smaller than 1 μm. MPs are particles between 1 μm and 5 mm. Mesoplastics are plastic particles between 5 and 25 mm. Macroplastics are plastic particles between 25 mm and 1 m, while megaplastics are plastic particles larger than 1 m (GESAMP 2019) (Table 1.1).

The morphology of plastic pieces, like their sizes, also varies. Filaments are thin thread-like particles composed of fibers and threads. Fragments are irregularly shaped plastics with round or angular fractures. Films are composed of thin film sheets. Foams are flexible particles. Pellets, granules or beads are spherical, flat on one side or cylindrical in shape (EU 2023). MPs are especially available in foam, fiber, bread and fragment structures (Filgueiras et al. 2019).

Another morphological feature is the classification of MPs according to their colors. Color perception is a subjective situation since it depends on the observer in visual determination and is a parameter that limits the accuracy and objective classification of the colors of MPs (EU 2023). However, when MPs are generally classified according to their colors, they are divided into black, gray, white, red, orange, yellow, green, blue, purple, pink, brown, multicolored and colorless (Yu et al. 2019; Ahmed et al. 2022; EU 2023). MPs designated as transparent are examined in two parts, opaque (not allowing the light to pass) and transparent (allowing the light to pass), depending on the degree to which light passes through the particle (EU 2023).

The most abundant MPs have the chemical structures of PET, PVC, PP, PS, polyethylene (PE), polyester (PES), polyamide (PA), and polyurethane (PUR) (Yu et al. 2019; Ahmed et al. 2022).

1.3 THE ENVIRONMENTAL DISTRIBUTION OF MICROPLASTICS

MPs can be transported between aquatic, terrestrial and atmospheric environments, with the main transport occurring via rivers, surface runoff waters and wind (Haque and Fan 2023; Kurniawan et al. 2023a, 2023b) (Figure 1.3). The presence of MPs has been detected all over the world, including aquatic environments such as Arctic (Adams et al. 2021), Antarctica (Reed et al. 2018), the Pacific Ocean (Desforges et al. 2014), the Atlantic Ocean (Enders et al. 2015), the Caspian Lake (Mehdinia et al. 2020), Tibetan Plateau (Wang et al. 2023) and atmosphere of New Zealand (Fan et al. 2022). They have also been detected in different types of organisms such as mussel (Li et al. 2016), fish (Khan et al. 2020), watermelon (Isari et al. 2021), drinking water (Acarer 2023) and sheep (Beriot et al. 2021). Therefore, MPs have become a rapidly increasing source of pollution worldwide.

As mentioned above, due to the increasing plastic pollution, especially MPs, scientific studies are being conducted on how abiotic and biotic compartments are affected. In addition, national and international policies have begun to be developed against plastic pollution (Hettiarachchi and Meegoda

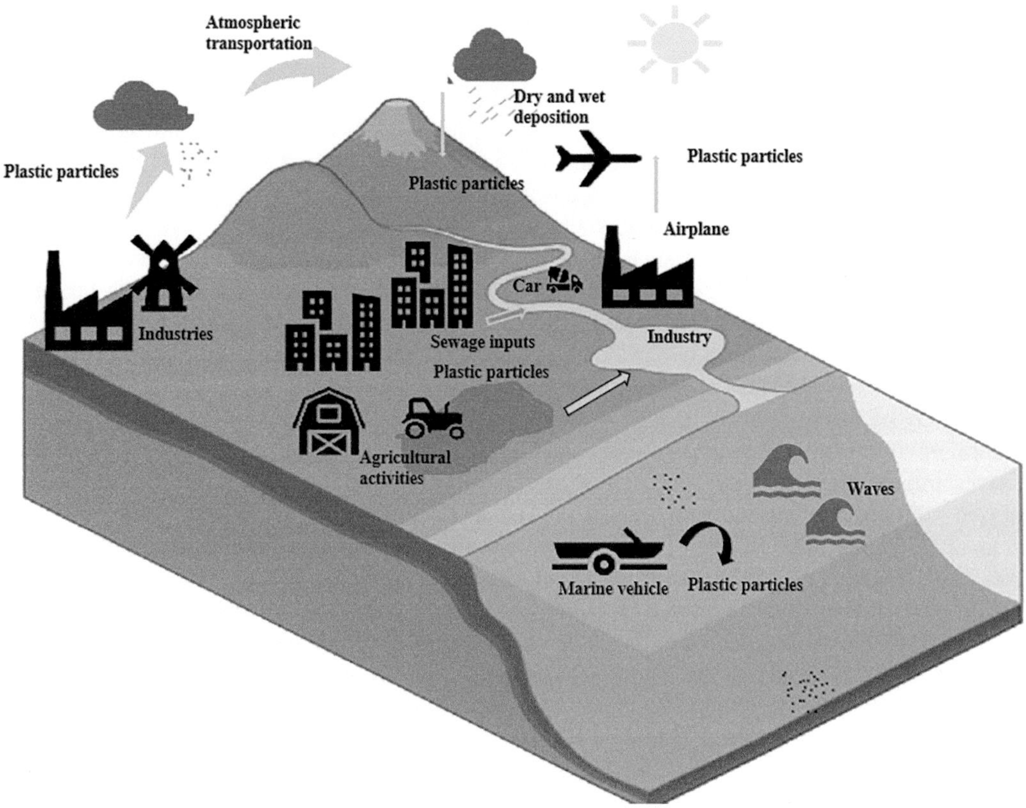

FIGURE 1.3 The global cycle of MPs.

2023; Kurniawan et al. 2023b; Pathak 2023). These studies need to be made more permanent through a structure such as the Stockholm Convention for POPs. For example, the Philippines, located in the Pacific Ocean, has started studies to investigate the sources of plastics, characterize plastics and determine socio-environmental factors within the framework of its plastic regulation policy (Van Ryan Kristopher et al. 2021). A more comprehensive report example was made by Plastics Europe. It is one of the international organizations that examines the production, use and distribution of plastics and provides a lot of information about plastics in European countries with Fact reports every year. According to 2022 data, the number of people working in the plastics sector in Europe is more than 1.5 million and there are more than 52 thousand industrial establishments. While there is a cash turnover of over 400 billion euros in the plastics sector, plastic trade is over 14.4 billion euros. In plastic production imports, the United States, South Korea and the United Kingdom are in the top three, while in exports, the United Kingdom, Turkey and China are in the top three. In plastic transformation, China, the United Kingdom and the United States are the top three import countries, while the United Kingdom, the United States and Switzerland are the top three export countries. The Netherlands, Norway, Spain and Germany are countries with plastic recycling rates over 40% (Plastics Europe, 2022). In this context, it is necessary to know the status of plastics in the aquatic, terrestrial and atmospheric environment in the global transformation or cycle.

1.3.1 Aquatic Environments

Aquatic ecosystems are reservoirs of environmental pollutants. Scientific studies conducted in aquatic ecosystems have shown that environmental pollutants can be detected in abiotic and biotic layers, as well as being toxic to living organisms in aquatic ecosystems (Serra and Colomer 2023). Like environmental pollutants, MPs are one of the most important and emerging pollutants of the aquatic ecosystem and end up in rivers, wastewater treatment plants (WWTPs) and oceans sediment (Zhang et al. 2020; Groffen et al. 2021; Rocha et al. 2021).

Plastic products used in areas such as medical, industrial, entertainment in the terrestrial ecosystem or fishing in the freshwater ecosystem reach freshwater resources as microplastic inputs (Gopalakrishnan et al. 2023). MPs used in terrestrial ecosystems degrade in the soil and enter freshwater ecosystems or enter freshwater ecosystems as macroplastics and turn into MP particles (Gopalakrishnan et al. 2023; Zhang et al. 2023a, 2023b).

In many of the studies, the presence of MPs in water and sediment samples taken from the natural environment was examined and characteristic features such as the amount of MPs, their shapes and chemical structure were determined. Neelavannan et al. (2022) determined an average of 606 ± 360 MP/kg in sediment samples of Anchar Lake (Northern Himalaya, India). Morgado et al. (2022) found 1,090 MP/kg in sediment samples of Aveiro River, 488 MP/kg in sediment samples of Mira River and 460 MP/kg in sediment samples of Mondego River. A study in the Paraíba do Sul River Basin (Brazil) reported high MP concentrations in the confluence areas of the rivers. 0.71 ± 0.25 MP/m^3 were detected in the confluence areas of the Paraíba do Sul and Muriaé rivers in the basin, 0.42 ± 0.23 MP/m^3 in the confluence areas of the Paraíba do Sul and Dois Rios rivers and 0.38 ± 0.14 MP/m^3 in the confluence areas of the Paraíba do Sul and Pomba rivers (da Costa et al. 2023). In the Vembanand Lake (India), a Ramsar site, 26.79 ± 3.74 items/L MPs was detected in surface water samples and 52.70 ± 5.43 items/L MPs was detected in bottom water samples (Anagha et al. 2023). The overall concentration of MPs in Lake Tai Basin (China) is 0–142 particles/L, 0 –18.6 particles/L in lake water samples, 1.56–1.42 × 10^2 particles/L in river water samples and 0.16–0.7 particles/L in estuary water samples (Bhutto et al. 2024).

On the other hand, wastewater treatment facilities, where industrial, municipal and domestic wastes accumulate, are one of the important places where MPs accumulate. While macroplastic pieces are separated in WWTPs, MPs and nanoplastics remain in the water. Therefore, methods such as chemical removal, membrane filtration and biodegradation are recommended for MPs remaining in water and sludge (Golmohammadi et al. 2023). Since long-term solutions to eliminate MPs from environmental compartments have not yet been fully established, WWTPs used in water and sediment treatment are being used (Tang and Hadibarata 2021). According to Qaiser et al. (2023), WWTP can remove more than 90% of MPs. Bioremediation, one of the technologies in these facilities, is a sustainable method that removes pollutants in wastewater and sediment using microbial activity. For bioremediation of MPs, MPs should be broken down using a suitable species and adapted to treatment conditions (Ahmed et al. 2024). Another method, the membrane bioreactor, filters MPs from wastewater, reducing them by 90%. Additionally, the application of rapid sand filtration and reverse osmosis processes also ensures the retention of MPs (Xu et al. 2021). Due to the use of WWTP in developing and developed countries, MPs accumulate in water and sediment. For example, in a study conducted in Vietnam, a developing country, it was determined that while there were 24.3 × 10^3 units/m^3 of MPs in the 17,000 m^3/day water entering the WWTP, the number of MPs in the water mass exiting after treatment was 0.81 x 10^3 units/m^3 (Nguyen et al. 2023). In a study conducted in Turkey, another developing country, the MP input to WWTP was determined as 4.9 × 10^{11} MP/day while the MP output was determined as 1.18 × 10^{10} MP/day (Koyuncuoğlu and Erden 2023). In a study conducted in the United Kingdom, an example of a developed country, the MPs concentration entering the WWTP was 9.9 × 10^{13} MPs/L, while the MPs concentration in the outlet water was 7.8 × 10^{10} MPs/day (Horton et al. 2021).

Marine ecosystems act as a sink for MPs because the vast majority of plastic produced reaches freshwater ecosystems, terrestrial ecosystems and atmospheric transport (Ikenoue et al. 2023; Verma et al. 2024). Moreover, marine ecosystems are also vulnerable to MPs pollution, which can originate from different anthropogenic impacts such as coastal tourism, fishing and shipping industry (Cole et al. 2011). MPs that enter the aquatic ecosystem in these ways may be carried to the water surface or to more distant parts by waves in the water part, while they may accumulate in the sediment part. In addition, particles coming from atmospheric convection and particles coming from runoff of soil surfaces are also ways of MPs entry into the marine ecosystem. MPs ingested by biota living in the marine ecosystem may accumulate in the tissue and pass to higher level organisms through the food chain when these organisms are eaten (Au et al. 2017; Miller et al. 2020).

There are many studies investigating MPs pollution in marine ecosystems, starting with Thompson et al. (2004). The studies conducted in marine ecosystems, as in freshwater ecosystems, include studies such as detecting MP concentrations in water and sediment samples and separating them according to their different characteristics. There are also studies in which MPs are detected in biota living in marine ecosystems and their effects are observed. In a study conducted in the Pacific Ocean, 0.26 ± 0.08 microplastic/m^3 was determined in continental water samples, 0.36 ± 0.10 microplastic/m^3 in water samples taken from stations in international waters and 0.24 ± 0.09 microplastic/m^3 and 0.22 ± 0.08 microplastic/m^3 in Eastern and Western Galapagos coastal water samples, respectively (Alfaro-Núñez et al. 2021). In a study conducted in the Mediterranean Sea, which is the sixth largest accumulation area for marine plastic litter pollution, 1.30 MPs/m^3 were detected in surface sea waters, 32.4 MPs/kg in sediment and 10.7 MPs/kg in beach sand in the Tarragona (Spain) region (Expósito et al. 2021). Ikenoue et al. (2023) reported that in the surface water samples from Chukchi Sea, Western Arctic Ocean, the 39 pieces/samples were detected.

In aquatic ecosystems, whether freshwater or marine, MPs have been detected in water and sediment compartments, as well as in living organisms habituating in these systems. MPs pose a threat to aquatic life because they are mistaken for food (Cole et al. 2011). Seven beluga whale (*Delphinapterus leucas*) samples caught by fishermen off Tuktoyaktuk (Northwest Territories, Canada) had an average of 97 ± 42 MPs/individual detected in the gastrointestinal tract (Moore et al. 2020). An average of 2.93 items/individual MPs was detected in oysters (*Crassostrea gigas, C. angulat, C. hongkongensis* and *C. sikamea*) collected from local fish markets in China (Teng et al. 2019). In a study with mussels (*Mytilus galloprovincialis*) collected from the Adriatic Sea, MPs were found as 3.1 ± 1.2 items/individual in mussels from the Dobrota region and 2.5 ± 1.1 items/individual in mussels from the Cogi region (Bošković et al. 2023). Harikrishnan et al. (2023) reported that fewer MPs were detected in pelagic fish species than in demersal fish species. Marine plastics have been detected in the carcasses of bird species on the North Sea coast of Schleswig-Holstein (Bange et al. 2023). Plastics have also been detected in marine debris in 37 of 73 seabird species in the Santos Basin in the South Atlantic Ocean (Baes et al. 2024).

MPs have also been shown to have negative effects on organisms (Koelmans 2015). It has been observed that MPs affect the immune system in mussels (*M. galloprovincialis*) exposed to microplastic fragments for a long time, as they cause toxic effects in mussels together with pathogenic bacteria. Thus, it has been determined that pathogenic bacteria attached to MPs increase the effects of MPs with the Trojan horse effect (Sun et al. 2023). Juvenile rainbow trout (*Oncorhynchus mykiss*) individuals exposed to MPs exhibited an immune system response with an increase in the number of neutrophils and a decrease in the number of lymphocytes. In addition, a decrease in the plasma protein ceruloplasmin was observed with the development of liver damage (Hollerova et al. 2023). In the hematological parameters of *Pseudobagrus fulvidraco* individuals exposed to MPs, there was a significant decrease in calcium, magnesium and total protein, while there was an increase in parameters such as glucose and cholesterol. In addition, a significant increase was observed in the enzymatic antioxidant systems (superoxide dismutase, catalase and glutathione-*S*-transferase), which are the most important defense mechanisms against oxidative stress (Lee et al. 2023).

1.3.2 Terrestrial Ecosystems

The first studies on MPs started in aquatic ecosystems and have been increasingly carried out in these systems every year. Studies on soil, one of the other ecosystems, have started to be carried out in recent years (Zhang et al. 2022). Although plastics, and therefore MPs, are produced, used, distributed and disposed of on land, there are few studies on soil ecosystems (Gopalakrishnan et al. 2023). The most important reason for this is the lack of analytical procedures used to detect MPs in soils and the insufficient number of monitoring data on microplastic formation and distribution in soil. Since it is estimated that there may be more MPs in all soil ecosystems than in all aquatic ecosystems, together with the microplastic input formed by the application of sewage sludge or biosolids to agricultural areas, soil ecosystems act as a pool or storage for MPs (Wang et al. 2020; Sajjad et al. 2022) (Figure 1.4).

MPs input may occur in soil ecosystems due to different agricultural activities. MP input sources may include biosolids, compost application, sewage overflow, different irrigation techniques, fertilizers, pesticides and atmospheric sediments (Shorobi et al. 2023; Xiang et al. 2023). MPs in soil are called primary and secondary sources. While primary sources come from cosmetics, pharmaceuticals and industrial production, secondary sources come from physical, biological or chemical processes such as atmospheric deposition and industrial production destruction. Especially secondary sources are the main cause of soil pollution. These MPs can migrate horizontally or vertically due to effects such as plant roots, soil fauna and wind (Zhang et al. 2022). When terrestrial plants absorb MPs from the soil by their roots, they experience problems such as hindering their growth and development and remaining in plant bodies. As a result of their accumulation in plants, they accumulate in the food chain, reach the human organism and cause deterioration in human health (Xiang et al. 2023).

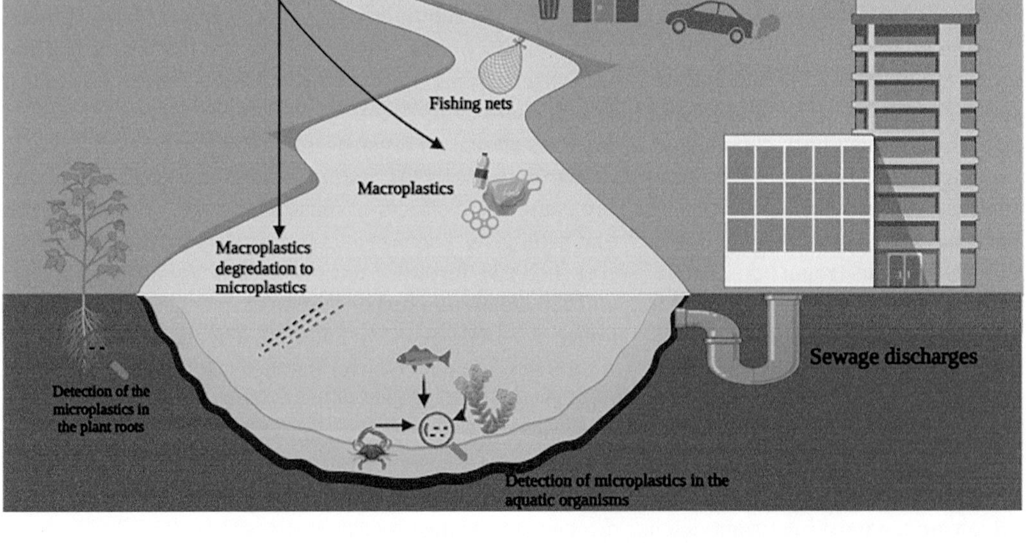

FIGURE 1.4 The contamination of aquatic and terrestrial ecosystems with microplastic.

The presence of MPs has been detected in industrial products produced or processed in the terrestrial environment, such as paddy fields (mean 1,952.86 ± 114.36 particles/kg) (Amirhosseini et al. 2023), sheep milk (128 MPs/L) (Rbaibi Zipak et al. 2024), honey (mean 314 ± 353 MP/kg) (Basaran et al. 2024), table salt (55.2 ± 43.7 MP/kg) (Makhdoumi et al. 2023) and sugar (57.7 ± 20.6 particles/kg) (Makhdoumi et al. 2023).

Studies on the presence of MPs in soil ecosystems are focused on studies conducted in agricultural areas. The presence of MPs in agricultural soils is affected by many factors such as the chemical structure of the soil, the planting period of agricultural products and irrigation methods (Yu et al. 2021). It has been reported that surface irrigation, one of these factors, causes high levels of MPs input into soils, followed by sprinkler and drip irrigation methods (Zhang 2020). In samples collected from agricultural lands in Shouguang City, one of the major vegetable production regions of China, MPs were detected between 310 and 5,698 items/kg (average: 1,444 ± 986 items/kg) (Yu et al. 2021). In Urumqi, another agricultural region of China, an average of 7,763 ± 2,773 units/kg of MPs was reported in greenhouse soils, 4,128 ± 2,235 units/kg in vegetable growing soils and 3,178 ± 3,172 units/kg in crop fields (Li et al. 2023). According to Harms et al. (2021), 3.7 ± 11.9 MPs/kg were reported in agricultural soils in the northern part of Germany. Medyńska-Juraszek and Szczepańska (2023), in their study on different soil types in Poland, detected an average of 1,540 ± 912 particles/kg in loamy soils, an average of 933 ± 682 particles/kg in clay soils, an average of 1,200 ± 234 particles/kg in flooded soils and an average of 4,050 ± 2,831 particles/kg in sewage-treated soils. In soil samples taken from agricultural and urban areas in different regions of Türkiye, an average of 192.7 ± 14.2 particles/kg was detected in agricultural areas and an average of 127.3 ± 21.6 particles/kg in urban areas, and it was reported that the presence of MPs in agricultural areas was significantly higher than in urban areas (Akca et al. 2024).

Plants, one of the fundamental biotas of terrestrial ecosystems, may be affected by MPs in their growth and development processes (Gan et al. 2023). Lian et al. (2024) found that MPs had negative effects on lettuce growth parameters. There was a decrease in the shoot length and weight of lettuce. They showed that there was a decrease in photosynthesis rate due to the decrease in chlorophyll content. Zhang et al. (2024) found that the effects of MPs on the growth parameters of maize plants depended on the exposure dose, particle size and application soil structure. They showed that MPs application in calcareous soils led to an increase in chlorophyll pigment, but MPs application in silty soils reduced plant growth. In a study conducted on the presence of MPs in vegetables and fruits purchased from local markets, it was determined that carrots among vegetables and apples among fruits contained high levels of MPs (Conti et al. 2020). It has been determined that MPs cause differences in plant properties such as amino acids, oils and colors (Kim et al. 2022). The detection of MPs in consumable vegetables and fruits poses a health threat to organisms in the food web including humans (Zhou et al. 2023). Farm animals ingest MPs through contaminated soil, water and plants. For example, livestock grazing in vegetable fields are exposed primarily to MPs through packaging residues and agricultural mulching materials left on fields (Bahrani et al. 2024).

Microbial communities in soil are also affected by MPs. It has been reported that different types of MPs have microbial communities that form biofilms on their surfaces. While MP biofilms caused a decrease in the microbial community consisting of Acidobacteriota, Myxococcota, Gemmatimonadota and Chloroflexi, tire wear MPs caused an increase in Bacteroidota and PP MPs caused an increase in Firmicutes and Actinobacteriota groups (Ding et al. 2024). In addition, the presence of MPs in soil showed an increase in the number of some bacterial groups involved in the nitrogen cycle (Zhang et al. 2023a).

1.3.3 Atmospheric Compartments

Studies on the presence of MPs in the atmosphere are not as intensive as in aquatic ecosystems (Akanyange et al. 2021). It has been reported that MPs can be carried by wind and mixed into terrestrial or aquatic ecosystems with dry or wet precipitation (Knobloch et al. 2021; Wei et al. 2023).

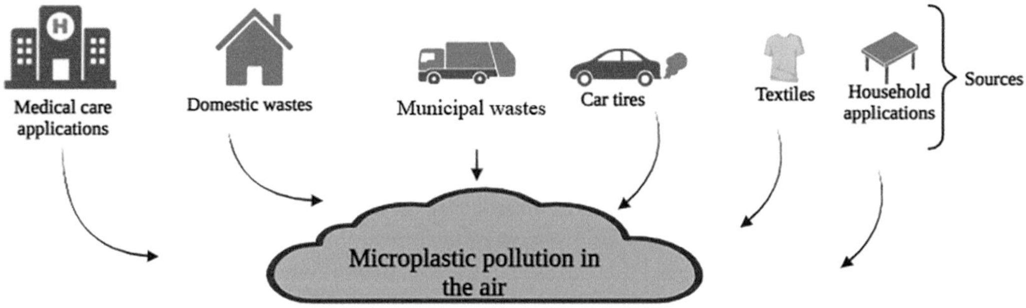

FIGURE 1.5 Microplastic pollution in the atmosphere.

Thus, MPs cycle is provided between water, land and atmospheric environments (Wei et al. 2023) (Figure 1.5). Wind is one of the main factors in the transport of MPs (Akanyange et al. 2021). Meteorological events such as rain are also responsible for the transport of MPs (Wu et al. 2023). In addition, the source of MPs in indoor environments is the wear of textile products and furniture, while the main sources in outdoor areas are car tires, urban waste and improper waste management in the textile industry (Akanyange et al. 2021).

Atmospheric MPs exist in various matrices due to long-distance travel and meteorological transport. The number of atmospheric MP samples from the northwest of the South China Sea falling dry into the ocean in autumn was 1.4×10^3 t (Ding et al. 2021). An average of 463 MP/L particles was detected in cloud water samples collected at 1,545 m altitude in eastern China, and 0.21 MP/m^3 particles were detected in air samples (Xu et al. 2023). In another study conducted in China, the seasonal distribution of atmospheric MPs was found to be high in spring (0.19 ± 0.02 particle/m^3), summer (0.18 ± 0.02 particle/m^3) and autumn (0.20 ± 0.02 particle/m^3) and low in winter (0.10 ± 0.01 particle/m^3) (Yuan et al. 2023). Knobloch et al. (2021) detected approximately 80–330 particles m^2/day in dry air samples in a study conducted in New Zealand. Celik-Saglam et al. (2023), in their seasonal atmospheric MPs study in Turkey, found that the ranges of MPs were 7,207–21,042 MP/m^3 in summer, 7,245–32,950 MP/m^3 in autumn, 4,035–58,270 MP/m^3 in winter and 7,275–37,094 MP/m^3 in spring. The mean number of MPs recovered from urban areas near the Patna Airport in India and from the peri-urban areas at the Indian Institute of Technology Patna are $1,959.63 \pm 204.8$ MPs/m^2/day and $1,320.4 \pm 125.7$ MPs/m^2/day, respectively (Parashar and Hait 2023).

In living organisms that breathe through lungs, such as humans, via atmospheric means, the MPs pass through the lung membrane and mix into the blood, which can ultimately cause significant health effects (Akanyange et al. 2021). For example, a study in Shanghai (China) found that indoor air inhaled by people contained 43 ± 16 particles/m^3 and exhaled air contained 12 ± 5 particles/m^3 (Geng et al. 2023). In a study conducted in Wuxi People's Hospital and Wuxi Children's Hospital (China), an average of 14.19 ± 14.57 particles/g was found in human lung samples (Zhu et al. 2024).

1.4 CONCLUSION

The biotic and abiotic environment is in a constant state of pollution, and the effects of the factors that cause this pollution on the environment are revealed more rapidly with developing technology. MPs, one of these factors, occur as a result of plastic materials breaking down into smaller pieces. Regardless of their chemical structure, shape or color, whether they are of primary or secondary origin, they have the ability to be found in water, soil, air and living tissue. Studies conducted in the scientific world reveal the status of these pollutants in water, soil and air and their effects on biota. These studies can be a source of understanding, preventing and stopping microplastic pollution in the future, as the main subject of various regulations and conferences at national and international levels.

REFERENCES

Acarer, S. (2023). Abundance and characteristics of microplastics in drinking water treatment plants, distribution systems, water from refill kiosks, tap waters and bottled waters. *Science of the Total Environment*, 163866.

Adams, J. K., Dean, B. Y., Athey, S. N., Jantunen, L. M., Bernstein, S., Stern, G., Diamond, M. L., & Finkelstein, S. A. (2021). Anthropogenic particles (including microfibers and microplastics) in marine sediments of the Canadian Arctic. *Science of the Total Environment*, 784, 147155.

Ahmed, R., Hamid, A. K., Krebsbach, S. A., He, J., & Wang, D. (2022). Critical review of microplastics removal from the environment. *Chemosphere*, 293, 133557.

Ahmed, S. F., Islam, N., Tasannum, N., Mehjabin, A., Momtahin, A., Chowdhury, A. A., Almomani, F., & Mofijur, M. (2024). Microplastic removal and management strategies for wastewater treatment plants. *Chemosphere*, 347, 140648.

Akanyange, S. N., Lyu, X., Zhao, X., Li, X., Zhang, Y., Crittenden, J. C., Anning, C., Chen, T., Jiang, T. & Zhao, H. (2021). Does microplastic really represent a threat? A review of the atmospheric contamination sources and potential impacts. *Science of the Total Environment*, 777, 146020.

Akca, M. O., Gündoğdu, S., Akca, H., Delialioğlu, R. A., Aksit, C., Turgay, O. C., & Harada, N. (2024). An evaluation on microplastic accumulations in Turkish soils under different land uses. *Science of the Total Environment*, 911, 168609.

Alfaro-Núñez, A., Astorga, D., Cáceres-Farías, L., Bastidas, L., Soto Villegas, C., Macay, K., & Christensen, J. H. (2021). Microplastic pollution in seawater and marine organisms across the Tropical Eastern Pacific and Galápagos. *Scientific Reports*, 11(1), 6424.

Alharbi, O. M., Khattab, R. A., & Ali, I. (2018). Health and environmental effects of persistent organic pollutants. *Journal of Molecular Liquids*, 263, 442–453.

Almeida, C. M. R., Sáez-Zamacona, I., Silva, D. M., Rodrigues, S. M., Pereira, R., & Ramos, S. (2023). The role of estuarine wetlands (saltmarshes) in sediment microplastics retention. *Water*, 15(7), 1382.

Amirhosseini, K., Haghani, Z., & Alikhani, H. A. (2023). Microplastics pollution in rice fields: A case study of Pir Bazar rural district of Gilan, Iran. *Environmental Monitoring and Assessment,* 195(12), 1473.

Anagha, P. L., Viji, N. V., Devika, D., & Ramasamy, E. V. (2023). Distribution and abundance of microplastics in the water column of Vembanad Lake–A Ramsar site in Kerala, India. *Marine Pollution Bulletin*, 194, 115433.

Arslan, P., Gül, G., & Günal, A. Ç. (2023). How does the persistent organic pollutant perfluorooctane sulfonate (PFOS) affect health of the common carp? *Water, Air, & Soil Pollution*, 234(10), 634.

Arslan, P., & Ozeren, S. C. (2022a). A case study of 38 micro-organic pollutants contamination in Kirmir Stream, Turkey. *Environmental Quality Management*, 31(4), 283–290.

Arslan, P., & Ozeren, S. C. (2022b). Physiological and histopathological alterations in *Capoeta baliki* and *Squalius pursakensis* after caused by some environmental pollutants. *Environmental Monitoring and Assessment*, 194(3), 183.

Askar, M. K., Al-Kamaki, Y. S., & Hassan, A. (2023). Utilizing polyethylene terephthalate PET in concrete: A review. *Polymers*, 15(15), 3320.

Au, S. Y., Lee, C. M., Weinstein, J. E., van den Hurk, P., & Klaine, S. J. (2017). Trophic transfer of microplastics in aquatic ecosystems: Identifying critical research needs. *Integrated Environmental Assessment and Management*, 13(3), 505–509.

Baes, L., Santiago, C. D., Roman, L., dos Santos Costa, P. C., Pugliesi, É., & Reigada, C. (2024). Beached seabirds as plastic biomonitors in Brazil from the Beach Monitoring Project of the Santos Basin (PMP-BS). *Marine Pollution Bulletin*, 199, 115847.

Bahrani, F., Mohammadi, A., Dobaradaran, S., De-la-Torre, G. E., Arfaeinia, H., Ramavandi, B., Saeedi, R., & Tekle-Röttering, A. (2024). Occurrence of microplastics in edible tissues of livestock (cow and sheep). *Environmental Science and Pollution Research*, 31(14), 22145–22157.

Bai, R., Liu, H., Cui, J., Wu, Y., Guo, X., Liu, Q., Liu, Q., Gao, H., Yan, H., & He, W. (2024). The characteristics and influencing factors of farmland soil microplastic in Hetao Irrigation District, China. *Journal of Hazardous Materials*, 465, 133472.

Bajt, O. (2021). From plastics to microplastics and organisms. *FEBS Open Bio*, 11(4), 954–966.

Bange, A., Backes, A., Garthe, S., & Schwemmer, P. (2023). Prey choice and ingestion of microplastics by common shelducks and common eiders in the Wadden Sea World Heritage Site. *Marine Biology*, 170, 54.

Bao, L. J., Maruya, K. A., Snyder, S. A., & Zeng, E. Y. (2012). China's water pollution by persistent organic pollutants. *Environmental Pollution*, 163, 100–108.

Basaran, B., Özçifçi, Z., Akcay, H. T., & Aytan, Ü. (2023). Microplastics in branded milk: Dietary exposure and risk assessment. *Journal of Food Composition and Analysis*, 123, 105611.

Basaran, B., Özçifçi, Z., Kanbur, E. D., Akçay, H. T., Gül, S., Bektaş, Y., & Aytan, Ü. (2024). Microplastics in honey from Türkiye: Occurrence, characteristic, human exposure, and risk assessment. *Journal of Food Composition and Analysis*, 135, 106646.

Bataineh, K. M. (2020). Life-cycle assessment of recycling postconsumer high-density polyethylene and polyethylene terephthalate. *Advances in Civil Engineering*, 2020(1), 8905431.

Beriot, N., Peek, J., Zornoza, R., Geissen, V., & Lwanga, E. H. (2021). Low density-microplastics detected in sheep faeces and soil: A case study from the intensive vegetable farming in Southeast Spain. *Science of the Total Environment*, 755, 142653.

Bhutto, S. U. A., Akram, M., & You, X. Y. (2024). Probabilistic risk assessment of microplastics in Tai Lake, China. *Science of the Total Environment*, 914, 169965.

Bizzotto, E. C., Villa, S., & Vighi, M. (2009). POP bioaccumulation in macroinvertebrates of alpine freshwater systems. *Environmental Pollution*, 157(12), 3192–3198.

Boitsov, S., Frantzen, S., Bruvold, A., & Grøsvik, B. E. (2024). Varying temporal trends in the levels of six groups of legacy persistent organic pollutants (POPs) in liver of three gadoid species from the North Sea. *Chemosphere*, 349, 140939.

Bošković, N., Joksimović, D., & Bajt, O. (2023). Microplastics in mussels from the Boka Kotorska Bay (Adriatic Sea) and impact on human health. *Food and Chemical Toxicology*, 173, 113641.

Carpenter, E. J., Anderson, S. J., Harvey, G. R., Miklas, H. P., & Peck, B. B. (1972). Polystyrene spherules in coastal waters. *Science*, 178(4062), 749–750.

Carpenter, E. J., & Smith Jr., K. L. (1972). Plastics on the Sargasso Sea surface. *Science*, 175, 1240–1241.

Carson, R. (1962). *Silent Spring*. Houghton Mifflin Company.

Celik-Saglam, I., Yurtsever, M., Civan, M., Yurdakul, S., & Cetin, B. (2023). Evaluation of levels and sources of microplastics and phthalic acid esters and their relationships in the atmosphere of highly industrialized and urbanized Gebze, Türkiye. *Science of the Total Environment*, 881, 163508.

Chai, B., Wei, Q., She, Y., Lu, G., Dang, Z., & Yin, H. (2020). Soil microplastic pollution in an e-waste dismantling zone of China. *Waste Management*, 118, 291–301.

Chen, M., Tian, J., Gan, Z., Wu, J., Ding, S., & Su, S. (2024). Tissue distribution and trophic transfer of organophosphate triesters and diesters in three marine mammals of the Liaodong Bay and the Northern Yellow Sea. *Journal of Hazardous Materials*, 461, 132694.

Chiesa, L. M., Nobile, M., Malandra, R., Pessina, D., Panseri, S., Labella, G. F., & Arioli, F. (2018). Food safety traits of mussels and clams: Distribution of PCBs, PBDEs, OCPs, PAHs and PFASs in sample from different areas using HRMS-Orbitrap® and modified QuEChERS extraction followed by GC-MS/MS. *Food Additives & Contaminants: Part A*, 35(5), 959–971.

Cho, Y., Jeon, H. J., Lee, S. E., Kim, C., Kim, G., Kim, K., Kim, Y.-G., & Lee, S. R. (2024). Microplastic accumulation dynamics in Hanriver headwaters: Sediment interactions and environmental implication. *Journal of Hazardous Materials*, 472, 134445.

Cole, M., Lindeque, P., Halsband, C., & Galloway, T. S. (2011). Microplastics as contaminants in the marine environment: A review. *Marine Pollution Bulletin*, 62(12), 2588–2597.

Conti, G. O., Ferrante, M., Banni, M., Favara, C., Nicolosi, I., Cristaldi, A., Fiore, M., & Zuccarello, P. (2020). Micro- and nano-plastics in edible fruit and vegetables. The first diet risks assessment for the general population. *Environmental Research*, 187, 109677.

da Costa, I. D., Costa, L. L., & Zalmon, I. R. (2023). Microplastics in water from the confluence of tropical rivers: Overall review and a case study in Paraiba do Sul River basin. *Chemosphere*, 338, 139493.

Desforges, J. P. W., Galbraith, M., Dangerfield, N., & Ross, P. S. (2014). Widespread distribution of microplastics in subsurface seawater in the NE Pacific Ocean. *Marine Pollution Bulletin*, 79(1–2), 94–99.

Ding, J., Liang, Z., Lv, M., Li, X., Lu, S., Ren, S., Yang, X., Li, X., Tu, C., Zhu, D., & Chen, L. (2024). Aging in soil increases the disturbance of microplastics to the gut microbiota of soil fauna. *Journal of Hazardous Materials*, 461, 132611.

Ding, Y., Zou, X., Wang, C., Feng, Z., Wang, Y., Fan, Q., & Chen, H. (2021). The abundance and characteristics of atmospheric microplastic deposition in the northwestern South China Sea in the fall. *Atmospheric Environment*, 253, 118389.

Dissanayake, P. D., Kim, S., Sarkar, B., Oleszczuk, P., Sang, M. K., Haque, M. N., Ahn, J. H., Bank, M. S., & Ok, Y. S. (2022). Effects of microplastics on the terrestrial environment: A critical review. *Environmental Research*, 209, 112734.

Duttagupta, S., Mukherjee, A., Bhattacharya, A., & Bhattacharya, J. (2020). Wide exposure of persistent organic pollutants (PoPs) in natural waters and sediments of the densely populated Western Bengal basin, India. *Science of the Total Environment*, 717, 137187.

Egessa, R., Nankabirwa, A., Ocaya, H., & Pabire, W. G. (2020). Microplastic pollution in surface water of Lake Victoria. *Science of the Total Environment*, 741, 140201.

Enders, K., Lenz, R., Stedmon, C. A., & Nielsen, T. G. (2015). Abundance, size and polymer composition of marine microplastics ≥ 10 μm in the Atlantic Ocean and their modelled vertical distribution. *Marine Pollution Bulletin*, 100(1), 70–81

EU. 2023. Guidance on the Monitoring of Marine Litter in European Seas. https://publications.jrc.ec.europa.eu/repository/handle/JRC133594. Accessed: 07.10.2024.

Expósito, N., Rovira, J., Sierra, J., Folch, J., & Schuhmacher, M. (2021). Microplastics levels, size, morphology and composition in marine water, sediments and sand beaches. Case study of Tarragona coast (western Mediterranean). *Science of the Total Environment*, 786, 147453.

Fan, W., Salmond, J. A., Dirks, K. N., Cabedo Sanz, P., Miskelly, G. M., & Rindelaub, J. D. (2022). Evidence and mass quantification of atmospheric microplastics in a coastal New Zealand city. *Environmental Science & Technology*, 56(24), 17556–17568.

Ferguson, L., Awe, A., & Sparks, C. (2024). Microplastic concentrations and risk assessment in water, sediment and invertebrates from Simon's Town, South Africa. *Heliyon*, 10(7), e28514.

Filgueiras, A. V., Gago, J., Campillo, J. A., & León, V. M. (2019). Microplastic distribution in surface sediments along the Spanish Mediterranean continental shelf. *Environmental Science and Pollution Research*, 26(21), 21264–21273.

Gan, Q., Cui, J., & Jin, B. (2023). Environmental microplastics: Classification, sources, fates, and effects on plants. *Chemosphere*, 313, 137559.

Geng, Y., Zhang, Z., Zhou, W., Shao, X., Li, Z., & Zhou, Y. (2023). Individual exposure to microplastics through the inhalation route: Comparison of microplastics in inhaled indoor aerosol and exhaled breath air. *Environmental Science & Technology Letters*, 10(6), 464–470.

GESAMP. (2019). Guidelines or the monitoring and assessment of plastic litter and microplastics in the ocean (Kershaw P. J., Turra A. and Galgani F. Editors), (IMO/FAO/UNESCO-IOC/UNIDO/WMO/IAEA/UN/UNEP/UNDP/ISA Joint Group of Experts on the Scientific Aspects of Marine Environmental Protection). Rep. Stud. GESAMP No. 99, 130p.

Geyer, H. J., Rimkus, G. G., Scheunert, I., Kaune, A., Schramm, K. W., Kettrup, A., Zeeman, M., Muir, D. C. G., Hansen, L. G., & Mackay, D. (2000). Bioaccumulation and occurrence of endocrine-disrupting chemicals (EDCs), persistent organic pollutants (POPs), and other organic compounds in fish and other organisms including humans. In: B. Beek (Eds.), *Bioaccumulation–New Aspects and Developments* (pp. 1–166). Berlin, Heidelberg: Springer.

Goerke, H., Weber, K., Bornemann, H., Ramdohr, S., & Plötz, J. (2004). Increasing levels and biomagnification of persistent organic pollutants (POPs) in Antarctic biota. *Marine Pollution Bulletin*, 48(3–4), 295–302.

Golmohammadi, M., Musavi, S. F., Habibi, M., Maleki, R., Golgoli, M., Zargar, M., Dumee, L. F., Baroutian, S., & Razmjou, A. (2023). Molecular mechanisms of microplastics degradation: A review. *Separation and Purification Technology*, 309, 122906.

Gómez-Gutiérrez, A., Garnacho, E., Bayona, J. M., & Albaigés, J. (2007). Assessment of the Mediterranean sediments contamination by persistent organic pollutants. *Environmental Pollution*, 148(2), 396–408.

Gopalakrishnan, K. K., Sivakumar, R., & Kashian, D. (2023). The microplastics cycle: An in-depth look at a complex topic. *Applied Sciences*, 13(19):10999.

Groffen, T., Rijnders, J., van Doorn, L., Jorissen, C., De Borger, S. M., Luttikhuis, D. O., de Deyn, L., Covaci, A., & Bervoets, L. (2021). Preliminary study on the distribution of metals and persistent organic pollutants (POPs), including perfluoroalkylated acids (PFAS), in the aquatic environment near Morogoro, Tanzania, and the potential health risks for humans. *Environmental Research*, 192, 110299.

Gündoğdu, S., Cevik, C., & Ataş, N. T. (2020). Occurrence of microplastics in the gastrointestinal tracts of some edible fish species along the Turkish coast. *Turkish Journal of Zoology*, 44(4), 312–323.

Haque, F., & Fan, C. (2023). Fate and impacts of microplastics in the environment: Hydrosphere, pedosphere, and atmosphere. *Environments*, 10(5), 70.

Harikrishnan, T., Janardhanam, M., Sivakumar, P., Sivakumar, R., Rajamanickam, K., Raman, T., Thangavule, M., Muthusamy, G., & Singaram, G. (2023). Microplastic contamination in commercial fish species in southern coastal region of India. *Chemosphere*, 313, 137486.

Harms, I. K., Diekötter, T., Troegel, S., & Lenz, M. (2021). Amount, distribution and composition of large microplastics in typical agricultural soils in Northern Germany. *Science of the Total Environment*, 758, 143615.

Hartmann, C., Kaiser, A. M., Moche, W., Weiss, S., Raffesberg, W., Scharf, S., Graf-Rohrmeister, K., Thanhaeuser, M., Haiden, N., & Uhl, M. (2024). Persistent organic pollutants in Austrian human breast milk collected between 2013 and 2016. *Journal of Xenobiotics*, 14(1), 247–266.

Herceg-Romanić, S., Kljaković-Gašpić, Z., Klinčić, D., & Ujević, I. (2014). Distribution of persistent organic pollutants (POPs) in cultured mussels from the Croatian coast of the Adriatic Sea. *Chemosphere*, 114, 69–75.

Hettiarachchi, H., & Meegoda, J. N. (2023). Microplastic pollution prevention: The need for robust policy interventions to close the loopholes in current waste management practices. *International Journal of Environmental Research and Public Health*, 20(14), 6434.

Hollerova, A., Hodkovicova, N., Blahova, J., Faldyna, M., Franc, A., Pavlokova, S., Tichy, F., Postulkova, E., Mares, J., Medkova, D., Kyllar, M., & Svobodova, Z. (2023). Polystyrene microparticles can affect the health status of freshwater fish–Threat of oral microplastics intake. *Science of the Total Environment*, 858, 159976.

Horton, A. A., Cross, R. K., Read, D. S., Jürgens, M. D., Ball, H. L., Svendsen, C., Vollertsen, J., & Johnson, A. C. (2021). Semi-automated analysis of microplastics in complex wastewater samples. *Environmental Pollution*, 268, 115841.

Ikenoue, T., Nakajima, R., Fujiwara, A., Onodera, J., Itoh, M., Toyoshima, J., Watanabe, E., Murata, A., Nishino, S., & Kikuchi, T. (2023). Horizontal distribution of surface microplastic concentrations and water-column microplastic inventories in the Chukchi Sea, western Arctic Ocean. *Science of the Total Environment*, 855, 159564.

Isari, E. A., Papaioannou, D., Kalavrouziotis, I. K., & Karapanagioti, H. K. (2021). Microplastics in agricultural soils: A case study in cultivation of watermelons and canning tomatoes. *Water*, 13(16), 2168.

Jeon, S., Jeon, J. H., Jeong, J., Kim, G., Lee, S., Kim, S., Maruthupandy, M., Lee, K., Yang, S. I., & Cho, W. S. (2023). Size-and oxidative potential-dependent toxicity of environmentally relevant expanded polystyrene Styrofoam microplastics to macrophages. *Journal of Hazardous Materials*, 459, 132295.

Kabir, M., Habiba, U., Iqbal, M. Z., Shafiq, M., Farooqi, Z. R., Shah, A., & Khan, W. (2023, July). Impacts of anthropogenic activities & climate change resulting from increasing concentration of carbon dioxide on environment in 21st century; A critical review. *IOP Conference Series: Earth and Environmental Science*, 1194(1), 012010. IOP Publishing.

Kaloyianni, M., Bobori, D. C., Xanthopoulou, D., Malioufa, G., Sampsonidis, I., Kalogiannis, S., Kastrinaki, G., Dimitriadi, A., Koumoundouros, G., Lambropoulo, D. A., Kyzas, G. Z., & Bikiaris, D. N. (2021). Toxicity and functional tissue responses of two freshwater fish after exposure to polystyrene microplastics. *Toxics*, 9(11), 289.

Khan, F. R., Shashoua, Y., Crawford, A., Drury, A., Sheppard, K., Stewart, K., & Sculthorp, T. (2020). 'The plastic Nile': First evidence of microplastic contamination in fish from the Nile River (Cairo, Egypt). *Toxics*, 8(2), 22.

Kim, D., An, S., Kim, L., Byeon, Y. M., Lee, J., Choi, M. J., & An, Y. J. (2022). Translocation and chronic effects of microplastics on pea plants (*Pisum sativum*) in copper-contaminated soil. *Journal of Hazardous Materials*, 436, 129194.

Klánová, J., Matykiewiczová, N., Máčka, Z., Prošek, P., Láska, K., & Klán, P. (2008). Persistent organic pollutants in soils and sediments from James Ross Island, Antarctica. *Environmental Pollution*, 152(2), 416–423.

Knobloch, E., Ruffell, H., Aves, A., Pantos, O., Gaw, S., & Revell, L. E. (2021). Comparison of deposition sampling methods to collect airborne microplastics in Christchurch, New Zealand. *Water, Air, & Soil Pollution*, 232(4), 133.

Koelmans, A. A. (2015). Modeling the role of microplastics in bioaccumulation of organic chemicals to marine aquatic organisms. A critical review. In: M. Bergmann, L. Gutow, & M. Klages (Eds.), *Marine Anthropogenic Litter* (pp. 309–324). Cham: Springer Open.

Koyuncuoğlu, P., & Erden, G. (2023). Microplastics in municipal wastewater treatment plants: A case study of Denizli/Turkey. *Frontiers of Environmental Science & Engineering*, 17(8), 99.

Kurniawan, T. A., Haider, A., Ahmad, H. M., Mohyuddin, A., Aslam, H. M. U., Nadeem, S., Javed, M., Othman, M. H. D., & Chew, K. W. (2023a). Source, occurrence, distribution, fate, and implications of microplastic pollutants in freshwater on environment: A critical review and way forward. *Chemosphere*, 325, 138367.

Kurniawan, T. A., Haider, A., Mohyuddin, A., Fatima, R., Salman, M., Shaheen, A., Ahmad, H. M., Al-Hazmi, H. E., Othman, M. H. D., Aziz, F., Anouzla, A., & Ali, I. (2023b). Tackling microplastics pollution in global environment through integration of applied technology, policy instruments, and legislation. *Journal of Environmental Management*, 346, 118971.

Latifi, M. R., Biricik, Ö., & Mardani Aghabaglou, A. (2022). Effect of the addition of polypropylene fiber on concrete properties. *Journal of Adhesion Science and Technology*, 36(4), 345–369.

Lee, J. H., Kang, J. C., & Kim, J. H. (2023). Toxic effects of microplastic (Polyethylene) on fish: Accumulation, hematological parameters and antioxidant responses in Korean Bullhead, Pseudobagrus fulvidraco. *Science of the Total Environment*, 877, 162874.

Li, J., Qu, X., Su, L., Zhang, W., Yang, D., Kolandhasamy, P., Li, D., & Shi, H. (2016). Microplastics in mussels along the coastal waters of China. *Environmental Pollution*, 214, 177–184.

Li, W., Wang, S., Wufuer, R., Duo, J., & Pan, X. (2023). Distinct soil microplastic distributions under various farmland-use types around Urumqi, China. *Science of the Total Environment*, 857, 159573.

Lian, Y., Shi, R., Liu, J., Zeb, A., Wang, Q., Wang, J., Yu, M., Li, J., Zheng, Z., Ali, N., Bao, Y., & Liu, W. (2024). Effects of polystyrene, polyethylene, and polypropylene microplastics on the soil-rhizosphere-plant system: Phytotoxicity, enzyme activity, and microbial community. *Journal of Hazardous Materials*, 465, 133417.

Lieberzeit, P., Bekchanov, D., & Mukhamediev, M. (2022). Polyvinyl chloride modifications, properties, and applications. *Polymers for Advanced Technologies*, 33(6), 1809–1820.

Makhdoumi, P., Pirsaheb, M., Amin, A. A., Kianpour, S., & Hossini, H. (2023). Microplastic pollution in table salt and sugar: Occurrence, qualification and quantification and risk assessment. *Journal of Food Composition and Analysis*, 119, 105261.

Medyńska-Juraszek, A., & Szczepańska, A. (2023). Microplastic pollution in EU farmland soils: Preliminary findings from agricultural soils (southwestern Poland). *Agriculture*, 13(9), 1733.

Mehdinia, A., Dehbandi, R., Hamzehpour, A., & Rahnama, R. (2020). Identification of microplastics in the sediments of southern coasts of the Caspian Sea, north of Iran. *Environmental Pollution*, 258, 113738.

Menzel, T., Meides, N., Mauel, A., Mansfeld, U., Kretschmer, W., Kuhn, M., Herzig, E. M., Altstodt, V., Strohriegl, P., Senker, J., & Ruckdäschel, H. (2022). Degradation of low-density polyethylene to nanoplastic particles by accelerated weathering. *Science of the Total Environment*, 826, 154035.

Miller, M. E., Hamann, M., & Kroon, F. J. (2020). Bioaccumulation and biomagnification of microplastics in marine organisms: A review and meta-analysis of current data. *PloS One*, 15(10), e0240792.

Mintoo, M. J., & Kousar, R. (2024). Reports of microplastic in human blood. In: M. Shahnawaz, C. O. Adetunji, M. A. Dar, & D. Zhu (Eds.), *Microplastic Pollution* (pp. 263–270). Singapore: Springer Nature Singapore.

Moore, R. C., Loseto, L., Noel, M., Etemadifar, A., Brewster, J. D., MacPhee, S., Bendell, L., & Ross, P. S. (2020). Microplastics in beluga whales (*Delphinapterus leucas*) from the Eastern Beaufort Sea. *Marine Pollution Bulletin*, 150, 110723.

Morgado, V., Gomes, L., da Silva, R. J. B., & Palma, C. (2022). Microplastics contamination in sediments from Portuguese inland waters: Physical-chemical characterisation and distribution. *Science of the Total Environment*, 832, 155053.

Munno, K., Helm, P. A., Rochman, C., George, T., & Jackson, D. A. (2022). Microplastic contamination in Great Lakes fish. *Conservation Biology*, 36(1), e13794.

Muruganandam, M., Rajamanickam, S., Sivarethinamohan, S., Reddy, M. K., Velusamy, P., Gomathi, R., Ravindiran, G., Rao Gurugubelli, T., & Munisamy, S. K. (2023). Impact of climate change and anthropogenic activities on aquatic ecosystem–A review. *Environmental Research*, 238, 117233.

Neelavannan, K., Sen, I. S., Lone, A. M., & Gopinath, K. (2022). Microplastics in the high-altitude Himalayas: Assessment of microplastic contamination in freshwater lake sediments, Northwest Himalaya (India). *Chemosphere*, 290, 133354.

Nguyen, P. D., Tran, Q. V., Le, T. T., Nguyen, Q. H., Kieu-Le, T. C., & Strady, E. (2023). Evaluation of microplastic removal efficiency of wastewater-treatment plants in a developing country, Vietnam. *Environmental Technology & Innovation*, 29, 102994.

Parashar, N., & Hait, S. (2023). Plastic rain—Atmospheric microplastics deposition in urban and peri-urban areas of Patna City, Bihar, India: Distribution, characteristics, transport, and source analysis. *Journal of Hazardous Materials*, 458, 131883.

Pathak, G. (2023). Plastic politics: Industry stakeholders and the navigation of plastic control policy in India. *Environmental Politics*, 32(1), 135–156.

Pereira, A. D. S., da Silva, J., Taniguchi, S., Montone, R. C., & Lourenço, R. A. (2024). Persistent organic pollutants and polycyclic aromatic hydrocarbons in livers of stranded Arctocephalus australis in southern Brazilian beaches. *Marine Pollution Bulletin*, 200, 116129.

Plastics Europe. (2022). Plastics- The Fact 2022. https://plasticseurope.org/knowledge-hub/plastics-the-facts-2022/. Accessed: 01.06.2024.

Plastics Europe. (2024). Plastics- The Fact 2023. https://plasticseurope.org/knowledge-hub/plastics-the-fast-facts-2023/. Accessed: 01.06.2024.

Qaiser, Z., Aqeel, M., Sarfraz, W., Rizvi, Z. F., Noman, A., Naeem, S., & Khalid, N. (2023). Microplastics in wastewaters and their potential effects on aquatic and terrestrial biota. *Case Studies in Chemical and Environmental Engineering*, 8, 100536.

Rbaibi Zipak, S., Muratoglu, K., & Buyukunal, S. K. (2024). Microplastics in raw milk samples from the Marmara region in Turkey. *Journal of Consumer Protection and Food Safety*, 19, 1–12.

Reed, S., Clark, M., Thompson, R., & Hughes, K. A. (2018). Microplastics in marine sediments near Rothera research station, Antarctica. *Marine Pollution Bulletin*, 133, 460–463.

Reid, B. J., Jones, K. C., & Semple, K. T. (2000). Bioavailability of persistent organic pollutants in soils and sediments—A perspective on mechanisms, consequences and assessment. *Environmental Pollution*, 108(1), 103–112.

Ren, X., Zeng, G., Tang, L., Wang, J., Wan, J., Liu, Y., Yu, J., Yi, H., Ye, S., & Deng, R. (2018). Sorption, transport and biodegradation–An insight into bioavailability of persistent organic pollutants in soil. *Science of the Total Environment*, 610, 1154–1163.

Rocha, M. J., Ribeiro, A. B., Campos, D., & Rocha, E. (2021). Temporal-spatial survey of PAHs and PCBs in the Atlantic Iberian northwest coastline, and evaluation of their sources and risks for both humans and aquatic organisms. *Chemosphere*, 279, 130506.

Sajjad, M., Huang, Q., Khan, S., Khan, M. A., Liu, Y., Wang, J., Lian, F., Wang, Q., & Guo, G. (2022). Microplastics in the soil environment: A critical review. *Environmental Technology & Innovation*, 27, 102408.

Sakan, S., Ostojić, B., & Đorđević, D. (2017). Persistent organic pollutants (POPs) in sediments from river and artificial lakes in Serbia. *Journal of Geochemical Exploration*, 180, 91–100.

Serra, T., & Colomer, J. (2023). Scavenging of polystyrene microplastics by sediment particles in both turbulent and calm aquatic environments. *Science of the Total Environment*, 884, 163720.

Sheriff, I., Debela, S. A., & Mans-Davies, A. (2022). The listing of new persistent organic pollutants in the Stockholm Convention: Its burden on developing countries. *Environmental Science & Policy*, 130, 9–15.

Shorobi, F. M., Vyavahare, G. D., Seok, Y. J., & Park, J. H. (2023). Effect of polypropylene microplastics on seed germination and nutrient uptake of tomato and cherry tomato plants. *Chemosphere*, 329, 138679.

Şimşek, İ., & Bilgili, A. (2022). Investigation of monthly residues of polycyclic aromatic hydrocarbons in water and sediment samples from Kirikkale Kizilirmak River Basin. *Environmental Monitoring and Assessment*, 194(10), 683.

Steer, M., Cole, M., Thompson, R. C., & Lindeque, P. K. (2017). Microplastic ingestion in fish larvae in the western English Channel. *Environmental Pollution*, 226, 250–259.

Sun, C., Teng, J., Wang, D., Zhao, J., Shan, E., & Wang, Q. (2023). The adverse impact of microplastics and their attached pathogen on hemocyte function and antioxidative response in the mussel *Mytilus galloprovincialis*. *Chemosphere*, 325, 138381.

Sun, Y. X., Zhang, Z. W., Xu, X. R., Hu, Y. X., Luo, X. J., Cai, M. G., & Mai, B. X. (2015). Bioaccumulation and biomagnification of halogenated organic pollutants in mangrove biota from the Pearl River Estuary, South China. *Marine Pollution Bulletin*, 99(1–2), 150–156.

Suteja, Y., Purwiyanto, A. I. S., Purbonegoro, T., & Cordova, M. R. (2024). Spatial and temporal trends of microplastic contamination in surface sediment of Benoa Bay: An urban estuary in Bali-Indonesia. *Marine Pollution Bulletin*, 202, 116357.

Tan, H., Othman, M. H. D., Chong, W. T., Kek, H. Y., Wong, S. L., Nyakuma, B. B., Mong, G. R., Wahab, R. A., & Wong, K. Y. (2024). Turning plastics/microplastics into valuable resources? Current and potential research for future applications. *Journal of Environmental Management*, 356, 120644.

Tang, K. H. D., & Hadibarata, T. (2021). Microplastics removal through water treatment plants: Its feasibility, efficiency, future prospects and enhancement by proper waste management. *Environmental Challenges*, 5, 100264.

Tavşanoğlu, Ü. N., Başaran Kankılıç, G., Akca, G., Çırak, T., & Erdoğan, Ş. (2020). Microplastics in a dam lake in Turkey: Type, mesh size effect, and bacterial biofilm communities. *Environmental Science and Pollution Research*, 27, 45688–45698.

Teng, J., Wang, Q., Ran, W., Wu, D., Liu, Y., Sun, S., Liu, H., Cao, R., & Zhao, J. (2019). Microplastic in cultured oysters from different coastal areas of China. *Science of the Total Environment*, 653, 1282–1292.

Thompson, R. C., Olsen, Y., Mitchell, R. P., Davis, A., Rowland, S. J., John, A. W., McGonigle, D., & Russell, A. E. (2004). Lost at sea: Where is all the plastic? *Science*, 304(5672), 838.

UNEP. (2009). Handbook for effective participation in thework of the POPs Review Committee, (February). *Secretariat of the Stockholm Convention on Persistent Organic Pollutants United Nations Environment Programme International Environment House 11–13*, chemin des Anémones CH-1219, Châtelaine, Geneva, Switzerland.

Van Ryan Kristopher, R. G., Jaraula, C. M. B., & Paler, M. K. O. (2021). The nexus of macroplastic and microplastic research and plastic regulation policies in the Philippines marine coastal environments. *Marine Pollution Bulletin*, 167, 112343.

Verma, A., Chand, N., Upadhyay, P., Sharma, S., & Prajapati, S. K. (2024). Systematic review on microplastics as a threat to terrestrial and aquatic eco-environment. *Sustainable Chemistry One World*, 3, 100013.

Wang, F., Lei, A., Huang, S., Su, L., Li, M., Craig, N. J., & Lai, Z. (2024). Impact of typhoon events on microplastic distribution in offshore sediments in Leizhou Peninsula of the South China Sea. *Environmental Pollution*, 348, 123817.

Wang, T., Qu, L., Luo, D., Ji, X., Ma, Z., Wang, Z., Dahlgren, R. A., Zhang, M., & Shang, X. (2023). Microplastic pollution characteristics and its future perspectives in the Tibetan Plateau. *Journal of Hazardous Materials*, 457, 131711.

Wang, W., Ge, J., Yu, X., & Li, H. (2020). Environmental fate and impacts of microplastics in soil ecosystems: Progress and perspective. *Science of the Total Environment*, 708, 134841.

Wang, X. P., Yao, T. D., Cong, Z. Y., Yan, X. L., Kang, S. C., & Zhang, Y. (2007). Distribution of persistent organic pollutants in soil and grasses around Mt. Qomolangma, China. *Archives of Environmental Contamination and Toxicology*, 52, 153–162.

Wei, L., Yue, Q., Chen, G., & Wang, J. (2023). Microplastics in rainwater/stormwater environments: Influencing factors, sources, transport, fate, and removal techniques. *TrAC Trends in Analytical Chemistry*, 165, 117147.

Wu, J., Ye, Q., Sun, L., Liu, J., Huang, M., Wang, T., Wu, P., & Zhu, N. (2023). Impact of persistent rain on microplastics distribution and plastisphere community: A field study in the Pearl River, China. *Science of the Total Environment*, 879, 163066.

Xiang, P., Zhang, T., Wu, Q., & Li, Q. (2023). Systematic review of degradation processes for microplastics: Progress and prospects. *Sustainability*, 15(17), 12698.

Xu, X., Li, T., Zhen, J., Jiang, Y., Nie, X., Wang, Y., Yuan, X.-Z., Mao, X., Xue, L., & Chen, J. (2023). Characterization of microplastics in clouds over Eastern China. *Environmental Science & Technology Letters*, 11(1), 16–22.

Xu, Z., Bai, X., & Ye, Z. (2021). Removal and generation of microplastics in wastewater treatment plants: A review. *Journal of Cleaner Production*, 291, 125982.

Yu, J., Wang, P., Ni, F., Cizdziel, J., Wu, D., Zhao, Q., & Zhou, Y. (2019). Characterization of microplastics in environment by thermal gravimetric analysis coupled with Fourier transform infrared spectroscopy. *Marine Pollution Bulletin*, 145, 153–160.

Yu, L., Zhang, J., Liu, Y., Chen, L., Tao, S., & Liu, W. (2021). Distribution characteristics of microplastics in agricultural soils from the largest vegetable production base in China. *Science of the Total Environment*, 756, 143860.

Yuan, Z., Pei, C., Li, H., Lin, L., Liu, S., Hou, R., Ligo, R., & Xu, X. (2023). Atmospheric microplastics at a southern China metropolis: Occurrence, deposition flux, exposure risk and washout effect of rainfall. *Science of the Total Environment*, 869, 161839.

Zhang, S., Pei, L., Zhao, Y., Shan, J., Zheng, X., Xu, G., Sun, Y., & Wang, F. (2023a). Effects of microplastics and nitrogen deposition on soil multifunctionality, particularly C and N cycling. *Journal of Hazardous Materials*, 451, 131152.

Zhang, S., Wang, W., Yan, P., Wang, J., Yan, S., Liu, X., & Aurangzeib, M. (2023b). Microplastic migration and distribution in the terrestrial and aquatic environments: A threat to biotic safety. *Journal of Environmental Management*, 333, 117412.

Zhang, X., Robson, M., Jobst, K., Pena-Abaurrea, M., Muscalu, A., Chaudhuri, S., Marvin, C., Brindle, I. D., Reiner, E. J., & Helm, P. (2020). Halogenated organic contaminants of concern in urban-influenced waters of Lake Ontario, Canada: Passive sampling with targeted and non-targeted screening. *Environmental Pollution*, 264, 114733.

Zhang, Y., Song, M., Zhu, Y., Li, H., Zhang, Y., Wang, G., Chen, X., Zhang, W., Wang, H., Wang, Y., Shao, R., Guo, J., & Yang, Q. (2024). Impact of microplastic particle size on physiological and biochemical properties and rhizosphere metabolism of *Zea mays* L.: Comparison in different soil types. *Science of the Total Environment*, 908, 168219.

Zhang, Z. (2020). Monitoring and Evaluation of Irrigation Water Utilization Coefficient under Different Irrigation Methods in Xiliaohe Well Irrigation District (Doctoral dissertation, Changchun Institute of Technology).

Zhang, Z., Zhao, S., Chen, L., Duan, C., Zhang, X., & Fang, L. (2022). A review of microplastics in soil: Occurrence, analytical methods, combined contamination and risks. *Environmental Pollution*, 306, 119374.

Zhou, J., Jia, R., Brown, R. W., Yang, Y., Zeng, Z., Jones, D. L., & Zang, H. (2023). The long-term uncertainty of biodegradable mulch film residues and associated microplastics pollution on plant-soil health. *Journal of Hazardous Materials*, 442, 130055.

Zhu, L., Kang, Y., Ma, M., Wu, Z., Zhang, L., Hu, R., Xu, Q., Zhu, J., Gu, X., & An, L. (2024). Tissue accumulation of microplastics and potential health risks in human. *Science of the Total Environment*, 915, 170004.

2 Plastic-Associated Contaminants

Omowunmi H. Fred-Ahmadu, Frederick O. Ahmadu, and Oluwafemi T. Ogunmodede

2.1 INTRODUCTION

Synthetic plastics are a cocktail of many chemicals, some of which are contaminants and others are classified as pollutants (Koelmans et al., 2016). For the purpose of proper understanding, it is important to differentiate between contaminants and pollutants. A contaminant is a substance present where it should not be or at concentrations above background. A pollutant is a substance that results in or can result in adverse biological effects to resident communities. All pollutants are contaminants but not all contaminants are pollutants (Chapman, 2007). The chemical composition of plastics includes their monomeric units and additives such as softeners, fillers, UV stabilizers, antioxidants, pigments, fire retardants, lubricants, impact modifiers, cross-linking agents, blowing agents, antistatic agents and so on. The additives are chemical compounds added to plastic during production to improve performance or functionality of the products (Ambrogi et al., 2016; Fred-Ahmadu et al., 2020). Plastics are non-biodegradable and some of the chemical additives are classified as environmental pollutants because of their persistence, bioaccumulation and toxic (PBT) characteristics. Throughout the plastic life cycle, from the extraction of oil and gas to the synthesis of polymers and chemicals during production, to the usage and disposal of plastics, a large number of hazardous chemicals are used, released and emitted into various environmental matrices. The growing environmental awareness and increasing amount of scientific evidence about the presence of synthetic polymers in aqueous and terrestrial environments have triggered the interest and concern of the general public and scientific community about this relatively new class of contaminants. The term "plastics" includes a variety of materials, which contain a diversity of additive chemicals, some required by law and some added by industry, and many of which can leach and extract off the polymer (Gulizia et al., 2023; Luo et al., 2019).

Plastics have been found to act as a vector for chemicals in the marine environment, concentrating and transporting toxic pollutants to organisms which ingest them (Koelmans et al., 2016). When plastic particles are abundant in seas, marine life encounter the chemical-laden particles and ingest them leading to potentially harmful effects (Rochman et al., 2013). Since small particles (micro(nano)plastics) are proven to make up a significant portion of plastic pollution in the environment, understanding the interaction of these plastic particles with contaminants and pollutants will be important for the overall picture of the effects of plastic and ultimately for assessing the ecological impact of plastic debris. This also has implications for the disposal of plastics and our management of both freshwater and the oceans. Due to environmental persistence and the increasing rates of plastics disposal, it is conceivable that plastics will continue to serve as a source of chemicals in both the terrestrial and aquatic environment for the foreseeable future.

Much work has been done on the effects of plastic debris in the marine, freshwater and terrestrial environments and the adsorption of metals and persistent organic pollutants (POPs) such as dichlorodiphenyltrichloroethane (DDT), polycyclic aromatic hydrocarbons (PAHs), organochlorine pesticides (OCPs) and polychlorinated biphenyls (PCBs) to these materials (Bakir et al., 2014; Dobaradaran et al., 2018; Fred-Ahmadu et al., 2022). This is different from the chemical interaction

DOI: 10.1201/9781032717814-2

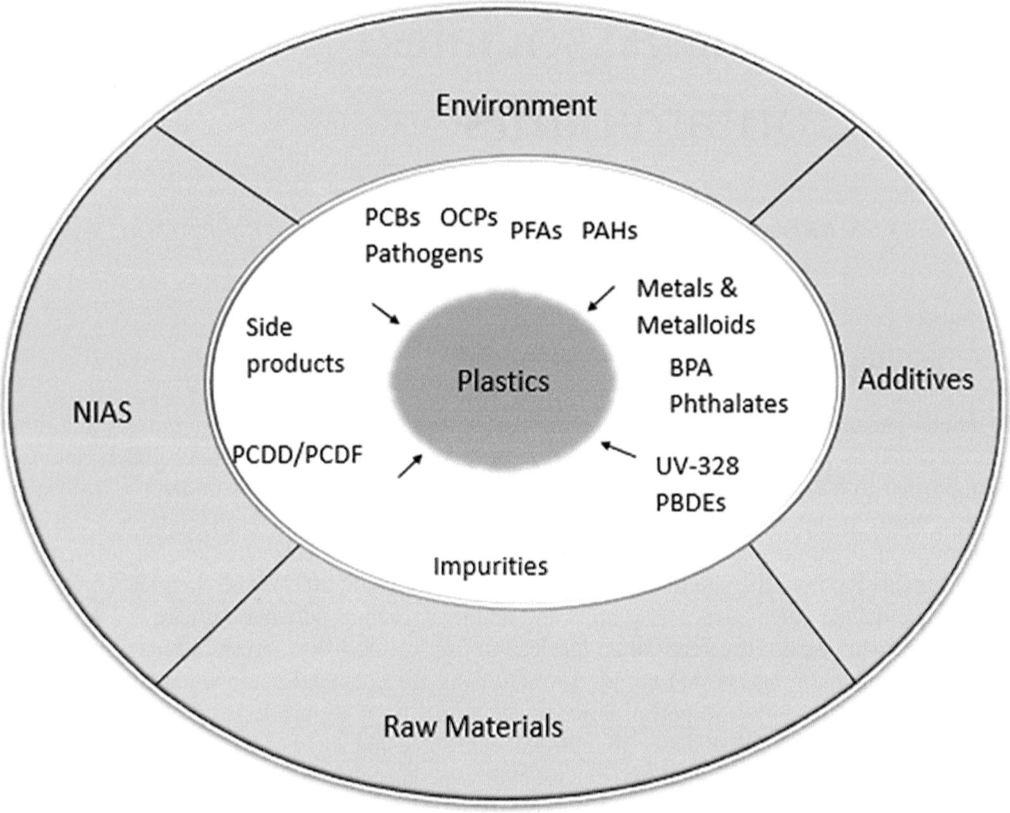

FIGURE 2.1 Sources of plastic contaminants.

of a plastic particle and a given chemical, in which the chemical can move into the polymer, and may even be a catalyst for degrading the plastic. Simulation studies and laboratory experiments conducted at a range of temperatures and pressures have shown that conditions are often a key factor in the extent of uptake of a chemical into a polymer (Town et al., 2023). With regard to conditions in the marine environment, the global distribution of microplastic (MP) particles has been modelled using a variety of transport mechanisms, indicating that there are more than 5.25×10^{12} plastics in the surface seawater with a weight of 35,000 tonnes (Auta et al., 2017; Jambeck et al., 2015; Uzun et al., 2022). The statistical distribution of temperature and salinity of the sea in the models may allow for assessment of the most probable locations for chemical uptake into plastic in the marine environment. The major classes of contaminants associated with plastics are organic contaminants, inorganic contaminants (metals and metalloids) and pathogens. The plastics derive their toxicity potential from their ability to sorb/desorb chemical contaminants and pathogens that originate from chemical additives, raw materials, the environment and non-intentionally added substances (NIAS) as shown in Figure 2.1.

2.1.1 Chemical Contaminants Associated with Plastics

Concerns about the ubiquitous distribution of plastics throughout the global environment have fostered an increased awareness of the adverse effects of plastics on human health and wildlife. This concern over plastics initially focused on the physical damage caused by plastic debris on wildlife and the environment. More recently, the potential for plastics to cause chemical damage has

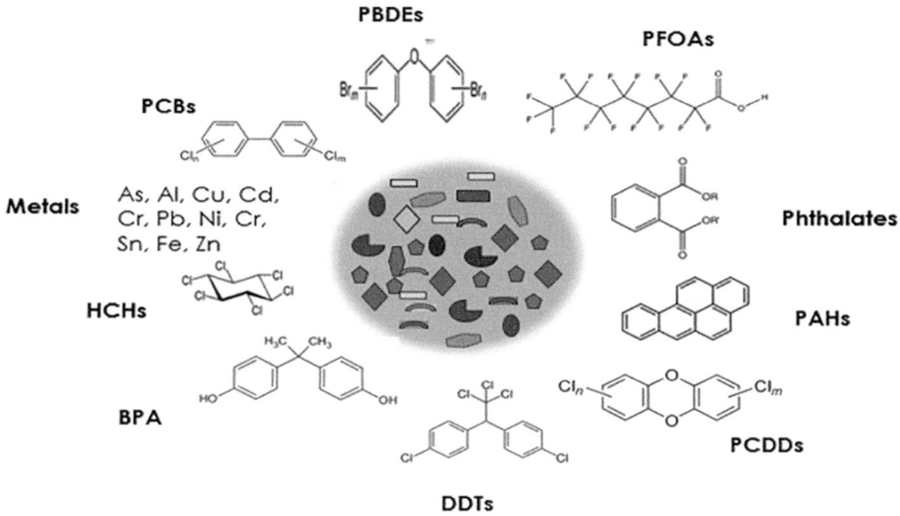

FIGURE 2.2 Chemicals associated with plastics.

received greater attention because their pervasiveness and small sizes as microplastics and nanoplastics allow them to penetrate the biogeochemical cycles and food webs, spreading from the surface layer of coastal waters to the sediment compartment of the open oceans (Benson et al., 2022; Rodrigues et al., 2022). According to a United Nations Environment Programme technical report, more than 13,000 chemicals have been identified with plastics, over 7,000 of which have been analysed for their hazardous properties and more than 3,200 of them are of potential concern (UNEP, 2023). These chemicals include but are not limited to additives such as flame retardants (PBDEs), plasticisers (phthalates), stabilizers (BPA, Cd and Pb compounds), heat stabilizers (Cd and Pb compounds), slip agents (zinc stearate), biocides (arsenic and organotin compounds), organic pigments (cobalt (II)diacetate), inorganic pigments (titanium oxide, Cd, Cr and Pb compounds), UV stabilizers (UV-328), per- and polyfluoroalkyl substances (PFASs) and certain alkylphenols and alkylphenol ethoxylates (APEOs). PAHs, OCPs and PCBs are sorbed from the environment while NIAS include polychlorinated dibenzo-p-dioxins and polychlorinated dibenzofurans (PCDD/PCDFs), PCBs, PCNs and hexachlorobenzene (HCB) (Hahladakis et al., 2013; UNEP, 2023). The structures of some of the common contaminants are depicted in Figure 2.2. Plastic-associated chemical contaminants are further discussed in the following sections.

2.1.1.1 Hydrophobic Organic Compounds (HOCs) Associated with Plastics

Many studies have reported the adsorption of various organic pollutants, such as PAHs, PCBs, OCPs, DDTs, phthalates and antibiotics, by plastic particles (Fred-Ahmadu et al., 2022; Liu et al., 2019). Both chemical (chemisorption) and physical (physisorption) associations between the sorbate and the sorbent are relevant when discussing sorption. The solid phase and the chemicals interact during chemisorption by the formation of covalent bonds; non-covalent van der Waals forces are necessary for physisorption. When chemisorption is involved, it is difficult to desorb compounds from the solid phase because chemisorption is usually irreversible unless covalent connections are broken (Endo & Koelmans, 2016). However, physisorbed chemicals are much easily released from the solid phase matrix. Many chemical additives of environmental concern such as phthalates, PBDEs and BPA have been shown to leach from plastics because they are loosely held within to the plastic structure by weak van der Waals forces (physisorption). PBDEs (up to 2,158 ng/g) were found to be associated with plastic particles (0.315–1 mm) collected from the Pacific Ocean in Japan (Yeo et al., 2019). In an experiment simulating marine conditions, the leaching of plasticisers

(BPA and DEHP) from PVC was quantified using gel permeation chromatography and thermal gravimetric analysis (TGA). It was reported that both additives leached and leaching was enhanced by increased temperature confirming that PVC particles contribute to plasticizer pollution in the aquatic environment (Gulizia et al., 2023). A study modelled the importance of MPs as a pathway for the transfer of HOCs to marine life, taking the influence of gut surfactants, pH and temperature into consideration. The results suggested that ingestion of MPs does not provide a quantitatively important additional pathway for the transfer of adsorbed chemicals from seawater to biota via the gut (Bakir et al., 2016). In another study, modelling of the bioaccumulation and biomagnification potential of MPs in a cetacean food web was carried out. In comparison to high biomagnification behaviour of HOCs in marine food webs, low biomagnification capacity of MP was predicted in the cetacean food webs by the model (Alava, 2020). However, a critical review of many experimental studies suggests that the role of MPs in HOCs accumulation depends on the chemical partitioning between MPs and biota, when MP is ingested by an organism. So that MP ingestion may either adsorb HOCs or contaminate the organism by desorption of HOCs, depending on the chemical fugacity gradient between ingested plastic and organism tissue (Koelmans, 2015). Therefore, it is possible that marine organisms experience a trade-off between negative effects of chemical transfer from additives to the organism and positive effects of attenuation of POP bioaccumulation, upon ingestion of MP (Koelmans, 2015). However, more research investigating these chemical exchanges in realistic scenarios, considering complex mixtures of chemicals and different exposure pathways are needed.

2.1.1.2 Metals and Metalloids

Metal pollutants that may be adsorbed from the aquatic environment by plastics have been identified as a potential transport medium to biota upon ingestion of the plastic particle. Studies have found that various potentially toxic metals and metalloids such as aluminium (Al), arsenic (As), cadmium (Cd), nickel (Ni), iron (Fe), mercury (Hg), copper (Cu), lead (Pb), chromium (Cr), tin (Sn), titanium (Ti) and zinc (Zn) have been adsorbed from the surrounding environment by plastics, often at concentrations several orders of magnitude greater than the ambient environment (Li et al., 2020; Vedolin et al., 2018). Plastic surfaces are susceptible to metal(loids) adsorption when weathering by mechanical, UV and chemical (hydrolysis) processes occurs. Weathering modifies the plastics by creating reactive surfaces/binding sites for metal(loids) (Brennecke et al., 2016). Fragmentation and surface modification lead to an increase in the surface-to-volume ratio and the creation of new functional groups (such as esters, keto groups and quinone-related structures) on the surface of polymers (Yu et al., 2019). The sorption and desorption pathways of metal–plastic association could have long-term effects that are currently ill understood. For instance, the potential to release adsorbed metals by plastics has been investigated by some studies. Pb adsorbed onto conventional and biodegradable polyethylene (PE) plastics was reported to completely desorb and release into simulated digestive systems within 10 hours, indicating the potential toxicity to organisms if ingested (Lin et al., 2023). Similarly, MPs facilitated the bioaccumulation and toxicity of Cd, resulting in oxidative damage and inflammation in *Danio rerio* (adult fish), due to the long-term co-exposure of Cd and MPs (Lu et al., 2018). On the contrary, a study reported the exposure of mussels to Hg via MPs, microalgae and water and found that Hg ingested via MPs was comparable to the levels of Hg accumulated in microalgae or dissolved in water. The tested mussels excreted more than 70% of the MP-associated Hg over the depuration time. This observation was attributed to a strong affinity between Hg and MPs and the non-nutritiveness of the plastics (Rivera-Hernández et al., 2019). This study suggests that plastic particles could be used for eliminating ingested Hg from organisms.

The adverse effects of metals associated with plastic may be due to the additive and local effects of the metals. For many metals, the release into the gastrointestinal tract after ingestion of the plastic may cause an increase in metal ion bioavailability compared with ingestion of the free metal, due to the metal ion being released from the plastic in a different complex form. Potentially toxic metals can cause carcinogenic, teratogenic and allergenic effects and also induce behavioural alterations at neurological levels (Tchounwou et al., 2012).

2.1.2 Biological Contaminants Associated with Plastics

Plastics have the ability to adsorb and transport organic matter, biomolecules, nutrients and bacteria in environmental compartments, particularly in aquatic ecosystems. They can act as a novel surface for the spread of nonindigenous species (NIS) into new habitats (Audrézet et al., 2021). When a novel microbial community is attached to plastic distinct from the surroundings, it is described as plastisphere (Zettler et al., 2013). Apart from the direct physical effects on organisms such as entanglements, gut blockage and abrasion, plastics (including MPs) could act as vectors for biological agents such as microbes when biofilms are formed on their surfaces (GESAMP, 2016). These biofilms can also promote chemical or physical adsorption of other contaminants through trapping, accumulation and multiple binding sites (Tumwesigye et al., 2023). According to Arias-Andres et al. (2018), biofilms can capture and spread the eggs of invasive species, including invertebrates, to new areas. These invasive species' eggs have the potential to seriously threaten the recipient system's ecology when they spread to other water bodies. MPs cover large distances as they float on the sea surface, which could facilitate diseases spread to distant places from their initial source. The chance of pathogen ingestion by filter-feeding marine invertebrate species is further increased by the possibility that sinking plastic particles will concentrate these pathogens in the benthos, which may lead to a significant impact on organisms (Zhang et al., 2022).

Research in this area is still very limited, and more attention is needed considering the issue of biological invasion. Although the extent of the participation of plastics (including MPs) in the transfer of pathogens to organisms is still a subject of investigation, a number of studies have reported the ability of plastics to act as a substrate to bacteria, viruses, antibiotic resistance genes (ARGs) and certain invertebrates. Some of the findings are discussed in the following sections.

2.1.2.1 Bacteria, Fungi and Antibiotic Resistance Genes (ARGs)

Plastics may be regarded as emerging habitats, otherwise called "plastisphere" that facilitate the dispersion of microbial communities, which can sometimes include pathogenic bacteria. Unlike autochthonous substrates, plastics have a longer half-life than most naturally occurring floating marine substrates and a hydrophobic surface that encourages microbial colonization, biofilm and viruses (Meng et al., 2021; Zettler et al., 2013). Scanning electron microscopy (SEM) data by Zettler et al. (2013) demonstrated that MP surfaces are linked to biofilms, and high-throughput sequencing technology identified the bacteria within the biofilms. The SEM data showed that cyanobacteria, proteobacteria and bacteroides were among the bacteria that accumulated on MPs. In a recent study, researchers found that pristine and weathered plastics (LDPE and PP) submerged for 7 days in a river harboured opportunistic "microbial hitchhikers" such as *Pseudomonas aeruginosa* and *Acinetobacter baumannii* as well as a distinct set of ARGs, indicating that environmental plastics could serve as reservoirs of harmful pathogenic organisms (Zadjelovic et al., 2023). Another study by Lear et al. (2022) showed that various human pathogens cling to plastics in seawater. Plastic strips (made of LDPE, HDPE, PP, PVC and PET) were submerged in the ocean off the coast of Falmouth, England. A thin layer of biofilm was formed on the plastics within 7 days. *Galleria mellonella*, or wax moth larvae, were injected with the biofilm in order to determine the extent to which plastic can be colonized by potentially pathogenic bacteria. Over the course of 5 weeks, the number of pathogenic bacteria increased significantly and the mortality rate of Galleria jumped from 4% in week one to 65% in week five. Serratia and Enterococcus species were among the pathogens that were extracted from plastic biofilms using Galleria enrichment, and they also carried a variety of ARGs. In another laboratory study by Arias-Andres et al. (2018), the permissiveness of aquatic bacteria towards a model antibiotic resistance plasmid was carried out by comparing communities that form biofilms on MPs against those that are free-living. An exogenous and red-fluorescent *E. coli* donor strain was used to introduce the green-fluorescent broad-host-range plasmid, which encodes for trimethoprim resistance. An increased frequency of plasmid transfer in bacteria associated with MPs was observed compared to bacteria that are free-living or in natural aggregates.

Table 2.1 presents further evidence of pathogenic bacteria, fungi and ARGs association with plastics of various types. Most of the published reports available concerning the colonization of plastics by biological agents relate to MPs. MPs have been shown to function as selective artificial microhabitats that not only attract specific fungal communities but also accumulate certain opportunistic human pathogens, such as cryptococcal and Phoma-like species (Gkoutselis et al., 2021). The available reports also showed that bacterial colonization is independent of plastic polymer type and microbial community structures of plastics-associated bacteria were different from that in surrounding waters. However, whether pathogens carried by MPs can affect the health of organisms that ingest them is yet to be clearly understood and requires further investigation.

2.1.2.2 Viruses

Viral pathogens are currently the greatest risk for zoonotic spillover and human host infection, as demonstrated by the recent COVID-19 outbreak. Even though humans are always in contact with animal viruses, it is uncommon for them to spread from an animal host to a human. When it does, though, the quick development and spread of viruses means that a large-scale epidemic or possibly a pandemic may result. Zoonotic virus transmission from animals to humans can happen through direct or indirect contact, necessitating an animal–human point of contact or environmental source contact (Nandi & Allen, 2021; Ormsby et al., 2024). The magnitude of the involvement of plastics as environmental source contact to organisms and humans needs to be further investigated. However, some studies have reported the participation of plastics in the retention of viral pathogens. SARS-CoV-2 was found to be more stable on plastic and stainless steel than on copper and cardboard, and viable virus was detected up to 72 hours after application to these surfaces. The longest viability of both SARS-CoV-1 and SARS-CoV-2 viruses was on stainless steel and plastic; the estimated median half-life of SARS-CoV-2 was approximately 5.6 hours on stainless steel and 6.8 hours on plastic (van Doremalen et al., 2020). In another study by Moresco et al. (2022), it was reported that the presence of an envelope may limit virus interaction with plastisphere; however, the ability to recover both enveloped and non-enveloped infectious viruses from colonized MP pellets highlights the potential public health risk of surface waters becoming contaminated with pathogen-laden MPs. Table 2.2 presents some of the studies that have reported the association of different plastic polymers with viruses, with persistence ranging from 48 to 216 hours at temperatures between 21°C and 27°C. Pathogenic bacteria adsorption on plastics is more frequently reported than other biological agents.

2.2 FACTORS INFLUENCING PLASTICS INTERACTION WITH CHEMICAL CONTAMINANTS AND PATHOGENIC AGENTS

The capacity of MPs to transport chemical contaminants and pathogens on their surfaces depends on their adsorption strength for such chemical contaminants/pathogens, the physical and chemical properties of the plastic and environmental conditions. The major mechanisms by which plastic particles adsorb contaminants include hydrophobic interactions, partition and electrostatic interactions. Others are pi–pi interactions, van der Waals forces, hydrogen bonding and other non-covalent interactions (Liu et al., 2016). In addition, plastic particles have the potential to hinder pathogenic bacteria from contacting porous media or competing with them for deposition sites, which could result in the long-distance spread of harmful microorganisms in the aquatic environment. Since pathogens and plastic particles have surfaces that are net negatively charged, electrostatic repulsion between

TABLE 2.1
Evidence of Pathogenic Bacteria, Fungi and Antibiotic Resistance Genes on Plastic

Plastic Type	Name of Bacteria/Fungi	Size of Plastic	Environmental Matrix	Location	References
PE & PP	*Exiguobacterium and Eubacterium*	0.01–6.93 mm	River water	Japan	(Guruge et al. (2024)
HDPE & PP	*Salmonella enteritidis, S. typhimurium and S. typhi*	20 μm	Surface water	Malawi and the U.K.	Ormsby et al. (2024)
LDPE & PP (pristine and weathered)	*Pseudomonas aeruginosa, Acinetobacter aeromonas, ARGs*	Square films 4 cm^2	River water	River Sowe, United Kingdom	Zadjelovic et al. (2023)
EPS, PE, PP & PET	*Gammaproteobacteria, Alphaproteobacteria; Bacilli and Clostridia*		Coastal waters	South Korea	Jeong et al. (2023)
PE, PP (fragments & pellets)	*Escherichia coli* & *Vibrio* spp.	1–5 mm	Beach sand	Canary Islands, Spain	Hernández-Sánchez et al. (2023)
LDPE, HDPE, PP, PVC, PET	*Serratia and Enterococcus species and ARGs*	85 mm by 25 mm strips	Coastal waters	Cornwall, UK.	Lear et al. (2022)
PE microbeads & polyester fibres	*Toxoplasma gondii, Cryptosporidium parvum and Giardia enterica*	100 and 500 μm; 400–700 μm and 800–1,200 μm	Seawater	USA	Zhang et al. (2022)
PE & PS	*Vibrio* spp.	<5 mm	Sea water	Baltic Sea, Germany	Kesy et al. (2021)
PS	*Ochrobactrum anthropic*, *Acinetobacter lwoffii*, *Afipia broomeae*, *A. baumannii* and *P. aeruginosa*	≤1 mm	Landfill leachate	Laogang landfill, Shanghai, China	Shi et al. (2021)
PP straws	*Planktothrix agardhii* and *P. rubescens*	<5 mm	Freshwater lake	Hungary	Szabó et al. (2021)
Plastic waste	*P. herbarum*, *R. anthropophila*, *C. lunata*, *N. diffluens* and *Rh. mucilaginosa*, cryptococcal and Phoma-like species	<5 mm	Terrestrial soil	Siaya, Western Kenya	Gkoutselis et al. (2021)
Food-related marine plastic litter	*Achromobacter xylosoxidans*, *Lelliottia amnigena*, *Neisseria flava/perflava/subflava*, *Pantoea agglomerans*, *Providencia rettgeri*, *Pseudomonas fluorescens*, *Rahnella aquatilis*, and *Serratia fonticola*	>5 mm	Sea water	North Atlantic Ocean, the North Channel and the Irish Sea	Moore et al. (2020)
Plastic pellets	*Escherichia coli* & *Vibrio* spp.	3–5 mm	Public bathing beaches	Forth Estuary, Scotland, UK	Rodrigues et al. (2019)

TABLE 2.2
Instances of Pathogenic Virus Loading on Plastic Surfaces

Polymer Type	Name of Virus	Size of Plastic	Persistence (hours)	Temperature (°C)	Location	References
PP pellets	Rotavirus (RV) SA11 (a nonenveloped enteric virus) and the enveloped bacteriophage Phi6	2 mm	48	25–27	Scotland	Moresco et al. (2022)
NA	SARS-CoV-1 (AY274119.3) SARS-CoV-2 (MN985325.1)	NA	~72	21–23	USA	van Doremalen et al. (2020)
NA	SARS-CoV-2	NA	~96	22	China	Chin et al. (2020)
PTFE, PVC	Human coronavirus 229E	NA	~120	21	UK	Warnes et al. (2015)
NA	SARS-CoV-1 (HKU39849)	NA	~120	22–25	Hong Kong	Chan et al. (2011)
PS	SARS-CoV-1 (FFM1)	NA	~216	21–25	Germany	Rabenau et al. (2005)

them occurs most of the time. Nevertheless, the pathogen flagella's hydrophobicity, proteins, surface charges and intricacy all work together to overcome the repulsion (Rühs et al., 2014). The factors that shape the interaction of plastics with chemical contaminants and pathogenic biological agents are summarized in Figure 2.3.

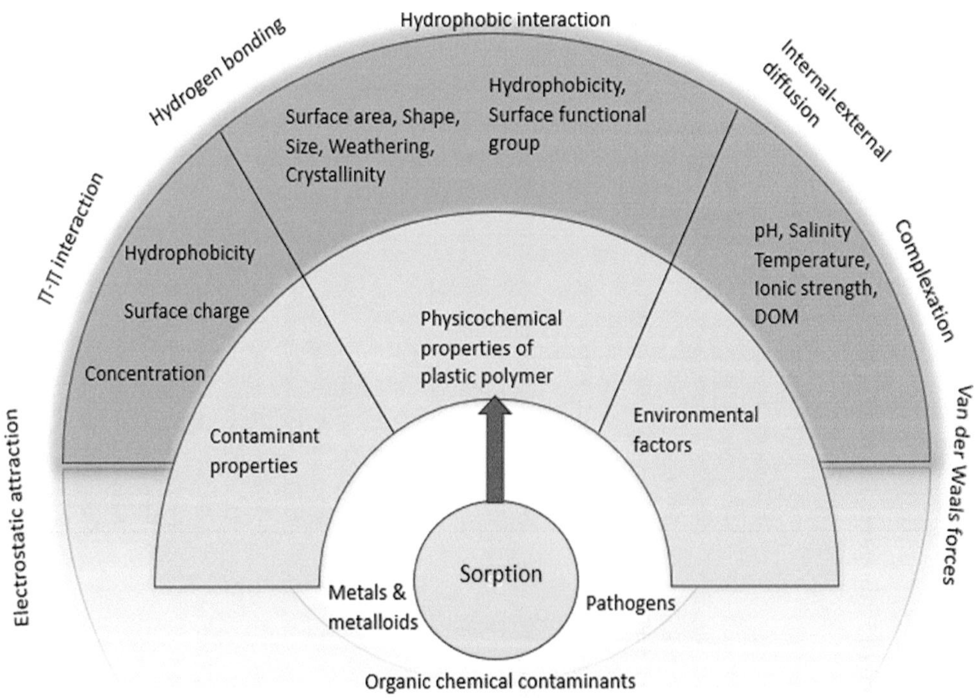

FIGURE 2.3 Factors influencing the sorption of contaminants to plastics and sorption mechanisms.

2.2.1 Physical Properties of Plastics

Physical characteristics such as surface area, particle size, shape (fibres or spherical), degree of crystallinity and surface weathering have been shown to play a role in the association of plastics with contaminants and pollutants. MPs provide the large surface-to-volume ratio and small particle size required for the adsorption and transport of species in the aquatic environment (Agboola & Benson, 2021). In a study to investigate the interaction of zoonotic protozoan parasites *Toxoplasma gondii, Cryptosporidium parvum* and *Giardia enterica* with PE microbeads and polyester microfibres in seawater, it was reported that microfibres showed a higher potential for associating with pathogens than microbeads. Furthermore, in terms of particle size, more pathogens showed affinity for the large microfibres (800–1,200 μm) when compared to the other MP types. The microbeads (100–500 μm) and small microfibres (400–700 μm) were comparable in their ability to associate with pathogens, with the exception of *G. enterica* where there was a statistical difference between the small microfibres and the microbeads (Zhang et al., 2022). In another study, Bhagwat et al. (2021) demonstrated that weathered microfibres exhibited higher surface areas for adsorption of perfluorooctane sulfonate (PFOS) and lead (Pb) at environmentally relevant concentrations due to biomass accumulation compared to pristine samples. Another study indicated that the adsorption behaviour of PE showed that as the surface morphology changed, hydrogen bonding increased and enhanced the adsorption capacity of weathered MPs (Yu et al., 2024). When virgin plastics undergo weathering, they form oxygen-containing functional groups and microcracks, which enhance their sorption capacity (G. Liu et al., 2019).

The presence of straight chains with regularly spaced side groups promotes crystallization in polymer molecules. Amorphous polymers, on the other hand, are characterized by chains that are incapable of forming a rigid arrangement. They form an amorphous glassy solid in which the molecular chains are randomly arranged. The sorption isotherm of a rubbery polymer is essentially linear and the absorption process is reversible and non-competitive. Conversely, glassy polymers undergo adsorption, showing competition with any coexisting chemicals and their isotherms are nonlinear (Endo & Koelmans, 2019). The crystalline region within a polymer does not favour the sorption of chemicals due to the high energy that is required to destabilize the strongly ordered polymer chains. However, the amorphous portion of the polymer which has randomly oriented polymeric chains is more available for greater sorption of chemical compounds (Endo & Koelmans, 2019; Fred-Ahmadu et al., 2020). Guo et al. (2012) carried out a study of the adsorption of HOCs (i.e., phenanthrene, naphthalene, lindane and 1-naphthol) by three types of polymers namely PE, polystyrene (PS) and polyphenyleneoxide (PPO) and found that the sorption of HOCs was enhanced as the mobile and rubbery domains on the polymers increased. This is an indication that crystallinity regulated the sorption capacities of HOCs by plastics.

2.2.2 Chemical Properties of Plastics

Apart from physical properties, chemical characteristics such as hydrophobicity and polymer surface functional groups have been shown to influence the adsorption behaviour of plastics towards chemical contaminants and pathogens. Most plastics are synthesized from chemicals sourced from fossil fuels and they possess intrinsic hydrophobic properties. Hydrophobic interaction is one of the predominant mechanisms influencing the sorption of hydrophobic organic contaminants (HOCs) to plastics (Yu et al., 2019). Generally, hydrophobic interaction increases with the organic contaminant–water partition coefficient of the sorbent (Cui et al., 2023). Surface polarity of MPs determines how strongly an organic contaminant interacts with it; HOCs preferably adhere to non-polar surfaces (Mato et al., 2001). However, the presence of oxygen-containing functional groups (ether moieties) on a polar polymer like PPO could make them act as hydrogen bond acceptor and interact with water molecules (H bond donor), thus forming water clusters at its surfaces through hydrogen bonding. The water clusters would reduce the accessibility of organic contaminants to the sorption domains and compete with them for sorption sites, thereby reducing their sorption capacity (Guo et al., 2012).

Strong hydrophobicity promotes the nonspecific adsorption of germs on abiotic surfaces such as plastics, but hydrophilic viruses and bacteria have a higher migratory potential in aquatic environments (Liu et al., 2021). Some bacteria produce surfactants during growth, which may lead to the replacement of certain proteins, reducing the surface tension and altering viscosity moduli (Liu et al., 2020). Thus, the formation of protein networks with lower surface tension facilitates bacterial adsorption, especially on the surfaces of MPs with low interfacial tension.

2.2.3 Contaminant Properties

Concentration of sorbate, hydrophobicity and octanol–water partition coefficient are some of the characteristics that determine the adsorption of chemical contaminants to plastic surface. The adsorption capacity of PP and PVC towards Pb and Mn was found to strongly correlate with the metal concentration in seawater (Gao et al., 2019). A study demonstrated the adsorption of metal ions of Cr, Cu, Ag, Cd, Hg, Ni, Co, Pb and Zn in fresh water on virgin and weathered PE plastic particles and reported that the difference in ion concentration between the liquid phase (metal solution) and the solid phase (PE particles) was the main factor controlling the adsorption process and the metal ion concentration eventually reached equilibrium (Turner & Holmes, 2015). Hydrophobic nonpolar contaminants also show high affinity for hydrophobic plastic surfaces (Yu et al., 2019). Many bacteria have short appendages called fimbriae that they employ to adhere to both organic and inorganic substrates as well as one another (Zhao et al., 2024).

2.2.4 Environmental Factors

In aquatic environments, factors such as pH, salinity, temperature, ionic strength and dissolved organic matter (DOM) content have been shown to influence the sorption behaviour of plastics. Soil organic matter also influences the sorption capacity of plastics when its mass fraction exceeds 0.1% in the terrestrial environment (Guo et al., 2012). Previous research findings suggest that as pH decreases, the surfaces of PE and PS particles undergo surface modifications and protonation, which increases the sorption of organic contaminants such as PFOS and chlorinated phenols (CPs) (Puckowski et al., 2021). In another study, the results obtained for the adsorption of CPs on MPs in a water matrix indicated a lower sorption affinity of CPs with MPs at pH 4 and pH 10, with maximum affinity at pH 7; whereas the pH value of the water matrix did not affect the adsorption of halogenated benzenes and PAHs on MPs (Vujić et al., 2023). These reported variations in the affinity of contaminants at different pH values are possibly due to other factors. Also, ion concentration and DOM compete with MPs for adsorption, resulting in lower adsorption performance of weathered MPs in complex natural environments (Yu et al., 2024). In an experiment to evaluate the synergistic effects of salinity and temperature on the sorption behaviour of sulfamethoxazole (an antibiotic), findings revealed that increasing salinity from 0.05% to 3.5% reduced the sorption capacities of all MPs by $53.3\% \pm 5\%$. Furthermore, temperature reduction had a more significant effect on sorption than temperature rise (Kuang et al., 2023). Due to the complexity of the environment, many factors are at play at the same time during the sorption process, and it can only be concluded that the most dominant factor(s) will eventually prevail.

2.3 EFFECTS OF CONTAMINANT-LADEN PLASTICS ON ORGANISMS AND ENVIRONMENT

Risk is a function of exposure (dose) and hazard. To assess the risks associated with plastics, it is important to understand the level of exposure (threshold levels and dose–response relationships) and hazards (such as the possibility that MPs will have negative effects through implausible mechanisms). Exposure levels are the amounts of MPs found in the environment, including in living organisms. The hazard characterization phase of the risk assessment of MPs in the marine environment is still ongoing because there is a lack of data on established effect levels and exposure levels (GESAMP, 2016). Furthermore, there is currently no risk assessment methodology that considers the multidimensionality of MP particles against the background of numerous natural particles, which collectively exhibit an infinite combination of sizes, shapes, densities and chemical signatures (Koelmans et al., 2022).

Nonetheless, the reported effects of contaminant-laden plastic particles in organisms and the environment could range from physical, chemical to biological damages. Physically, plastic debris in aquatic or terrestrial environment is unsightly and harbours organic materials that produce foul odour and reduce air quality. In the open sea, large plastic materials have caused entanglement, injuries and death of organisms. Ingestion of small plastic particles has been shown to result in gut blockage, inflammation, oxidative stress and reduced body weight (Bhuyan, 2022). A study demonstrated that sea urchins (*Paracentrotus lividus*) and filter feeders both have a tendency to ingest biofilm colonized weathered MPs more frequently than pristine particles, suggesting the risk of the plastisphere exposure in the interaction (Murano et al., 2021). These marine invertebrates scout for and adsorb particulate matter using chemical sensing (Liu et al., 2021). Over time, biofilm colonization may have altered MPs' characteristics making them more enticing for ingestion, which may lead to increased pathogen virulence through environmental MPs. However, accurate and comprehensive quantitative risk assessment of plastic-associated pathogens is still lacking. Furthermore, organism or human exposure to plastic-associated pathogens leading to infection is yet to be established to our knowledge.

2.4 HUMAN HEALTH IMPACTS OF PLASTIC-ASSOCIATED CONTAMINANTS

Humans are exposed to plastics on a daily basis by ingestion, inhalation or dermal contact due to the ubiquitous presence of plastics in air, soil, sediment, water and food products (Alijagic et al., 2024; Wang et al., 2019). Many consumer goods such as packaging materials, electronics, medical equipment, utensils, toys and clothes are made from plastic (Völker et al., 2022). Humans can have direct exposure to plastic particles by ingestion through the use of plastic food materials or consuming plastic-contaminated foods and drinks such as seafood, bottled water, beer and milk (Diaz-Basantes et al., 2020; Loyo-Rosales et al., 2004; Ribeiro et al., 2020). The additives included in plastic materials to improve their functionality often leach and migrate into the surrounding media such as food resulting in potential human exposure during use. The most prevalent of these chemical additives are flame retardants (PBDEs), phthalates (plasticizers), BPA and PFASs (Seewoo et al., 2023). These substances are categorized as persistent organic pollutants because they do not readily degrade and remain in the environment for a long time. Because they are fat-soluble, they can bioaccumulate in animal and marine tissues and biomagnify in the food chain leading to human exposure (Ábalos et al., 2019). BPA and phthalates have been identified as metabolism-disrupting compounds (MDCs), which promote obesity through adipogenic activity in cell and animal models (Völker et al., 2022). Although micro- and nanoplastics have been detected in different parts of the human body including blood (Leslie et al., 2022), placenta (Ragusa et al., 2022) and testis (Hu et al., 2024), the extent of their human health effect is still unclear, but a study reported a correlation between the concentration of MPs in the faeces of patients with severity of inflammatory bowel disease (Yan et al., 2022). However, the cause-and-effect relationship between MPs and inflammatory bowel disease is still subject to investigation. Furthermore, human studies conducted in high-MP exposure work environments, like factories that produce synthetic textiles, have revealed a rise in respiratory symptoms (Turcotte et al., 2013) and stomach and oesophageal cancers (Gallagher et al., 2015). Because plastics are made of so many different kinds of chemicals, humans are exposed to multiple chemicals at the same time in real-world situations. Current knowledge is limited about how these chemicals (and/or their metabolites) interact across different biological pathways to cause negative effects on human health (Quiros-Alcala & Barr, 2023). It is unclear whether they do so antagonistically, synergistically or independently (Bakir et al., 2012). Regulating these chemicals based on individual thresholds does not necessarily protect against combination effects and mixture toxicity of multiple plastic-associated chemicals (Savitz & Hattersley, 2023). More research is required to tackle plastic chemical exposure problems based on mixture toxicity information and inform regulatory policy that will address the co-occurrence of chemical mixtures that leaches from common consumer plastic products (Savitz & Hattersley, 2023). Given the evidence of human exposure to plastic and associated chemicals, there is a need to develop measurement and mitigation strategies to prevent any public health crisis.

2.5 CONCLUSION AND FUTURE PERSPECTIVES

The contaminants associated with plastics originate from the plastic raw materials, production additives, NIAS and the environment. The widespread use of plastics in virtually all spheres of economic and environmental life has led to the global ubiquity of the material. Plastic hydrophobicity and surface area provide a suitable binding substrate to both chemical and biological contaminants and pollutants with great potential to cause harmful effects on organisms and ecosystems exposed to them. A number of risks have been associated with plastics and their associated contaminants; however, further studies are needed to elucidate the contribution of environmental plastic particles in the spread of pathogenic biological agents, leading to infection and whether plastic particles can be used to eliminate harmful chemicals ingested by organisms or as adsorbents for environmental remediation. In addition, risk assessment methodologies that consider the multidimensionality of MP particles against the background of natural particles need to be developed.

REFERENCES

Ábalos, M., Barceló, D., Parera, J., la Farré, M., Llorca, M., Eljarrat, E., Giulivo, M., Capri, E., Paunović, M., Milačič, R., & Abad, E. (2019). Levels of regulated POPs in fish samples from the Sava River Basin. Comparison to legislated quality standard values. *Science of the Total Environment*, *647*, 20–28. https://doi.org/10.1016/j.scitotenv.2018.07.371

Agboola, O. D., & Benson, N. U. (2021). Physisorption and chemisorption mechanisms influencing micro (nano) plastics-organic chemical contaminants interactions: A review. *Frontiers in Environmental Science*, *9*. https://doi.org/10.3389/fenvs.2021.678574

Alava, J. J. (2020). Modeling the bioaccumulation and biomagnification potential of microplastics in a cetacean foodweb of the Northeastern Pacific: A prospective tool to assess the risk exposure to plastic particles. *Frontiers in* Marine *Science*, *7*. https://doi.org/10.3389/fmars.2020.566101

Alijagic, A., Suljević, D., Fočak, M., Sulejmanović, J., Šehović, E., Särndahl, E., & Engwall, M. (2024). The triple exposure nexus of microplastic particles, plastic-associated chemicals, and environmental pollutants from a human health perspective. *Environment International*, 108736. https://doi.org/10.1016/j.envint.2024.108736

Ambrogi, V., Carfagna, C., Cerruti, P., & Marturano, V. (2016). Additives in polymers. In C. F. Jasso-Gastinel, & J. M. Kenny (Eds.), *Modification of Polymer Properties* (pp. 87–108). Oxford: William Andrew Publishing. https://doi.org/10.1016/B978-0-323-44353-1.00004-X

Arias-Andres, M., Klümper, U., Rojas-Jimenez, K., & Grossart, H.-P. (2018). Microplastic pollution increases gene exchange in aquatic ecosystems. *Environmental Pollution*, *237*, 253–261. https://doi.org/10.1016/j.envpol.2018.02.058

Audrézet, F., Zaiko, A., Lear, G., Wood, S. A., Tremblay, L. A., & Pochon, X. (2021). Biosecurity implications of drifting marine plastic debris: Current knowledge and future research. *Marine Pollution Bulletin*, *162*, 111835. https://doi.org/10.1016/j.marpolbul.2020.111835

Auta, H. S., Emenike, C. U., & Fauziah, S. H. (2017). Distribution and importance of microplastics in the marine environment: A review of the sources, fate, effects, and potential solutions. *Environment International*, *102*, 165–176. https://doi.org/10.1016/j.envint.2017.02.013

Bakir, A., O'Connor, I. A., Rowland, S. J., Hendriks, A. J., & Thompson, R. C. (2016). Relative importance of microplastics as a pathway for the transfer of hydrophobic organic chemicals to marine life. *Environmental Pollution*, *219*, 56–65. https://doi.org/10.1016/j.envpol.2016.09.046

Bakir, A., Rowland, S. J., & Thompson, R. C. (2012). Competitive sorption of persistent organic pollutants onto microplastics in the marine environment. *Marine Pollution Bulletin*, *64*(12), 2782–2789. https://doi.org/10.1016/j.marpolbul.2012.09.010

Bakir, A., Rowland, S. J., & Thompson, R. C. (2014). Transport of persistent organic pollutants by microplastics in estuarine conditions. *Estuarine, Coastal and Shelf Science*, *140*, 14–21. https://doi.org/10.1016/j.ecss.2014.01.004

Benson, N. U., Agboola, O. D., Fred-Ahmadu, O. H., De-la-Torre, G. E., Oluwalana, A., & Williams, A. B. (2022). Micro(nano)plastics prevalence, food web interactions, and toxicity assessment in aquatic organisms: A review. *Frontiers in Marine Science*, *9*(March), 1–19. https://doi.org/10.3389/fmars.2022.851281

Bhagwat, G., Tran, T. K. A., Lamb, D., Senathirajah, K., Grainge, I., O'Connor, W., Juhasz, A., & Palanisami, T. (2021). Biofilms enhance the adsorption of toxic contaminants on plastic microfibers under environmentally relevant conditions. *Environmental Science & Technology, 55*, 8877–8887. https://doi.org/10.1021/acs.est.1c02012

Bhuyan, S. (2022). Effects of microplastics on fish and in human health. *Frontiers in Environmental Science*, *10*(March), 1–17. https://doi.org/10.3389/fenvs.2022.827289

Brennecke, D., Duarte, B., Paiva, F., Caçador, I., & Canning-Clode, J. (2016). Microplastics as vector for heavy metal contamination from the marine environment. *Estuarine, Coastal and Shelf Science*, *178*, 189–195. https://doi.org/10.1016/j.ecss.2015.12.003

Chan, K. H., Peiris, J. S. M., Lam, S. Y., Poon, L. L. M., Yuen, K. Y., & Seto, W. H. (2011). The effects of temperature and relative humidity on the viability of the SARS Coronavirus. *Advances in Virology*, *2011*, 734690. https://doi.org/10.1155/2011/734690

Chapman, P. M. (2007). Determining when contamination is pollution—Weight of evidence determinations for sediments and effluents. *Environment International*, *33*(4), 492–501. https://doi.org/10.1016/j.envint.2006.09.001

Chin, A. W. H., Chu, J. T. S., Perera, M. R. A., Hui, K. P. Y., Yen, H.-L., Chan, M. C. W., Peiris, M., & Poon, L. L. M. (2020). Stability of SARS-CoV-2 in different environmental conditions. *The Lancet. Microbe*, *1*(1), e10. https://doi.org/10.1016/S2666-5247(20)30003-3

Cui, W., Hale, R. C., Huang, Y., Zhou, F., Wu, Y., Liang, X., Liu, Y., Tan, H., & Chen, D. (2023). Sorption of representative organic contaminants on microplastics: Effects of chemical physicochemical properties, particle size, and biofilm presence. *Ecotoxicology and Environmental Safety*, *251*, 114533. https://doi.org/10.1016/j.ecoenv.2023.114533

Diaz-Basantes, M. F., Conesa, J. A., & Fullana, A. (2020). Microplastics in honey, beer, milk and refreshments in Ecuador as emerging contaminants. *Sustainability*, *12*(14), Article 14. https://doi.org/10.3390/su12145514

Dobaradaran, S., Schmidt, T. C., Nabipour, I., Khajeahmadi, N., Tajbakhsh, S., Saeedi, R., Javad Mohammadi, M., Keshtkar, M., Khorsand, M., & Faraji Ghasemi, F. (2018). Characterization of plastic debris and association of metals with microplastics in coastline sediment along the Persian Gulf. *Waste Management*, *78*, 649–658. https://doi.org/10.1016/j.wasman.2018.06.037

Endo, S., & Koelmans, A. A. (2016). Sorption of hydrophobic organic compounds in marine environments: Equilibrium. In H. Takada, & H. K. Karapanagioti (Eds.), *Hazardous Chemicals Associated with Plastics in the Marine Environment* (pp. 1–20). Cham: Springer. https://doi.org/10.1007/698_2016_11

Endo, S., & Koelmans, A. A. (2019). Sorption of hydrophobic organic compounds to plastics in the marine environment: Equilibrium. In O. Hutzinger (ed.), *Handbook of Environmental Chemistry*. Berlin: Springer-Verlag. https://doi.org/10.1007/698_2016_11

Fred-Ahmadu, O. H., Bhagwat, G., Oluyoye, I., Benson, N. U., Ayejuyo, O. O., & Palanisami, T. (2020). Interaction of chemical contaminants with microplastics: Principles and perspectives. *Science of the Total Environment*, *706*, 135978. https://doi.org/10.1016/j.scitotenv.2019.135978

Fred-Ahmadu, O. H., Tenebe, I. T., Ayejuyo, O. O., & Benson, N. U. (2022). Microplastics and associated organic pollutants in beach sediments from the Gulf of Guinea (SE Atlantic) coastal ecosystems. *Chemosphere*, *298*(March), 134193. https://doi.org/10.1016/j.chemosphere.2022.134193

Gallagher, L. G., Li, W., Ray, R. M., Romano, M. E., Wernli, K., Gao, D., Thomas, D., & Checkoway, H. (2015). Occupational exposures and risk of stomach and esophageal cancers: Update of a cohort of female textile workers in Shanghai, China. *American Journal of Industrial Medicine*, *58*(3), 267–275.

Gao, F., Li, J., Sun, C., Zhang, L., Jiang, F., Cao, W., & Zheng, L. (2019). Study on the capability and characteristics of heavy metals enriched on microplastics in marine environment. *Marine Pollution Bulletin, 144*, 61–67. https://doi.org/10.1016/j.marpolbul.2019.04.039

GESAMP. (2016). Sources, fate and effects of microplastics in the marine environment: Part 2 of a global assessment—Science for sustainable Oceans. *International Maritime Organization, 93*, 1–217.

Gkoutselis, G., Rohrbach, S., Harjes, J., Obst, M., Brachmann, A., Horn, M. A., & Rambold, G. (2021). Microplastics accumulate fungal pathogens in terrestrial ecosystems. *Scientific Reports, 11*(1), 13214. https://doi.org/10.1038/s41598-021-92405-7

Gulizia, A. M., Philippa, B., Zacharuk, J., Motti, C. A., & Vamvounis, G. (2023). Plasticiser leaching from polyvinyl chloride microplastics and the implications for environmental risk assessment. *Marine Pollution Bulletin, 195*, 115392. https://doi.org/10.1016/j.marpolbul.2023.115392

Guo, X., Wang, X., Zhou, X., Kong, X., Tao, S., & Xing, B. (2012). Sorption of four hydrophobic organic compounds by three chemically distinct polymers: Role of chemical and physical composition. *Environmental Science & Technology, 46*(13), 7252–7259. https://doi.org/10.1021/es301386z

Guruge, K. S., Goswami, P., Kanda, K., Abeynayaka, A., Kumagai, M., Watanabe, M., & Tamamura-Andoh, Y. (2024). Plastiome: Plastisphere-enriched mobile resistome in aquatic environments. *Journal of Hazardous Materials, 471*, 134353. https://doi.org/10.1016/j.jhazmat.2024.134353

Hahladakis, J., Smaragdaki, E., Vasilaki, G., & Gidarakos, E. (2013). Use of sediment quality guidelines and pollution indicators for the assessment of heavy metal and PAH contamination in Greek surficial sea and lake sediments. *Environmental Monitoring and Assessment, 185*(3), 2843–2853. https://doi.org/10.1007/s10661-012-2754-2

Hernández-Sánchez, C., Pestana-Ríos, Á. A., Villanova-Solano, C., Domínguez-Hernández, C., Díaz-Peña, F. J., Rodríguez-Álvarez, C., Lecuona, M., & Arias, Á. (2023). Bacterial colonization of microplastics at the beaches of an Oceanic Island, Tenerife, Canary Islands. *International Journal of Environmental Research and Public Health, 20*(5), 3951. https://doi.org/10.3390/ijerph20053951

Hu, C. J., Garcia, M. A., Nihart, A., Liu, R., Yin, L., Adolphi, N., Gallego, D. F., Kang, H., Campen, M. J., & Yu, X. (2024). Microplastic presence in dog and human testis and its potential association with sperm count and weights of testis and epididymis. *Toxicological Sciences, 200*(2), 235–240. https://doi.org/10.1093/toxsci/kfae060

Jambeck, J. R., Geyer, R., Wilcox, C., Siegler, T. R., Perryman, M., Andrady, A., Narayan, R., & Law, K. L. (2015). Plastic waste inputs from land into the ocean. *Science, 347*(6223), 768–771. https://doi.org/10.1126/science.1260352

Jeong, G., Kim, H.-J., Kim, K. E., Kim, Y. J., Lee, T.-K., Shim, W. J., & Jung, S. W. (2023). Selective attachment of prokaryotes and emergence of potentially pathogenic prokaryotes on four plastic surfaces: Adhesion study in a natural marine environment. *Marine Pollution Bulletin, 193*, 115149. https://doi.org/10.1016/j.marpolbul.2023.115149

Kesy, K., Labrienz, M., Scales, B., Kreikemeyer, B., & Oberbeckmann, S. (2021). Vibrio colonization is highly dynamic in early microplastic-associated biofilms as well as on field-collected microplastics. *Microorganisms, 9*(76). https://doi.org/10.3390/microorganisms9010076

Koelmans, A. A. (2015). Modeling the role of microplastics in bioaccumulation of organic chemicals to marine aquatic organisms. A critical review. In M. Bergmann, L. Gutow, & M. Klages (Eds.), *Marine Anthropogenic Litter* (pp. 309–324). Cham: Springer International Publishing. https://doi.org/10.1007/978-3-319-16510-3_11

Koelmans, A. A., Bakir, A., Burton, G. A., & Janssen, C. R. (2016). Microplastic as a vector for chemicals in the aquatic environment: Critical review and model-supported reinterpretation of empirical studies. *Environmental Science and Technology, 50*(7), 3315–3326. https://doi.org/10.1021/acs.est.5b06069

Koelmans, A. A., Redondo-Hasselerharm, P. E., Nor, N. H. M., de Ruijter, V. N., Mintenig, S. M., & Kooi, M. (2022). Risk assessment of microplastic particles. *Nature Reviews Materials, 7*(2), Article 2. https://doi.org/10.1038/s41578-021-00411-y

Kuang, B., Chen, X., Zhan, J., Zhou, L., Zhong, D., & Wang, T. (2023). Interaction behaviors of sulfamethoxazole and microplastics in marine condition: Focusing on the synergistic effects of salinity and temperature. *Ecotoxicology and Environmental Safety, 259*, 115009. https://doi.org/10.1016/j.ecoenv.2023.115009

Lear, L., Padfield, D., Dowsett, T., Jones, M., Kay, S., Hayward, A., & Vos, M. (2022). Bacterial colonisation dynamics of household plastics in a coastal environment. *Science of the Total Environment, 838*, 156199. https://doi.org/10.1016/j.scitotenv.2022.156199

Leslie, H. A., van Velzen, M. J. M., Brandsma, S. H., Vethaak, A. D., Garcia-Vallejo, J. J., & Lamoree, M. H. (2022). Discovery and quantification of plastic particle pollution in human blood. *Environment International, 163*, 107199. https://doi.org/10.1016/j.envint.2022.107199

Li, W., Lo, H.-S., Wong, H.-M., Zhou, M., Wong, C.-Y., Tam, N. F.-Y., & Cheung, S.-G. (2020). Heavy metals contamination of sedimentary microplastics in Hong Kong. *Marine Pollution Bulletin*, *153*, 110977. https://doi.org/10.1016/J.MARPOLBUL.2020.110977

Lin, W.-H., Wang, H.-Y., Kuo, J., & Lo, S.-L. (2023). Adsorption and desorption characteristics of heavy metals onto conventional and biodegradable plastics. *Chemosphere*, *333*, 138920. https://doi.org/10.1016/j.chemosphere.2023.138920

Liu, J., Qi, S., Yao, J., Yang, D., Xing, X., Liu, H., & Qu, C. (2016). Contamination characteristics of organochlorine pesticides in multimatrix sampling of the Hanjiang River Basin, southeast China. Chemosphere, *163*, 35–43. https://doi.org/10.1016/j.chemosphere.2016.07.040

Liu, F., Liu, G., Zhu, Z., Wang, S., & Zhao, F. (2019). Interactions between microplastics and phthalate esters as affected by microplastics characteristics and solution chemistry. *Chemosphere*, *214*, 688–694. https://doi.org/10.1016/j.chemosphere.2018.09.174

Liu, G., Fang, Z., Zhong, H., Shi, L., Yang, X., & Liu, Z. (2020). Transport of Pseudomonas aeruginosa in porous media mediated by low-concentration surfactants: The critical role of surfactant to change cell surface hydrophobicity. *Water Resources Research*, *56*. https://doi.org/10.1029/2019WR026103

Liu, G., Zhu, Z., Yang, Y., Sun, Y., Yu, F., & Ma, J. (2019). Sorption behavior and mechanism of hydrophilic organic chemicals to virgin and aged microplastics in freshwater and seawater. *Environmental Pollution*, *246*, 26–33. https://doi.org/10.1016/j.envpol.2018.11.100

Liu, S., Shi, J., Wang, J., Dai, Y., Li, H., Li, J., Liu, X., Chen, X., Wang, Z., & Zhang, P. (2021). Interactions between microplastics and heavy metals in aquatic environments: A review. *Frontiers in Microbiology*, *12*. https://www.frontiersin.org/articles/10.3389/fmicb.2021.652520

Loyo-Rosales, J. E., Rosales-Rivera, G. C., Lynch, A. M., Rice, C. P., & Torrents, T. (2004). Migration of nonylphenol from plastic containers to water and a milk surrogate. *Journal of Agricultural and Food Chemistry*, *52*(7), 2016–2020. https://doi.org/10.1021/JF0345696

Lu, K., Qiao, R., An, H., & Zhang, Y. (2018). Influence of microplastics on the accumulation and chronic toxic effects of cadmium in zebrafish (*Danio rerio*). *Chemosphere*, *202*, 514–520. https://doi.org/10.1016/j.chemosphere.2018.03.145

Luo, H., Xiang, Y., He, D., Li, Y., Zhao, Y., Wang, S., & Pan, X. (2019). Leaching behavior of fluorescent additives from microplastics and the toxicity of leachate to Chlorella vulgaris. *Science of the Total Environment*, *678*, 1–9. https://doi.org/10.1016/J.SCITOTENV.2019.04.401

Mato, Y., Isobe, T., Takada, H., Kanehiro, H., Ohtake, C., & Kaminuma, T. (2001). Plastic resin pellets as a transport medium for toxic chemicals in the marine environment. *Environmental Science and Technology*, *35*(2), 318–324. https://doi.org/10.1021/es0010498

Meng, J., Zhang, Q., Zheng, Y., He, G., & Shi, H. (2021). Plastic waste as the potential carriers of pathogens. *Current Opinion in Food Science*, *41*, 224–230. https://doi.org/10.1016/j.cofs.2021.04.016

Moore, R. E., Millar, B. C., & Moore, J. E. (2020). Antimicrobial resistance (AMR) and marine plastics: Can food packaging litter act as a dispersal mechanism for AMR in oceanic environments? *Marine Pollution Bulletin*, *150*, 110702. https://doi.org/10.1016/j.marpolbul.2019.110702

Moresco, V., Charatzidou, A., Oliver, D. M., Weidmann, M., Matallana-Surget, S., & Quilliam, R. S. (2022). Binding, recovery, and infectiousness of enveloped and non-enveloped viruses associated with plastic pollution in surface water. *Environmental Pollution*, *308*, 119594. https://doi.org/10.1016/j.envpol.2022.119594

Murano, C., Donnarumma, V., Corsi, I., Casotti, R., & Palumbo, A. (2021). Impact of microbial colonization of polystyrene microbeads on the toxicological responses in the sea urchin Paracentrotus lividus. *Environmental Science & Technology*, *55*, 7990–8000. https://doi.org/10.1021/acs.est.1c00618

Nandi, A., & Allen, L. J. S. (2021). Probability of a zoonotic spillover with seasonal variation. *Infectious Disease Modelling*, *6*, 514–531. https://doi.org/10.1016/j.idm.2021.01.013

Ormsby, M. J., White, H. L., Metcalf, R., Oliver, D. M., Feasey, N. A., & Quilliam, R. S. (2024). Enduring pathogenicity of African strains of *Salmonella* on plastics and glass in simulated peri-urban environmental waste piles. *Journal of Hazardous Materials*, *461*, 132439. https://doi.org/10.1016/j.jhazmat.2023.132439

Puckowski, A., Cwięk, W., Mioduszewska, K., Stepnowski, P., & Białk-Bielińska, A. (2021). Sorption of pharmaceuticals on the surface of microplastics. *Chemosphere*, *263*, 127976. https://doi.org/10.1016/j.chemosphere.2020.127976

Quiros-Alcala, L., & Barr, D. B. (2023). Invited perspective: Mixtures-are they worth the risk (assessment)? *Environmental Health Perspectives, 131*, 41301. https://doi.org/10.1289/EHP12596

Rabenau, H. F., Cinatl, J., Morgenstern, B., Bauer, G., Preiser, W., & Doerr, H. W. (2005). Stability and inactivation of SARS coronavirus. *Medical Microbiology and Immunology*, *194*(1), 1–6. https://doi.org/10.1007/s00430-004-0219-0

Ragusa, A., Notarstefano, V., Svelato, A., Belloni, A., Gioacchini, G., Blondeel, C., Zucchelli, E., De Luca, C., D'Avino, S., Gulotta, A., Carnevali, O., & Giorgini, E. (2022). Raman microspectroscopy detection and characterisation of microplastics in human breastmilk. *Polymers*, *14*(13), 2700. https://doi.org/10.3390/polym14132700

Ribeiro, F., Oko, E. D., Brien, J. W. O., Fraissinet-tachet, S., Brien, S. O., Gallen, M., Samanipour, S., Kaserzon, S., Mueller, J. F., Galloway, T., & Thomas, K. V. (2020). Quantitative analysis of selected plastics in high-commercial-value Australian seafood by pyrolysis gas chromatography mass spectrometry. *Environmental Science & Technology*, *54*, 9408–9417. https://doi.org/10.1021/acs.est.0c02337

Rivera-Hernández, J. R., Fernández, B., Santos-Echeandia, J., Garrido, S., Morante, M., Santos, P., & Albentosa, M. (2019). Biodynamics of mercury in mussel tissues as a function of exposure pathway: Natural vs microplastic routes. *Science of the Total Environment*, *674*, 412–423. https://doi.org/10.1016/j.scitotenv.2019.04.175

Rochman, C. M., Hoh, E., Kurobe, T., & Teh, S. J. (2013). Ingested plastic transfers hazardous chemicals to fish and induces hepatic stress. *Scientific Reports*, *3*, 3263. https://doi.org/10.1038/srep03263

Rodrigues, A., Oliver, D. M., McCarron, A., & Quilliam, R. S. (2019). Colonisation of plastic pellets (nurdles) by *E. coli* at public bathing beaches. *Marine Pollution Bulletin*, *139*, 376–380. https://doi.org/10.1016/j.marpolbul.2019.01.011

Rodrigues, J. P., Duarte, A. C., & Santos-Echeandía, J. (2022). Interaction of microplastics with metal(oid)s in aquatic environments: What is done so far? *Journal of Hazardous Materials Advances*, *6*, 100072. https://doi.org/10.1016/j.hazadv.2022.100072

Rühs, P. A., Böcker, L., Inglis, R. F., & Fischer, P. (2014). Studying bacterial hydrophobicity and biofilm formation at liquid–liquid interfaces through interfacial rheology and pendant drop tensiometry. *Colloids and Surfaces B: Biointerfaces*, *117*, 174–184. https://doi.org/10.1016/j.colsurfb.2014.02.023

Savitz, D. A., & Hattersley, A. M. (2023). Evaluating chemical mixtures in epidemiological studies to inform regulatory decisions. *Environmental Health Perspectives*, *131*(4), 045001. https://doi.org/10.1289/EHP11899

Seewoo, B. J., Goodes, L. M., Mofflin, L., Mulders, Y. R., Wong, E. V., Toshniwal, P., Brunner, M., Alex, J., Johnston, B., Elagali, A., Gozt, A., Lyle, G., Choudhury, O., Solomons, T., Symeonides, C., & Dunlop, S. A. (2023). The plastic health map: A systematic evidence map of human health studies on plastic-associated chemicals. *Environment International*, *181*, 108225. https://doi.org/10.1016/j.envint.2023.108225

Shi, J., Wu, D., Su, Y., & Xie, B. (2021). Selective enrichment of antibiotic resistance genes and pathogens on polystyrene microplastics in landfill leachate. *Science of the Total Environment*, *765*, 142775. https://doi.org/10.1016/j.scitotenv.2020.142775

Szabó, I., Al-Omari, J., Szerdahelyi, G. S., Farkas, M., Al-Omari, Y., Szabó, P. M., Sebők, R., Griffitts, J., Kriszt, B., & Szoboszlay, S. (2021). In situ investigation of plastic-associated bacterial communities in a freshwater lake of Hungary. *Water, Air, & Soil Pollution*, *232*(12), 493. https://doi.org/10.1007/s11270-021-05445-0

Tchounwou, P. B., Yedjou, C. G., Patlolla, A. K., & Sutton, D. J. (2012). Heavy metals toxicity and the environment. *EXS*, *101*, 133–164. https://doi.org/10.1007/978-3-7643-8340-4_6

Town, R. M., van Leeuwen, H. P., & Duval, J. F. L. (2023). Effect of polymer aging on uptake/release kinetics of metal ions and organic molecules by micro- and nanoplastics: Implications for the bioavailability of the associated compounds. *Environmental Science & Technology*, *57*(43), 16552–16563. https://doi.org/10.1021/acs.est.3c05148

Tumwesigye, E., Felicitas Nnadozie, C., C Akamagwuna, F., Siwe Noundou, X., William Nyakairu, G., & Odume, O. N. (2023). Microplastics as vectors of chemical contaminants and biological agents in freshwater ecosystems: Current knowledge status and future perspectives. *Environmental Pollution*, *330*, 121829. https://doi.org/10.1016/j.envpol.2023.121829

Turcotte, S. E., Chee, A., Walsh, R., Grant, F. C., Liss, G. M., Boag, A., Forkert, L., Munt, P. W., & Lougheed, M. D. (2013). Flock worker's lung disease: Natural history of cases and exposed workers in Kingston, Ontario. *Chest*, *143*(6), 1642–1648. https://doi.org/10.1378/chest.12-0920

Turner, A., & Holmes, L. A. (2015). Adsorption of trace metals by microplastic pellets in fresh water. *Environmental Chemistry*, *12*(5), 600–610. https://doi.org/10.1071/EN14143

UNEP. (2023). *Chemicals in Plastics—A Technical Report*. https://www.unep.org/resources/report/chemicals-plastics-technical-report

Uzun, P., Farazande, S., & Guven, B. (2022). Mathematical modeling of microplastic abundance, distribution, and transport in water environments: A review. *Chemosphere*, *288*, 132517. https://doi.org/10.1016/j.chemosphere.2021.132517

van Doremalen, N., Bushmaker, T., Morris, D. H., Holbrook, M. G., Gamble, A., Williamson, B. N., Tamin, A., Harcourt, J. L., Thornburg, N. J., Gerber, S. I., Lloyd-Smith, J. O., de Wit, E., & Munster, V. J. (2020). Aerosol and surface stability of SARS-CoV-2 as compared with SARS-CoV-1. *The New England Journal of Medicine*, *382*, 1564–1567. https://doi.org/10.1056/NEJMc2004973

Vedolin, M. C., Teophilo, C. Y. S., Turra, A., & Figueira, R. C. L. (2018). Spatial variability in the concentrations of metals in beached microplastics. *Marine Pollution Bulletin*, *129*(2), 487–493. https://doi.org/10.1016/J.MARPOLBUL.2017.10.019

Völker, J., Ashcroft, F., Vedøy, Å., Zimmermann, L., & Wagner, M. (2022). Adipogenic activity of chemicals used in plastic consumer products. *Environmental Science & Technology*, *56*(4), 2487–2496. https://doi.org/10.1021/acs.est.1c06316

Vujić, M., Vasiljević, S., Nikić, J., Kordić, B., Agbaba, J., & Tubić, A. (2023). Sorption behavior of organic pollutants on biodegradable and nondegradable microplastics: pH effects. *Applied Sciences*, *13*(23), Article 23. https://doi.org/10.3390/app132312835

Wang, W., Gao, H., Jin, S., Li, R., & Na, G. (2019). The ecotoxicological effects of microplastics on aquatic food web, from primary producer to human: A review. *Ecotoxicology and Environmental Safety*, *173*, 110–117. https://doi.org/10.1016/J.ECOENV.2019.01.113

Warnes, S. L., Little, Z. R., & Keevil, C. W. (2015). Human Coronavirus 229E remains infectious on common touch surface materials. *mBio*, *6*(6), e01697–15. https://doi.org/10.1128/mBio.01697-15

Yan, Z., Liu, Y., Zhang, T., Zhang, F., Ren, H., & Zhang, Y. (2022). Analysis of microplastics in human feces reveals a correlation between fecal microplastics and inflammatory bowel disease status. *Environmental Science & Technology*, *56*(1), 414–421. https://doi.org/10.1021/acs.est.1c03924

Yeo, B. G., Takada, H., Yamashita, R., Okazaki, Y., Uchida, K., Tokai, T., Tanaka, K., & Trenholm, N. (2019). PCBs and PBDEs in microplastic particles and zooplankton in open water in the Pacific Ocean and around the coast of Japan. *Marine Pollution Bulletin*, *151*, 110806. https://doi.org/10.1016/j.marpolbul.2019.110806

Yu, F., Qin, Q., Zhang, X., & Ma, J. (2024). Characteristics and adsorption behavior of typical microplastics in long-term accelerated weathering simulation. *Environmental Science: Processes & Impacts, 26*, 882–890. https://doi.org/10.1039/D4EM00062E

Yu, F., Yang, C., Zhu, Z., Bai, X., & Ma, J. (2019). Adsorption behavior of organic pollutants and metals on micro/nanoplastics in the aquatic environment. *Science of the Total Environment*, *694*, 133643. https://doi.org/10.1016/J.SCITOTENV.2019.133643

Zadjelovic, V., Wright, R. J., Borsetto, C., Quartey, J., Cairns, T. N., Langille, M. G. I., Wellington, E. M. H., & Christie-Oleza, J. A. (2023). Microbial hitchhikers harbouring antimicrobial-resistance genes in the riverine plastisphere. *Microbiome*, *11*(1), 225. https://doi.org/10.1186/s40168-023-01662-3

Zettler, E. R., Mincer, T. J., & Amaral-Zettler, L. A. (2013). Life in the "plastisphere": Microbial communities on plastic marine debris. *Environmental Science & Technology*, *47*(13), 7137–7146. https://doi.org/10.1021/es401288x

Zhang, E., Kim, M., Rueda, L., Rochman, C., VanWormer, E., Moore, J., & Shapiro, K. (2022). Association of zoonotic protozoan parasites with microplastics in seawater and implications for human and wildlife health. *Scientific Reports*, *12*(1), 6532. https://doi.org/10.1038/s41598-022-10485-5

Zhao, H., Hong, X., Chai, J., Wan, B., Zhao, K., Han, C., Zhang, W., & Huan, H. (2024). Interaction between microplastics and pathogens in subsurface system: What we know so far. *Water*, *16*(3), Article 3. https://doi.org/10.3390/w16030499

3 Microbiome in Atmospheric Microplastics and Microplastic Deposition on Symbiotic Organisms

Kulathi Nishshankage, Ishani Gunasekara, and Meththika Vithanage

3.1 INTRODUCTION

Atmospheric microplastics (AMPs) have become a pressing environmental concern due to their widespread presence in the air we breathe. The atmosphere includes ambient air, deposition, and dust capable of resuspension into the air (O'Brien et al., 2023). Microplastics (MPs) are any synthetic solid particle or polymeric matrix, with regular or irregular shape and size ranging from 1 µm to 5 mm, of either primary or secondary manufacturing origin, which are insoluble in water (Frias and Nash, 2019). These tiny plastic particles finally end up in different environmental compartments including the atmosphere. The studies have reported the presence of MPs in the air of different regions and in air masses over the water bodies demonstrating the capability of long-range transport and wide spatial distribution of MPs (Habibi et al., 2022). The impact of AMPs on the environment and human health is significant, with studies highlighting their potential to contaminate soil, water, and even the food chain (Yao et al., 2022).

The interactions between AMPs and microbial communities involve several aspects. AMPs can form biofilms, creating a unique and dynamic ecosystem, referred to as atmospheric plastisphere. Microorganisms colonize the surface of MPs, which can provide a suitable substrate for growth and important nutrients for their survival. The surface characteristics of MPs, such as size, shape, roughness, and hydrophobicity, influence the initial interaction through electrostatic forces. Environmental factors such as temperature, pH, and ionic strength also impact the interactions between Microorganisms and MPs (Yang et al., 2023). For example, the rough surface of weathered MPs can facilitate bacterial growth and biofilm formation. The microbial communities on MPs, known as the "plastisphere," are distinct from those in the surrounding environment. These communities can include various bacteria such as Aeromonas, Rhodococcus, Pseudomonas, Enterobacter, Halomonas, Mycobacterium, Photobacterium, and Shigella, as well as fungi. MPs can be toxic to both autotrophic and heterotrophic bacterial growth, mainly due to toxic organic additives leaking from the MPs (Yang et al., 2023). Studying the interactions of AMPs and microbial communities is essential for safeguarding environmental health, human well-being, and economic stability. It provides a comprehensive understanding that informs both scientific knowledge and practical solutions to address the challenges posed by MP pollution.

Despite the growing awareness of AMPs, there remains a significant knowledge gap regarding the interactions between these particles and microbial communities. Most of the research has focused on the effects of MPs on aquatic ecosystems and terrestrial environments, with relatively little attention given to the atmospheric context (Kumar et al., 2023). This lack of understanding is

DOI: 10.1201/9781032717814-3

particularly concerning given the ubiquity of AMPs and their potential to influence microbial communities in various ecosystems.

Several key research gaps and emerging topics require further investigation to fully understand the connections between the microbiome and AMPs. The composition and diversity of microbial communities associated with atmospheric aerosols and dust particles remain poorly understood. The formation, structure, and function of microbial communities on AMPs are largely unexplored. The mechanisms by which microbial communities interact with AMPs and influence their fate and effects are not well characterized. Furthermore, the distribution and diversity of microbial communities on AMPs across different ecosystems and regions are largely unknown. The ecological implications of AMPs and their associated microbial communities on ecosystems are not well understood (Rao et al., 2020).

This chapter focuses on two major aspects of the atmospheric microplastisphere. The atmospheric microplastisphere represents an important pathway for the global distribution of MP pollution. Therefore, it is important to understand the aspects of it. The microbiome in the atmosphere section explores the composition and diversity of microbial communities associated with atmospheric aerosols and dust particles. MP deposition on symbiotic organisms' section examines the deposition of AMPs on symbiotic organisms and the impacts on these relationships.

Understanding the interactions between AMPs and microbial communities is crucial for several reasons. AMPs can contaminate soil, water, and even the food chain, posing significant environmental and health risks. The deposition of AMPs on ecosystems can have long-term ecological consequences, including changes to microbial communities and ecosystem functioning. The interactions between AMPs and microbial communities are a critical area of research that requires further investigation and attention. This chapter aims to provide a comprehensive overview of the current state of knowledge on the microbiome in the atmosphere and MP deposition on symbiotic organisms. The exploration of these topics will contribute to a deeper understanding of the ecological and environmental implications of AMPs.

3.2 UNDERSTANDING THE MICROBIOME IN THE ATMOSPHERE

Atmospheric aerosols and dust particles are complex mixtures of natural and anthropogenic substances that can influence the Earth's climate, air quality, and ecosystem health. These particles are often associated with microbial communities that play critical roles in various ecological processes. The composition and diversity of microbial communities in atmospheric aerosols and dust particles are influenced by a range of factors such as source, transport, and chemical composition. The source is the origin of the aerosol or dust particle, such as soil, vegetation, or industrial activities. Transport means the distance and pathways by which the aerosol or dust particle is transported through the atmosphere. Chemical composition of the aerosol or dust particle can influence microbial growth and survival. Studies have shown that atmospheric aerosols and dust particles can harbor a wide range of microbial species. These Microorganisms can influence the chemical and physical properties of the aerosol or dust particle, affecting its fate and effects on the environment (Rao et al., 2020; Buseck and Pósfai, 1999)

The atmosphere is a vast and dynamic habitat that contains a diverse array of Microorganisms, collectively known as the atmospheric microbiome. These microbes, which include bacteria, archaea, fungi, and viruses, are transported through the air via atmospheric processes such as wind, turbulence, and precipitation (Šantl-Temkiv et al., 2022; Archer et al., 2023). The atmospheric microbiome plays a crucial role in various ecological and biogeochemical processes, influencing cloud formation, precipitation patterns, and even human health (Šantl-Temkiv et al., 2022).

One of the key components of the atmospheric microbiome is the microbiome associated with atmospheric aerosols and dust particles. These particles, which range in size from a few nanometers to tens of micrometers, serve as carriers for the transport of Microorganisms through the atmosphere (Šantl-Temkiv et al., 2022; Naumova and Kabilov, 2022). The composition and diversity of the microbiome within these particles are influenced by various factors, including the source of the particles, meteorological conditions, and atmospheric chemistry.

Previous studies have shown that the bacterial microbiome in atmospheric aerosols is highly diverse, with members from various phyla such as Proteobacteria, Firmicutes, Actinobacteria, Bacteroidetes, and Cyanobacteria. The relative abundance of these phyla can vary depending on the location and environmental conditions. For example, Firmicutes can make up a significant portion (20%–30%) of the bacterial microbiome in urban areas, while Cyanobacteria, Bacteroidetes, and other phyla are present in lower abundances (1%–5%). The composition of the bacterial microbiome can also display diurnal and seasonal dynamics, as well as respond to meteorological events such as dust storms, showers, and fogs (Naumova and Kabilov, 2022).

The fungal component of the atmospheric microbiome, known as the mycobiome, is also an important aspect of the atmospheric ecosystem. Studies have shown that the mycobiome in atmospheric aerosols is dominated by members of the phylum Ascomycota, which includes many common airborne fungi such as Cladosporium and Alternaria. The diversity and composition of the mycobiome can vary depending on factors such as location, season, and air pollution levels (Naumova and Kabilov, 2022).

One of the key challenges in studying the atmospheric microbiome is the low biomass and high diversity of Microorganisms present in the air. This requires the use of advanced sampling and analytical techniques, such as high-volume sampling and metagenomic sequencing, to accurately characterize the composition and diversity of the microbiome (Dommergue et al., 2019). Additionally, the atmospheric microbiome is influenced by a complex interplay of physical, chemical, and biological factors, making it difficult to establish causal relationships between the microbiome and its environment (Šantl-Temkiv et al., 2022).

Despite these challenges, research on the atmospheric microbiome has yielded important insights into the role of Microorganisms in atmospheric processes. For example, certain bacteria and fungi can act as ice nuclei, facilitating the formation of ice crystals in clouds and influencing precipitation patterns (Šantl-Temkiv et al., 2022). Additionally, the atmospheric transport of Microorganisms can have implications for human health, as pathogens and allergens can be dispersed through the air (Naumova and Kabilov, 2022).

In conclusion, the atmospheric microbiome, including the microbiome associated with atmospheric aerosols and dust particles, is a complex and dynamic ecosystem that plays a crucial role in various ecological and biogeochemical processes. Understanding the composition, diversity, and function of the atmospheric microbiome is essential for understanding the global distribution of Microorganisms and their impact on the environment and human health. Future research in this field will likely focus on developing more advanced sampling and analytical techniques, as well as integrating data from various disciplines to gain a more comprehensive understanding of the atmospheric microbiome.

3.3 THE ATMOSPHERIC MICROPLASTISPHERE

Dr. Erik Zettler and his associates first proposed the idea of the "plastisphere," a unique microbial population that sticks to plastic surfaces and differs from its surroundings (Sun et al., 2023). Subsequently, a highly populated community of metabolically active Microorganisms is developed on the surfaces of MPs, which is known as the "microplastisphere" (Shruti et al., 2023). This chapter addresses "atmospheric microplastisphere," specifically Microorganisms attached to MP particles that are atmospheric in origin.

3.3.1 Factors Affecting Microbial Growth on Atmospheric Microplastics

Microbial growth may be impacted by the physical properties of AMPs, such as their size, shape, surface roughness, type of polymer, and hydrophobicity. Due to their large specific surface area, MPs encourage microbial cells to adhere to one another and change the composition of microbial communities according to the type of polymer they colonize (Frère et al., 2018). Polyethylene (PE), polypropylene (PP), polystyrene (PS), and polyvinyl stearate (PVS) are some of the main polymer types that could be found in MPs (Yukioka et al., 2020).

Additionally, AMPs can serve as a platform for Microorganisms, offering an advantageous microhabitat on which Microorganisms are easily attracted and form biofilms on their hydrophobic surfaces. This could promote colony protection and reproduction through the release of a polysaccharide matrix (Luo et al., 2022). Further, AMPs are exposed to various environmental factors, such as temperature, humidity, and nutrient availability, which may have an impact on the process of microbial development. Similar to certain aquatic habitats, favorable geographic locations may encourage faster microbial colonization and activity on AMPs (Stabnikova et al., 2021).

UV radiation accelerates the aging of suspended MPs in the atmosphere because it initiates photo-oxidation reactions that break polymer chains and weaken materials (Biber et al., 2019). Due to the increased temperatures, levels of oxygen, and UV radiation, atmospheric conditions accelerate this degradation more than terrestrial or aquatic habitats (Mao et al., 2020). Aging causes surface alterations in MPs, such as increased hydrophobicity and the creation of rough surfaces, which improve the adsorption capacity of contaminants (Cai et al., 2018). More sites for microbial adhesion and growth are provided by this potential increase in effective surface area. Furthermore, the organic matter attached to MPs may encourage microbial development by making trace elements easier to adsorb and move through (Wijesekara et al., 2018). Additionally, the duration of exposure combined with the adsorption of substances such as proteins, carbohydrates, lipids, fatty acids, and nucleic acids onto AMPs may accelerate the process of microbial colonization (Mammo et al., 2020). Figure 3.1 depicts the factors that could affect the growth of microbes on AMPs.

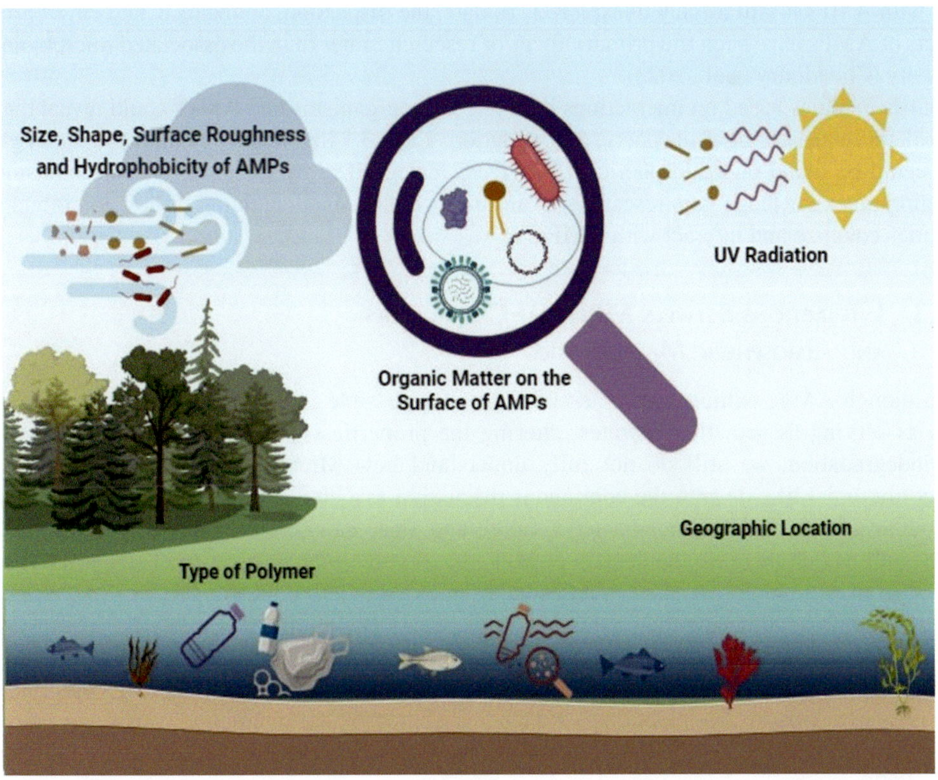

FIGURE 3.1 Factors that could affect the growth of microbes on atmospheric microplastics (AMPs) (Created with Biorender.com).

TABLE 3.1
Potential Microbial Communities That Could Present on Atmospheric Microplastic Surfaces

Microorganism	Survival Mechanisms	References
Bacillus subtilis NASA8	Sporulation, potential microniches in surfaces providing shielding from harmful radiation and enhancing survival	Smith et al. (2011)
Bacillus pumilus	Sporulation, UV tolerance, novel DNA repair genes	Khodadad et al. (2017) and Tirumalai et al. (2013)
Deinococcus aetherius	Sporulation, high tolerance to desiccation and γ-radiation, aggregation for shielding against radiation	Yang et al. (2010)
Halobacterium sp. NRC-1	Cold shock response with extra copies of cold-shock genes, higher radiation tolerance	Leuko et al. (2015)
Methicilin-resistant *Staphylococcus aureus* (MRSA)	Alteration of metabolic pathways and antibiotic resistance patterns to UV radiation	Chudobova et al. (2015)

3.3.2 Diversity of Microbial Communities in Atmospheric Microplastics

Microorganisms, including bacteria, archaea, fungi, and various other eukaryotes, have the capability to colonize MPs (Sun et al., 2023). The identification of the microbial communities associated with AMPs is still largely unexplored. To date, the dispersion, abundance, and environmental effects of AMPs have been the primary focus of research rather than the associated microbiological diversity (Choudhury et al., 2023).

Further studies based on interactions between Microorganisms and AMPs could reveal the ecological consequences of atmospheric MP pollution. Table 3.1 illustrates potential microbial groups that could be found on atmospheric MP surfaces, categorized by their survival mechanisms. By bridging this knowledge gap, researchers can improve their understanding of how microbial communities colonize and interact with AMPs.

3.3.3 Interactions Between Microbial Communities and Atmospheric Microplastics

Even though AMPs exhibit favorable conditions for possible interactions with Microorganisms, such as serving as growth substrates, altering the properties of Microorganisms, and assisting in biodegradation, we still do not fully understand how Microorganisms adhering to AMPs affect microbial life, despite the continuous discussion associated with plastic and MP pollution (Stabnikova et al., 2021). Since bacteria, fungi, and algae can colonize plastic surfaces and produce biofilms that increase the density of MPs, the distribution of plastic and MPs in water layers may shift over time. For instance, studies have demonstrated that plastic particles may settle as a result of biofilms biofouling them (Harrison et al., 2011). By relating this to AMPs, we can deduce that similar mechanisms could take place in the atmospheric conditions as well. As shown in Figure 3.2, the primary pathways for interaction and accumulation between AMPs and microbes may include bioadsorption, biouptake, and biodegradation. These interactions raise concerns about the possible emergence of antibiotic-resistant microbial communities within biofilms (Mammo et al., 2020).

Research has demonstrated that plastic surfaces serve as both carriers and protective environments for viruses, as seen by the 72-hour persistence of SARS-CoV-2 on plastic surfaces

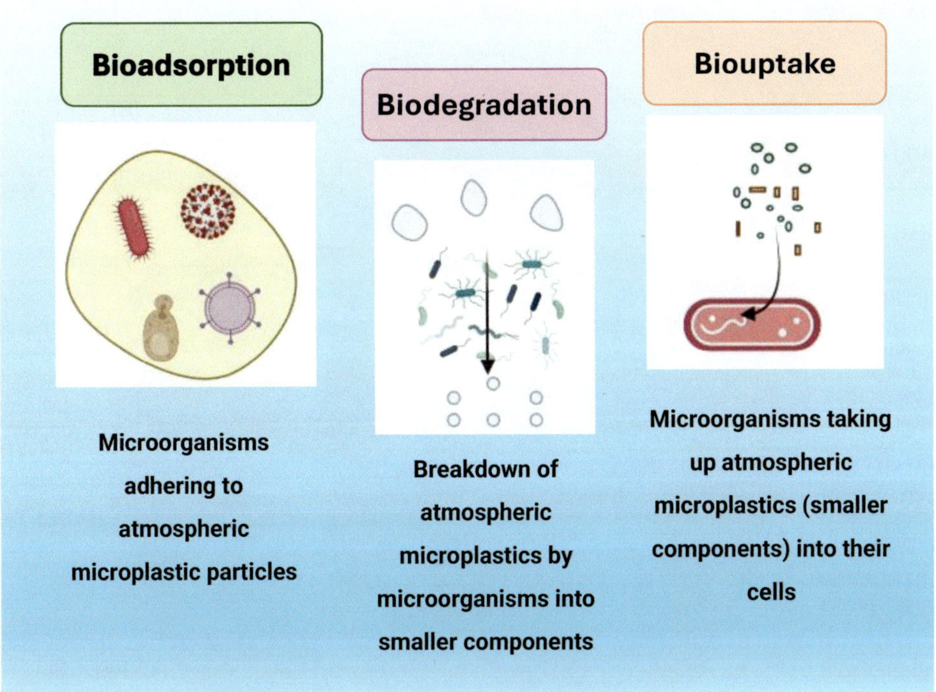

FIGURE 3.2 Pathways of interaction and accumulation between atmospheric microplastics and microorganisms (Created with Biorender.com).

(Van Doremalen et al., 2020). To limit the risks that AMPs raise to human health and the sustainability of the environment, managing these interactions requires a full understanding of the difficulties involved and the implementation of specialized measures. Additionally, photodegradation, which is caused by solar radiation, can change mechanical, physicochemical, thermal, and surface properties of AMPs, particularly if chromophores are present. As a result, the photodegradation process may produce lower molecular weight MP particles, gaseous products, and organic compounds that either promote or inhibit microbial growth (López et al., 2023).

3.3.4 Biogeography of the Atmospheric Microplastisphere

MP dispersal in the atmosphere could be studied using the same "biogeographic" techniques that are used to examine the spread of organisms in relation to their surroundings. When released into the atmosphere, MPs can travel great distances due to wind currents and atmospheric circulation patterns, which causes them to become widely distributed in both terrestrial and marine habitats (Haque and Fan, 2023). According to studies, AMPs can move from low-latitude ocean surfaces to far-off places, such as the Arctic zones (Zhang et al., 2019).

The land surface of Earth is characterized by specified biogeographic regions, which are generally classified into six regions (Womack et al., 2010). Nevertheless, no solid research has been done so far to definitively establish the existence of biogeographic areas in the atmosphere. However, more broadly, six separate "cells" of air surround the Earth because of the combination of Earth's rotation and the differential heating of air at the tropics and poles (Klein, 2015) as shown in Figure 3.3 (Shackford, 2005). Thus, it is possible that distinct microbial communities may reside

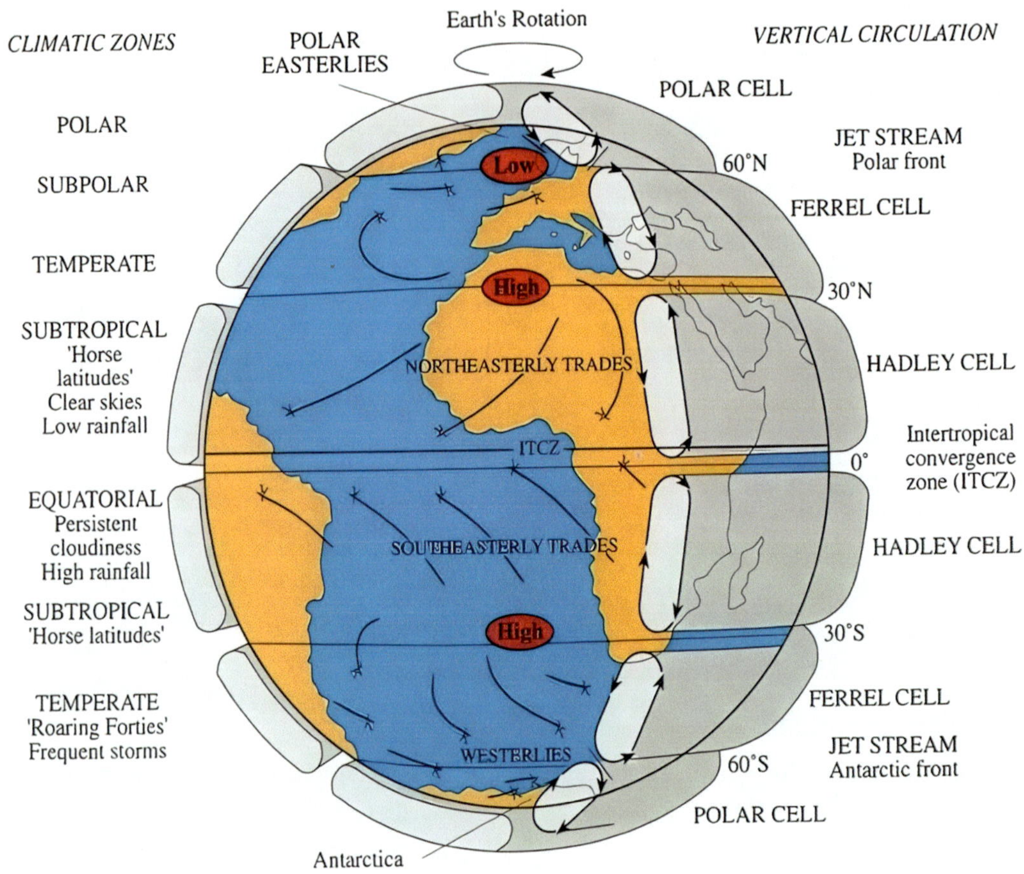

FIGURE 3.3 The six-cell model of global air circulation (From Charnook, 1996).

in these air cells, possibly adhering to surfaces of AMPs. Additional research employing both culture-dependent and culture-independent molecular techniques may provide insight into determining the Microorganisms linked to AMPs in all atmospheric cells worldwide.

3.3.5 Impacts on Ecosystems

AMPs are considered emerging pollutants, and inhaling them can be harmful to one's health (Perera et al., 2023). During the COVID-19 pandemic, surgical masks were a new source of MPs, which raised concerns about the attachment of pathogenic bacteria and the potential health risks that could arise from breathing them directly (Haque and Fan, 2023). During an experiment, the emission of nano- and microplastic particles ranging from 5 to 600 nm was detected from face masks (Ma et al., 2021).

There are concerns that ingesting MPs could cause gastrointestinal inflammations in people, particularly if the particles contain chemicals from the environment that could have harmful consequences and damage organs, raising the risk of cancer (Kirstein et al., 2016). We can assume that ingestion of AMPs to organisms may pose similar risks as in MP deposits in terrestrial and aquatic habitats. Furthermore, the movement of MPs through the atmosphere may make it easier for them to enter different ecosystems, increasing the risk of environmental pollution and its resulting impact on ecosystem health and biodiversity (Ghosh et al., 2023). In a study, the relationship between the

AMPs and the phyllosphere bacterial community (PBC) was examined. Genus *Sphingomonas* was majorly discovered to contribute to the pathogenesis of PBC caused by AMPs. This demonstrates the risks to the environment and the rise in pathogenicity within the phyllosphere microbial community (Xu et al., 2024).

Moreover, macromolecules derived from bacteria, such as proteins, lipids, nucleic acids, amino acids, ribonucleic acid, and deoxyribonucleic acid (Kurepa et al., 2020), may combine to create bio-coronas on AMPs. Such coronas may have detrimental consequences on ecosystems, including the generation of reactive oxygen species (ROS), cellular uptake, and ultimately cell death (Cao et al., 2022). The interactions between different microbial populations on AMPs may give rise to concerns about the possible establishment of antibiotic-resistant microbial communities within biofilms (Mammo et al., 2020). Therefore, there is a chance that the presence of AMPs in drug-exposed habitats will accelerate the development of antibiotic resistance, endangering both ecological integrity and public health.

3.4 MICROPLASTIC DEPOSITION ON SYMBIOTIC ORGANISMS

Most of the MPs originate from fresh waters and terrestrial sources, for instance, stormwater outflow, sewage effluent, or possibly through the deposition of airborne MPs (Zhang, 2017). These might affect the mutualism in relation to terrestrial and aquatic biomes. As there is limited data regarding the AMPs, particularly the effects on symbiotic relationships, the next sections will discuss on atmospheric deposition of MPs, possible stresses in the symbiotic organisms, and approaches to determine the impacts on the biology of the organisms and in the long run on the ecosystem.

3.4.1 Atmospheric Deposition of Microplastics on Ecosystems

It has been suggested that MP pollution throughout large areas of more than 100 km could be transported by atmospheric deposition (Allen et al., 2019). Figure 3.4 depicts the possible pathways of MP transportation and deposition. Studies indicate that amounts of anthropopressure, dominant

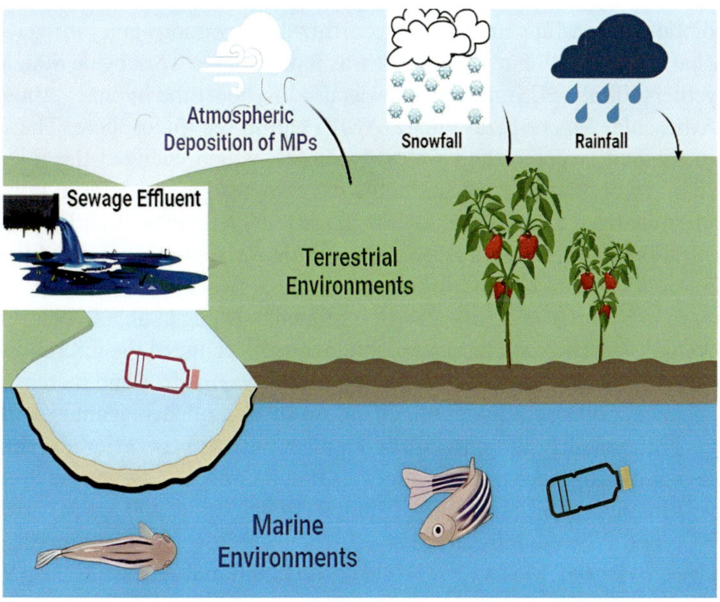

FIGURE 3.4 Pathways of microplastic transport and deposition (Created with Biorender.com).

land cover, and distance to the first sources are necessary factors influencing MP deposition in the atmosphere (Brahney et al., 2020; Roblin et al., 2020). In general, fibers and fragments together with thermoplastic polymers are the most frequently identified MP types deposited in the air (Zhang et al., 2020). Among them, fibers have been classified as the most prominent type due to having vast production, easy analysis attributed to the large size, and high release rates (Szewc et al., 2021). The deposition factor works in a way that slows down the transportation rates of MP and also increases their in situ breakdown. According to the research conducted by Allen et al. (2019), snowfall poses the possibility of contributing more AMPs deposition compared to rainfall.

3.4.2 Microplastic Exposure as a Stressor for Symbiotic Relationships

Scleractinian corals that create reefs and algae have symbiotic connections. Although they are negatively impacted by temperature changes and bleaching, research suggests that MPs may have an impact on the photosynthesis ability of the photosymbiont and the heterotrophic feeding of the corals (Allen et al., 2017). Further, research has demonstrated that MP spheres can be ingested by corals and encased in mesenterial tissues within their stomach chambers (Hall et al., 2015). Also, MPs are likely to disrupt the mutualistic association between anthozoa and algae (Okubo et al., 2018). It has been reported that sea anemones and scleractinian corals respond to plastic pollution in a similar manner. Therefore, sea anemones may be useful as a model species for evaluating coral reaction to MP exposure (Grajales and Rodríguez, 2016). According to an experiment by de Orte et al. (2019), bleached sea anemones that have expelled their symbiotic algae due to stress retained plastic microfibers in their bodies for much longer compared to healthy anemones with symbionts. The proportion of microfibers ingested was depended on the chemical cues present in the plastic biofilms or the surrounding environment. Furthermore, the researchers asserted that they suffered higher magnitudes of adverse effects when subjected to bleaching. This implies that similarly several stressors, such as pollution from MPs, hazardous chemicals, and ocean warming (Anthony, 2016), may also operate as stressors for the coral reef systems.

In terrestrial environments, MPs have the potential to damage the soil–plant–microbe system and ultimately have an impact on the environment since they may serve as a vector for the spread of pollutants and Microorganisms. By changing plant growth and nutrient levels, this may have an impact on symbiotic relationships including mycorrhizal connections and nitrogen-fixing bacterial communities (Zhai et al., 2024). For example, it was reported in a work by de Souza Machado et al. (2019) that polyethersulfone (PES) microfiber was used to boost the hyphae, arbuscules, and coils that are off in Arbuscular Mycorrhizal Fungi (AMF) within the rhizosphere. These had an impact on alterations in the soil structure and water dynamics, which changed the microbial diversity, activity, and role in the community. Additionally, the composition of bacterial populations such as *Phaseolus vulgaris* changed in response to low-density polyethylene (LDPE) and polylactic acid (PLA) at concentrations of 0.5%, 1.0%, and 2.5%. While *Rhizobiaceae* showed the highest occurrence in the 2.5% LDPE treatment and the lowest in the 2.5% PLA treatment, *Comamonadaceae* showed an increase in abundance across all MP treatments (Meng et al., 2023).

In a different study, starch-based biodegradable polymers changed the microbial composition of the rhizosphere by encouraging the growth of *Bacillus* and *Variovorax* and increasing the synthesis of the volatile chemical dodecanal (Qi et al., 2018). As such, it impacts plant growth processes in a number of ways. These modify the mutualistic connections between Microorganisms and plants. MPs are capable of realtering the enzyme processes that are responsible for the cycling of nutrients and the decomposition of organic matter. For example, in one study, 200 μm particles of PE, polyvinyl chloride (PVC), and PS were introduced at a 2% concentration, and the impact on wheat rhizosphere bacteria was measured. Most of the levels of the functional categories drop when exposed to PS. PVC demonstrated positive effects in xenobiotic metabolism and degradation and PE and PVC exposure enhanced specific functional categories. Figure 3.5 depicts the impact of MPs on symbiotic relationships in marine and terrestrial ecosystems.

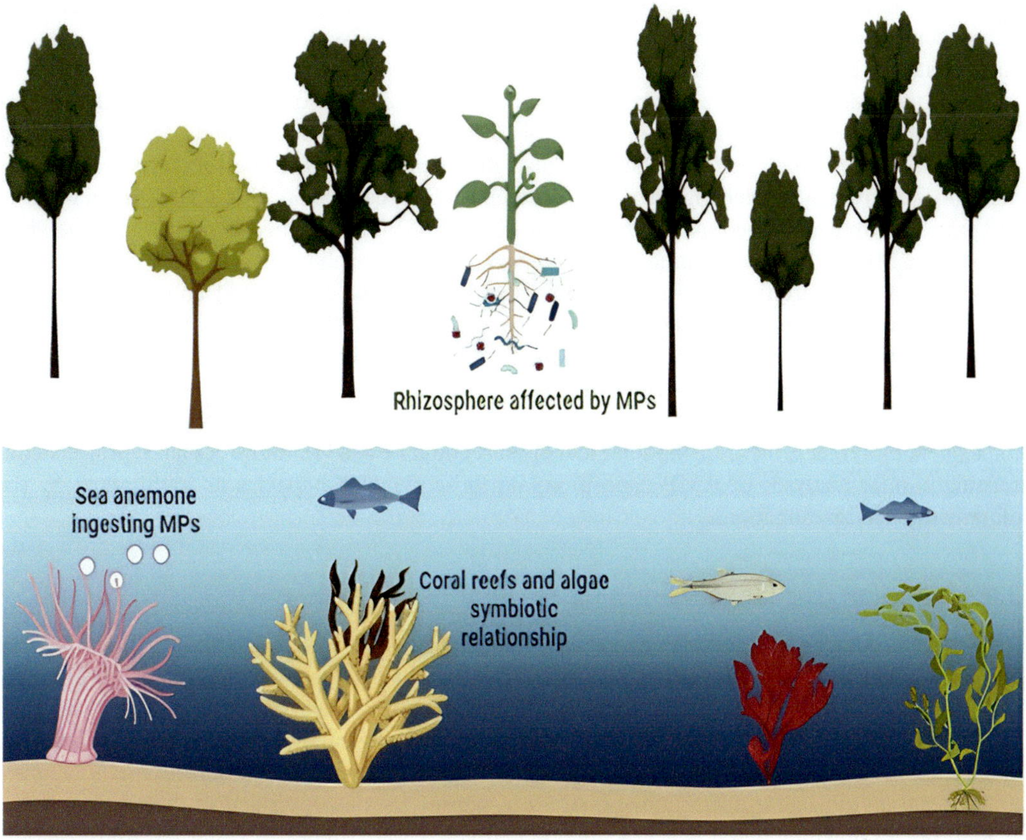

FIGURE 3.5 Impact of microplastics on symbiotic relationships in marine and terrestrial ecosystems (Created with Biorender.com).

3.4.3 Assessment of Microplastic Deposition on Symbiotic Organisms: Methodological Approaches

Different methods, depending on the investigation, could potentially be used to measure the amount of MP deposition on symbiotic organisms. Although there are no direct investigations on the effects of AMP deposition on symbiotic species, future research can benefit from the approaches of already available studies. According to a study by de Orte et al. (2019), plastic microfibers made of nylon, polyester, and PP were given to the sea anemone *Aiptasia pallida*. Fluorescent ropes that are sold commercially were used to create these plastic microfibers. These polymers were introduced to bleached (without algal symbionts) and symbiotic (with algal symbionts) anemones, which were cultivated under carefully monitored conditions and fed once a week. Raman spectroscopy was used to characterize the microfibers before they were combined with shrimp homogenate or artificial saltwater. Microscopy was used to track the ingestion of microfibers, and R software was used to statistically assess the ingestion and egestion rates in relation to different polymer types and the presence of shrimp homogenate.

Another study demonstrated the relationship between coral symbiotic density and MP concentration. The results showed apoptotic mechanisms as well as their capabilities of influencing the balance and relationship between coral-Symbiodiniaceae and also the changes in the distribution of corals. Stereo microscopy has been used, as well as the Micro Fourier Transform Infrared Spectroscopy (m-FTIR) while conducting this investigation to identify and characterize the MPs.

Utilizing commercial kits and spectroscopic assays, the density of coral symbionts was determined. Subsequently, statistical analysis and quality control procedures were carried out to determine sample correlations (Tang et al., 2021).

Pocillopora damicornis is a scleractinian coral that relies on a zooxanthellae intracellular symbiosis. In this symbiotic relationship, the corals offer inorganic nutrients and carbon dioxide to the zooxanthellae, which perform photosynthesis organically in return; the corals receive organic nutrients and oxygen in exchange from zooxanthellae. The reactions of this scleractinian coral to MP exposure were examined in a study by Tang et al. (2018). The findings demonstrated that while exposure to MPs had minimal immediate effect on zooxanthellae abundance and key photosynthate translocation transporters, however, it could affect the immune system and ability to withstand stress in corals. Spectrophotometers were employed to compare the density and chlorophyll concentration of zooxanthellae that form symbiotic relationships with the host organism. Commercial kits and microplate spectrophotometers were used to measure the enzyme activities. IBM® SPSS® Statistical software was used for statistical analysis, and the Illumina HiseqX platform was used for sequencing. Hence, these methodological options could be used in future research projects concerning similar analyses of AMP depositions on those symbiotic species as highlighted by the aforementioned investigations.

3.4.4 Microplastic-Induced Stress on the Biology of Symbiosis

Research has documented that MPs can cause biological stress in organisms, ranging from physical damage to fatal consequences (Lei et al., 2018). Translocation may occur through the digestive system after passing through biological barriers like cell membranes and affect organisms (Ma et al., 2020). Internal exposure may cause oxidative stress, immunological problems, metabolic disorders, and variations in gene expression (Qiao et al., 2019). The exposure of coral *Pocillopora damicornis* to MPs caused molecular and metabolic stress responses, as was described in the section above. A total of 215 downregulated and 134 upregulated genes were identified using the transcriptome analysis data. JNK signaling pathways and the stress response were caused by upregulated genes. An increase in antioxidant enzyme activity (SOD and CAT) led to a rise in ROS levels. The lower GST activity suggested suppressed detoxification. A decrease in AKP activity suggested a weakened immune system. While the coral-zooxanthella symbiosis did not appear to be disrupted by short-term exposure to MPs, more research is required to determine long-term biological changes (Tang et al., 2018).

Corals typically have ciliary movement and mucus production as part of their natural cleaning process, which helps them get rid of non-nutritional particles (Lanctôt et al., 2020). Studies have shown that exposure to various plastic types, as opposed to food particles, can cause mucus development (Connors, 2017). High levels of MP exposure hindered the ability to develop and catch prey (Savinelli et al., 2020), interfered with symbiont uptake into host cells (Okubo et al., 2018), and altered stress responses and suppressed immune functions (Rocha et al., 2020). Thus, this might have an impact on the symbiotic relationship; researchers could anticipate comparable negative impacts from AMPs as well, even though there is not sufficient evidence of how modifications in atmospheric plastic deposition have affected the biology of symbiosis.

3.4.5 Long-Term Ecological Consequences

In marine invertebrates, gut obstruction, false satiation, inflammation, energy reserve depletion, decreased survival, growth and fecundity impairments, and larval development retardation are among the negative health effects that have been documented (de Orte et al., 2019). Also, the cumulative and long-term impacts include resistance to biotic and abiotic degradation processes as MPs accumulate in soil and degrade slowly. In terrestrial environments, MPs remain for long periods of time and accumulate gradually. Physical weathering processes like fragmentation and abrasion may produce smaller particles (Duan et al., 2021). While contact with environmental conditions like sunlight and moisture may cause some polymers to deteriorate over time, overall degradation rates are typically slow (Ding et al., 2022). MP breakdown is additionally assisted by biological degradation, which is facilitated by soil microbes. The effectiveness of this process varies based on a number of variables, including the type of plastic, the microbial community composition, and the surrounding environmental circumstances (Thakur et al., 2023).

The persistent use of MPs may cause a decrement in the ability of soil fertility and further affect all the processes that are naturally associated with nutrient cycling. Such changes may affect fundamental processes such as the decomposition of organic matter, nitrogen cycling, and carbon sequestration that occur in the soil (Chia et al., 2022). MPs may also cause changes to the microbial populations in the soil around the roots where they accumulate; this may affect root growth, nutrient acquisition, and water uptake and utilization, which may adversely affect plant health and productivity (Bouaicha et al., 2022). This disturbance may over time lead to changes in the composition of vegetation, reduced crop yield, and altered ecosystem dynamics (Rillig et al., 2018). If the soil environment has been dominated for a long time by MPs, it can have ecological or even evolutionary significance on microniches and the plant–microbe symbiotic interactions. Since MPs and their environment are habitable in most circumstances, MPs can be viewed as hotspots for microbial action that dramatically affects the evolutionary steps by horizontal genetic transfer between microbes (Wu et al., 2019).

3.5 CHALLENGES AND FUTURE DIRECTIONS

There are still questions and gaps in our knowledge about the research on AMPs (Ahmad et al., 2023). Initiatives for research are needed to completely comprehend the growing threat of AMPs. Marine MP pollution was first recognized as a major issue (Ng and Obbard, 2006), and now it is more apparent in both the atmosphere and the hydrosphere. Recent studies suggest that MPs may infiltrate the environment through a variety of sources and pathways. There could be a number of sources and activities that contribute to the production and release of AMPs, even though the precise amounts are unclear. According to González-Pleiter et al. (2021), building sites, dust, and synthetic garments are the main producers of MPs in the air. Furthermore, these MP particles migrate farther due to processes like precipitation, wind entrainment, and re-aerosolization from the sea surface (Tiernan et al., 2022) making management more difficult. It becomes more difficult because even tourists regularly carry these particles long distances, such as to the summit of Mount Everest (Napper et al., 2020; Wright et al., 2020). To save our ecosystems, it is imperative that we consider future plans for lowering the pollution brought on by AMPs.

3.5.1 Future Research on Atmospheric Microplastisphere

It is anticipated that one or more of the following approaches, which are listed in the description, will be applied in future research aimed at reducing the risks associated with AMP pollution. At first, a systematic investigation could be conducted to determine and categorize the chemical and physical constituents of the AMP sources in terms of the type of polymer, quantities, characteristics, and orders of magnitude of primary and secondary MPs in diverse environments (Noorimotlagh et al., 2024). Additionally, a thorough inventory of the primary sources such as synthetic textiles, construction dust, and other urban sources that lead to the generation of AMPs must be created (Prasittisopin et al., 2023; Jahandari, 2023). Further, it is important to understand the basic processes by which these MP particles are transported through air, rainfall, and other processes and how they are being deposited and distributed (Jahandari, 2023).

More studies exploring MP deposition in remote and pristine sites (e.g., high-altitude regions and pole areas) are hence necessary to determine the levels of pollution and where MPs may have the greatest impact on such fragile ecosystems (Huang et al., 2021). Future microbiome research focusing on these particles and diverse species should be directed toward the interactions among symbiotic species and MPs, and their potential behavioral, health, and ecological impacts (Cui et al., 2022). Moreover, interdisciplinary research could be utilized to conduct effective mitigation strategies and supply a complete understanding. Therefore, the provision of sound scientific evidence to policymakers is very important for solving this environmentally destructive issue through policies and laws. This includes encouraging environmentally friendly behavior, enhancing waste management through waste collection strategies, and educating folks about the origins and impacts of MPs (Belontz et al., 2019). Working on solutions to these challenges and exploring these lines of research present an opportunity to reduce their impact on the environment.

According to Zhang et al. (2020), AMP research is still in its infancy, and there is not enough comparable data on characterization and abundance. Due to the current nonstandardized operation techniques for MP sampling and detection, this is particularly the case in remote locations and with regard to the composition of MPs. Standardized procedures for AMP sampling and measurement will enable results to be repeated and compared, as well as provide high-quality data required for risk evaluations. Also, there is a need to expand global study on the temporal and spatial variability of AMP depositions. Research is also required to comprehend the relationship between human exposure, ecosystems, and other substances and AMPs (Zhang et al., 2020). With substantial unmet research objectives and difficulties in defining and characterizing AMP transport dynamics, atmospheric particle transport modeling is a key area of future concentration for AMP transport studies.

3.5.2 Methodological Challenges in Studying Atmospheric Microplastisphere

There is a significant risk of contamination at every stage of sample collection, processing, and analysis since MPs are widely distributed and originate from a wide range of sources in both indoor and outdoor environments (Brander et al., 2020). Therefore, it is essential to evaluate potential contamination throughout the sample collecting and analysis procedure using field and laboratory blanks as well as equipment. To comprehend MP heterogeneity and sample representativeness, replicated samples are also required (Lenaker et al., 2019). Quality-control measures pertaining to sample collection have not always been consistent in research conducted thus far. Therefore, a thorough evaluation of quality assurance pertaining to (1) optimizing sample representativeness and reproducibility in varied media, (2) limiting bias, and (3) minimizing contamination is required in order to help guide the development of robust sample-collection procedures (Koelmans et al., 2019). To reduce unpredictable outcomes from partial digestion of organic materials, undesired MP degradation, and digestion reactions, separation procedure improvements are required (Sutkar et al., 2023). Additionally, validation of characterization techniques should be conducted for a wide range of MP sizes, polymer compositions, and weathering and aging circumstances (Hurley et al., 2018).

In the past, to verify the presence and quantification, the majority of MP research relied on basic microscopic visualization, which works well for large particles (Silva et al., 2018). However, the quality and resolution of the microscope, and human bias, all have a significant impact on this procedure (Li et al., 2018). Therefore, spectroscopic techniques are more accurate in investigating particles that are specifically smaller than 500 μm (Zhang et al., 2020). Raman and FTIR spectroscopy can be used to measure the chemical reactions of the bonds in MP material. Fluorescence is the primary problem with short-wavelength Raman lasers. The conventional Raman wavelengths are extremely near to several maximum excitation wavelengths, which induces fluorescence in the particle. The signal is obscured by this fluorescence, making it impossible to analyze the chemical bonds. To overcome this, research is currently employing a 785 nm near-infrared laser (Allen et al., 2019). Characterizing small MPs (<20 μm) and nanoplastics (<1 μm), which are increasingly identified as posing a significant concern to environmental health, requires specific techniques (Lei et al., 2018). Furthermore, to enable focused research in regions with elevated MP concentrations, quick and affordable analyses of field-collected samples are required. These analyses must be able to quickly determine the relative amount of MP contamination over time and geography (Iwanowicz et al., 2024).

3.5.3 From Research to Conservation and Policy

Making decisions on complicated subjects is difficult because there are usually many parties involved, each with competing goals, trade-offs, and preferred courses of action (Lyons et al., 2008). When an issue is new and poorly studied, as airborne MP contaminants are, then another level of complexity appears. Many quantitative, qualitative, and hybrid approaches to decision-making have been put forth; of these, structured decision-making, or "SDM," as shown in Figure 3.6 has proven to be a successful strategy for handling challenging issues. Thus, future studies could focus on answering these quantitative research questions as well as finding and filling knowledge gaps in supply networks, human behavior, and the danger that AMPs pose to the environment (Tiernan et al., 2022).

The SDM principle addresses the following concerns. The stakeholders' main objectives should be based on AMPs, their level of attention on this crisis, approach in decision-making, steps that need to be taken, advantages and disadvantages of those actions, and areas that lack understanding of AMPs. Stakeholders can counteract the pollution generated by AMPs by implementing conservation measures and laws and providing solutions to these questions.

3.5.4 Bridging the Gaps: Building Interdisciplinary Partnerships

Solving complex issues of plastic pollution requires bridging disciplinary boundaries among generations with different age groups who may have different perspectives regarding environmental issues and between and within the arts, sciences, social sciences, and humanities. The *Synthetic Collective* (a group of scientists, artists, and humanities researchers) employs rigorously interdisciplinary practices grounded in scientific data to produce resonant knowledge as artifacts, artworks, curatorial interventions, articles, and other forms that are legible to scholarly audiences, policy decision-makers, people of art, and the wider public. Humanities researchers, particularly those focused on visual culture and media studies, can collaborate with scientists from fields such as geology, environmental science, biology, and chemistry. This interdisciplinary approach will

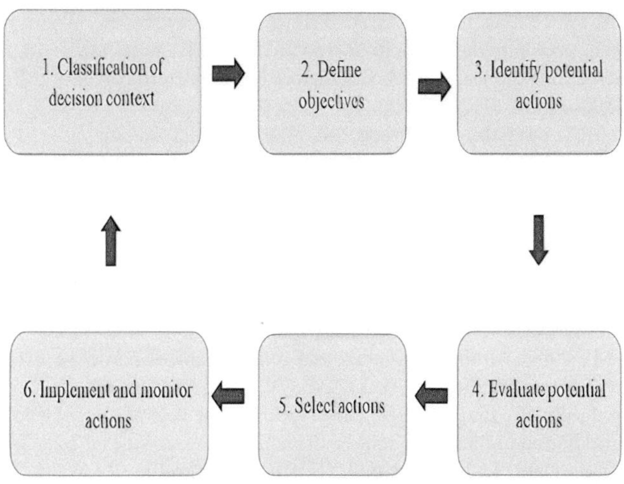

FIGURE 3.6 Overview of a structured decision-making (SDM) process (Tiernan et al., 2022).

provide insights into how plastic waste is introduced, into the environment, its impacts, and how it is archived. Several examples will be given on the related concept of citizen science, outreach, and media channels; another aspect is a reflection on an artwork that covers plastic waste in various artistic disciplines by scientists and the interdisciplinary approach adopted to tackle the global issue of plastic pollution (Belontz et al., 2019).

The case of plastiglomerate – an agglomerating mass of plastics and natural elements as described in an influential scientific paper, during international art exhibitions, public lectures, and the media – is a great example of a collaborative work on a visualization of plastic contamination (Corcoran et al., 2014). Based on the idea that land use is influenced by culture, the nonprofit Center for Land Use Interpretation (CLUI) has followed the "life" of plastics in Texas, from petroleum to trash in rivers. As a method of research, they find ways to relate local meanings and practices involving land-use phenomena to scientific explanations, without devaluing their own scientific work. Also, using text, art, photography, artifacts, and video, the polymedia installations in both the permanent exhibition space in Los Angeles and traveling exhibition sites around the United States create a "thick description" of the political and scientific dimensions of a suite of land-use problems. These arts are not in a vacuum but in a wider world that also produces literature writing on ecocriticism, human and nonhuman condition, plastic pollution, etc. (Gabrys et al., 2013; Robertson, 2016; Huang, 2017). Therefore, in order to bridge the gaps and develop solutions for the pollution caused by MPs in the atmosphere, interdisciplinary work like these could be employed.

REFERENCES

Ahmad, M., Chen, J., Khan, M.T., Yu, Q., Phairuang, W., Furuuchi, M., Ali, S.W., Nawab, A., Panyametheekul, S., 2023. Sources, analysis, and health implications of atmospheric microplastics. *Emerg. Contam.* 9, 100233.

Allen, A.S., Seymour, A.C., Rittschof, D., 2017. Chemoreception drives plastic consumption in a hard coral. *Mar. Pollut. Bull.* 124, 198–205.

Allen, S., Allen, D., Phoenix, V.R., Le Roux, G., Durántez Jiménez, P., Simonneau, A., Binet, S., Galop, D., 2019. Atmospheric transport and deposition of microplastics in a remote mountain catchment. *Nat. Geosci.* 12, 339–344.

Anthony, K.R., 2016. Coral reefs under climate change and ocean acidification: challenges and opportunities for management and policy. *Annu. Rev. Environ. Resour.* 41(1), 59–81.

Archer, S.D., Lee, K.C., Caruso, T., Alcami, A., Araya, J.G., Cary, S.C., Cowan, D.A., Etchebehere, C., Gantsetseg, B., Gomez-Silva, B., Hartery, S., 2023. Contribution of soil bacteria to the atmosphere across biomes. *Sci. Total Environ.* 871, 162137.

Belontz, S.L., Corcoran, P.L., Davis, H., Hill, K.A., Jazvac, K., Robertson, K., Wood, K., 2019. Embracing an interdisciplinary approach to plastics pollution awareness and action. *Ambio* 48, 855–866.

Biber, N.F., Foggo, A., Thompson, R.C., 2019. Characterising the deterioration of different plastics in air and seawater. *Mar. Pollut. Bull.* 141, 595–602.

Bouaicha, O., Mimmo, T., Tiziani, R., Praeg, N., Polidori, C., Lucini, L., Vigani, G., Terzano, R., Sanchez-Hernandez, J.C., Illmer, P., 2022. Microplastics make their way into the soil and rhizosphere: A review of the ecological consequences. *Rhizosphere* 22, 100542.

Brahney, J., Hallerud, M., Heim, E., Hahnenberger, M., Sukumaran, S., 2020. Plastic rain in protected areas of the United States. *Science* 368, 1257–1260.

Brander, S.M., Renick, V.C., Foley, M.M., Steele, C., Woo, M., Lusher, A., Carr, S., Helm, P., Box, C., Cherniak, S., 2020. Sampling and quality assurance and quality control: A guide for scientists investigating the occurrence of microplastics across matrices. *Appl. Spectrosc.* 74, 1099–1125.

Buseck, P.R., Pósfai, M., 1999. Airborne minerals and related aerosol particles: Effects on climate and the environment. *Proc. Nat. Acad. Sci.* 96(7), 3372–3379.

Cai, L., Wang, J., Peng, J., Wu, Z., Tan, X., 2018. Observation of the degradation of three types of plastic pellets exposed to UV irradiation in three different environments. *Sci. Total Env.* 628, 740–747.

Cao, J., Yang, Q., Jiang, J., Dalu, T., Kadushkin, A., Singh, J., Fakhrullin, R., Wang, F., Cai, X., Li, R., 2022. Coronas of micro/nano plastics: A key determinant in their risk assessments. *Particle Fibre Toxicol.* 19(1), 55.

Charnook, H., 1996. The Atmosphere and the Ocean. In: *Oceanography, an illustrated guide,* edited by C.P. Summerhayes and S.A. Thorpe. New York: John Wiley & Sons.
Chia, R.W., Lee, J.-Y., Jang, J., Kim, H., Kwon, K.D., 2022. Soil health and microplastics: A review of the impacts of microplastic contamination on soil properties. *J. Soils Sediments* 22, 2690–2705.
Choudhury, A., Simnani, F.Z., Singh, D., Patel, P., Sinha, A., Nandi, A., Ghosh, A., Saha, U., Kumari, K., Jaganathan, S.K., Kaushik, N.K., 2023. Atmospheric microplastic and nanoplastic: The toxicological paradigm on the cellular system. *Ecotoxicol. Environ. Safety* 259, 115018.
Chudobova, D., Cihalova, K., Jelinkova, P., Zitka, J., Nejdl, L., Guran, R., Klimanek, M., Adam, V., Kizek, R., 2015. Effects of stratospheric conditions on the viability, metabolism and proteome of prokaryotic cells. *Atmosphere* 6, 1290–1306.
Connors, E.J., 2017. Distribution and biological implications of plastic pollution on the fringing reef of Mo'orea, French Polynesia. *PeerJ* 5, e3733.
Corcoran, P.L., Moore, C.J., Jazvac, K., 2014. An anthropogenic marker horizon in the future rock record. *GSA Today* 24, 4–8.
Cui, J., Chen, C., Gan, Q., Wang, T., Li, W., Zeng, W., Xu, X., Chen, G., Wang, L., Lu, Z., 2022. Indoor microplastics and bacteria in the atmospheric fallout in urban homes. *Sci. Total Environ.* 852, 158233.
de Orte, M.R., Clowez, S., Caldeira, K., 2019. Response of bleached and symbiotic sea anemones to plastic microfiber exposure. *Environ. Pollut.* 249, 512–517.
de Souza Machado, A.A., Lau, C.W., Kloas, W., Bergmann, J., Bachelier, J.B., Faltin, E., Becker, R., Görlich, A.S., Rillig, M.C., 2019. Microplastics can change soil properties and affect plant performance. *Environ. Sci. Technol.* 53, 6044–6052.
Ding, L., Ouyang, Z., Liu, P., Wang, T., Jia, H., Guo, X., 2022. Photodegradation of microplastics mediated by different types of soil: The effect of soil components. *Sci. Total Environ.* 802, 149840.
Dommergue, A., Amato, P., Tignat-Perrier, R., Magand, O., Thollot, A., Joly, M., Bouvier, L., Sellegri, K., Vogel, T., Sonke, J.E., Jaffrezo, J.L., 2019. Methods to investigate the global atmospheric microbiome. *Front. Microbiol.* 10, 243.
Duan, J., Bolan, N., Li, Y., Ding, S., Atugoda, T., Vithanage, M., Sarkar, B., Tsang, D.C., Kirkham, M., 2021. Weathering of microplastics and interaction with other coexisting constituents in terrestrial and aquatic environments. *Water Res.* 196, 117011.
Frère, L., Maignien, L., Chalopin, M., Huvet, A., Rinnert, E., Morrison, H., Kerninon, S., Cassone, A.L., Lambert, C., Reveillaud, J., Paul-Pont, I., 2018. Microplastic bacterial communities in the Bay of Brest: Influence of polymer type and size. *Environ. Pollut.* 242, 614–625.
Frias, J.P., Nash, R., 2019. Microplastics: Finding a consensus on the definition. *Mar. Pollut. Bull.* 138, 145–147.
Gabrys, J., Hawkins, G., Michael, M., 2013. *Accumulation: The Material Politics of Plastic*. Routledge.
Ghosh, S., Sinha, J.K., Ghosh, S., Vashisth, K., Han, S., Bhaskar, R., 2023. Microplastics as an emerging threat to the global environment and human health. *Sustainability* 15(14), 10821.
González-Pleiter, M., Edo, C., Aguilera, Á., Viúdez-Moreiras, D., Pulido-Reyes, G., González-Toril, E., Osuna, S., de Diego-Castilla, G., Leganés, F., Fernández-Piñas, F., 2021. Occurrence and transport of microplastics sampled within and above the planetary boundary layer. *Sci. Total Environ.* 761, 143213.
Grajales, A., Rodríguez, E., 2016. Elucidating the evolutionary relationships of the Aiptasiidae, a widespread cnidarian–dinoflagellate model system (Cnidaria: Anthozoa: Actiniaria: Metridioidea). *Mol. Phylogenet. E* 94, 252–263.
Habibi, N., Uddin, S., Fowler, S.W., Behbehani, M., 2022. Microplastics in the atmosphere: A review. *J. Environ. Expo. Assess* 1(6), 10–20517.
Hall, N., Berry, K., Rintoul, L., Hoogenboom, M., 2015. Microplastic ingestion by scleractinian corals. *Mar. Biol.* 162, 725–732.
Haque, F., Fan, C., 2023. Fate and impacts of microplastics in the environment: Hydrosphere, pedosphere, and atmosphere. *Environments* 10(5), 70.
Harrison, J.P., Sapp, M., Schratzberger, M., Osborn, A.M., 2011. Interactions between microorganisms and marine microplastics: A call for research. *Mar. Technol. Soc. J.* 45(2), 12–20.
Huang, M.N., 2017. Ecologies of entanglement in the great Pacific garbage patch. *J. Asian Am. Stud.* 20, 95–117.
Huang, Y., He, T., Yan, M., Yang, L., Gong, H., Wang, W., Qing, X., Wang, J., 2021. Atmospheric transport and deposition of microplastics in a subtropical urban environment. *J. Hazard. Mater.* 416, 126168.
Hurley, R.R., Lusher, A.L., Olsen, M., Nizzetto, L., 2018. Validation of a method for extracting microplastics from complex, organic-rich, environmental matrices. *Environ. Sci. Technol.* 52, 7409–7417.
Iwanowicz, D.D., Baldwin, A.K., Barber, L.B., Blazer, V.S., Corsi, S.R., Duris, J.W., Fisher, S.C., Focazio, M., Janssen, S.E., Jasmann, J.R., 2024. Integrated science for the study of microplastics in the environment—A strategic science vision for the US Geological Survey (No. 1411345541). US Geological Survey.

Jahandari, A., 2023. Microplastics in the urban atmosphere: Sources, occurrences, distribution, and potential health implications. *J. Hazard. Mater. Adv.* 12, 100346.

Khodadad, C.L., Wong, G.M., James, L.M., Thakrar, P.J., Lane, M.A., Catechis, J.A., Smith, D.J., 2017. Stratosphere conditions inactivate bacterial endospores from a Mars spacecraft assembly facility. *Astrobiology* 17(4), 337–350.

Kirstein, I.V., Kirmizi, S., Wichels, A., Garin-Fernandez, A., Erler, R., Löder, M., Gerdts, G., 2016. Dangerous hitchhikers? Evidence for potentially pathogenic Vibrio spp. on microplastic particles. *Mar. Environ. Res.* 120, 1–8.

Klein, A.M., 2015. *Microbial biodiversity of the atmosphere* (Doctoral dissertation, University of Oregon).

Koelmans, A.A., Nor, N.H.M., Hermsen, E., Kooi, M., Mintenig, S.M., De France, J., 2019. Microplastics in freshwaters and drinking water: Critical review and assessment of data quality. *Water Res.* 155, 410–422.

Kumar, A., Yu, Z.G., Thakur, T.K., 2023. Microplastic pollutants in terrestrial and aquatic environment. *Environ. Sci. Pollut. Res.* 30(49), 107296–107299.

Kurepa, J., Shull, T.E., Smalle, J.A., 2020. Metabolomic analyses of the bio-corona formed on TiO 2 nanoparticles incubated with plant leaf tissues. *J. Nanobiotechnology*, *18*, 1–10.

Lanctôt, C.M., Bednarz, V.N., Melvin, S., Jacob, H., Oberhaensli, F., Swarzenski, P.W., Ferrier-Pagès, C., Carroll, A.R., Metian, M., 2020. Physiological stress response of the scleractinian coral Stylophora pistillata exposed to polyethylene microplastics. *Environ. Pollut.* 263, 114559.

Lei, L., Wu, S., Lu, S., Liu, M., Song, Y., Fu, Z., Shi, H., Raley-Susman, K.M., He, D., 2018. Microplastic particles cause intestinal damage and other adverse effects in zebrafish Danio rerio and nematode Caenorhabditis elegans. *Sci. Total Environ.* 619, 1–8.

Lenaker, P.L., Baldwin, A.K., Corsi, S.R., Mason, S.A., Reneau, P.C., Scott, J.W., 2019. Vertical distribution of microplastics in the water column and surficial sediment from the Milwaukee River Basin to Lake Michigan. *Environ. Sci. Technol.* 53, 12227–12237.

Leuko, S., Domingos, C., Parpart, A., Reitz, G., Rettberg, P., 2015. The survival and resistance of Halobacterium salinarum NRC-1, Halococcus hamelinensis, and Halococcus morrhuae to simulated outer space solar radiation. *Astrobiology* 15(11), 987–997.

Li, J., Liu, H., Chen, J.P., 2018. Microplastics in freshwater systems: A review on occurrence, environmental effects, and methods for microplastics detection. *Water Res.* 137, 362–374.

López, A.P.A., Trilleras, J., Arana, V.A., Garcia-Alzate, L.S., Grande-Tovar, C.D., 2023. Atmospheric microplastics: Exposure, toxicity, and detrimental health effects. *RSC Adv.* 13(11), 7468–7489.

Luo, H., Liu, C., He, D., Xu, J., Sun, J., Li, J., Pan, X., 2022. Environmental behaviors of microplastics in aquatic systems: A systematic review on degradation, adsorption, toxicity and biofilm under aging conditions. *J. Hazard. Mater.* 423, 126915.

Lyons, J.E., Runge, M.C., Laskowski, H.P., Kendall, W.L., 2008. Monitoring in the context of structured decision-making and adaptive management. *J. Wildl. Manag.* 72, 1683–1692.

Ma, H., Pu, S., Liu, S., Bai, Y., Mandal, S., Xing, B., 2020. Microplastics in aquatic environments: Toxicity to trigger ecological consequences. *Environ. Pollut.* 261, 114089.

Ma, J., Chen, F., Xu, H., Jiang, H., Liu, J., Li, P., Chen, C.C., Pan, K., 2021. Face masks as a source of nanoplastics and microplastics in the environment: Quantification, characterization, and potential for bioaccumulation. *Environ. Pollut.* 288, 117748.

Mammo, F.K., Amoah, I.D., Gani, K.M., Pillay, L., Ratha, S.K., Bux, F., Kumari, S., 2020. Microplastics in the environment: Interactions with microbes and chemical contaminants. *Sci. Total Environ.*, 743, 140518.

Mao, R., Lang, M., Yu, X., Wu, R., Yang, X., Guo, X., 2020. Aging mechanism of microplastics with UV irradiation and its effects on the adsorption of heavy metals. *J. Hazard. Mater.* 393, 122515.

Meng, F., Harkes, P., van Steenbrugge, J.J., Geissen, V., 2023. Effects of microplastics on common bean rhizosphere bacterial communities. *Appl. Soil Ecol.* 181, 104649.

Napper, I.E., Davies, B.F., Clifford, H., Elvin, S., Koldewey, H.J., Mayewski, P.A., Miner, K.R., Potocki, M., Elmore, A.C., Gajurel, A.P., 2020. Reaching new heights in plastic pollution—Preliminary findings of microplastics on Mount Everest. *One Earth* 3, 621–630.

Naumova, N.B., Kabilov, M.R., 2022. About the biodiversity of the air microbiome. *Acta Nat.* 14(4), 50.

Ng, K., Obbard, J.P., 2006. Prevalence of microplastics in Singapore's coastal marine environment. *Mar. Pollut. Bull.* 52, 761–767.

Noorimotlagh, Z., Hopke, P.K., Mirzaee, S.A., 2024. A systematic review of airborne microplastics emissions as emerging contaminants in outdoor and indoor air environments. *Emerg. Contam.* 10, 100372.

O'Brien, S., Rauert, C., Ribeiro, F., Okoffo, E.D., Burrows, S.D., O'Brien, J.W., Wang, X., Wright, S.L., Thomas, K.V., 2023. There's something in the air: A review of sources, prevalence and behaviour of microplastics in the atmosphere. *Sci. Total Environ.* 874, 162193.

Okubo, N., Takahashi, S., Nakano, Y., 2018. Microplastics disturb the anthozoan-algae symbiotic relationship. *Mar. Pollut. Bull.* 135, 83–89.

Prasittisopin, L., Ferdous, W., Kamchoom, V., 2023. Microplastics in construction and built environment. *Dev. Built Environ.* 15, 100188.

Perera, K., Ziajahromi, S., Nash, S.B., Leusch, F.D., 2023. Microplastics in Australian indoor air: abundance, characteristics, and implications for human exposure. *Sci. Total Environ*, 889, p.164292.

Qi, Y., Yang, X., Pelaez, A.M., Lwanga, E.H., Beriot, N., Gertsen, H., Garbeva, P., Geissen, V., 2018. Macro- and micro-plastics in soil-plant system: Effects of plastic mulch film residues on wheat (Triticum aestivum) growth. *Sci. Total Environ.* 645, 1048–1056.

Qiao, R., Sheng, C., Lu, Y., Zhang, Y., Ren, H., Lemos, B., 2019. Microplastics induce intestinal inflammation, oxidative stress, and disorders of metabolome and microbiome in zebrafish. *Sci. Total Environ.* 662, 246–253.

Rao, Y., Li, H., Chen, M., Fu, Q., Zhuang, G., Huang, K., 2020. Characterization of airborne microbial aerosols during a long-range transported dust event in Eastern China: Bacterial community, influencing factors, and potential health effects. *Aerosol Air Qual. Res.* 20(12), 2834–2845.

Rillig, M.C., de Souza Machado, A.A., Lehmann, A., Klümper, U., 2018. Evolutionary implications of microplastics for soil biota. *Environ. Chem.* 16, 3–7.

Robertson, K., 2016. Plastiglomerate. *E-flux J.* 86, 1–11.

Roblin, B., Ryan, M., Vreugdenhil, A., Aherne, J., 2020. Ambient atmospheric deposition of anthropogenic microfibers and microplastics on the western periphery of Europe (Ireland). *Environ. Sci. Technol.* 54, 11100–11108.

Rocha, R., Rodrigues, A., Campos, D., Cícero, L., Costa, A., Silva, D., Oliveira, M., Soares, A., Silva, A.P., 2020. Do microplastics affect the zoanthid Zoanthus sociatus? *Sci. Total Environ.* 713, 136659.

Šantl-Temkiv, T., Amato, P., Casamayor, E.O., Lee, P.K., Pointing, S.B., 2022. Microbial ecology of the atmosphere. *FEMS Microbiol. Rev.* 46(4), fuac009.

Savinelli, B., Fernández, T.V., Galasso, N.M., D'Anna, G., Pipitone, C., Prada, F., Zenone, A., Badalamenti, F., Musco, L., 2020. Microplastics impair the feeding performance of a Mediterranean habitat-forming coral. *Mar. Environ. Res.* 155, 104887.

Shackford, J.K., 2005. *Continuous late pleistocene paleoclimate record from the Southwest African Margin: A multi-proxy approach* (MS thesis). Georgia State University. https://doi.org/10.57709/1059586

Shruti, V.C., Kutralam-Muniasamy, G., Pérez-Guevara, F., 2023. Microplastisphere antibiotic resistance genes: A bird's-eye view on the plastic-specific diversity and enrichment. *Sci. Total Environ.* 912, 169316.

Silva, A.B., Bastos, A.S., Justino, C.I., da Costa, J.P., Duarte, A.C., Rocha-Santos, T.A., 2018. Microplastics in the environment: Challenges in analytical chemistry-A review. *Anal. Chim. Acta* 1017, 1–19.

Smith, D.J., Griffin, D.W., McPeters, R.D., Ward, P.D., Schuerger, A.C., 2011. Microbial survival in the stratosphere and implications for global dispersal. *Aerobiologia* 27, 319–332.

Stabnikova, O., Stabnikov, V., Marinin, A., Klavins, M., Klavins, L., Vaseashta, A., 2021. Microbial life on the surface of microplastics in natural waters. *Appl. Sci.* 11(24), 11692.

Sun, Y., Wu, M., Zang, J., Du, L., Huang, M., Chen, C., Wang, J., 2023. Plastisphere microbiome: Methodology, diversity, and functionality. *iMeta* 2, e101.

Sutkar, P.R., Gadewar, R.D., Dhulap, V.P., 2023. Recent trends in degradation of microplastics in the environment: A state-of-the-art review. *J. Hazard. Mater. Adv.* 11, 100343.

Szewc, K., Graca, B., Dołęga, A., 2021. Atmospheric deposition of microplastics in the coastal zone: Characteristics and relationship with meteorological factors. *Sci. Total Environ.* 761, 143272.

Tang, J., Ni, X., Zhou, Z., Wang, L., Lin, S., 2018. Acute microplastic exposure raises stress response and suppresses detoxification and immune capacities in the scleractinian coral Pocillopora damicornis. *Environ. Pollut.* 243, 66–74.

Tang, J., Wu, Z., Wan, L., Cai, W., Chen, S., Wang, X., Luo, J., Zhou, Z., Zhao, J., Lin, S., 2021. Differential enrichment and physiological impacts of ingested microplastics in scleractinian corals in situ. *J. Hazard. Mater.* 404, 124205.

Thakur, B., Singh, J., Singh, J., Angmo, D., Vig, A.P., 2023. Biodegradation of different types of microplastics: Molecular mechanism and degradation efficiency. *Sci. Total Environ.* 877, 162912.

Tiernan, H., Friedman, S., Clube, R.K., Burgman, M.A., Castillo, A.C., Stettler, M.E., Kazarian, S.G., Wright, S., De Nazelle, A., 2022. Implementation of a structured decision-making framework to evaluate and advance understanding of airborne microplastics. *Environ. Sci. Policy* 135, 169–181.

Tirumalai, M.R., Rastogi, R., Zamani, N., O'Bryant Williams, E., Allen, S., Diouf, F., Kwende, S., Weinstock, G.M., Venkateswaran, K.J., Fox, G.E., 2013. Candidate genes that may be responsible for the unusual resistances exhibited by Bacillus pumilus SAFR-032 spores. *PLoS One* 8(6), e66012.

Van Doremalen, N., Bushmaker, T., Morris, D.H., Holbrook, M.G., Gamble, A., Williamson, B.N., Tamin, A., Harcourt, J.L., Thornburg, N.J., Gerber, S.I., Lloyd-Smith, J.O., 2020. Aerosol and surface stability of SARS-CoV-2 as compared with SARS-CoV-1. *New Engl. J. Med.* 382(16), 1564–1567.

Wijesekara, H., Bolan, N.S., Bradney, L., Obadamudalige, N., Seshadri, B., Kunhikrishnan, A., Dharmarajan, R., Ok, Y.S., Rinklebe, J., Kirkham, M.B., Vithanage, M., 2018. Trace element dynamics of biosolids-derived microbeads. *Chemosphere* 199, 331–339.

Womack, A.M., Bohannan, B.J., Green, J.L., 2010. Biodiversity and biogeography of the atmosphere. *Philos. Trans. R. Soc. B, Biol. Sci.* 365(1558), 3645–3653.

Wright, S.L., Ulke, J., Font, A., Chan, K.L.A., Kelly, F.J., 2020. Atmospheric microplastic deposition in an urban environment and an evaluation of transport. *Environ. Int.* 136, 105411.

Wu, X., Pan, J., Li, M., Li, Y., Bartlam, M., Wang, Y., 2019. Selective enrichment of bacterial pathogens by microplastic biofilm. *Water Res.* 165, 114979.

Xu, L., Li, K., Bai, X., Zhang, G., Tian, X., Tang, Q., Zhang, M., Hu, M., Huang, Y., 2024. Microplastics in the atmosphere: Adsorb on leaves and their effects on the phyllosphere bacterial community. *J. Hazard. Mater.* 462, 132789.

Yang, W., Li, Y., Boraschi, D., 2023. Association between microorganisms and microplastics: How does it change the host–pathogen interaction and subsequent immune response? *Int. J. Mol. Sci.* 24(4), 4065.

Yang, Y., Itoh, T., Yokobori, S.I., Shimada, H., Itahashi, S., Satoh, K., Ohba, H., Narumi, I., Yamagishi, A., 2010. Deinococcus aetherius sp. nov., isolated from the stratosphere. *Int. J. Syst. Evol. Microbiol.* 60(4), 776–779.

Yao, X., Luo, X.S., Fan, J., Zhang, T., Li, H., Wei, Y., 2022. Ecological and human health risks of atmospheric microplastics (MPs): A review. *Environ. Sci. Atmos.* 2(5), 921–942.

Yukioka, S., Tanaka, S., Nabetani, Y., Suzuki, Y., Ushijima, T., Fujii, S., Takada, H., Van Tran, Q., Singh, S., 2020. Occurrence and characteristics of microplastics in surface road dust in Kusatsu (Japan), Da Nang (Vietnam), and Kathmandu (Nepal). *Environ. Pollut.* 256, 113447.

Zhai, Y., Bai, J., Chang, P., Liu, Z., Wang, Y., Liu, G., Cui, B., Peijnenburg, W., Vijver, M.G., 2024. Microplastics in terrestrial ecosystem: Exploring the menace to the soil-plant-microbe interactions. *TrAC Trends Anal. Chem.* 174, 117667.

Zhang, H., 2017. Transport of microplastics in coastal seas. *Estuar. Coast. Shelf Sci.* 199, 74–86.

Zhang, Y., Gao, T., Kang, S., Sillanpää, M., 2019. Importance of atmospheric transport for microplastics deposited in remote areas. *Environ. Pollut.* 254, 112953.

Zhang, Y., Kang, S., Allen, S., Allen, D., Gao, T., Sillanpää, M., 2020. Atmospheric microplastics: A review on current status and perspectives. *Earth Sci. Rev.* 203, 103118.

4 Phyllosphere as a Signature of Microplastic Deposition in Peri-Urban Environments

Manish Chaudhary, Ayushi, Kothari, Anjali Mishra, and Surindra Suthar

4.1 INTRODUCTION

Microplastic (MP) pollution is an emerging environmental threat, especially in the peri-urban area of convergence between the ecosystems of the urban and rural areas. These regions are hybrids and quite often characterize a very unique set of environmental challenges that emerge from the interaction of anthropogenic activities, agricultural activities, and natural ecosystems. MP refers to plastic pieces less than 5 mm in size (Iqbal et al., 2020). MPs have become a ubiquitous pollutant entering the terrestrial, freshwater, and marine ecosystems (Borah et al., 2023). This widespread occurrence of MPs is very alarming as these microscopic particles come from various sources, including large fragmentation from debris and synthetic textiles and wear and tear of vehicle tyres, among others (Qiu et al., 2020). Parts of the plant aboveground called the phyllosphere, may be considered a sink for MP deposition (Adomako & Yu, 2023; Dris et al., 2016). According to Adomako & Yu (2023), the assimilation of deposited particles on plant surfaces may be associated with types of microorganisms, insects, and other living organisms that exist in and around the plant zone. This might have a radical influence not only on plant health and growth but also on the dynamics of ecology networking at the peri-urban level. The present chapter, therefore, summarizes the complex relationship among MP, plant-associated microbiota such as phytobiomes, and environmental health at the peri-urban level. Indeed, the study of such interactions is intended to lead to a proper interpretation of the many different impacts due to MP pollution in those regions, shedding light on how to reduce it and encourage sustainable urban development in harmony with nature.

4.2 UNDERSTANDING THE PHYLLOSPHERE

4.2.1 Definition and Composition

The phyllosphere—a complex, dynamic environment—encompasses leaf surfaces, stem surfaces, and other aerial parts of plants (De Mandal & Jeon, 2023). Various bacterial, fungal, algal, and yeast species inhabit this unique ecological niche, supporting the overall health and functioning of the plant (Leonard et al., 2023; Liu et al., 2022). The phyllosphere microbiome contributes significantly to nutrient cycling, plant growth promotion, and disease resistance (Ali et al., 2023). The structure and diversity of microbial communities in the phyllosphere are influenced by multiple factors. Due to the fluctuations in habitats and available resources within plant species, the morphology of the leaf surface can support a diverse microbial community for its physical and chemical characteristics (Gong & Xin, 2021). In the same vein, intensive changes by anthropogenic activities such as agriculture, and land use change could strongly affect phyllosphere microbiota (Perreault & Laforest-Lapointe, 2022). The use of fertilizer and pesticides can indeed alter the

DOI: 10.1201/9781032717814-4

functional behaviour of microbial communities, which in turn affects plant health and ecosystem stability (Kapoor et al., 2024; Yu et al., 2023). There is an intense need to study the phyllosphere concerning its biodiversity and species interactions.

4.2.2 Phyllosphere Habitat

The phyllosphere ecosystem is a unique group of plants and their parts. The terrestrial phyllosphere covers almost 6.4×10^5 km^2 and has distinct microhabitats, which makes it a suitable area for microbes to live (Catania et al., 2022), such as yeasts, bacteria, fungi, protozoa, and worms (Bashir et al., 2022; Adomako & Yu, 2023). These microbes have different connections with the plants they live on, such as commensalism, mutualism (symbionts), and hostility. Various global approaches benefit from these phyllospheric microbial associations, such as carbon and nitrogen biogeochemical cycles and environmental effects (Adomako & Yu, 2023). According to Praeg et al. (2020), the microenvironment and leaf physiology are determined by the arrangement of epidermal cells. This microenvironment allows for the number and dispersion of microorganisms on the leaf surface. Mostly, epiphytes grow in a biofilm-like fashion; nevertheless, the areas of the leaf that exudate nutrients—the trichomes, veins, and epidermal cell groves—are ideal for forming more extensive bacterial aggregation. The outer cuticle of the leaf facilitates microbial colonization. Permeability, wettability, and the presence or absence of aliphatic chemicals in the cuticle layer are physicochemical features of the leaf surface that microbes can cling to.

The epiphytes' ability to thrive and expand may depend critically on the site's water permeability. Additionally, epiphytes use the phyllosphere for colonization and nutrient leaching. Bacterial communities are densely populated on the surface of leaves that receive more water and nutrients. For example, *Pseudomonas syringae* secretes syringafactin on the leaf's cuticle layer, which helps make sucrose available for sustained epiphytic growth. Bacteria generally maintain cuticular permeability by secreting biosurfactants (Furtak et al., 2019). *Pantoea agglomerans* and *Pantoea eucalypti* 299R both have fructose availability. Epiphytic microorganisms use the phyllosphere microbial assemblage for specialized microbial dispersal, survival, and blooming (colonization). Concurrently, surface microbes alter the phyllosphere's chemical composition and give rise to the epiphytic community's varied oligotrophic lifestyle. Additionally, microbes create specialized spaces on leaf surfaces through their interacting lifestyle, which allows for the continuous maintenance of microbial populations (Scholier et al., 2023).

4.3 MP SOURCES AND CHARACTERISTICS

4.3.1 Urbanization and MP Generation

Peri-urban regions are often referred to as situated between the urban and rural environments, creating an ecological and social landscape. These are mainly characterized by their proximity to the urban centres, which may dictate their environmental and socio-economic activities. The increase in the production and appearance of MPs (>5 mm) (Laskar & Kumar, 2019) has been the most common problem in peri-urban areas. This phenomenon is mainly enhanced by industrial activities such as processes of manufacturing and the release of harmful substances that overwhelm the urban areas (Mandal & Mishra, 2023). Additionally, vehicular emissions significantly contribute to MP pollution by causing friction and abrasion of rubber fragments into the ambient air (Arole et al., 2023; Luo et al., 2021). Another critical contributor to MP production involves synthetic textiles, which have now become frequent products including clothes and domestic materials. Microfibres formed during the washing of clothes can be broken into the drainage systems, which often lack proper filtration procedures that can retain such fine pollutants (De Falco et al., 2019). Further, the bigger fragments of plastic material break down into MP from littered items and municipal waste, which spread throughout the earth and water environments. Besides this, urban runoff and atmospheric deposition contribute significantly to the pool of MP in the local peri-urban zones (Smyth et al., 2021). Heavy rainfall may wash off contaminants that stick on urban surfaces into nearby water bodies, potentially transporting MP and accumulating them into aquatic ecosystems (Zhang et al., 2022a). Similar could be the presence of atmospheric deposition containing airborne particles that settle down to the ground and water bodies, thus contributing to the widespread phenomenon of MP in such environments. The reinforcing nature of such cumulative factors underlines the urgency to integrate solutions to tackle MP pollution in peri-urban areas and give importance to sustainable waste management, innovative industrial regulations, and popular public awareness programmes to tackle such growing environmental issues.

4.3.2 Types of MPs

They can also be classified into two types, according to their source and technology of production: primary MPs and secondary MPs (see Figure 4.1). Primary MPs are the plastics produced on a small scale for commerce, such as microbeads used in cosmetics, toothpaste, and exfoliating scrubs (Anagnosti et al., 2021). Such small-sized particles have been engineered to make these products more textured and effective, but are now considered to pose a problem due to not being efficiently filtered out in the processes of waste treatment, which may as a result cause their abundance in aquatic ecosystems (Anagnosti et al., 2021). These, instead of primary MPs, are the MP particles resulting from the breakages and deterioration of larger plastic products like bottles, bags, and fishing nets (Kalogerakis et al., 2017).

Environmental effects such as mechanical abrasion, chemical weathering, and exposure to UV radiation fragment large plastic particles in the long run into smaller particle sizes, which range from a few micrometres to a few millimetres (Sun et al., 2022). This change might also be contributing to the widespread accumulation of MPs in any system, and environmental systems including rivers, oceans, and soils. The size, shape, and size of MPs further characterized by polymer type or composition determine the distribution, behaviour, and interactions occurring between them and their environment (Alfonso et al., 2021; Kameda et al., 2021). For example, the small particles can be suspended in the water columns, and the big ones settle at the bottom of the water bodies. Similarly, the dimension and shape of MPs may make it easier for animals to ingest as they are commonly mistaken for food by marine animals due to their shapes (Coppock et al., 2019). The chemical nature of MPs controls the degree to which MP can adsorb toxic contaminants from their environment, which implies that it eventually leads to the agent's bioaccumulation in the web of food. All these varied perceptions are thus required to understand the environmental implications of MPs and to formulate an appropriate strategy regarding their management and mitigation.

4.4 PHYLLOSPHERE AS A SINK FOR MPs

4.4.1 Mechanisms of Deposition

In the phyllosphere, MPs can easily deposit on the surface of the leaves through various transport mechanisms. Atmospheric deposition is the most favourable route through which the MPs in the atmosphere accumulate on the surface of the leaf through gravity, precipitation, wind, vehicle movement and other environmental factors (Adomako & Yu, 2023; Leitão et al., 2024). Likewise, MPs can enter the environment through rainwater, as raindrops transport these particles from the atmosphere onto the surfaces of plants (Roblin et al., 2020). Moreover, agricultural and human actions play a role in depositing MPs on plants by introducing them via dust particles in the air (Nizzetto et al., 2016). Figure 4.2 depicts the various ways in which MPs are deposited on leaves. The shape and texture of leaves play a role in their ability to retain and accumulate MPs. Differences in leaf size, shape, and composition can affect how well they hold onto MPs (Leitão et al., 2024). For instance, surfaces that repel water might make it easier for MPs to stick since less moisture allows

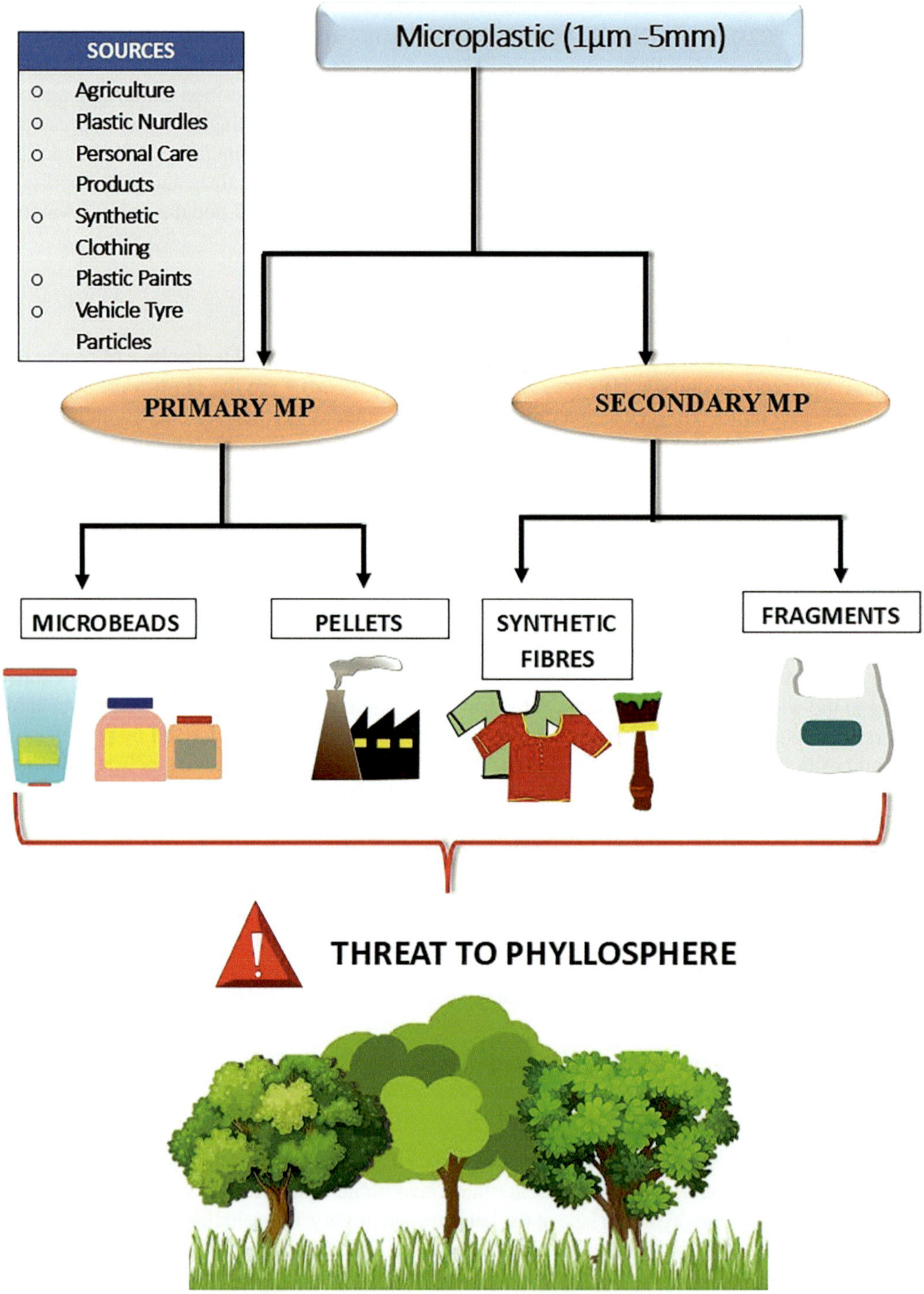

FIGURE 4.1 Sources of phyllospheric microplastic based on their types.

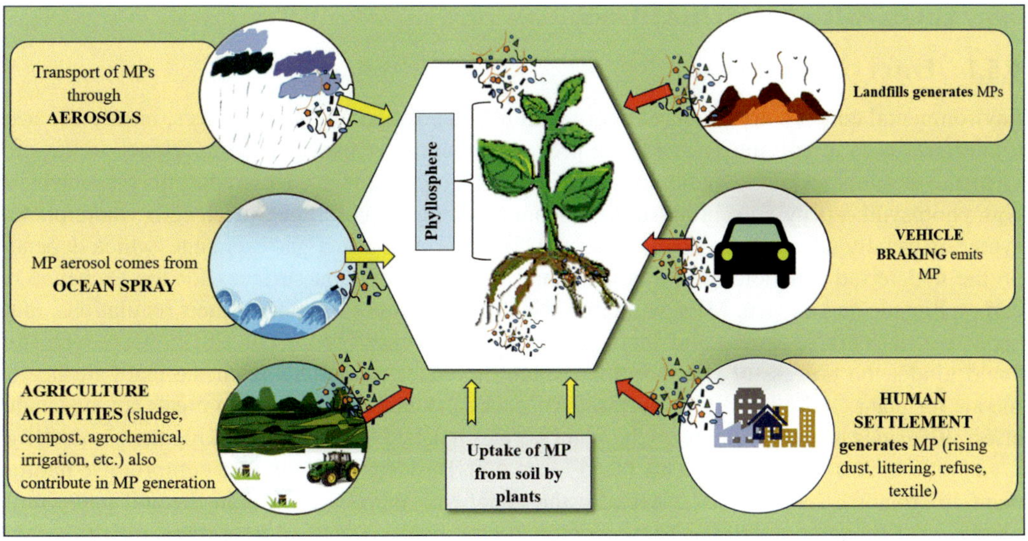

FIGURE 4.2 MPs possible routes to the phyllosphere: red arrow indicates sources and yellow arrows indicate transport mechanism.

them to adhere more readily to surfaces. Additionally, features like trichomes (tiny hair-like projections) and other specific leaf characteristics can create spots for trapping MPs, thereby improving their retention (Perera et al., 2024; Xing et al., 2017). Furthermore, the natural waxy coatings on leaves that help reduce water loss and protect against pathogens may also influence MP adherence by either promoting or hindering their accumulation (Perera et al., 2024). Examining the connection between traits and MP behaviour is essential for understanding the ecological effects of MPs, in terrestrial environments.

4.4.2 MP Accumulation Patterns

Some of the recorded plant species in the literature indicate distinct accumulation patterns of MPs in the phyllosphere (group of leaves, flowers, and stems) (Leitão et al., 2024; Li et al., 2019). They are quantitative in their accumulative patterns, and these two factors are quite peculiar to them: the extent of MPs degree and the method of deposition, contact at, or interaction with the plant surface (Liu et al., 2022). The age of the leaves may determine the accumulation patterns of MPs due to different surface structures and associated metabolic activities between old and young leaves (Kanojia et al., 2021; Mechaber et al., 1996). Here, the orientation of leaves is of paramount importance. This varied direction of leaves towards wind or rain can enhance the different deposition sites for MPs (Leitão et al., 2024; Xu et al., 2024). The settling of particles relates to flow and water characteristics (Xiong et al., 2018). In addition, other physical properties related to the leaf architecture, such as surface roughness, cuticular wax, and trichomes can affect MPs retention and adhesion (Leonard et al., 2023; Xing et al., 2017). The outcomes can be further tuned by environmental factors, such as humidity, temperature, and air contaminants that can disrupt these interactions (Adomako & Yu, 2023) by changing the MPs' physical characteristics and modifying plants' physiological properties (Gan et al., 2023). For instance, high levels of humidity favour the entry of MPs into foliage, but heavy rain or strong winds can wash it away or sweep it along great distances (Zhang et al., 2022b). Collectively, all these interacting variables formed in a complex way and form a web of networks determining MP deposition on plant surfaces, showing the need for more research into the ecological significance of MP accumulation in terrestrial ecosystems.

4.5 ECOLOGICAL IMPLICATIONS

4.5.1 Effect on Plant Health

Environmental concern arises increasingly with MP deposition on plant surfaces, mostly because it interferes with important physiological processes that are of paramount importance to the survival and development of the plant. In such cases, MP deposition on leaves and stems may interfere with photosynthesis by preventing sunlight from reaching the chlorophyll-rich cells where photosynthesis usually takes place (Zhang et al., 2022b). MP even affect gas exchange, which depends on the undeterred functioning of stomata-small openings on plant surfaces for the exchange of carbon dioxide and oxygen. MPs, while blocking the stomata or disrupting their regulations, may hence restrict carbon uptake and inefficiently release oxygen (Wu et al., 2019). Moreover, this could hinder the process of transpiration, where the water evaporates from the plant surface to the air for plant cooling and transportation of necessary nutrients. This can create an avalanche effect of various negative effects on the plant's water balance, potentially making the plants dehydrated or deficient in some nutrients (see Figure 4.3). Besides, MPs can also compromise natural plant–microbiome interactions, especially the beneficial microorganisms in the soil and within plant roots (Zhang et al., 2021). These microbial communities play an important role in nutrient uptake, disease resistance, and thus overall plant health (Adomako & Yu, 2023). MPs can induce changes in the structure and function of such microbial communities, with loss to plant health and therefore agricultural productivity as a potential consequence. The increasing presence of MPs in the environment has increasingly disturbing implications for ecosystems and food security and thus calls for further research and awareness of their probable long-term implications for plant physiology and health.

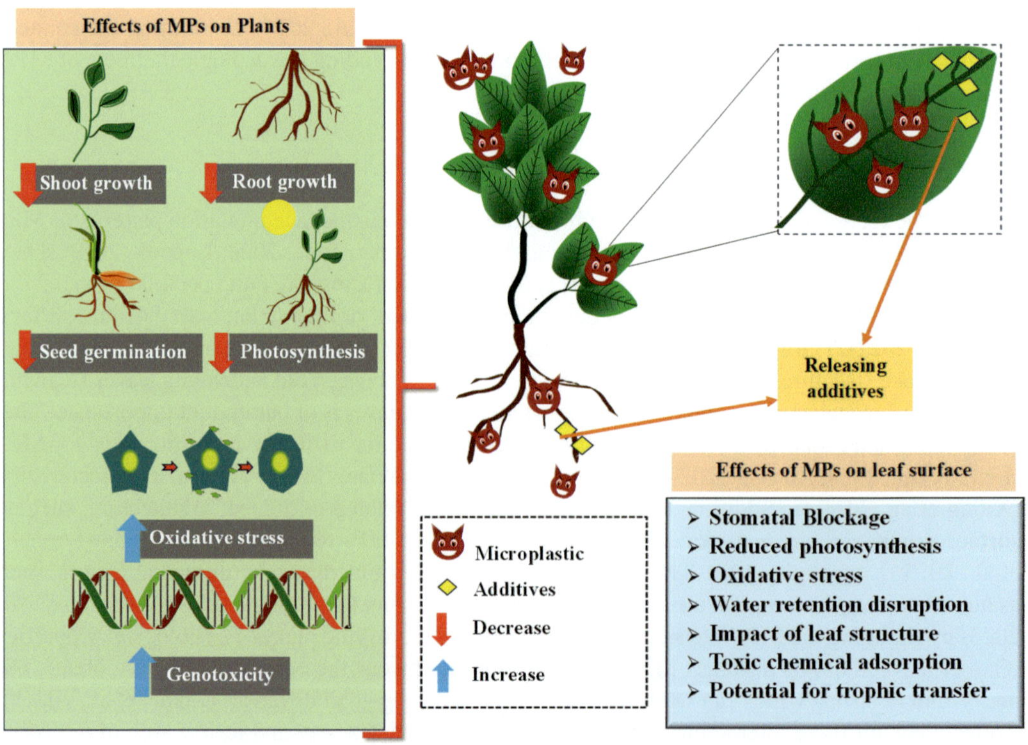

FIGURE 4.3 Phyllospheric MP effect on plant and leaf surfaces.

4.5.2 The Effects of MP on the Microbial Community of the Phyllosphere

MP interaction with microbial communities can cause extreme changes in species diversity and abundance within these communities (Xu et al., 2024). MPs introduced to multiple ecosystems provide a new substrate to which microorganisms attach themselves (Adomako & Yu, 2023). This colonization may alter the microbial community composition, favouring certain species that are MPs hospitable and potentially repressing others less adaptable (Amaral-Zettler et al. 2020). Thus, the microbial diversity could be decreased and affect the delicate balance that occurs in such ecosystems. Yet, the addition of MPs could also interfere with nutrient cycling, which is an essential function in maintaining ecosystem health. Some microbes are imperative decomposers and help in nutrient recycling. The effect of MPs, especially their functions, becomes less efficient, during the community shift of microbes(Zhou et al., 2021). Thus, the functioning might compromise nutrient cycles, thereby changing plant growth and the bigger food webs. These effects aside, the ability of microbes to form biofilm could have implications for the circulation of pathogens and harmful chemicals within the ecosystem via MPs (Ren et al., 2020; Rong et al., 2021). MPs adsorb various kinds of pollutants, including heavy metals and organic toxins, and it could be that microorganisms carry such cargoes to other locations as they migrate or are carried through water currents (Rafa et al., 2024). It raises concern not only for the health of aquatic life but also for human health through food chain contamination. In general, the interactions of MPs and microbial communities have profound implications extending beyond an immediate environmental interest to possibly profound structural and functional changes in ecosystems.

4.6 CONCLUSIONS

The phyllosphere, being the leaves and stems of plants above ground, can be a very critical ecosystem of MP deposition within peri-urban environments where the effects of urbanization are most strongly expressed. As the urban areas have been extending further, encroaching upon natural ecosystems and transforming knowledge about the complex interrelations existing between MPs and the phyllosphere. The understanding developed herein will not only contribute to assessing ecological risks but also provide valuable insight towards efficient strategies for environmental management—a process that intends to safeguard biodiversity and ecosystem health. In peri-urban areas, where transitions between fully urbanized, suburban, and rural environments become gradual, plant surfaces can serve both as a sink and as a potential source of MPs, which are so prevalent in the modern landscape.

Future research needs to be oriented more towards the quantification of MP accumulation in different plant species, relating studies to species diversity, plant morphology, and habitat type. Moreover, the long-term impacts of MP deposition on plant–microbe interactions need to be understood, as their roles are crucial in nutrient cycling, plant health, and ecosystem resilience. Understanding how these associations are disrupted by MP enables the identification of the broader ecological implications of pollution. In addition, there is a requirement to develop targeted approaches that mitigate MP pollution within urban and peri-urban landscapes. These include better waste management practices, community awareness, and green infrastructure to filter and trap MPs before they reach the local ecosystems. This calls for effective collaboration among researchers, policymakers, and community stakeholders because of successfully addressing this timely environmental issue. Centring research efforts and action in this area will help us nurture healthy ecosystems resilient to the negative impact of urbanization and pollution, with guaranteed healthy life for future generations by sustainably using natural resources.

4.7 FUTURE DIRECTIONS

4.7.1 Research Gaps and Recommendations

Further research will be needed to fully explain the complex, often interconnected relationships between phyllosphere microbiomes—the microbial communities on the surface of leaves—and MPs in our environment. Current studies indicate that MPs may influence microbiome composition and functionality, with potential measurable impacts on ecosystem processes important to nutrient cycling and plant health. It will also be very important to understand the long-term effects of microbial communities and broader environmental systems and human health. Enhanced monitoring of peri-urban areas being a very important zone of contact between urban and rural ecosystems can help to effectively address this growing concern. These are often hotspots of pollution, including the accumulation of MPs, hence with special challenges and opportunities for research. Second, there is an urgent need for standardized protocols concerning the sampling and analysis of MPs, so that consistency in data collection can be achieved and comparisons are enabled across various studies. Such complex dynamics will need interdisciplinary approaches by ecologists, microbiologists, environmental scientists, and experts in public health. Allowing different views and expertise to converge will permit researchers to grasp the implications of MPs on microbial life in the phyllosphere and beyond, thus yielding better-informed management and mitigation of their impacts on ecosystems and human well-being. An integrated approach would not only help advance our scientific knowledge but also breed effective policy-making aimed at safeguarding both environmental and public health in an increasingly plastic-laden world.

4.7.2 Policy and Management Implications

Since mitigation of MP pollution is extremely important, there is a dire need for effective policy measures. Ambulatory regulations governing plastic production, waste management, and promoting societal awareness in peri-urban environments can significantly reduce MP origin. Incorporation of scientific findings into urban development will ensure a more robust and sustainable ecosystem. This chapter aims to add to the ever-growing literature knowledge on the most severe impacts in the natural environment, more importantly in the peri-urban area's MP. Becoming an important and integral part of understanding the broader environmental impacts caused by any pollution is phyllosphere.

REFERENCES

Adomako, M. O., & Yu, F. H. (2023). Potential effects of micro- and nanoplastics on phyllosphere microorganisms and their evolutionary and ecological responses. *Science of the Total Environment*, *884*, 163760. https://doi.org/10.1016/J.SCITOTENV.2023.163760

Alfonso, M. B., Arias, A. H., Ronda, A. C., & Piccolo, M. C. (2021). Continental microplastics: Presence, features, and environmental transport pathways. *Science of the Total Environment*, *799*, 149447. https://doi.org/10.1016/J.SCITOTENV.2021.149447

Ali, S., Tyagi, A., & Bae, H. (2023). Plant microbiome: An ocean of possibilities for improving disease resistance in plants. *Microorganisms*, *11*(2). MDPI. https://doi.org/10.3390/microorganisms11020392

Amaral-Zettler LA, Zettler ER, Mincer TJ. Ecology of the plastisphere. *Nature Reviews Microbiology*, *18*(3), 139–151. https://doi.org/10.1038/s41579-019-0308-0. Epub 2020 Jan 14. PMID: 31937947.

Anagnosti, L., Varvaresou, A., Pavlou, P., Protopapa, E., & Carayanni, V. (2021). Worldwide actions against plastic pollution from microbeads and microplastics in cosmetics focusing on European policies. Has the issue been handled effectively? *Marine Pollution Bulletin*, *162*, 111883. https://doi.org/10.1016/j.marpolbul.2020.111883

Arole, K., Velhal, M., Tajedini, M., Xavier, P. G., Bardasz, E., Green, M. J., & Liang, H. (2023). Impacts of particles released from vehicles on environment and health. *Tribology International*, *184*, 108417. https://doi.org/10.1016/J.TRIBOINT.2023.108417

Bashir, I., War, A.F., Rafiq, I., Reshi, Z.A., Rashid, I., Shouche, Y.S., 2022. Phyllosphere microbiome: Diversity and functions. *Microbiological Research, 254*, 126888.

Borah, S. J., Gupta, A. K., Gupta, A., Bhawna, Kumar, S., Sharma, R., Kumar, R., Kumar, P., Dubey, K. K., Kaushik, S., Mishra, A. K., & Kumar, V. (2023). Grasping the supremacy of microplastic in the environment to understand its implications and eradication: A review. *Journal of Materials Science*, *58*(32), 12899–12928. https://doi.org/10.1007/s10853-023-08806-8

Catania, V., Bueno, R. S., Alduina, R., Grilli, E., La Mantia, T., Castaldi, S., & Quatrini, P. (2022). Soil microbial biomass and bacterial diversity in southern European regions vulnerable to desertification. *Ecological Indicators*, *145*, 109725. https://doi.org/10.1016/J.ECOLIND.2022.109725

Coppock, R. L., Galloway, T. S., Cole, M., Fileman, E. S., Queirós, A. M., & Lindeque, P. K. (2019). Microplastics alter feeding selectivity and faecal density in the copepod, Calanus helgolandicus. *Science of the Total Environment*, *687*, 780–789. https://doi.org/10.1016/J.SCITOTENV.2019.06.009

De Falco, F., Di Pace, E., Cocca, M., & Avella, M. (2019). The contribution of washing processes of synthetic clothes to microplastic pollution. *Scientific Reports*, *9*(1), 6633. https://doi.org/10.1038/s41598-019-43023-x

De Mandal, S., & Jeon, J. (2023). Phyllosphere microbiome in plant health and disease. *Plants, 12*(19). Multidisciplinary Digital Publishing Institute (MDPI). https://doi.org/10.3390/plants12193481

Dris, R., Gasperi, J., Saad, M., Mirande, C., & Tassin, B. (2016). Synthetic fibers in atmospheric fallout: A source of microplastics in the environment? *Marine Pollution Bulletin*, *104*(1–2), 290–293. https://doi.org/10.1016/J.MARPOLBUL.2016.01.006

Furtak, K., Grzadziel, J., Gałazka, A., & Niedźwiecki, J. (2019). Analysis of soil properties, bacterial community composition, and metabolic diversity in fluvisols of a floodplain area. *Sustainability 2019, 11*(14), 3929. https://doi.org/10.3390/SU11143929

Gan, Q., Cui, J., & Jin, B. (2023). Environmental microplastics: Classification, sources, fates, and effects on plants. *Chemosphere*, *313*, 137559. https://doi.org/10.1016/J.CHEMOSPHERE.2022.137559

Gong, T., & Xin, X.-F. (2021). Phyllosphere microbiota: Community dynamics and its interaction with plant hosts. *Journal of Integrative Plant Biology*, *63*(2), 297–304. https://doi.org/https://doi.org/10.1111/jipb.13060

Iqbal, S., Xu, J., Allen, S. D., Khan, S., Nadir, S., Arif, M. S., & Yasmeen, T. (2020). Unraveling consequences of soil micro- and nano-plastic pollution on soil-plant system: Implications for nitrogen (N) cycling and soil microbial activity. *Chemosphere*, *260*, 127578. https://doi.org/10.1016/J.CHEMOSPHERE.2020.127578

Kalogerakis, N., Karkanorachaki, K., Kalogerakis, G. C., Triantafyllidi, E. I., Gotsis, A. D., Partsinevelos, P., & Fava, F. (2017). Microplastics generation: Onset of fragmentation of polyethylene films in marine environment mesocosms. *Frontiers in Marine Science*, *4*(MAR). https://doi.org/10.3389/fmars.2017.00084

Kameda, Y., Yamada, N., & Fujita, E. (2021). Source- and polymer-specific size distributions of fine microplastics in surface water in an urban river. *Environmental Pollution*, *284*, 117516. https://doi.org/10.1016/J.ENVPOL.2021.117516

Kanojia, A., Shrestha, D. K., & Dijkwel, P. P. (2021). Primary metabolic processes as drivers of leaf ageing. *Cellular and Molecular Life Sciences*, *78*(19), 6351–6364. https://doi.org/10.1007/s00018-021-03896-6

Kapoor, D., Sharma, P., Sharma, M. M. M., Yadav, S., & Husen, A. (2024). Exploring soil microbiota and their role in plant growth, stress tolerance, disease control and nutrient immobilizer. *Biocatalysis and Agricultural Biotechnology*, 103358. https://doi.org/10.1016/J.BCAB.2024.103358

Laskar, N., & Kumar, U. (2019). Plastics and microplastics: A threat to environment. *Environmental Technology & Innovation*, *14*, 100352. https://doi.org/10.1016/J.ETI.2019.100352

Leitão, I. A., Van Schaik, L., Iwasaki, S., Ferreira, A. J. D., & Geissen, V. (2024). Accumulation of airborne microplastics on leaves of different tree species in the urban environment. *Science of the Total Environment*, *948*, 174907. https://doi.org/https://doi.org/10.1016/j.scitotenv.2024.174907

Leonard, J., Borthakur, A., Koutnik, V. S., Brar, J., Glasman, J., Cowger, W., Dittrich, T. M., & Mohanty, S. K. (2023). Challenges of using leaves as a biomonitoring system to assess airborne microplastic deposition on urban tree canopies. *Atmospheric Pollution Research*, *14*(2), 101651. https://doi.org/10.1016/J.APR.2023.101651

Li, L., Zhou, Q., Yin, N., Tu, C., & Luo, Y. (2019). Uptake and accumulation of microplastics in an edible plant. *Chinese Science Bulletin*, *64*(9), 928–934. https://doi.org/https://doi.org/10.1360/N972018-00845

Liu, X., Lu, J., He, S., Tong, Y., Liu, Z., Li, W., & Xiayihazi, N. (2022). Evaluation of microplastic pollution in Shihezi city, China, using pine needles as a biological passive sampler. *Science of the Total Environment*, *821*, 153181. https://doi.org/10.1016/J.SCITOTENV.2022.153181

Luo, Z., Zhou, X., Su, Y., Wang, H., Yu, R., Zhou, S., Xu, E. G., & Xing, B. (2021). Environmental occurrence, fate, impact, and potential solution of tire microplastics: Similarities and differences with tire wear particles. *Science of the Total Environment*, *795*, 148902. https://doi.org/10.1016/J.SCITOTENV.2021.148902

Mandal, G., & Mishra, S. (2023). A review on emerging micro and nanoplastic pollutants, heavy metals and their remediation techniques. *Nanofabrication*, *8*. https://doi.org/10.37819/nanofab.008.302

Mechaber, W. L., Marshall, D. B., Mechabers, R. A., Jobeii, R. T., Chew, F. S., & Harold Mooney, C. A. (1996). Mapping leaf surface landscapes (epicuticular lipids/phylloplane/ecological scale/microtopography/atomic force microscopy). *Proceedings of the National Academy of Sciences*, *93*(10), 4600–4603.

Nizzetto, L., Futter, M., & Langaas, S. (2016). Are agricultural soils dumps for microplastics of urban origin? *Environmental Science & Technology*, *50*(20). https://doi.org/10.1021/acs.est.6b04140

Perera, K., Ziajahromi, S., Nash, S. B., & Leusch, F. D. L. (2024). Evaluating the retention of airborne microplastics on plant leaf: Influence of leaf morphology. *Environmental Pollution*, *346*, 123673. https://doi.org/10.1016/J.ENVPOL.2024.123673

Perreault, R., & Laforest-Lapointe, I. (2022). Plant-microbe interactions in the phyllosphere: Facing challenges of the anthropocene. *ISME Journal, 16*(2), 339–345. Springer Nature. https://doi.org/10.1038/s41396-021-01109-3

Praeg, N., Seeber, J., Leitinger, G., Tasser, E., Newesely, C., Tappeiner, U., & Illmer, P. (2020). The role of land management and elevation in shaping soil microbial communities: Insights from the Central European Alps. *Soil Biology and Biochemistry*, *150*, 107951. https://doi.org/10.1016/J.SOILBIO.2020.107951

Qiu, R., Song, Y., Zhang, X., Xie, B., & He, D. (2020). Microplastics in urban environments: Sources, pathways, and distribution. In D. He & Y. Luo (Eds.), *Microplastics in Terrestrial Environments: Emerging Contaminants and Major Challenges* (pp. 41–61). Springer International Publishing. https://doi.org/10.1007/698_2020_447

Rafa, N., Ahmed, B., Zohora, F., Bakya, J., Ahmed, S., Ahmed, S. F., Mofijur, M., Chowdhury, A. A., & Almomani, F. (2024). Microplastics as carriers of toxic pollutants: Source, transport, and toxicological effects. *Environmental Pollution*, *343*, 123190. https://doi.org/10.1016/j.envpol.2023.123190

Roblin, B., & Aherne, J. (2020). Moss as a biomonitor for the atmospheric deposition of anthropogenic microfibres. *Science of The Total Environment*, *715*, 136973.

Ren, X., Tang, J., Liu, X., & Liu, Q. (2020). Effects of microplastics on greenhouse gas emissions and the microbial community in fertilized soil. *Environmental Pollution*, *256*, 113347. https://doi.org/10.1016/J.ENVPOL.2019.113347

Rong, L., Zhao, L., Zhao, L., Cheng, Z., Yao, Y., Yuan, C., Wang, L., & Sun, H. (2021). LDPE microplastics affect soil microbial communities and nitrogen cycling. *Science of the Total Environment*, *773*, 145640. https://doi.org/10.1016/J.SCITOTENV.2021.145640

Scholier, T., Lavrinienko, A., Brila, I., Tukalenko, E., Hindström, R., Vasylenko, A., Cayol, C., Ecke, F., Singh, N. J., Forsman, J. T., Tolvanen, A., Matala, J., Huitu, O., Kallio, E. R., Koskela, E., Mappes, T., & Watts, P. C. (2023). Urban forest soils harbour distinct and more diverse communities of bacteria and fungi compared to less disturbed forest soils. *Molecular Ecology*, *32*(2), 504–517. https://doi.org/10.1111/MEC.16754

Smyth, K., Drake, J., Li, Y., Rochman, C., Van Seters, T., & Passeport, E. (2021). Bioretention cells remove microplastics from urban stormwater. *Water Research*, *191*, 116785. https://doi.org/10.1016/j.watres.2020.116785

Sun, J., Zheng, H., Xiang, H., Fan, J., & Jiang, H. (2022). The surface degradation and release of microplastics from plastic films studied by UV radiation and mechanical abrasion. *Science of the Total Environment*, *838*, 156369. https://doi.org/10.1016/J.SCITOTENV.2022.156369

Wu, Y., Guo, P., Zhang, X., Zhang, Y., Xie, S., & Deng, J. (2019). Effect of microplastics exposure on the photosynthesis system of freshwater algae. *Journal of Hazardous Materials*, *374*, 219–227. https://doi.org/10.1016/J.JHAZMAT.2019.04.039

Xing, Z., Liu, Y., Cai, W., Huang, X., Wu, S., & Lei, Z. (2017). Efficiency of trichome-based plant defense in Phaseolus vulgaris depends on insect behavior, plant ontogeny, and structure. *Frontiers in Plant Science*, *8*. https://doi.org/10.3389/fpls.2017.02006

Xiong, P., Chen, Z., Jia, Z., Wang, Z., Palta, J. A., & Xu, B. (2018). Variability in leaf wettability and surface water retention of main species in semiarid Loess Plateau of China. *Ecohydrology*, *11*(8), e2021. https://doi.org/10.1002/ECO.2021

Xu, L., Li, K., Bai, X., Zhang, G., Tian, X., Tang, Q., Zhang, M., Hu, M., & Huang, Y. (2024). Microplastics in the atmosphere: Adsorb on leaves and their effects on the phyllosphere bacterial community. *Journal of Hazardous Materials*, *462*, 132789. https://doi.org/10.1016/J.JHAZMAT.2023.132789

Yu, Z., Lu, T., & Qian, H. (2023). Pesticide interference and additional effects on plant microbiomes. *Science of the Total Environment*, *888*, 164149. https://doi.org/10.1016/J.SCITOTENV.2023.164149

Zhang, J., Ding, W., Zou, G., Wang, X., Zhao, M., Guo, S., & Chen, Y. (2022a). Urban pipeline rainwater runoff is an important pathway for land-based microplastics transport to inland surface water: A case study in Beijing. *The Science of the Total Environment*, *861*, 160619. https://api.semanticscholar.org/CorpusID:252774437

Zhang, X., Li, Y., Ouyang, D., Lei, J., Tan, Q., Xie, L., Li, Z., Liu, T., Xiao, Y., Farooq, T. H., Wu, X., Chen, L., & Yan, W. (2021). Systematical review of interactions between microplastics and microorganisms in the soil environment. *Journal of Hazardous Materials*, *418*, 126288. https://doi.org/10.1016/J.JHAZMAT.2021.126288

Zhang, Z., Cui, Q., Chen, L., Zhu, X., Zhao, S., Duan, C., Zhang, X., Song, D., & Fang, L. (2022b). A critical review of microplastics in the soil-plant system: Distribution, uptake, phytotoxicity and prevention. *Journal of Hazardous Materials*, *424*, 127750. https://doi.org/10.1016/J.JHAZMAT.2021.127750

Zhou, J., Wen, Y., Marshall, M. R., Zhao, J., Gui, H., Yang, Y., Zeng, Z., Jones, D. L., & Zang, H. (2021). Microplastics as an emerging threat to plant and soil health in agroecosystems. *Science of the Total Environment*, *787*, 147444. https://doi.org/10.1016/J.SCITOTENV.2021.147444

5 Exploring the Microbiome of the Marine Microplastisphere

Implications for Microplastic Fate and Ecological Impacts

Wahappulige Madhushan Wichakshana Perera and Praveen Abhishek

5.1 INTRODUCTION

5.1.1 The Marine Microplastisphere (MMPL): An Overview

Most plastic waste released into the environment is resistant to corrosion, chemically stable, and difficult for microbes to break down (Lambert and Wagner, 2016; Yang et al., 2020). Therefore, those plastic are the reason for a lot of environmental issues (Wright et al., 2020). Between 4.8 and 12.7 million tonnes of plastic debris are expected to enter the oceans each year, and this number is likely to rise by an order of magnitude over the next 10 years (Wright et al., 2020; Jambeck et al., 2015). Plastic pollution in our oceans has been a growing concern since the 1970s, with microplastic (MP) emerging as a particularly troubling issue in recent years (Carpenter and Smith, 1972; Thompson et al., 2004). These tiny plastic fragments, no larger than 5 mm, have infiltrated marine ecosystems worldwide (Andrady, 2011; Marsay et al., 2023).

As MP persists in marine environments, they undergo continuous physical and chemical transformations. These particles experience fragmentation due to various environmental factors, including UV radiation, mechanical abrasion, and oxidative processes (Andrady, 2011; Gewert et al., 2017; Gewert et al., 2015). These MPs can benefit the colonizing microbial communities by offering them shelters that let microorganisms withstand extreme environmental conditions, nutrient supplies, and physical support (Oberbeckmann et al., 2015; Shen et al., 2019; Ahmed et al., 2022). Therefore, their surfaces become colonized by diverse microbial communities, creating a novel ecological niche referred to as the "Plastisphere" (Zettler et al., 2013; Jacquin et al., 2024; Roager and Sonnenschein, 2019); this process happens in the marine environment to combine with MP and same microbes known as MMPL. The plastisphere may be regarded as a new microecosystem in pelagic waters, and includes primary producers, predators, symbionts, and decomposers (Amaral-Zettler et al., 2020). This dynamic microenvironment hosts a complex assemblage of bacteria, archaea, fungi, and other microorganisms (Figure 5.1), significantly altering the properties and behavior of the plastic particles in aquatic ecosystems (Amaral-Zettler et al., 2020).

Numerous marine creatures easily colonize MPs at sea, forming an ecosystem that is different from the surrounding water (Zettler et al., 2013). It was demonstrated that the composition of the plastic microbiome varies depending on the geographic location (Oberbeckmann et al., 2015; Amaral-Zettler et al., 2015; Coons et al., 2021) and that environmental elements like salinity,

 DOI: 10.1201/9781032717814-5

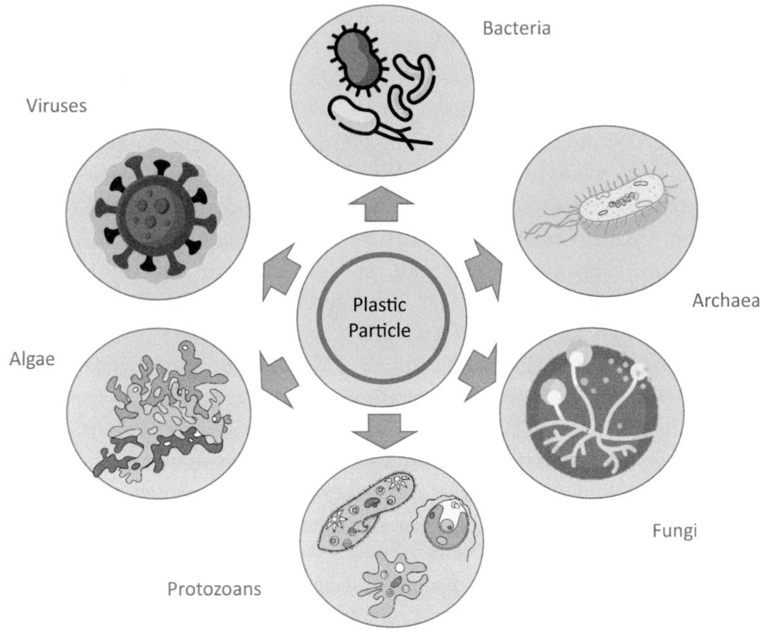

FIGURE 5.1 Microorganisms, which contribute to the formation of marine microplastisphere (MMPL).

water temperature, light, and nutrient availability have an impact on it (Amaral-Zettler et al., 2015; Oberbeckmann et al., 2015; Misic et al., 2019). What is particularly intriguing is how these communities differ from one plastic particle to another. The type of plastic, the surrounding environment, and even the season can all influence which microbes call a particular piece of MP home (Amaral-Zettler et al., 2015; Oberbeckmann et al., 2015). It is a dynamic ecosystem, constantly shifting and adapting to its environment.

5.2 COMPOSITION AND COLONIZATION OF THE MARINE MICROPLASTISPHERE (MMPL)

5.2.1 Microbial Diversity (Bacteria, Archaea, Fungi, Viruses)

A number of investigations showed notable variations from microhabitats in the same setting (Bryant et al., 2016; Ogonowski et al., 2018). This differs from a variety of other substrates, including cloth, glass, rubber, metal, and seawater (Ogonowski et al., 2018). Bacteria are perhaps the most abundant inhabitants of this plastic-based ecosystem (Nikolopoulou et al., 2023). The primary bacteria that are abundant in the plastisphere are typically Proteobacteria, Bacteroidetes, Firmicutes, and Cyanobacteria. Studies have revealed a wide variety of bacterial species, from common marine bacteria to more specialized organisms. For instance, Zettler et al. (2013) found that members of the bacterial genus *Vibrio* spp. were particularly abundant in MP samples from the North Atlantic. Some of these genera can cause diseases in humans and marine animals.

Archaea, once thought to inhabit only extreme environments, have also been found in the MMPL. While less abundant than bacteria, their presence adds another layer of complexity to this ecosystem (De Tender et al., 2015). Fungi, the often-overlooked members of the microbial world, also play a role in the MMPL. These eukaryotic microorganisms have been observed forming intricate networks on plastic surfaces, potentially contributing to the degradation of the plastic itself (Kettner et al., 2017). Even viruses, which exist on the borderline of what we consider “life”, are present in the microPL. These tiny genetic entities can infect the other microorganisms present, potentially influencing the composition and dynamics of the entire community (Oberbeckmann et al., 2015).

The plastisphere exhibits selective properties for specific microorganisms, creating a unique ecological niche in marine environments (Amaral-Zettler et al., 2020; Lacerda et al., 2024). This selectivity arises from a combination of factors related to the physicochemical properties of plastics and the dynamics of microbial colonization (Yang et al., 2020; Zhai et al., 2023). The hydrophobic nature and specific chemical composition of plastic surfaces favor the attachment of certain microbial species adapted to these conditions (Rummel et al., 2017). Additionally, plastics may leach organic compounds that serve as carbon sources for specific microbial groups, promoting their growth (Romera-Castillo et al., 2018). The plastisphere can also provide protection against environmental stressors, such as UV radiation and desiccation, selecting microorganisms that can exploit this protective environment (Oberbeckmann et al., 2015; Sun et al., 2023). As biofilms develop, early colonizers may alter the surface conditions, influencing subsequent microbial succession and community composition (Dang and Lovell, 2016). Furthermore, different plastic polymers may be selected for distinct microbial communities due to variations in surface chemistry and degradability (Kirstein et al., 2019). These selective pressures collectively result in microbial assemblages on MP that differ significantly from those in the surrounding water column, contributing to the plastisphere's distinct ecological characteristics and potential impacts on marine ecosystems. Different types of plastics can attract different microbes based on their surface properties, with some bacteria thriving on certain plastics by producing substances that help them stick. Research has shown that bacterial families such as *Rhodobacteraceae* and *Flavobacteriaceae* can dominate on certain plastics due to their ability to produce extracellular polymeric substances that enhance adhesion (Mu et al., 2023; Dussud et al., 2018). Microorganisms are often more likely to colonize biodegradable plastics, as these materials can serve as a carbon source. Studies have identified specific bacteria capable of degrading polyolefins like polyethylene (PE) and polypropylene (PP), which suggests that these microbes preferentially colonize plastics that they can metabolize (de Vogel et al., 2024; Sharma et al., 2022).

5.2.2 Biofilm Formation on Microplastic Surfaces

The microbial colonization process on MP follows a complex and dynamic pattern of biofilm formation, rather than occurring randomly. This systematic colonization transforms the initially abiotic plastic surface into a diverse microbial ecosystem. The formation of these biofilms involves a series of stages, including initial attachment, microcolony formation, maturation, and dispersion (Figure 5.2), each characterized by distinct microbial community compositions and activities (Flemming and Wuertz, 2019). Microorganisms attach to the MP surface and biofilm formation might occur within a few days (De Sales-Ribeiro et al., 2020; Debroas et al., 2017).

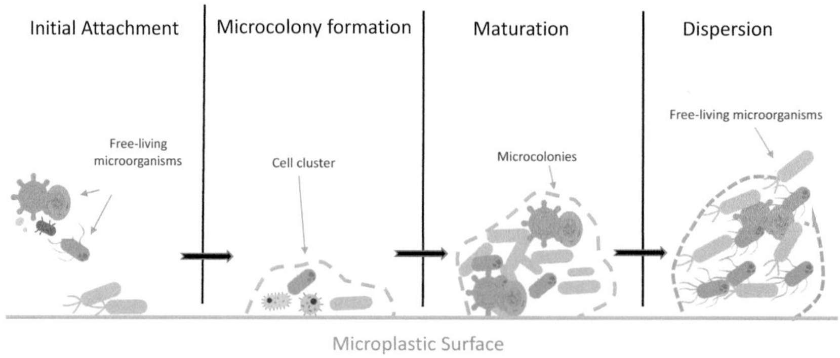

FIGURE 5.2 Biofilm formation process on a microplastic surface (initial attachment, microcolony formation, maturation, and dispersion).

A variety of factors influence the formation of these biofilms. Environmental factors have a significant impact on microbial diversity and function, which in turn affects the structure and formation of MP biofilms (Yang et al., 2020; Zhai et al., 2023). Spatial and seasonal factors play a significant role (Oberbeckmann et al., 2015). The type of polymer, its surface texture, and even the size of the plastic particle can all affect which microorganisms colonize it and also it affects the biofilm development (Oberbeckmann et al., 2015). The microbial communities found on plastics in the warm waters of the tropics, for instance, can differ significantly from those in colder, polar regions. Similarly, the community composition can shift with the changing seasons, reflecting the dynamic nature of marine ecosystems.

The process of biofilm formation typically begins with the attachment of pioneering bacteria to the plastic surface. These early colonizers modify the surface, paving the way for other microorganisms to join the community. Over time, a complex three-dimensional structure develops, complete with channels for nutrient flow and areas of specialized activity (Flemming et al., 2016; Qian et al., 2024). As the biofilm matures, it can dramatically alter the properties of the plastic surface. The once-smooth surface becomes rougher and more hospitable to a wider range of organisms. This change can have implications for how plastic behaves in the marine environment, potentially affecting its buoyancy, degradation rate, and interactions with marine life (Rummel et al., 2017).

Understanding the composition and colonization patterns of the marine MMP is crucial for several reasons. First, it elucidates how plastic pollution is altering the microbial landscape of oceanic ecosystems (Amaral-Zettler et al., 2020). Second, it raises important questions about the potential for MP to act as vectors for the transport of harmful microorganisms across marine environments (Kirstein et al., 2019). Finally, it may provide insights into potential bioremediation strategies, as some of these microorganisms might possess the capacity to degrade plastic polymers (Yoshida et al., 2016; Jacquin et al., 2019).

5.3 FUNCTIONS AND INTERACTIONS OF THE MICROPLASTISPHERE MICROBIOME

The microorganisms colonizing MP in marine environments are not merely passive hitchhikers. They form complex communities that interact with the plastic substrate and the surrounding environment, significantly influencing the fate and impact of MP in marine ecosystems (Zettler et al., 2013; Amaral-Zettler et al., 2020). Two key aspects of these interactions are MP degradation and transport (Jacquin et al., 2019) and bioaccumulation processes (Vethaak and Leslie, 2016).

The microbial colonization of MP presents both potential benefits and risks for marine ecosystems, with complex and sometimes contradictory implications for oceanic health and function (Wright et al., 2020). On one hand, some microorganisms might be capable of breaking down plastics, offering a glimmer of hope for natural bioremediation (Yoshida et al., 2016). Certain bacteria, for instance, have shown the ability to degrade polyethylene terephthalate (PET), a common plastic found in water bottles (Yoshida et al., 2016). However, some microbes (e.g., *Sakeiensis* sp.) could potentially make MP more harmful. They might facilitate the transport of MP through marine ecosystems or even concentrate toxic substances on the plastic surface (Vedolin et al., 2018; Zhang et al., 2022).

The ecological implications of MMPL become more complex when examining the interactions between MP-associated microbial communities and marine organisms. The bioaccumulation of MP is one issue. When marine life consumes these microscopic particles, they may absorb contaminants and pathogens that stick to the MPs' surfaces (Issac and Kandasubramanian, 2021). The pollutants can biomagnify through the food chain when larger organisms eat the contaminated prey, leading to potentially dangerous amounts in apex predators, including animals that humans consume. The safety of seafood and the possible movement of pollutants up the food chain to human consumers are seriously called into question by this (Walkinshaw et al., 2020; Huang et al., 2021).

These interactions introduce additional layers of complexity to marine life. Research indicates that biofilms developing on MP surfaces can alter the chemical signature and visual characteristics of these particles, potentially increasing their attractiveness to marine organisms (Zhang et al., 2022). For instance, studies demonstrated that plastic debris colonized by algae emits dimethyl sulfide, a volatile organic compound that serves as a foraging cue for many marine species (Savoca et al., 2016). This biogenic infochemical production may inadvertently attract organisms to MP, leading to higher ingestion rates. Additionally, the biofilm coating can change the color, texture, and buoyancy of MP, causing them to more closely resemble natural food particles in the water column. These alterations in physical and chemical properties due to microbial colonization may contribute to increased consumption of MP by various marine species, from planktonic organisms to larger predators. This could have far-reaching consequences for marine food webs and, ultimately, human health. Moreover, the MMPL isn't isolated from the rest of the marine ecosystem (Arias-Andres et al., 2018; Rogers et al., 2020; Oberbeckmann et al., 2015).

This chapter explores the intricate world of the MMPL, examining its composition, formation, and functions, as well as its broader ecological implications. The discussion synthesizes current research findings and identifies critical knowledge gaps in the field. By presenting a comprehensive overview of the MMPL, this chapter aims to highlight the urgent need for interdisciplinary approaches to address the complex issue of MP pollution in oceanic environments. The subsequent sections will delve into the microbial ecology of plastic-associated communities and their interactions with marine ecosystems. This analysis underscores the importance of collaborative scientific efforts in understanding and mitigating the impacts of MP pollution on marine systems.

5.3.1 Microplastic Degradation and Transport

Some members of the plastisphere community have shown the potential to break down plastic polymers, promoting natural bioremediation of plastic pollution. They identified a bacterium, *Ideonella sakaiensis*, capable of using PET as its primary carbon and energy source (Yoshida et al., 2016). This organism degrades the PET using two enzymes, PETase and MHETase, to terephthalate (TPA) and ethylene glycol (EG), which serve as carbon sources for the bacterium (Tanaka et al., 2024). This discovery sparked excitement about the potential for microbial degradation of plastics in the environment.

Table 5.1 shows a selection of known plastic degraders in marine habitats, and this also shows the species of the microbes and type of the polymer they degrade. The rate of microbial plastic degradation in natural environments is typically very slow, and the process is influenced by various factors including polymer type, environmental conditions, and the specific microbial community present (Jacquin et al., 2019). General biofilm properties suggest potential mechanisms for accelerating polymer breakdown, such as trapping enzymes close to surfaces and creating localized chemical gradients (Flemming et al., 2016). On the other hand, biofilms can also form a protective layer, potentially slowing down photodegradation and other abiotic degradation processes (Rummel et al., 2017).

The transport of MP through marine ecosystems is another area where the plastisphere plays a crucial role. As microorganisms colonize plastic particles, they can alter the particle's buoyancy, leading to changes in its vertical distribution in the water column (Gaylarde et al., 2023). Lobelle and Cunliffe (2011) observed that biofilm formation on plastic pieces in seawater could lead to a loss of buoyancy, causing the particles to sink (Lobelle and Cunliffe, 2011). This microbially mediated transport can influence the distribution of MP in marine environments, potentially increasing their availability to different groups of organisms. Furthermore, the plastisphere can facilitate the horizontal transport of microorganisms, potentially including pathogens or invasive species. MP can act as rafts, allowing microbes to travel vast distances and potentially colonize new environments (Kirstein et al., 2019). This "plastic vector effect" adds another layer of complexity to the ecological implications of marine plastic pollution.

TABLE 5.1
A Selection of Known Plastic Degraders in Marine Habitats

Plastic Type	Microorganism	Type	Habitat	Location	References
LDPE	*Kocuria palustris, Bacillus pumilus, Bacillus subtilis, Bacillus sphaericus, Bacillus cereus*	Bacteria	Pelagic water	India	Harshvardhan and Jha (2013) and Sudhakar et al. (2008)
HDPE	*Bacillus sphaericus, Bacillus cereus, Brevibacillus borstelensis, Aspergillus tubingensis, Aspergillus flavus*	Bacteria, fungi	Pelagic water, seawater, coastal sediment	India	Sudhakar et al. (2008), Mohanrasu et al. (2018), and Devi et al. (2015)
PE	*Zalerion maritimum, Rhodococcus ruber*	Fungi, bacteria	Marine water, coastal sediment	Portugal, Japan	Yoshida et al. (2016) and Paço et al. (2017)
PET	*Bacillus cereus, Bacillus gottheilii, Ideonella sakaiensis*	Bacteria	Mangrove sediment, marine water	Nigeria, Japan	Yoshida et al. (2016) and Auta et al. (2017)
PP	*Rhodococcus sp., Bacillus sp., Acinetobacter baumannii*	Bacteria	Mangrove sediment, marine water	Nigeria, Saudi Arabia	Auta et al. (2018) and Al-Sabagh et al. (2020)
PS	*Bacillus cereus, Bacillus gottheilii*	Bacteria	Mangrove sediment	Nigeria	Januar et al. (2017)
PCL	*Pseudomonas* sp.	Bacteria	Plastic in coastal	Japan	Suzuki et al. (2018)
PVC	*Pseudomonas citronellolis*	Bacteria	Marine water	China	Yang et al. (2020)

5.3.2 Bioaccumulation Processes of Contaminants

The interaction between MP and their associated microbiomes can significantly influence bioaccumulation processes in marine ecosystems. One key aspect of this is the role of the plastisphere in altering the surface properties of MP, which can affect their ability to adsorb and concentrate various pollutants from the surrounding water (Rochman et al., 2013). Many persistent organic pollutants (POPs) and heavy metals have a high affinity for plastic surfaces, and MP has been shown to concentrate these contaminants to levels far exceeding those in the surrounding seawater (Mato et al., 2001). Biofilms can either enhance or inhibit the sorption of pollutants, depending on the specific compounds and microorganisms involved (Richard et al., 2019).

Biofilms on MP have been found to enhance the sorption of mercury, potentially increasing the risk of mercury bioaccumulation in marine food webs (Graca et al., 2024). The plastisphere can also influence bioaccumulation processes by altering how marine organisms interact with MP. As discussed earlier, biofilms can make MP more attractive to some marine grazers, potentially increasing their ingestion rates. This "trojan horse" effect could lead to increased bioaccumulation of both the plastic particles themselves and any associated pollutants (Zettler et al., 2013).

Moreover, the biofilms can bioaccumulate pollutants themselves, adding another layer to the transfer of these contaminants through marine food webs. Also, certain plastic-degrading microorganisms in biofilms have been observed to break down some organic pollutants, suggesting a potential mitigating effect on pollutant bioaccumulation (Johansen et al., 2018; Sudhakar et al., 2008). When plastic-associated microbes are consumed by larger organisms, they may serve as an additional vector for pollutant transfer (Avio et al., 2017). Bioaccumulation processes associated with the plastisphere are highly complex and can vary significantly depending on the specific plastics, pollutants, microorganisms, and environmental conditions involved. For example, Kettner et al. (2017) found that different types of plastic (PE, polystyrene, and polylactic acid) showed distinct patterns of microbial colonization and pollutant adsorption, highlighting the need for polymer-specific assessments of bioaccumulation risks.

In conclusion, the interactions of the MMPL microbiome play a crucial role in shaping the environmental fate and impacts of MP on marine ecosystems. These microscopic communities exert far-reaching effects, from influencing degradation and transport processes to mediating complex bioaccumulation pathways (Amaral-Zettler et al., 2020). A deeper understanding of the MMPL's dynamics will be instrumental in developing effective strategies to address this pervasive environmental issue. Future studies should focus on elucidating the long-term ecological consequences of MP–microbe interactions and their broader implications for marine ecosystem health.

5.4 ECOLOGICAL IMPLICATIONS

5.4.1 Effects on Marine Organisms and Food Webs

The proliferation of MP in marine environments has far-reaching consequences for marine organisms and food webs. As these tiny plastic particles become ubiquitous in our oceans, they are interacting with a wide range of species across trophic levels. One of the primary concerns is the ingestion of MP by marine organisms. From zooplankton to large predatory fish, numerous species have been observed consuming MP, often mistaking them for food particles (Valdez-Cibrián et al., 2024). Figure 5.3 shows the impacts of MP ingestion on marine food webs. This ingestion can lead to physical harm, such as intestinal blockage or damage to the digestive tract (Wright et al., 2013; Malik et al., 2023). However, the implications of MP proliferation extend beyond mere physical obstruction. The studies demonstrated that microbial communities are diverse and can differ significantly from those in the surrounding water (Zettler et al., 2013; Liu et al., 2024). Therefore, when marine organisms ingest MP, they are not just consuming inert plastic particles but also introducing novel microbial assemblages into their digestive systems. This could potentially disrupt the delicate

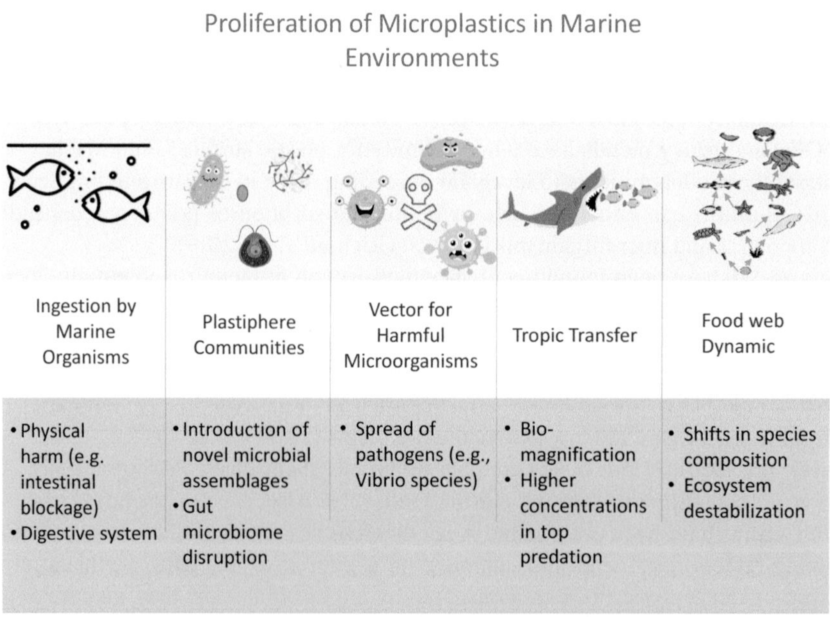

FIGURE 5.3 Impacts of MP ingestion on marine food webs.

balance of their gut microbiomes, with cascading effects on their health and metabolism (Pan et al., 2024). Furthermore, MPs are vectors of harmful microorganisms. Some studies have found potential pathogens, such as *Vibrio* spp., thriving on MP surfaces (Kirstein et al., 2019). As these plastic particles move through food webs, they could facilitate the spread of these pathogens to a diverse array of marine life.

The trophic transfer of MP is another critical concern for ecosystem health, as it poses risks to marine organisms and potentially impacts human health through the food chain (Saikumar et al., 2024). As smaller organisms ingest MP, these particles can accumulate in predators higher up the food chain. This biomagnification could lead to higher concentrations of MP in top predators, including commercially important fish species consumed by humans (Setälä et al., 2014). Moreover, the presence of MP could alter food web dynamics in subtle, yet significant ways. For instance, if certain species are more adversely affected by MP ingestion, it could lead to shifts in species composition and abundance, potentially destabilizing entire food webs (Galloway et al., 2017; Olmo-Gilabert et al., 2024).

5.4.2 Impacts on Biogeochemical Cycling

The introduction of MP into marine environments is not just a matter of pollution; it's fundamentally altering biogeochemical cycling in our oceans (Liu et al., 2024). These tiny plastic particles, along with their associated plastisphere communities, are becoming active players in the cycling of carbon, nutrients, and other elements (Rogers et al., 2020; Dodhia et al., 2023). One of the most significant impacts is on carbon cycling. As Romera-Castillo et al. (2018) discovered, plastics release dissolved organic carbon (DOC) into seawater as they degrade. This additional source of carbon could fuel microbial metabolism, potentially altering the balance of carbon fixation and respiration in marine ecosystems. Furthermore, the sinking of MP particles could accelerate the export of carbon to the deep ocean, a process known as the biological carbon pump (Cole et al., 2016).

The plastisphere communities themselves are hotspots of microbial activity, often showing higher metabolic rates than surrounding waters. This enhanced microbial activity could lead to localized zones of nutrient depletion or enrichment, depending on the specific metabolic processes occurring on the plastic surfaces (Arias-Andres et al., 2018). Also, MP interacts with other biogeochemical cycles in complex ways. For instance, they can adsorb and concentrate trace metals from seawater. Holmes et al. (2014) found that weathered plastic pellets adsorbed significantly more trace metals (such as copper and arsenic) than virgin pellets. This could potentially alter the bioavailability of these elements to marine organisms (Du et al., 2024).

The impact on nitrogen cycling is another area of concern. Some studies have suggested that plastic surfaces could provide ideal conditions for nitrogen-fixing bacteria, potentially enhancing nitrogen-fixation rates in certain marine environments (Bryant et al., 2016; Zhou et al., 2024). Conversely, if MP is ingested by zooplankton or other marine organisms, it could disrupt their role in nutrient recycling through processes like excretion and mortality. Perhaps most concerning is the potential for MP to disrupt the delicate balance of microbial communities that drive many of these biogeochemical processes. As Kettner et al. (2017) observed, MP can alter the composition of fungal communities in aquatic ecosystems. Similar shifts in bacterial communities could have far-reaching consequences for nutrient cycling and energy flow in marine food webs (Collaboration et al., 2018).

5.5 FUTURE DIRECTIONS AND MITIGATION STRATEGIES

5.5.1 Research Challenges and Opportunities

Research into marine MP continues to reveal complex challenges and unanswered questions. A primary obstacle in this field is the comprehensive characterization of the plastisphere community. The development of a global database cataloging plastic-associated microorganisms would significantly advance the understanding of the core microbiome of the plastisphere and its functional properties, inducing a resource would facilitate more systematic and comparative studies of plastisphere ecology across different marine environments (Roager and Sonnenschein, 2019).

As MPs are globally transportable by wind or ocean currents, they can be found in practically every ecosystem, including the most distant parts of our planet (Dey et al., 2022). We are also realizing that the journey of an MP particle is far from simple. As it travels through different environments—from the sunlit surface waters to the dark ocean depths and eventually to the seafloor—it encounters a whole range of chemical conditions. These changing environments could dramatically influence how microbes interact with the plastic. It's like watching a tiny spaceship navigate through different alien worlds, each with its own unique challenges (Brandon et al., 2019). While previous studies have explored MP transport, this research specifically evaluates drag models for irregularly shaped particles, enhancing understanding of settling velocity in marine environments (Coyle et al., 2023). One exciting frontier is the application of cutting-edge molecular techniques. Metagenomics, proteomics, and metabolomics could give us unprecedented insights into how these microbial communities function and interact with their plastic homes. It is like having a microscopic GoPro strapped to each microbe, allowing us to see exactly what they're doing and how they're doing it (Satta et al., 2024).

Fundamental technical challenges persist in the field of MP research. Key issues include the efficient separation of MP particles from seawater or sediment matrices and the accurate identification of various degraded plastic types (Lim et al., 2024). Current methodologies often prove time-consuming and labor-intensive, limiting large-scale environmental assessments. There is a critical need for the development and implementation of standardized, efficient techniques capable of addressing the scale and complexity of environmental samples. Such advancements would significantly enhance the ability to conduct comprehensive analyses of MP pollution in diverse marine environments (Wiggin and Holland, 2019; Araujo et al., 2018).

5.5.1.1 Potential Solutions and Conservation Efforts

While the number of challenges is increasing, they are encouraging us to find new solutions for that. As we better understand the plastisphere, we might discover microbes with an appetite for plastic in Section 5.3.1, which can contribute to the degradation of plastics. These could be employed for the bioremediation purposes of MP, transforming our plastic problem into a microbial feat. It is crucial to standardize the protocol for future study endeavors in plastisphere ecology. Understanding and evaluating this intricate environment would be accelerated if researchers from all around the world could agree on procedures for gathering, handling, and evaluating plastisphere communities (Roager and Sonnenschein, 2019).

Expanding the knowledge base on MP and associated microorganisms should be associated with regulatory frameworks, which is important to policy perspectives. In evaluating the effects of plastic pollutants, the potential for microbial dispersion over oceanic distances has to be factored in as well. Also, we might employ it for biosecurity measures and the purpose of ecosystem management; it would have great implications. Exposing the scientific publication to the public community is of paramount importance. It is important to note that with the progress of research and the development of our understanding of the complex relationships of the plastisphere, science communication becomes increasingly relevant and effective. Perhaps, the improved appreciation of the plastisphere's workings and its complexity will lead the general public to adjust their consumption choices. It may also help in promoting polity towards a reduction in plastic waste production.

As researchers elucidate the intricate dynamics of the plastisphere, effective science communication becomes essential. Enhancing the public knowledge on plastic complex lifecycle in ocean environments will give sustainable consumer habits and increase the support policies aimed at plastic waste reduction. Restoring marine MP is a task that requires a broader framework and a cross-disciplinary approach. To properly assess this problem, further research should be needed. Such research should be complemented by the creation of new detection and removal methods and better regulatory practices to prevent plastic from entering oceanic systems as well. This problem is a multiscaled problem, from processes occurring at the molecular level to large-scale interactions within the ecosystems in the world, which highlights the need for integration between scientists and across borders.

The way to go is the integration of knowledge from various disciplines. It is equally important to place the observations in the context of the environment. This approach enhances the understanding of the integration of MP–microbe interactions into marine ecosystems and biogeochemical cycles. Certainly, the complexity of this problem is high, but every new discovery enables us to see further one of the most severe environmental problems of our time in the prospects of its resolution.

REFERENCES

Ahmed, R., Hamid, A. K., Krebsbach, S. A., He, J. & Wang, D. 2022. Critical review of microplastics removal from the environment. *Chemosphere,* 293, 133557.

Al-Sabagh, A., Noah, A. Z., Hussein, A., El-Arabi, A. & El Safty, M. E. L. 2020. Enhancing oil based drilling fluids parameters using secondary synthetized emulsifiers and nanoparticles. *Egyptian Journal of Chemistry*, 63, 4119–4129.

Amaral-Zettler, L. A., Zettler, E. R. & Mincer, T. J. 2020. Ecology of the plastisphere. *Nature Reviews Microbiology,* 18, 139–151.

Amaral-Zettler, L. A., Zettler, E. R., Slikas, B., Boyd, G. D., Melvin, D. W., Morrall, C. E., Proskurowski, G. & Mincer, T. J. 2015. The biogeography of the plastisphere: implications for policy. *Frontiers in Ecology and the Environment,* 13, 541–546.

Andrady, A. L. 2011. Microplastics in the marine environment. *Marine Pollution Bulletin*, 62, 1596–1605.

Araujo, C. F., Nolasco, M. M., Ribeiro, A. M. & Ribeiro-Claro, P. J. 2018. Identification of microplastics using Raman spectroscopy: latest developments and future prospects. *Water Research*, 142, 426–440.

Arias-Andres, M., Klümper, U., Rojas-Jimenez, K. & Grossart, H. P. 2018. Microplastic pollution increases gene exchange in aquatic ecosystems. *Environmental Pollution,* 237, 253–261.

Auta, H. S., Emenike, C. U. & Fauziah, S. H. 2017. Distribution and importance of microplastics in the marine environment: a review of the sources, fate, effects, and potential solutions. *Environment International*, 102, 165–176.

Auta, H. S., Emenike, C. U., Jayanthi, B. & Fauziah, S. H. 2018. Growth kinetics and biodeterioration of polypropylene microplastics by Bacillus sp. and Rhodococcus sp. isolated from mangrove sediment. *Marine Pollution Bulletin*, 127, 15–21.

Avio, C. G., Gorbi, S. & Regoli, F. 2017. Plastics and microplastics in the oceans: from emerging pollutants to emerged threat. *Marine Environmental Research,* 128, 2–11.

Brandon, J. A., Jones, W. & Ohman, M. D. 2019. Multidecadal increase in plastic particles in coastal ocean sediments. *Science Advances,* 5, eaax0587.

Bryant, J. A., Clemente, T. M., Viviani, D. A., Fong, A. A., Thomas, K. A., Kemp, P., Karl, D. M., White, A. E. & Delong, E. F. 2016. Diversity and activity of communities inhabiting plastic debris in the North Pacific Gyre. *mSystems,* 1, e00024-16.

Carpenter, E. J. & Smith, K. L. 1972. Plastics on the Sargasso Sea surface. *Science,* 175, 1240–1241.

Cole, M., Lindeque, P. K., Fileman, E., Clark, J., Lewis, C., Halsband, C. & Galloway, T. S. 2016. Microplastics alter the properties and sinking rates of zooplankton faecal pellets. *Environmental Science & Technology,* 50, 3239–3246.

Collaboration, P., Aidala, C., Akiba, Y., Alfred, M., Andrieux, V., Aoki, K., Apadula, N., Asano, H., Ayuso, C., Azmoun, B., Babintsev, V., Bagoly, A., Bandara, N. S., Barish, K. N., Bathe, S., J., et al. 2018. Nuclear dependence of the transverse-single-spin asymmetry for forward neutron production in polarized p+A collisions at sqrt[s_{NN}]=200 GeV. *Physical Review Letters,* 120, 022001.

Coons, A. K., Busch, K., Lenz, M., Hentschel, U. & Borchert, E. 2021. Biogeography rather than substrate type determines bacterial colonization dynamics of marine plastics. *PeerJ*, 9, e12135.

Coyle, R., Service, M., Witte, U., Hardiman, G. & McKinley, J. 2023. Modeling microplastic transport in the marine environment: testing empirical models of particle terminal sinking velocity for irregularly shaped particles. *ACS ES&T Water,* 3, 984–995.

Dang, H. & Lovell, C. R. 2016. Microbial surface colonization and biofilm development in marine environments. *Microbiology and Molecular Biology Reviews,* 80, 91–138.

De Sales-Ribeiro, C., Brito-Casillas, Y., Fernandez, A. & Caballero, M. J. 2020. An end to the controversy over the microscopic detection and effects of pristine microplastics in fish organs. *Scientific Reports*, 10, 12434.

De Vogel, F. A., Goudriaan, M., Zettler, E. R., Niemann, H., Eich, A., Weber, M., Lott, C. & Amaral-Zettler, L. A. 2024. Biodegradable plastics in Mediterranean coastal environments feature contrasting microbial succession. *Science of the Total Environment,* 928, 172288.

Debroas, D., Domaizon, I., Humbert, J.-F., Jardillier, L., Lepère, C., Oudart, A. & Taïb, N. 2017. Overview of freshwater microbial eukaryotes diversity: a first analysis of publicly available metabarcoding data. *FEMS Microbiology Ecology*, 93, fix023.

De Tender, C.A., Devriese, L.I., Haegeman, A., Maes, S., Ruttink, T. and Dawyndt, P. 2015. Bacterial community profiling of plastic litter in the Belgian part of the North Sea. *Environmental science & technology*, *49*(16), 9629–9638. DOI: 10.1021/acs.est.5b01093.

Devi, R. S., Kannan, V. R., Nivas, D., Kannan, K., Chandru, S. & Antony, A. R. 2015. Biodegradation of HDPE by Aspergillus spp. from marine ecosystem of Gulf of Mannar, India. *Marine Pollution Bulletin*, 96, 32–40.

Dey, S., Rout, A. K., Behera, B. K. & Ghosh, K. 2022. Plastisphere community assemblage of aquatic environment: plastic-microbe interaction, role in degradation and characterization technologies. *Environmental Microbiome*, 17, 32.

Dodhia, M. S., Rogers, K. L., Fernández-Juárez, V., Carreres-Calabuig, J. A., Löscher, C. R., Tisserand, A. A., Keulen, N., Riemann, L., Shashoua, Y. & Posth, N. R. 2023. Microbe-mineral interactions in the plastisphere: coastal biogeochemistry and consequences for degradation of plastics. *Frontiers in Marine Science*, 10, 1134815.

Du, H., Chen, P., Lin, X., Zheng, J., Liu, H. & Wang, X. 2024. Adsorption of metals on aged microplastics in intensive mariculture areas: aggravating the potential ecological risks to marine organisms. *Science of the Total Environment,* 946, 173964.

Dussud, C., Hudec, C., George, M., Fabre, P., Higgs, P., Bruzaud, S., Delort, A.-M., Eyheraguibel, B., Meistertzheim, A.-L. & Jacquin, J. 2018. Colonization of non-biodegradable and biodegradable plastics by marine microorganisms. *Frontiers in Microbiology*, 9, 1571.

Flemming, H. C., Wingender, J., Szewzyk, U., Steinberg, P., Rice, S. A. & Kjelleberg, S. 2016. Biofilms: an emergent form of bacterial life. *Nature Reviews MicrobiologyNat Rev Microbiol,* 14, 563–575.

Flemming, H.-C. & Wuertz, S. 2019. Bacteria and archaea on Earth and their abundance in biofilms. *Nature Reviews Microbiology,* 17, 247–260.

Galloway, T. S., Cole, M. & Lewis, C. 2017. Interactions of microplastic debris throughout the marine ecosystem. *Nature Ecology & Evolution,* 1, 116.

Gaylarde, C. C., De Almeida, M. P., Neves, C. V., Neto, J. A. & Da Fonseca, E. M. 2023. The importance of biofilms on microplastic particles in their sinking behavior and the transfer of invasive organisms between ecosystems. *Micro*, 3, 320–337.

Gewert, B., Ogonowski, M., Barth, A. & MacLeod, M. 2017. Abundance and composition of near surface microplastics and plastic debris in the Stockholm Archipelago, Baltic Sea. *Marine Pollution Bulletin,* 120, 292–302.

Gewert, B., Plassmann, M. M. & MacLeod, M. 2015. Pathways for degradation of plastic polymers floating in the marine environment. *Environmental Science: Processes & Impacts,* 17, 1513–1521.

Graca, B., Rychter, A., Bełdowska, M. & Wojdasiewicz, A. 2024. Seasonality of mercury and its fractions in microplastics biofilms-comparison to natural biofilms, suspended particulate matter and bottom sediment. *Science of the Total Environment,* 948, 174814.

Harshvardhan, K. & Jha, B. 2013. Biodegradation of low-density polyethylene by marine bacteria from pelagic waters, Arabian Sea, India. *Marine Pollution Bulletin*, 77, 100–106.

Holmes, L.A., Turner, A. and Thompson, R.C. 2014. Interactions between trace metals and plastic production pellets under estuarine conditions. *Marine Chemistry*, 167, 25–32. https://doi.org/10.1016/j.marchem.2014.06.001

Huang, W., Song, B., Liang, J., Niu, Q., Zeng, G., Shen, M., Deng, J., Luo, Y., Wen, X. & Zhang, Y. 2021. Microplastics and associated contaminants in the aquatic environment: a review on their ecotoxicological effects, trophic transfer, and potential impacts to human health. *Journal of Hazardous Materials,* 405, 124187.

Issac, M. N. & Kandasubramanian, B. 2021. Effect of microplastics in water and aquatic systems. *Environmental Science and Pollution Research,* 28, 19544–19562.

Jacquin, J., Budinich, M., Chaffron, S., Barbe, V., Lombard, F., Pedrotti, M.-L., Gorsky, G., Ter Halle, A., Bruzaud, S., Kedzierski, M. & Ghiglione, J.-F. 2024. Niche partitioning and plastisphere core microbiomes in the two most plastic polluted zones of the world ocean. *Environmental Science and Pollution Research,* 31, 41118–41136.

Jacquin, J., Cheng, J., Odobel, C., Pandin, C., Conan, P., Pujo-Pay, M., Barbe, V., Meistertzheim, A. L. & Ghiglione, J. F. 2019. Microbial ecotoxicology of marine plastic debris: a review on colonization and biodegradation by the "plastisphere". *Frontiers in Microbiology,* 10, 865.

Jambeck, J. R., Geyer, R., Wilcox, C., Siegler, T. R., Perryman, M., Andrady, A., Narayan, R. & Law, K. L. 2015. Plastic waste inputs from land into the ocean. *Science*, 347, 768–771.

Januar, H. I., Zamani, N. P., Soedarma, D. & Chasanah, E. 2017. Bioactive cembranoid composition in the soft coral Sarcophyton glaccum on the response to changing pH. *ILMU KELAUTAN Indonesian Journal of Marine Sciences*, 22(1), 25.

Johansen, M. P., Prentice, E., Cresswell, T. & Howell, N. 2018. Initial data on adsorption of Cs and Sr to the surfaces of microplastics with biofilm. *Journal of Environmental Radioactivity,* 190–191, 130–133.

Kettner, M. T., Rojas-Jimenez, K., Oberbeckmann, S., Labrenz, M. & Grossart, H. P. 2017. Microplastics alter composition of fungal communities in aquatic ecosystems. *Environmental Microbiology,* 19(11), 4447–4459. https://doi.org/10.1111/1462-2920.13891.

Kirstein, I. V., Wichels, A., Gullans, E., Krohne, G. & Gerdts, G. 2019. The plastisphere - uncovering tightly attached plastic "specific" microorganisms. *PLoS One,* 14, e0215859.

Lacerda, A. L., Briand, J.-F., Lenoble, V., Oreste, E. Q., Kessler, F. & Pedrotti, M. L. 2024. Assessing the plastisphere from floating plastics in the Northwestern Mediterranean Sea, with emphasis on viruses. *Microorganisms*, 12, 444.

Lambert, S. & Wagner, M. 2016. Characterisation of nanoplastics during the degradation of polystyrene. *Chemosphere*, 145, 265–268.

Lim, S. J., Lee, K.-J., Nam, H., Kim, S. H., Kim, E.-J., Lee, S. & Chung, J. 2024. Progress and future directions bridging microplastics transport from pore to continuum scale: a comprehensive review for experimental and modeling approaches. *TrAC Trends in Analytical Chemistry,* 179, 117851.

Liu, S., Qiu, Y., He, Z., Shi, C., Xing, B. & Wu, F. 2024. Microplastic-derived dissolved organic matter and its biogeochemical behaviors in aquatic environments: a review. *Critical Reviews in Environmental Science and Technology,* 54, 865–882.

Lobelle, D. & Cunliffe, M. 2011. Early microbial biofilm formation on marine plastic debris. *Marine Pollution Bulletin,* 62, 197–200.

Malik, U. E., Schuster, M. & Richter, S. 2023. S2682 Extensive necrosis of the GI tract from caustic ingestion. *The American Journal of Gastroenterology*, 118(10S), S1837.

Marsay, K. S., Ambrosino, A. C., Koucherov, Y., Davidov, K., Figueiredo, N., Yakovenko, I., Itzahri, S., Martins, M., Sobral, P. & Oren, M. 2023. The geographical and seasonal effects on the composition of marine microplastic and its microbial communities: the case study of Israel and Portugal. *Frontiers in Microbiology*, 14, 1089926.

Mato, Y., Isobe, T., Takada, H., Kanehiro, H., Ohtake, C. & Kaminuma, T. 2001. Plastic resin pellets as a transport medium for toxic chemicals in the marine environment. *Environmental Science & Technology,* 35, 318–324.

Misic, C., Covazzi Harriague, A. & Ferrari, M. 2019. Hydrodynamic forcing and sand permeability influence the distribution of anthropogenic microparticles in beach sediment. *Estuarine, Coastal and Shelf Science,* 230, 106429.

Mohanrasu, K., Premnath, N., Prakash, G. S., Sudhakar, M., Boobalan, T., Arun, A. J. J. O. P. & Biology, P. B. 2018. Exploring multi potential uses of marine bacteria; an integrated approach for PHB production, PAHs and polyethylene biodegradation. *Journal of Photochemistry and Photobiology B: Biology*, 185, 55–65.

Mu, M., Liu, S., Deflorio, W., Hao, L., Wang, X., Salazar, K. S., Taylor, M., Castillo, A., Cisneros-Zevallos, L., Oh, J. K., Min, Y. & Akbulut, M. 2023. Influence of surface roughness, nanostructure, and wetting on bacterial adhesion. *Langmuir,* 39, 5426–5439.

Nikolopoulou, I., Piperagkas, O., Moschos, S. & Karayanni, H. 2023. Bacteria release from microplastics into new aquatic environments. *Diversity*, 15, 115.

Oberbeckmann, S., Löder, M. G. & Labrenz, M. 2015. Marine microplastic-associated biofilms–a review. *Environmental Chemistry*, 12, 551–562.

Ogonowski, M., Gerdes, Z. & Gorokhova, E. 2018. What we know and what we think we know about microplastic effects–a critical perspective. *Current Opinion in Environmental Science & Health*, 1, 41–46.

Olmo-Gilabert, R., Fagiano, V., Alomar, C., Rios-Fuster, B., Compa, M. & Deudero, S. 2024. Plastic webs, the new food: dynamics of microplastics in a Mediterranean food web, key species as pollution sources and receptors. *Science of the Total Environment,* 918, 170719.

Paço, A., Duarte, K., Da Costa, J. P., Santos, P. S., Pereira, R., Pereira, M., Freitas, A. C., Duarte, A. C. & Rocha-Santos, T. A. 2017. Biodegradation of polyethylene microplastics by the marine fungus Zalerion maritimum. *Science of the Total Environment*, 586, 10–15.

Pan, Y., Zhang, H., Zhu, L., Tan, J., Wang, B. & Li, M. 2024. The role of gut microbiota in MP/NP-induced toxicity. *Environmental Pollution,* 359, 124742.

Qian, Y., Huang, L., Yan, P., Wang, X. & Luo, Y. 2024. Biofilms on plastic debris and the microbiome. *Microorganisms*, 12, 1362.

Richard, H., Carpenter, E. J., Komada, T., Palmer, P. T. & Rochman, C. M. 2019. Biofilm facilitates metal accumulation onto microplastics in estuarine waters. *Science of the Total Environment,* 683, 600–608.

Roager, L. & Sonnenschein, E. C. 2019. Bacterial candidates for colonization and degradation of marine plastic debris. *Environmental Science & Technology*, 53, 11636–11643.

Rochman, C. M., Hoh, E., Kurobe, T. & Teh, S. J. 2013. Ingested plastic transfers hazardous chemicals to fish and induces hepatic stress. *Scientific Reports*, 3, 1–7.

Rogers, K. L., Carreres-Calabuig, J. A., Gorokhova, E., Posth, N. R. J. L. & Letters, O. 2020. Micro-by-micro interactions: how microorganisms influence the fate of marine microplastics. *Limnology and Oceanography Letters*, 5, 18–36.

Romera-Castillo, C., Pinto, M., Langer, T. M., Álvarez-Salgado, X. A. & Herndl, G. J. 2018. Dissolved organic carbon leaching from plastics stimulates microbial activity in the ocean. *Nature Communications,* 9, 1430.

Rummel, C. D., Jahnke, A., Gorokhova, E., Kühnel, D. & Schmitt-Jansen, M. 2017. Impacts of biofilm formation on the fate and potential effects of microplastic in the aquatic environment. *Environmental Science & Technology Letters,* 4, 258–267.

Saikumar, S., Mani, R., Ganesan, M., Dhinakarasamy, I., Palanisami, T. & Gopal, D. 2024. Trophic transfer and their impact of microplastics on estuarine food chain model. *Journal of Hazardous Materials,* 464, 132927.

Satta, A., Ghiotto, G., Santinello, D., Giangeri, G., Bergantino, E., Modesti, M., Raga, R., Treu, L., Campanaro, S. & Zampieri, G. 2024. Synergistic functional activity of a landfill microbial consortium in a microplastic-enriched environment. *Science of the Total Environment,* 947, 174696.

Savoca, M. S., Wohlfeil, M. E., Ebeler, S. E. & Nevitt, G. A. 2016. Marine plastic debris emits a keystone infochemical for olfactory foraging seabirds. *Science Advances,* 2, e1600395.

Setälä, O., Fleming-Lehtinen, V. and Lehtiniemi, M. 2014. Ingestion and transfer of microplastics in the planktonic food web. *Environmental pollution*, 185, 77–83. https://doi.org/10.1016/j.envpol.2013.10.013

Sharma, P., Mahongnao, S. & Nanda, S. 2022. Microbial degradation of micro-plastics. In A. Ahamad, P. Singh, D. Tiwary (eds.), *Plastic and Microplastic in the Environment* (pp. 222–244). John Wiley & Sons Ltd.

Shen, M., Zhang, Y., Zhu, Y., Song, B., Zeng, G., Hu, D., Wen, X. & Ren, X. 2019. Recent advances in toxicological research of nanoplastics in the environment: a review. *Environmental Pollution,* 252, 511–521.

Sudhakar, M., Doble, M., Murthy, P. S. & Venkatesan, R. 2008. Marine microbe-mediated biodegradation of low-and high-density polyethylenes. *International Biodeterioration & Biodegradation*, 61, 203–213.

Sun, Y., Wu, M., Zang, J., Du, L., Huang, M., Chen, C. & Wang, J. 2023. Plastisphere microbiome: methodology, diversity, and functionality. *Imeta,* 2, e101.

Suzuki, M., Tachibana, Y., Oba, K., Takizawa, R. & Kasuya, K.-I. 2018. Microbial degradation of poly (ε-caprolactone) in a coastal environment. *Polymer Degradation and Stability*, 149, 1–8.

Tanaka, Y., Hiraga, K. & Inui, M. 2024. Regulation of the expression of MHETase and TPA degradation genes involved in the degradation of PET in Ideonella sakaiensis. *The FEBS Journal,* 291, 4489–4500.

Thompson, R. C., Olsen, Y., Mitchell, R. P., Davis, A., Rowland, S. J., John, A. W., McGonigle, D. & Russell, A. E. 2004. Lost at sea: where is all the plastic? *Science,* 304, 838.

Valdez-Cibrián, A., Kozak, E. R. & Franco-Gordo, C. 2024. Microplastic ingestion in marine mesozooplankton species associated with functional feeding traits. *Marine Environmental Research,* 200, 106650.

Vedolin, M. C., Teophilo, C., Turra, A. & Figueira, R. C. L. 2018. Spatial variability in the concentrations of metals in beached microplastics. *Marine Pollution Bulletin*, 129, 487–493.

Vethaak, A. D. & Leslie, H. A. 2016. *Plastic Debris Is a Human Health Issue*. ACS Publications.

Walkinshaw, C., Lindeque, P. K., Thompson, R., Tolhurst, T. & Cole, M. 2020. Microplastics and seafood: lower trophic organisms at highest risk of contamination. *Ecotoxicology and Environmental Safety,* 190, 110066.

Wiggin, K. J. & Holland, E. B. 2019. Validation and application of cost and time effective methods for the detection of 3–500 μm sized microplastics in the urban marine and estuarine environments surrounding Long Beach, California. *Marine Pollution Bulletin*, 143, 152–162.

Wright, R. J., Erni-Cassola, G., Zadjelovic, V., Latva, M. & Christie-Oleza, J. A. 2020. Marine plastic debris: a new surface for microbial colonization. *Environmental Science & Technology*, 54, 11657–11672.

Wright, S. L., Thompson, R. C. & Galloway, T. S. 2013. The physical impacts of microplastics on marine organisms: a review. *Environmental Pollution*, 178, 483–492.

Yang, Y., Wang, J. & Xia, M. 2020. Biodegradation and mineralization of polystyrene by plastic-eating superworms Zophobas atratus. *Science of the Total Environment*, 708, 135233.

Yoshida, S., Hiraga, K., Takehana, T., Taniguchi, I., Yamaji, H., Maeda, Y., Toyohara, K., Miyamoto, K., Kimura, Y. & Oda, K. 2016. A bacterium that degrades and assimilates poly (ethylene terephthalate). *Science*, 351, 1196–1199.

Zettler, E. R., Mincer, T. J. & Amaral-Zettler, L. A. 2013. Life in the "plastisphere": microbial communities on plastic marine debris. *Environmental Science & Technology*, 47, 7137–7146.

Zhai, X., Zhang, X. H. & Yu, M. 2023. Microbial colonization and degradation of marine microplastics in the plastisphere: a review. *Frontiers in Microbiology,* 14, 1127308.

Zhang, S.-J., Zeng, Y.-H., Zhu, J.-M., Cai, Z.-H. & Zhou, J. 2022. The structure and assembly mechanisms of plastisphere microbial community in natural marine environment. *Journal of Hazardous Materials,* 421, 126780.

Zhou, Z., Hua, J., Xue, J. & Yu, C. 2024. Differential impacts of polyethylene microplastic and additives on soil nitrogen cycling: a deeper dive into microbial interactions and transformation mechanisms. *Science of the Total Environment,* 942, 173771.

6 Plastisphere Community Assemblage in Freshwater

Tatenda Dalu, Naicheng Wu, Collins Oduro, Matthew Elmo Gruwell, Edmond Sanganyado, Rana Al-Jaibachi, Franco Teixeira de Mello, Conrad Sparks, and Pule Mpopetsi

6.1 FRESHWATER PLASTISPHERE: A UNIQUE HABITAT

Plastic pollution and the consequent formation of micro- and nanoplastics are expected to continue to increase in aquatic environments (Harrison et al., 2018; Walker and Fequet, 2023). When plastic debris enters aquatic environments, it can be colonised by microbial communities, thus forming a plastisphere (Figure 6.1), which allows for the long-distance dispersal of microbes in the aquatic environment through travelling on microplastics or through dispersal (Dalu et al., 2019; Di Pippo et al., 2020; Barros and Seena, 2021). This novel ecological niche harbours distinct microbial communities, including bacteria, algae, fungi, and microeukaryotes, collectively representing a microecosystem markedly different from those found in the surrounding waters or on natural surfaces (Yang et al., 2020a). This microecosystem, while distinct, does not exist in isolation, but rather in dynamic interactions with its environmental context (Sauer et al., 2022; Bos et al., 2023).

The freshwater plastisphere is an emergent ecological niche within freshwater ecosystems. Their presence signifies the extensive impact of human activities, manifest in the pervasive distribution of plastics. These plastics are rapidly colonised by diverse microbial communities. Generally, in freshwater ecosystems (as in marine ecosystems), plastisphere bacterial composition or microbial communities appear to be different from those of the surrounding environment and from other substrate types (Hoellein et al., 2014; McCormick et al., 2014; Wang et al., 2021; Cao et al., 2022). For example, recent research has shown that, on average, freshwater microbial abundance on plastic debris (i.e., freshwater plastisphere) was higher than that on natural substrates in both lentic and

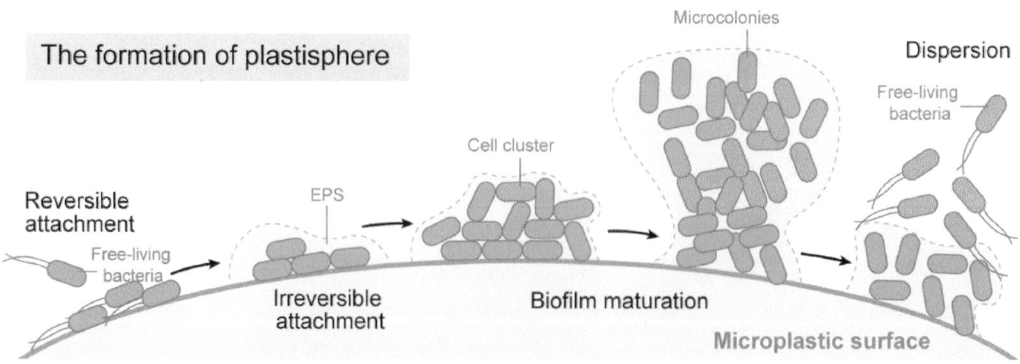

FIGURE 6.1 Plastisphere formation on a microplastic surface, starting with attachment and ending with the dispersal of free-living bacterial colonies. (Adapted from Shi et al., 2023.)

DOI: 10.1201/9781032717814-6

lotic environments (Miao et al., 2020). Furthermore, microbial communities on plastic substrates differed greatly from those on natural substrates (i.e., cobblestones) in terms of the carbon metabolic activities of the biofilms (Miao et al., 2020). A similar pattern is observed in marine environments (Zettler et al., 2013; Lear et al., 2021) and could imply that plastic substrates are unfavourable habitats for some microbial species, while at the same time providing a unique habitat that facilitates colonisation of other microbial species (Zettler et al., 2013; McCormick et al., 2014; Rummel et al., 2017). The colonisation process of microbial communities on plastic particles or substrates is influenced by the size, shape, and type of plastic, as well as by environmental factors such as temperature, nutrient availability, and contaminants (Oberbeckmann et al., 2015). These factors in turn create a habitat that is physically and chemically distinct from other aquatic surfaces, fostering unique microbial ecosystems (Barros and Seena, 2021; Miao et al., 2023). Thus, plastic particles or substrates, owing to their diverse chemical compositions and surface properties, provide various ecological niches that select for specific microbial assemblages (Bos et al., 2023). In addition, these substrates appear ideal for microbial colonisation (Wright et al., 2020a,b) and in creating conducive conditions for the formation of the plastisphere in aquatic environments. For example, in freshwater ecosystems, plastic debris fragments into small particles (<5 mm in size) called microplastics (Barros and Seena, 2021), which then provide suitable substrates for microbial communities to colonise, resulting in the formation of plastispheres. Therefore, coupled with their small sizes, microplastics that allow the formation of the plastisphere will be difficult to eradicate and will become a major challenge in freshwater ecosystems in the near future. Compared with other microbes, bacteria form a large component of microbial communities in freshwater ecosystems (Barros and Seena, 2021). However, this pattern could be due to bias centred towards the bacteria. For example, many studies on the plastisphere have focused on bacteria, with little attention paid to eukaryotes (Amaral-Zettler et al., 2020; Barros and Seena, 2021).

Focus on the plastisphere in freshwater began more recently, with a previous focus on the marine environment (McCormick et al., 2014; Barros and Seena, 2021) and with most of the research centred on microbial biodiversity and community structure assessments and not as much on the potential implications of the plastisphere on freshwater ecosystems. Thus, there is a paucity of information regarding the potential implications of plastisphere development for freshwater ecosystems (Barros and Seena, 2021). Notwithstanding, it is postulated that the driving factors behind microbial community assemblages on plastics in freshwater systems, and thus plastisphere development, are likely similar to those in marine systems (Harrison et al., 2018; Delacuvellerie et al., 2022). This chapter delves into the diverse forms and sizes of plastics, unique physicochemical properties affecting the plastisphere, and factors influencing their distribution and abundance.

6.2 DIVERSE FORMS AND SIZES PLASTIC DIVERSITY

The diversity of forms and sizes, particularly the range of microbes colonising nano-, micro-, and macroplastics, is evidence of the adaptability and resilience of microbes. This section discusses the intricate interactions and ecological effects stemming from the colonisation of plastics of varying sizes, compositions, and textures by diverse microbial communities, highlighting the complexity and multifaceted nature of the plastisphere. The plastisphere is prevalent on a wide range of plastic debris, from nano- and microplastics of less than 1 μm, several millimetres in size, and macroplastics visible to the naked eye (Cheng et al., 2021). In addition to size, plastics also vary in chemical composition, with polymers like polyethylene (PE), polypropylene (PP), and polystyrene (PS) offering different surface properties, such as hydrophobicity, which affect microbial colonisation (Cheng et al., 2021). Biodegradable plastics like polylactic acid (PLA) degrade faster, while durable polymers like PE and PP form long-lasting biofilms (Sun et al., 2022). Each category provides unique surfaces for microbial colonisation, creating a mosaic of habitats within the freshwater plastisphere.

6.2.1 Nano- and Microplastics

Owing to their minuscule size, nano- and microplastics have a high surface area-to-volume ratio, offering an extensive interface for microbial attachment, forming the plastisphere (Chouchene et al., 2023). These tiny particles can infiltrate various aquatic environments and biota, making them available to microorganisms through multiple pathways, including cellular uptake, where larger particles may be less likely to enter (Li et al., 2021). The small sizes of these plastics facilitate close interactions among microbial colonisers, potentially enhancing gene transfer events and fostering unique microbial consortia that include bacteria, fungi, and algae. Studies have shown that nano- and microplastics can harbour specialised microbial communities that differ significantly from those on macroplastic pieces and surrounding water, indicating a size-dependent colonisation process (Oberbeckmann et al., 2015). However, other factors may be important for the selection of plastisphere community (i.e., substratum chemistry, degradation state, redox conditions, and other physiochemical conditions). Thus, given the relative sizes of cells (i.e., say 1 μm length) and microplastics (up to 5 mm diameter), sizes make a large difference for further colonisation.

6.2.2 Macroplastics

Macroplastics, although more visually conspicuous, offer different ecological niches. The large surface areas of these plastics allow for more extensive and diverse microbial colonisation, supporting complex biofilms that can include a wider variety of species, from primary colonisers to high trophic levels, such as protists and small invertebrates. Biofilms can act as "microbial reefs", creating localised ecosystems that can affect the surrounding aquatic environment (Zettler et al., 2013). The colonisation patterns of macroplastics can be influenced by their physical structure, chemical composition, and environmental factors such as light, temperature, and nutrient availability.

6.3 Freshwater Plastisphere and Microbial Communities

Plastics found in freshwater systems are substrates for the development of the plastisphere by the microbial communities. Plastisphere development usually entails the transition of motile microorganisms to colonies (Figure 6.1). The transition may start with an individual cell forming tissue-like structures that create fluid channels between cells that are able to distribute nutrients (Amaral-Zettler et al., 2020). Community development is sequential in terms of complexity and community structure and is composed of diverse bacteria, archaea, and fungi. The initial colonisers were designed to make use of the limited sources of energy and synthetic habitat and pave the way for further degradation of the plastic as well as the development of more complex microbial communities (Kaushal et al., 2021; Di Pippo et al., 2022). Colonial microbes adhere to plastic and produce biofilms by secreting extracellular polymeric substances (EPSs) (Shi et al., 2023). The EPS creates a matrix that serves a protective function and at the same time can also contribute to the weathering of the surface, further contributing to the degradation of the plastic. These diverse phyla in the plastisphere include primary producers (e.g., cyanobacteria, diatoms, and microalgae), predators (e.g., choanoflagellates and hydroids), symbionts (e.g., sulfide-oxidising ectobionts on ciliates), and decomposers (e.g., fungi) (Amaral-Zettler et al., 2020). The plastisphere may also include pathogenic, invasive, and plastic-degrading species (Wright et al., 2020c). Community composition is influenced by factors such as polymer type, shape, size, and ambient environmental conditions (Figure 6.2). In freshwater environments (flowing or stagnant), the plastisphere contains distinct microbial communities that are determined and influenced by ambient conditions (Di Pippo et al., 2020, 2022; Barros and Seena, 2021).

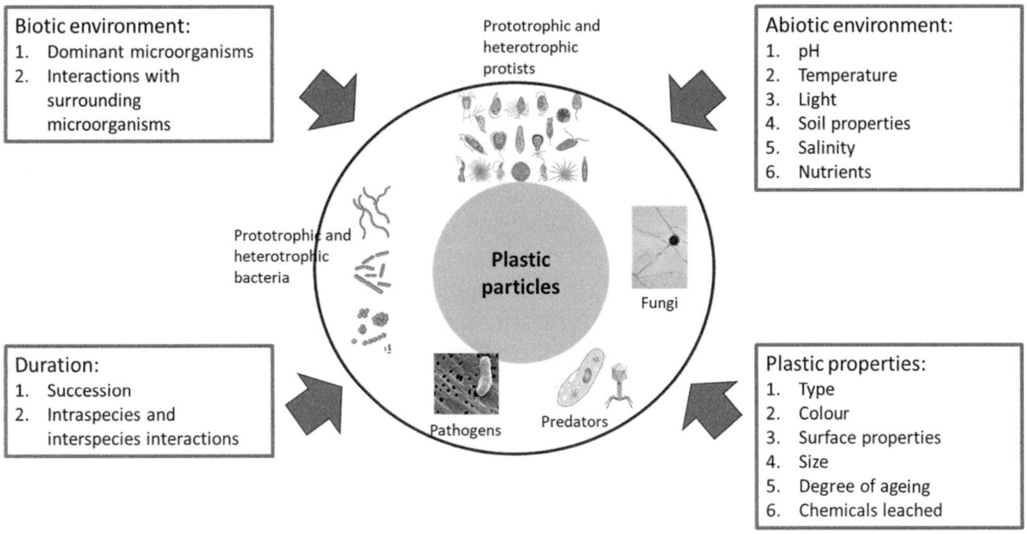

FIGURE 6.2 The plastisphere composition and the factors affecting it. (Adapted from Tang, 2024, published open access.)

The unique habitat of the plastisphere is profoundly influenced by physicochemical properties. These properties, encompassing both physical and chemical aspects, create an interactive environment that dictates the behaviour and impact of microplastics in freshwater systems (Badea et al., 2023; Dodhia et al., 2023). Understanding how these physicochemical characteristics shape the plastisphere provides crucial insights into its ecological role within freshwater ecosystems and its broader environmental impacts. The surface texture and topography of plastics are critical for their microbial adhesion and biofilm formation. Rough surfaces with grooves, pits, and cracks offer more niches and a higher surface area for microbial attachment than smooth surfaces do. These physical features can trap organic matter and nutrients, thereby providing a more conducive environment for microbial growth. Microbial communities exhibit preferential attachment to certain textures, which influences the initial colonisation and succession of microbial communities on plastics (Oberbeckmann et al., 2015).

The physical characteristics of plastics, including density, colour, shape, size, and crystalline structure, significantly influence their distribution and ecological interactions within freshwater ecosystems (Mendoza and Balcer, 2019; Wang et al., 2023). The density of the plastic material is particularly influential in determining its spatial distribution, with light plastics tending to float or remain suspended in the water column, whereas dense plastics are more likely to sink, thus differentially affecting their encounters with various microbial communities (Badea et al., 2023 and references therein). These variations in density are crucial for understanding how different types of plastics behave in aquatic environments, subsequently influencing their availability and interactions with microbial communities residing in these habitats. Studies have documented a distinct colour preference among organisms interacting with microplastics, suggesting that the colouration of these materials can be a critical factor in determining their ingestion and subsequent ecological interactions (Ren et al., 2020; Ríos et al., 2022; Badea et al., 2023). The colour of microplastics influences their visibility and attractiveness to organisms, including microbes with phototrophic capabilities and/or preferences. In aquatic ecosystems, where light penetration affects biological activity, coloured microplastics interact differently with microbial communities due to leachates from colour complexes, which affect microbial complexes more than transparent or neutrally coloured ones. This microbial behavioural response to colour underlines its potential influence on the ecological dynamics and interactions between microplastics and biota in freshwater environments.

6.4 MICROBIAL COMMUNITIES OF THE FRESHWATER PLASTISPHERE

Lentic (i.e., lakes, reservoirs, and wetlands) environments are thought to display high abundances of plastispheres compared to lotic (i.e., rivers and streams) environments in freshwater ecosystems because of their characteristics as sinks for plastic pollution (Tang et al., 2021). In lotic environments, the transportation and concentration of microplastics are highly influenced by river hydro-geomorphological factors and the physical properties of microplastics, such as size, shape, and density (Kiss et al., 2021; Kumar et al., 2021; Owowenu et al., 2023). In addition, the longitudinal flow of water in lotic systems could serve as an effective dispersal mechanism of plastispheres from headwaters to downstream reaches, especially during flooding periods. McCormick et al. (2014) highlighted that microplastics can be efficient vectors for transporting unique microbial communities to downstream river reaches. However, the survival of the plastisphere during long-term transportation in rivers (i.e., from upstream to downstream reaches) is not fully understood (Barros and Seena, 2021). Nonetheless, some studies have shown variation in the microbial community structure of plastics along different functional zones (McCormick et al., 2016).

Freshwater environments comprise diverse microbial taxa and communities, which are crucial components of nutrient recycling and plastic degradation. Plastisphere microbes start as aggregates of biofilms that are attached to or associated with plastics via polymeric matrices (Figure 6.1). Development starts with the aggregation and attachment of microbes such as bacteria, single-cell eukaryotes, larvae, spores, and dissolved organic matter. These matrices can influence ecological processes such as nutrient cycling, primary production, and ecosystem respiration (Figure 6.2; Sauer et al., 2022). Initially, research on plastispheres focused on bacteria, but there was a shift to include archaea and fungi when they were reported as plastispheres (Bryant et al., 2016; Yu et al., 2023; Zettler et al., 2013).

6.4.1 Bacteria

Bacteria are the most studied microbial group in freshwater plastispheres (Yu et al., 2023). Bacterial community composition depends on factors, such as geographical location, environmental conditions and stressors, taxa and community age, and plastic properties (Li et al., 2021). The Proteobacteria include a wide range of bacteria with various functionalities and are considered a dominant phylum in the freshwater plastisphere. Some *Pseudomonas* and Acinetobacteria are examples of *Proteobacteria* that can degrade plastics. Bacteriodetes belong to a phylum of bacteria of which some are known to degrade complex plastics. Examples include *Sphingomonas* and *Flexibacter*, which are known to occur in freshwater plastispheres (Shaw et al., 2023). *Sphingomonas* can degrade organic matter within the plastisphere, while *Flexibacter* contributes to biofilm formation and degradation of plastic, enabling further colonisation by other microbes (Shaw et al., 2023). Actinobacteria phyla microbes are associated with the freshwater plastisphere, including *Rhodococcus* and *Streptomyces*, which contribute to the degradation of plastics. Other bacterial phyla associated with the freshwater plastisphere include Firmicutes, Cyanobacteria, and Deinococcus-Thermus. The diverse types of bacteria in the freshwater plastisphere contribute to the added dynamics and impact of these microbes in plastic degradation, biofilm formation, and nutrient recycling.

6.4.2 Archaea

Archaea are less studied in freshwater systems than bacteria but are still found in the freshwater plastisphere. Due to challenges related to archaeal analysis, these microbes are possibly underestimated, but likely to play a significant role in freshwater plastispheres (Li et al., 2021). Crenarchaeota Archaea microbes are generally associated with the marine environment but have also been recorded in freshwater systems and plastispheres more frequently due to improved analytical developments and techniques (González-Pleiter et al., 2021).

6.4.3 Fungi

Fungi have filamentous growth forms (i.e., mycelia) and are a key constituent of freshwater plastispheres (Li et al., 2021). Although less abundant than bacteria, fungi play a key role in plastic degradation and biofilm formation. The Ascomycota phyla of fungi are diverse and include genera such as *Aspergillus* and *Penicillium*, which have been shown to degrade PE and various other types of plastic.

6.4.4 Co-occurrence and Multispecies Plastispheres

Plastisphere communities comprise numerous phyla, which may result in complex interactions (Chen et al., 2020a,b). The presence of microbial networks complicates the structure and function of the plastispheres. Given these complexities, there is a need to carry out co-occurrence network analyses to provide a comprehensive understanding of freshwater plastispheres (Yang et al., 2019). Interspecific interactions in plastispheres were analysed using topological statistics (e.g., node degree and clustering coefficient) to identify biological interactions, habitat preferences, and physiological characteristics (Chen et al., 2020c; Yang et al., 2019). Freshwater plastisphere interspecific interactions have shown that bacterial networks on polyvinyl chloride have a higher clustering coefficient than those on wood, and when compared to natural substrates, plastics can be colonised by more species with similar functions (Miao et al., 2021).

6.5 FACTORS INFLUENCING DISTRIBUTION AND ABUNDANCES OF FRESHWATER PLASTISPHERE

The physical and chemical characteristics of plastic debris, such as polymer type, surface texture, and the presence of additives, lay the foundation for the plastisphere and are universally acknowledged as the primary determinants of microbial colonisation and the subsequent development of the plastisphere (Zettler et al., 2013; Oberbeckmann et al., 2015). While these studies have made significant strides in identifying these relationships, there is a notable lack of mechanistic understanding of how specific plastic properties facilitate or inhibit microbial colonisation. Furthermore, Debroy et al. (2021) building on previous works introduced the critical role of biofilms in the degradation of microplastics, suggesting that the interplay between plastic properties and microbial biofilm formation could be pivotal in determining the fate of microplastics in aquatic environments. This perspective invites a deep investigation into biofilm–plastic interactions, a relatively underexplored domain with potential for mitigating microplastic pollution.

Environmental conditions, including temperature, precipitation, light availability, nutrient levels, geographic location, and water flow, including the physicochemical properties of the plastic substrate, heavily influence the distribution and abundance of the plastisphere (Barros and Seena, 2021). Rainfall and wind patterns act as natural forces that disperse and deposit these particles across various terrains (see Figure 6.2; Wright et al., 2021). Furthermore, water properties with nutrient elements such as carbon, phosphorus, and nitrogen have been shown to have a greater effect on microbial carbon metabolic functions than the plastic substrate type of the freshwater plastisphere (Chen et al., 2020c; Miao et al., 2020). These factors affect the viability and metabolism of microbial colonisers and influence the physical integrity and chemical leaching behaviour of plastics in freshwater environments. Studies have demonstrated that environmental variables markedly affect the microbial composition of plastics, often overshadowing the type of plastic material (e.g., Amaral-Zettler et al., 2020; Wright et al., 2020c). Thus, fluctuations in environmental conditions can lead to seasonal and spatial variations in plastisphere communities (Amaral-Zettler et al., 2020).

Evidence of the influence of biogeochemical conditions or redox conditions on plastisphere community structure has been observed in marine environments, in which plastisphere bacterial communities are related to locality-related physicochemical parameters such as salinity, pH, temperature, and dissolved oxygen (De Tender et al., 2015; Delacuvellerie et al., 2022; Dodhia et al., 2023). Thus, the geographical and hydrological features of river systems, particularly those in urbanised and industrial regions, play a pivotal role in the influx of microplastics into freshwater ecosystems, forming the bedrock of the plastisphere. In urbanised watersheds, the extent of urbanisation was found to correlate with changes in microbial community structure and antibiotic-resistance gene (ARG) profiles, highlighting the interaction between locality (i.e., human activity) and natural processes in shaping plastisphere communities (Zhu et al., 2023). To date, little research has been conducted to investigate the role of geographic location in relation to redox and (bio)geochemical conditions which define conditions such as pH, available ions, and oxygen, on the plastisphere community structure and/or diversity in freshwater plastisphere. Meijer et al. (2021) demonstrated that these areas are significant contributors to the global flux of microplastics, acting as conduits that channel waste from urban landscapes into the natural environment, subsequently shaping the biogeochemical conditions, redox, biological responses, spatial patterns, and density of the plastisphere. Although significant strides have been made to understand these dynamics, considerable gaps remain in our knowledge. Addressing these gaps through interdisciplinary research is essential to comprehensively understand the plastisphere and its implications for freshwater ecosystems and beyond.

The physicochemical properties of plastisphere substrates have been suggested to influence the distribution and microbial diversity of freshwater plastispheres (Barros and Seena, 2021; Miao et al., 2023). For example, different plastic types have been hypothesised to affect freshwater plastisphere bacterial and fungal community structures differently (Miao et al., 2023). The resistance of plastic debris to corrosion and microbial degradation is one of the factors that facilitate plastisphere formation in freshwater environments (Barros and Seena, 2021). In addition, the texture of plastic debris might determine the degree of its adsorption to certain microbial species, although the substrate is likely to change with age (Harrison et al., 2018). For example, recent findings have highlighted the influence of various types of plastics on forming distinct microbial communities within isolated freshwater ecosystems (Miao et al., 2023). In contrast, Wallbank et al. (2022) highlighted that immersion time, rather than substrate type and/or composition, was an important determinant of plastisphere community composition within marine environments, and this might likely be the same scenario within freshwater ecosystems. However, as in marine environments, it is possible that the freshwater plastisphere does not selectively thrive on plastic substrates to gain direct metabolic advantages but rather uses plastic surfaces as attachment substrates to form generalist microbial communities (Wallbank et al., 2022). While several studies have indicated that plastic size influences plastisphere communities (e.g., Debroas et al., 2017), other studies have shown that plastic size and shape have little influence on plastisphere abundance and diversity (e.g., Cheng et al., 2021; Delacuvellerie et al., 2022). Thus, further research is needed to better understand the relationship between substrate size and microbial communities in freshwater plastids. In contrast, substrate density may affect whether the plastisphere is transported (i.e., dispersed) through benthic or pelagic courses in freshwater systems (Barros and Seena, 2021). For example, biofouling by microorganisms can increase plastic density, causing the plastic (i.e., plastisphere) to sink to the bottom (Rummel et al., 2017).

Plastic debris can be beneficial to microbes by providing carbon and nutrients leached from organic additives that are added during the manufacturing of plastics (Wright et al., 2020c; Wallbank et al., 2022). Additives released during leaching can act as nutrient sources for plastisphere microbial communities, thus promoting their growth (Wen et al., 2015; Wallbank et al., 2022). However, plastic debris, particularly through the leaching of additives, can be a source of toxicants and thus negatively impact colonising microbes (Wallbank et al., 2022). This could also have negative implications for lower trophic levels because microbes constitute the base of the food web in aquatic environments. Therefore, the plastisphere of nutrient-rich microbial communities can be favoured by primary consumers, thereby facilitating the ingestion of plastisphere (Rummel et al., 2017). Although research on this type is most common for marine plastispheres (e.g., Wright et al., 2020a,b; Wallbank et al., 2022), there is also evidence of this pattern in freshwater (e.g., Wen et al., 2015).

6.6. PLASTIC COLONISATION AND BIODEGRADATION

The ubiquitous distribution of plastics is evident in all communities and societies worldwide (Hoseini and Bond, 2022). The presence and prevalence of plastic are not only bound by social or economic boundaries but rather by physical factors such as water and air movements. Although the sources of plastics maybe industrial and residential areas where there is poor waste management in place, the longer plastics remain in the environment, making them more susceptible to degradation. The degradation and denaturation of plastics lead to contamination and pollution of the natural environment, even in remote ecosystems.

6.6.1 Biodegradation in Freshwater Systems

The diversity of freshwater environments enables the formation of various ecological niches, including freshwater plastispheres, where organisms can thrive. Freshwater plastispheres and microbes have developed an array of methods not only to survive but also to thrive. While many plastics are not easily degraded, some freshwater microbes can degrade various types of plastics in various ways (Amaral-Zettler et al., 2020).

6.6.2 Freshwater Plastisphere and Pollutants

Freshwater plastics have the potential to adsorb and concentrate contaminants in the ambient environment. This, in turn, makes the contaminants bioavailable to microbes and the surrounding environment. Freshwater plastispheres are affected by organic compounds and pathogens.

6.6.2.1 Chemical Composition and Additives

Plastics are chemically synthesised polymers primarily composed of carbon and hydrogen atoms forming long chains that can incorporate various additives or adsorb environmental contaminants, thus affecting their interactions with microorganisms (Badea et al., 2023). These polymers are often infused with plasticizers, flame retardants, and ultraviolet stabilisers to modify their properties and enhance flexibility, durability, colour, and heat and environmental degradation resistance. Such modifications serve functional purposes and have implications for microbial colonisation. Polymers, such as PE, PP, and PS, exhibit diverse biodegradability levels, influencing the selection process of microbial assemblages that can utilise these materials as carbon sources (Bos et al., 2023). This biodegradability factor is crucial for determining the types of microbial communities that establish themselves on different plastics. Chouchene et al. (2023) elaborated on this aspect, highlighting that the interaction between microorganisms and plastics is not merely a surface-level phenomenon, but a complex process influenced by the inherent chemical nature of the polymers and their additives. These compounds, while enhancing the functional properties of plastics, also influence microbial colonisation and activity. Certain additives may serve as carbon sources for microbes, facilitating their growth and metabolic activity on plastic surfaces. However, as noted in the literature, leaching of these additives can introduce toxic substances into the aquatic environment, affecting the health and diversity of microbial communities (Zettler et al., 2013). Chouchene et al. (2023) further noted that these chemicals can act as selective pressures, shaping the microbial diversity of plastics and influencing the ecological roles of these communities in aquatic ecosystems.

6.6.2.2 Hydrophobic Organic Compounds (HOCs)

HOCs include an array of organic and inorganic chemicals, such as pesticides, herbicides, industrial chemicals, and pharmaceuticals. Their hydrophobic nature facilitates their partitioning from water to the plastic surface. Some bacteria can degrade HOCs via metabolic processes, making carbon available to themselves and breaking down the less readily degradable HOCs.

6.6.2.3 Microbial Pathogens

Pathogenic bacteria and protozoans from the ambient environment have the potential to adhere to plastics and be taken up by microbes. Microbes in freshwater plastispheres can prey on plastics-associated pathogens. The breakdown of pathogens releases nutrients into the ambient environment. One area of great interest in understanding the functionality of the freshwater plastisphere community is microplastic harbouring pathogenic microbes. Barros and Seena (2021) noted that we still lack a considerable understanding of the feasibility of freshwater plastisphere microplastics in transporting pathogenic microorganisms; however, in their review, they included a table citing 17 studies on species diversity in the freshwater plastisphere, eight of which found a significant assemblage of pathogenic bacteria or fungi (Barros and Seena, 2021). The breakdown of freshwater plastisphere water body types that detected pathogenic microbes on microplastics in that study was seven of nine river systems studied, one of four lake systems studied, and zero of three river and lake combinations studied (Barros and Seena, 2021). Freshwater plastisphere studies that have detected pathogenic bacteria on microplastics often find them with plastic-degrading bacteria (McCormick et al., 2014, 2016, Hoellein et al., 2017), in disproportionate numbers compared to the surrounding environment (Di Pippo et al., 2022; Kelly et al., 2021; Gong et al., 2019), or potentially using microplastics to vector pathogenic bacteria or antibiotic resistance (Amaral-Zettler et al., 2015; Di Pippo et al., 2022; Parathasarathy et al., 2019; Wu et al., 2019; Xue et al., 2020; Witsø et al., 2023). Xue et al. (2020) investigated freshwater plastisphere communities as a gradient along a single river and noted that the downstream microplastic communities were more heavily influenced by upstream microbes, whereas the downstream planktonic bacteria were more influenced by human activities. This demonstrated that microbes in the plastisphere could survive in water with high levels of antibiotics when non-plastisphere microbes could not (Xue et al., 2020). Balcázar et al. demonstrated a similar pattern by showing planktonic pathogens could survive antibiotic pressure when associated with biofilms but could not be outside of the biofilms (Balcázar et al., 2015). Such patterns enable microplastic communities to have broader functional diversity, evade pollutant antibiotics, and increase the survival of pathogenic bacteria downstream (Xue et al., 2020). They also found evidence that microplastics had pathogens associated with septicaemia, nosocomial infections, and diarrhoea in significantly higher numbers than the surrounding water (Xue et al., 2020). Witsø et al. (2023) took a different approach and sampled a river in Norway between June and September and found that the bacterial communities differed greatly depending on the month and location.

6.6.2.4 Pathogenic Microbes Found Associated with the Freshwater Plastisphere

Starting with eukaryotic pathogens, various fungal pathogens have been detected on microplastics from freshwater plastispheres, as there is evidence that freshwater microplastics typically host different flora of fungi than other substrates, such as rock and wood (Kettner et al., 2019), and the fungal communities on microplastics from the freshwater plastisphere differ significantly from the surrounding water (Xue et al., 2021). Kettner et al. (2019) found multiple fungi that are closely related to and therefore likely parasitic eukaryotes, zoosporic fungi from the genera *Pythium* and *Lagenidium*, and the fungus-like eukaryote Peronosporomycetes from the Oomycota. Xue et al. (2020) found strong colonial differentiation between plastic debris and the surrounding water for three types of pathogenic fungi, *Alternaria*, *Cladosporium*, and *Rhodotorul*a, in freshwater from an urban river system. Their study focused on plastic debris, not on microplastics; however, they discussed the threat of ingesting pathogenic fungi via ingestion of microplastics that are colonised by such fungi. Two types of dinoflagellates were found to be associated with microplastics by Kettner et al. (2019): a *Pfiesteria*-like species and *Heterocapsa* spp. Both dinoflagellates are associated with toxic tides and the fish die off.

To date, many studies have identified types of bacteria associated with pathogenicity in freshwater plastispheres living on microplastics. This list includes the following bacteria: (1) *Aeromonas*, *Arcobacter*, and Campylobacteraceae, which are known to cause diarrhoea and other gastrointestinal distress (McCormick et al., 2014); (2) *Pseudomonas* and *Acinetobacter* (McCormick et al., 2014; Xue et al., 2020); (3) known human pathogens *Pseudomonas monteilii* (*pneumonitis* and *bronchiectasis*) and *Pseudomonas mendocina* (spondyliscitis) (Wu et al., 2019); (4) *Pseudoanabaena*, *Leptolyngbya*, *Calothrix*, and *Phormidium* which are toxic cyanobacteria genera (Di Pippo et al., 2020); (5) the fish pathogen *Flavobacterium tructae* (Reisoglu et al., 2024); (6) Di Pippo et al. (2022) demonstrated that the freshwater plastispheres they studied from various Italian lakes showed microplastics with the presence of Legionella spp., *Pseudomonas aeruginosa*, and *Salmonella* spp., all of which are biofilm forming pathogens; and lastly (7) Witsø et al. (2023) used 16S sequencing to find the genera *Escherichia/Shigella*, *Listeria*, *Aeromona*, *Bacillus*, and *Enterobacter* and they confirmed pathogenicity using microbial identification techniques of the species *L. monocytogenes*, *E. coli*, *Providencia rettgeri*, *Citrobacter freundii*, *Mammaliicoccus vitulinus*, *Klebsiella pneumoniae*, and *Enterbacter* spp. This is likely just the beginning of our understanding of how bacteria can thrive on microplastics beyond environmental water, and the issue of microplastics vectoring pathogenic bacteria could become a serious problem in the future, especially in societies that lack purification technology or when the water is used for irrigation purposes (e.g., horticulture).

A few studies have investigated vectored pathogenic viruses in the freshwater plastisphere. Two viruses of concern contributing to foodborne illness were found to be associated with microplastics in the September sampling of the Leir River: Norovirus NoV GII and Adenovirus (Witsø et al., 2023). Li et al. (2022) studied viral load on microplastics in the Beilun River using metagenomic analysis and identified 1,719 species of viruses in five types of microplastics. Of the five microplastic types tested, PP had the most viral species at 1,372 *vs* 72 found in surface water and 84 found in sediment from the same area, demonstrating that microplastics are an important source of viral dispersal. Due to the nature of their study, Li et al. (2022) focused on viral metagenomic pathways, virulence factors, and ARGs. Among all the pathways identified, they identified genes that contribute to the following pathways of concern: endocrine and metabolic diseases, cancer, and infectious diseases (Li et al., 2022). More work is needed concerning the transmission of viruses in microplastics as part of the freshwater plastisphere.

Even smaller than viruses, a growing concern with the freshwater plastisphere is the ability of microplastics to transmit ARGs) and mobile genetic elements (MGE). Antibiotic resistance in the environment is one of the largest environmental health issues we face in this century (Yang et al., 2020a; Larsson and Flach, 2022) and anyway this is facilitated by horizontal gene transfer through the freshwater plastisphere is of concern (Yang et al., 2020a; Wang et al., 2020). Many studies have demonstrated the presence of ARGs in freshwater plastispheres (Wang et al., 2020; Li et al., 2022).

The functional role of freshwater plastisphere communities may also lead to positive outcomes. As different microbes are associated with microplastics, there exists the possibility of evolution towards a more efficient breakdown of plastics by microbes. In many of the broad genetic-based taxonomic surveys of microbes associated with microplastics in the freshwater plastisphere, they identify microbes known to degrade plastics (Mishra et al., 2024; Bocci et al. 2024, Reisoglu et al., 2024) in a new field some call "plastic-omics" (Viljakainen and Hug, 2021). One new shift in the study of microbial plastic degradation is the examination of the plastisphere as a "degradation unit" (Bocci et al., 2024) in which researchers take a broader look at the plastisphere to understand how microbial taxa can work together in enzyme production to increase polymer biodegradation (Li et al., 2023a, b; Mishra et al., 2024). A broader understanding of the possibility of a microbial degradation unit may allow for better management of plastic pollution. Mishra et al. (2024) suggested that the three future directions of this research should include a better understanding of the geographic distribution and structure of microbial communities in the freshwater plastisphere, and an understanding of how different microbes and communities contribute to plastic degradation.

6.7 RESEARCH CHALLENGES AND FUTURE DIRECTIONS

Plastisphere includes microbes that are persistent and can resist adverse conditions and develop complex 3D structures (i.e., extracellular polymeric structures). Total DNA extraction and sequencing are relative and not absolute measurements; hence, the measurements are compositional (Wright et al., 2020b). Although cell and biomass estimation is straightforward for planktonic systems, quantification of microbes has not been resolved for microbes growing in aggregates or on surfaces (Azeredo et al., 2017). These authors noted further challenges in biofilm and microbe analyses. Choosing an experimental platform for biofilm experiments requires careful consideration. The experiments will be influenced by platform selection, each with advantages and limitations.

- The Robbins device and drip flow reactors have little opportunity to follow biofilm development in real time.
- Biofilm mass and viability are measured differently and are dependent on microbial and molecular methods, as well as biofilm properties.
- Estimation of biofilm cell viability is based on colony-forming units on agar, which has drawbacks and limitations.
- Quantification of biofilms using quantitative polymerase chain reaction (qPCR) can overestimate the number of viable cells.

Despite the promising potential of biodegradation to mitigate plastic pollution in freshwater ecosystems, several challenges remain.

- **Limited Knowledge:** Our understanding of the specific microbial communities and metabolic pathways involved in plastic degradation in freshwater environments is still developing.
- **Environmental Variability:** The effectiveness of biodegradation can vary significantly depending on the specific freshwater ecosystem and the environmental conditions.
- **Upscaling Biodegradation Technologies:** Developing and scaling up biodegradation technologies for practical application in freshwater systems requires further research and investment.

Studying the plastisphere in freshwater environments presents a range of challenges due to the unique properties of both the plastics and the freshwater ecosystems. Several challenges exist that researchers face when investigating the plastisphere in freshwater environments such as the complexity of microbial communities, plastic composition and ageing, dynamic environmental conditions, contamination and sampling challenges, unknown ecological interactions, and limited baseline conditions. For example, freshwater ecosystems harbour highly diverse microbial communities, which vary significantly depending on location (e.g., rivers, lakes/reservoirs, and wetlands), thus identifying and characterising the microbial species that colonise plastic surfaces can be difficult due to the vast diversity of bacteria, algae, and fungi in these environments. Furthermore, in freshwater, biofilms form not only on plastic but also on various natural surfaces (i.e., rocks, sediments, and plants). Therefore, understanding the differences in colonisation between plastics and natural substrates is complex. Another important element is the diverse polymer types (e.g., PE and PS), each with varying degradation rates and surface characteristics that could potentially influence microbial colonisation and biofilm formation. It is also important to consider that rivers and streams have dynamic environmental conditions with strong currents and turbulence that can influence the transportation, breakdown, and biofilm formation of plastics. These conditions differ greatly from stagnant waters in lakes/reservoirs and/or wetlands, adding variability to studies. Plastisphere biofilms interact with freshwater flora and fauna in ways that are still poorly understood. Plastics may serve as artificial substrates for colonisation by organisms that would normally attach to natural surfaces, potentially affecting nutrient cycling and local food webs. Furthermore, plastics in freshwater can leach toxic additives, which may affect microbial colonisation and alter community composition. Therefore, understanding how these chemical interactions shape the plastisphere in different freshwater contexts is a key challenge.

REFERENCES

Amaral-Zettler, L.A., Zettler, E.R., and Mincer, T.J., 2020. Ecology of the plastisphere. *Nature Reviews Microbiology*, 18(3), 139–151.

Amaral-Zettler, L.A., Zettler, E.R., Slikas, B., Boyd, G.D., Melvin, D.W., Morrall, C.E., Proskurowski, G., and Mincer, T.J., 2015. The biogeography of the Plastisphere: implications for policy. *Frontiers in Ecology and the Environment*, 13(10), 541–546.

Azeredo, J., Azevedo, N.F., Briandet, R., Cerca, N., Coenye, T., Costa, A.R., Desvaux, M., Di Bonaventura, G., Hébraud, M., Jaglic, Z., and Kačániová, M., 2017. Critical review on biofilm methods. *Critical Reviews in Microbiology*, 43(3), 313–351.

Badea, M. A., Balas, M., Dinischiotu, A. 2023. Microplastics in freshwaters: implications for aquatic autotrophic organisms and fauna health. *Microplastics*, 2, 39–59.

Balcázar, J.L., Subirats, J., and Borrego, C.M. 2015. The role of biofilms as environmental reservoirs of antibiotic resistance. *Frontiers in Microbiology*, 6, 1216.

Barros, J., and Seena, S., 2021. Plastisphere in freshwaters: an emerging concern. *Environmental Pollution*, 290, 118123.

Bocci, V., Galafassi, S., Levantesi, C., Crognale, S., Amalfitano, S., Congestri, R., Matturro, B., Rossetti, S., and Di Pippo, F., 2024. Freshwater plastisphere: a review on biodiversity, risks, and biodegradation potential with implications for the aquatic ecosystem health. *Frontiers in Microbiology*, 15, 1395401.

Bos, R.P., Kaul, D., Zettler, E.R., Hoffman, J.M., Dupont, C.L., Amaral-Zettler, L.A., and Mincer, T.J., 2023. Plastics select for distinct early colonizing microbial populations with reproducible traits across environmental gradients. *Environmental Microbiology*, 25(12), 2761–2775.

Bryant, J.A., Clemente, T.M., Viviani, D.A., Fong, A.A., Thomas, K.A., Kemp, P., Karl, D.M., White, A.E., and DeLong, E.F., 2016. Diversity and activity of communities inhabiting plastic debris in the North Pacific Gyre. *mSystems*, 1(3), e00024.

Cao, J., Yang, Q., Jiang, J., Dalu, T., Kadushkin, A., Singh, J., Fakhrullin, R., Wang, F., Cai, X., and Li, R., 2022. Coronas of micro/nano plastics: a key determinant in their risk assessments. *Particle and Fibre Toxicology*, 19(1), 55.

Chen, C.C., Dai, L., Ma, L., and Guo, R.T., 2020a. Enzymatic degradation of plant biomass and synthetic polymers. *Nature Reviews Chemistry*, 4(3), 114–126.

Chen, J., Wang, P., Wang, C., Wang, X., Miao, L., Liu, S., Yuan, Q., and Sun, S., 2020b. Fungal community demonstrates stronger dispersal limitation and less network connectivity than bacterial community in sediments along a large river. *Environmental Microbiology*, 22(3), 832–849.

Chen, X., Chen, X., Zhao, Y., Zhou, H., Xiong, X., and Wu, C., 2020c. Effects of microplastic biofilms on nutrient cycling in simulated freshwater systems. *Science of the Total Environment*, 719, 137276.

Cheng, J., Jacquin, J., Conan, P., Pujo–Pay, M., Barbe, V., George, M., Fabre, P., Bruzaud, S., Ter Halle, A., Meistertzheim, A.L., and Ghiglione, J.F., 2021. Relative influence of plastic debris size and shape, chemical composition and phytoplankton–bacteria interactions in driving seawater plastisphere abundance, diversity and activity. *Frontiers in Microbiology*, 11, 610231.

Chouchene, K., Pinto da Costa, J., Chamkha, M., Ksibi, M., and Sayadi, S., 2023. Effects of microplastics' physical and chemical properties on aquatic organisms: state-of-the-art and future research trends. *Trends in Analytical Chemistry*, 156, 117192.

Dalu, T., Malesa, B., and Cuthbert, R.N., 2019. Assessing factors driving the distribution and characteristics of shoreline macroplastics in a subtropical reservoir. *Science of the Total Environment*, 696, 133992.

De Tender, C.A., Devriese, L.I., Haegeman, A., Maes, S., Ruttink, T., and Dawyndt, P., 2015. Bacterial community profiling of plastic litter in the Belgian part of the North Sea. *Environmental Science and Technology*, 49(16), 9629–9638.

Debroas, D., Mone, A., and Ter Halle, A., 2017. Plastics in the North Atlantic garbage patch: a boat-microbe for hitchhikers and plastic degraders. *Science of the Total Environment*, 599, 1222–1232.

Debroy, A., George, N., and Mukherjee, G., 2021. Role of biofilms in the degradation of microplastics in aquatic environments. *Journal of Chemical Technology and Biotechnology*, 97(12), 3271–3282.

Delacuvellerie, A., Ballerini, T., Frère, L., Matallana–Surget, S., Dumontet, B., and Wattiez, R., 2022. From rivers to marine environments: a constantly evolving microbial community within the plastisphere. *Marine Pollution Bulletin*, 179, 113660.

Di Pippo, F., Crognale, S., Levantesi, C., Vitanza, L., Sighicelli, M., Pietrelli, L., Di Vito, S., Amalfitano, S., and Rossetti, S., 2022. Plastisphere in lake waters: microbial diversity, biofilm structure, and potential implications for freshwater ecosystems. *Environmental Pollution*, 310, 119876.

Di Pippo, F., Venezia, C., Sighicelli, M., Pietrelli, L., Di Vito, S., Nuglio, S., and Rossetti, S., 2020. Microplastic-associated biofilms in lentic Italian ecosystems. *Water Research*, 187, 116429.

Dodhia, M.S., Rogers, K.L., Fernández-Juárez, V., Carreres-Calabuig, J.A., Löscher, C.R., Tisserand, A.A., Keulen, N., Riemann, L., Shashoua, Y., and Posth, N.R., 2023. Microbe-mineral interactions in the Plastisphere: coastal biogeochemistry and consequences for degradation of plastics. *Frontiers in Marine Science*, 10, 1134815.

Gong, M., Yang, G., Zhuang, L., and Zeng, E.Y., 2019. Microbial biofilm formation and community structure on low-density polyethylene microparticles in lake water microcosms. *Environmental Pollution*, 252, 94–102.

González-Pleiter, M., Velázquez, D., Casero, M.C., Tytgat, B., Verleyen, E., Leganés, F., Rosal, R., Quesada, A., and Fernández-Piñas, F., 2021. Microbial colonizers of microplastics in an Arctic freshwater lake. *Science of the Total Environment*, 795, 148640.

Harrison, J.P., Hoellein, T.J., Sapp, M., and Tagg, A.S., 2018. Microplastic-associated biofilms: a comparison of freshwater and marine environments. In: Wagner, M., and Lambert, S., (eds) *Freshwater Microplastics. The Handbook of Environmental Chemistry*, Volume 58. Springer, Cham.

Hoellein, T.J., McCormick, A.R., Hittie, J., London, M.G., Scott, J.W., and Kelly, J.J., 2017. Longitudinal patterns of microplastic concentration and bacterial assemblages in surface and benthic habitats of an urban river. *Freshwater Science*, 36(3), 491–507.

Hoellein, T.J., Rojas, M., Pink, A., Gasior, J., and Kelly, J.J., 2014. Anthropogenic litter in urban freshwater ecosystems: distribution and microbial interactions. *PLoS One*, 9(6), e98485.

Hoseini, M., and Bond, T., 2022. Predicting the global environmental distribution of plastic polymers. *Environmental Pollution*, 300, 118966.

Kaushal, J., Khatri, M., and Arya, S.K., 2021. Recent insight into enzymatic degradation of plastics prevalent in the environment: a mini-review. *Cleaner Engineering and Technology*, 2, 100083.

Kelly, J.J., London, M.G., McCormick, A.R., Rojas, M., Scott, J.W., and Hoellein, T.J., 2021. Wastewater treatment alters microbial colonization of microplastics. *PloS ONE*, 16(1), e0244443.

Kettner, M.T., Oberbeckmann, S., Labrenz, M., and Grossart, H.P., 2019. The eukaryotic life on microplastics in brackish ecosystems. *Frontiers in Microbiology*, 10, 538.

Kiss, T., Fórián, S., Szatmári, G., and Sipos, G., 2021. Spatial distribution of microplastics in the fluvial sediments of a transboundary river–a case study of the Tisza River in Central Europe. *Science of the Total Environment*, 785, 147306.

Kumar, R., Sharma, P., Verma, A., Jha, P.K., Singh, P., Gupta, P.K., Chandra, R., and Prasad, P.V., 2021. Effect of physical characteristics and hydrodynamic conditions on transport and deposition of microplastics in riverine ecosystem. *Water*, 13(19), 2710.

Larsson, D.G., and Flach, C.F., 2022. Antibiotic resistance in the environment. *Nature Reviews Microbiology*, 20(5), 257–269.

Lear, G., Kingsbury, J.M., Franchini, S., Gambarini, V., Maday, S.D.M., Wallbank, J.A., Weaver, L., and Pantos, O., 2021. Plastics and the microbiome: impacts and solutions. *Environmental Microbiome*, 16, 1–19.

Li, C., Wang, L., Ji, S., Chang, M., Wang, L., Gan, Y., and Liu, J., 2021. The ecology of the plastisphere: microbial composition, function, assembly, and network in the freshwater and seawater ecosystems. *Water Research*, 202, 117428.

Li, K., Jia, W., Xu, L., Zhang, M., and Huang, Y., 2023a. The plastisphere of biodegradable and conventional microplastics from residues exhibit distinct microbial structure, network and function in plastic-mulching farmland. *Journal of Hazardous Materials*, 442, 130011.

Li, R., Zhu, L., Cui, L., and Zhu, Y.G., 2022. Viral diversity and potential environmental risk in microplastic at watershed scale: Evidence from metagenomic analysis of plastisphere. *Environment International*, 161, 107146.

Li, W., Miao, L., Adyel, T.M., Wu, J., Yu, Y., and Hou, J., 2023b. Characterization of dynamic plastisphere and their underlying effects on the aging of biodegradable and traditional plastics in freshwater ecosystems. *Journal of Hazardous Materials*, 446, 130714.

McCormick, A., Hoellein, T.J., Mason, S.A., Schluep, J., and Kelly, J.J., 2014. Microplastic is an abundant and distinct microbial habitat in an urban river. *Environmental Science and Technology*, 48(20), 11863–11871.

McCormick, A.R., Hoellein, T.J., London, M.G., Hittie, J., Scott, J.W., and Kelly, J.J., 2016. Microplastic in surface waters of urban rivers: concentration, sources, and associated bacterial assemblages. *Ecosphere*, 7(11), e01556.

Meijer, L.J.J., van Emmerik, T., van der Ent, R., Schmidt, C., and Lebreton, L., 2021. More than 1000 rivers account for 80% of global riverine plastic emissions into the ocean. *Science Advances*, 7(18), eaaz5803.

Mendoza, L.M.R., and Balcer, M., 2019. Microplastics in freshwater environments: a review of quantification assessment. *TrAC Trends in Analytical Chemistry*, 113, 402–408.

Miao, L., Li, W., Adyel, T.M., Yao, Y., Deng, Y., Wu, J., Zhou, Y., Yu, Y., and Hou, J., 2023. Spatio-temporal succession of microbial communities in plastisphere and their potentials for plastic degradation in freshwater ecosystems. *Water Research*, 229, 119406.

Miao, L., Wang, C., Adyel, T.M., Wu, J., Liu, Z., You, G., Meng, M., Qu, H., Huang, L., Yu, Y., and Hou, J., 2020. Microbial carbon metabolic functions of biofilms on plastic debris influenced by the substrate types and environmental factors. *Environment International*, 143, 106007.

Miao, L., Wang, C., Adyel, T.M., Zhao, J., Yan, N., Wu, J., and Hou, J., 2021. Periphytic biofilm formation on natural and artificial substrates: comparison of microbial compositions, interactions, and functions. *Frontiers in Microbiology*, 12, 684903.

Mishra, S., Dash, D., and Das, A.P., 2024. Aquatic microbial diversity on plastisphere: colonization and potential role in microplastic biodegradation. *Geomicrobiology Journal*, 41(4), 312–323.

Oberbeckmann, S., Löder, M.G., and Labrenz, M., 2015. Marine microplastic-associated biofilms–a review. *Environmental Chemistry*, 12(5), 551–562.

Owowenu, E.K., Nnadozie, C.F., Akamagwuna, F., Noundou, X.S., Uku, J.E., and Odume, O.N., 2023. A critical review of environmental factors influencing the transport dynamics of microplastics in riverine systems: implications for ecological studies. *Aquatic Ecology*, 57(2), 557–570.

Reisoglu, Ş., Cati, C., Yurtsever, M., and Aydin, S., 2024. Evaluation of prokaryotic and eukaryotic microbial communities on microplastic-associated biofilms in marine and freshwater environments. *Engineering in Life Sciences*, 24(6), 2300249.

Ren, X., Sun, Y., Wang, Z., Barceló, D., Wang, Q., Zhang, Z., and Zhang, Y., 2020. Abundance and characteristics of microplastic in sewage sludge: a case study of yangling, Shaanxi Province, China. *Case Studies in Chemical and Environmental Engineering*, 2, 100050.

Ríos, J.M., Tesitore, G., and Teixeira de Mello, F., 2022. Does color play a predominant role in the intake of microplastics fragments by freshwater fish: an experimental approach with *Psalidodon eigenmanniorum*. *Environmental Science and Pollution Research*, 29(32), 49457–49464.

Rummel, C.D., Jahnke, A., Gorokhova, E., Kühnel, D., and Schmitt-Jansen, M., 2017. Impacts of biofilm formation on the fate and potential effects of microplastic in the aquatic environment. *Environmental Science and Technology Letters*, 4(7), 258–267.

Sauer, K., Stoodley, P., Goeres, D.M., Hall–Stoodley, L., Burmølle, M., Stewart, P.S., and Bjarnsholt, T., 2022. The biofilm life cycle: expanding the conceptual model of biofilm formation. *Nature Reviews Microbiology*, 20(10), 608–620.

Shaw, R., Halder, U., Chaudhuri, P., and Bandopadhyay, R., 2023. Antarctica microbial communities: ecological and industrial importance. In: Soni, R., Suyal, D.C., Morales–Oyervides, L., and Fouillaud, M., (eds) *Current Status of Marine Water Microbiology*. Springer Nature, Singapore.

Shi, X., Chen, Z., Wei, W., Chen, J., and Ni, B.J., 2023. Toxicity of micro/nanoplastics in the environment: roles of plastisphere and eco-corona. *Soil and Environmental Health*, 1(1), 100002.

Sun, Y., Shi, J., Wang, X., Ding, C., and Wang, J., 2022. Deciphering the mechanisms shaping the plastisphere microbiota in soil. *mSystems*, 7(4), e00352-22.

Tang, K.H.D., 2024. Terrestrial and aquatic plastisphere: formation, characteristics, and influencing factors. *Sustainability*, 16(5), 2163.

Tang, Y., Liu, Y., Chen, Y., Zhang, W., Zhao, J., He, S., Yang, C., Zhang, T., Tang, C., Zhang, C., and Yang, Z., 2021. A review: research progress on microplastic pollutants in aquatic environments. *Science of the Total Environment*, 766, 142572.

Viljakainen, V.R., and Hug, L.A., 2021. New approaches for the characterization of plastic-associated microbial communities and the discovery of plastic-degrading microorganisms and enzymes. *Computational and Structural Biotechnology Journal*, 19, 6191–6200.

Wallbank, J.A., Lear, G., Kingsbury, J.M., Weaver, L., Doake, F., Smith, D.A., Audrézet, F., Maday, S.D., Gambarini, V., Donaldson, L., and Theobald, B., 2022. Into the plastisphere, where only the generalists thrive: early insights in plastisphere microbial community succession. *Frontiers in Marine Science*, 9, 841142.

Walker, T.R., and Fequet, L., 2023. Current trends of unsustainable plastic production and micro (nano) plastic pollution. *TrAC Trends in Analytical Chemistry*, 160, 116984.

Wang, L., Tong, J., Li, Y., Zhu, J., Zhang, W., Niu, L., and Zhang, H., 2021. Bacterial and fungal assemblages and functions associated with biofilms differ between diverse types of plastic debris in a freshwater system. *Environmental Research*, 196, 110371.

Wang, S., Xue, N., Li, W., Zhang, D., Pan, X., and Luo, Y., 2020. Selectively enrichment of antibiotics and ARGs by microplastics in river, estuary and marine waters. *Science of the Total Environment*, 708, 134594.

Wang, Y., Liu, G., Wang, Y., Mu, H., Shi, X., Wang, C., and Wu, N., 2023. The global trend of microplastic research in freshwater ecosystems. *Toxics*, 11(6), 539.

Wen, G., Kotzsch, S., Vital, M., Egli, T., and Ma, J., 2015. BioMig–a method to evaluate the potential release of compounds from and the formation of biofilms on polymeric materials in contact with drinking water. *Environmental Science and Technology*, 49, 11659–11669.

Witsø, I.L., Basson, A., Vinje, H., Llarena, A.K., Bringas, C.S., Aspholm, M., Wasteson, Y., and Myrmel, M., 2023. Freshwater plastispheres as a vector for foodborne bacteria and viruses. *Environmental Microbiology*, 25(12), 2864–2881.

Wright, R.J., Bosch, R., Gibson, M.I., and Christie–Oleza, J.A., 2020a. Plasticizer degradation by marine bacterial isolates: a proteogenomic and metabolomic characterization. *Environmental Science and Technology*, 54, 2244–2256.

Wright, R.J., Erni-Cassola, G., Zadjelovic, V., Latva, M., and Christie-Oleza, J.A., 2020b. Marine plastic debris: a new surface for microbial colonization. *Environmental Science & Technology*, 54(19), 11657–11672.

Wright, R.J., Langille, M.G., and Walker, T.R., 2020c. Food or just a free ride? A meta-analysis reveals the global diversity of the plastisphere. *The ISME Journal*, 15(3), 789–806.

Wu, F., Ha, X., Wang, S., Li, J., and Gao, Y., 2024. Microplastic occurrences, transport, and quantification and associated effects on primary productivity and carbon cycling processes in freshwater ecosystems. *TrAC Trends in Analytical Chemistry*, 172, 117611.

Wu, X., Pan, J., Li, M., Li, Y., Bartlam, M., and Wang, Y., 2019. Selective enrichment of bacterial pathogens by microplastic biofilm. *Water Research*, 165, 114979.

Xue, N., Fang, Q., Pan, X., and Zhang, D., 2021. Distinct fungal plastisphere across different river functional zones: a watershed scale study. *Science of the Total Environment*, 752, 141879.

Xue, N., Wang, L., Li, W., Wang, S., Pan, X., and Zhang, D., 2020. Increased inheritance of structure and function of bacterial communities and pathogen propagation in plastisphere along a river with increasing antibiotics pollution gradient. *Environmental Pollution*, 265, 114641.

Yang, K., Chen, Q.L., Chen, M.L., Li, H.Z., Liao, H., Pu, Q., Zhu, Y.G., and Cui, L., 2020a. Temporal dynamics of antibiotic resistome in the plastisphere during microbial colonization. *Environmental Science and Technology*, 54(18), 11322–11332.

Yang, Y., Liu, G., Song, W., Ye, C., Lin, H., Li, Z., and Liu, W., 2019. Plastics in the marine environment are reservoirs for antibiotic and metal resistance genes. *Environment International*, 123, 79–86.

Yang, Y., Liu, W., Zhang, Z., Grossart, H.-P., and Gadd, G.M., 2020b. Microplastics provide new microbial niches in aquatic environments. *Applied Microbiology and Biotechnology*, 104(15), 6501–6511.

Yu, Y., Miao, L., Adyel, T.M., Waldschläger, K., Wu, J., and Hou, J., 2023. Aquatic plastisphere: interactions between plastics and biofilms. *Environmental Pollution*, 322, 121196.

Zettler, E.R., Mincer, T.J., and Amaral-Zettler, L.A., 2013. Life in the "plastisphere": microbial communities on plastic marine debris. *Environmental Science and Technology*, 47(13), 7137–7146.

Zhu, L., Li, R., Yang, K., Xu, F., Lin, C., Chen, Q., Zhu, D., Sun, Q., Zhu, Y.G., and Cui, L., 2023. Quantifying health risks of plastisphere antibiotic resistome and deciphering driving mechanisms in an urbanizing watershed. *Water Research*, 245, 120574.

7 Biofilm Dynamics and Environmental Implications on Plastic Surfaces

Ahamed Ashiq, Thilakshani Atugoda, and Yves Filion

7.1 INTRODUCTION

Microplastics (MPs), defined as plastic particles less than 5 mm in size, have become a pressing environmental concern due to their ubiquity and potential impacts on ecosystems. One crucial aspect of understanding MP pollution is the formation of biofilms and the subsequent microbial colonization on the plastic surfaces. Biofilms are complex communities of microorganisms, primarily bacteria and algae that attach to surfaces and encapsulate themselves in a self-produced, extracellular matrix. The formation of biofilms on MPs can have significant implications for the fate and transport of various pollutants in the environment. Microbes have the ability to adhere to various surfaces, such as natural materials, metals, and pharmaceuticals and personal care products (PPCPs) along with antibiotic resistance genes (ARGs) and pathogens, including synthetic polymeric surfaces of MPs, leading to the development of mature biofilms (Ramasamy & Lee, 2016). The process of MP biofouling, or the accumulation of organisms on the plastic surface, can increase the buoyancy of low-density MPs, causing them to sink or oscillate in the upper oceanic layers (Donlan, 2002; Munhoz et al., 2022). This has important consequences for the movement and distribution of MPs, as they may become incorporated into sinking zooplankton fecal pellets and marine snow, ultimately altering the pathways and fate of these pollutants (Kvale et al., 2023). Furthermore, the microbial colonization of MPs can have implications for the biodegradation and fragmentation of these materials. Certain microorganisms have been found to possess the ability to degrade various types of plastic, including MPs (Allemann et al., 2024). The biodegradation process leads to the breakdown of MPs into even smaller nanoplastics, exerting further impact on the environment and biota. Understanding the interactions between biofilms, MPs, and microbial communities is essential for developing effective strategies to mitigate the environmental impacts of MP pollution. Biofilm formation on the surfaces of MPs enhances their adhesiveness and density, largely due to the production of extracellular polymeric substances (EPSs) (Ghanadi et al., 2024). This process promotes the creation of aggregates and dense complexes that can sink through the water column, altering the distribution and fate of MPs (De-la-Torre, 2019; Kumar et al., 2021; Kvale et al., 2023; Munhoz et al., 2022). In summary, the interplay between biofilm formation and microbial colonization on MPs is a critical factor in determining the behavior, transport, and degradation of these pollutants in the environment. This chapter further explores the various aspects of this complex interaction. It discusses the process of biofilm formation on plastic surfaces, the role of biofilms in the biodegradation of MPs, the implications of biofilms as sinks and sources of pollutants, the facilitation of microbial transport by biofilms, and the challenges posed by antimicrobial and disinfectant resistance in biofilm–plastic systems. Continued research in this area is necessary to inform policies and responses to the growing problem of MP contamination.

DOI: 10.1201/9781032717814-7

7.2 BIOFILM–PLASTIC INTERACTIONS, MICROPLASTIC COLONIZATION, AND BIODEGRADATION RATES

Biofilms consist of diverse microorganisms embedded in an EPS matrix. These biofilms provide numerous benefits, including stable partnerships and protection from toxins and desiccation (Rummel et al., 2021; Tu et al., 2020). Biofilm development initiates with microbial adhesion to surfaces, transitioning from weak to strong interactions, forming irreversible bonds. As these cells grow, they create intricate biofilm communities. Outer biofilm layers can also facilitate cell dispersal, aiding the colonization of new environments (Renner & Weibel, 2011). The high surface area of MPs supports microbial attachment and growth (Zettler et al., 2013). This results in various microbial species coexisting closely, enhancing their interactions and symbiotic relationships through quorum sensing. This close coexistence of various microbial species enhances their interactions and symbiotic relationships, often facilitated by mechanisms such as quorum sensing. Quorum sensing allows microbes to communicate through signaling molecules, coordinating behaviors like biofilm formation and nutrient acquisition. Key factors such as surface roughness, charge, buoyancy, and the polymer composition of the plastic play a significant role in the early stages of biofilm formation. Research shows that the microbial diversity in these biofilms can vary considerably depending on the type of plastic polymer (Oberbeckmann et al., 2014, 2018). Other factors influencing biofilm establishment and development include surface availability, material morphology, and environmental conditions (Dussud et al., 2018; Jachimowicz et al., 2022; Xu et al., 2019).

Figure 7.1 illustrates the dynamic process of biofilm formation on plastic surfaces and the resultant changes in the properties of the plastic as it transitions from the early to the mature plastisphere. Initially, plastic in aquatic environments quickly acquires a conditioning film composed of inorganic and organic substances. This conditioning film facilitates the initial microbial attachment, which leads to biofilm formation. The conditioning film that forms on plastics or similar surfaces in aquatic environments is typically composed of a variety of organic and inorganic compounds, such as proteins, polysaccharides, lipids, and ions. This film is rapidly established within minutes

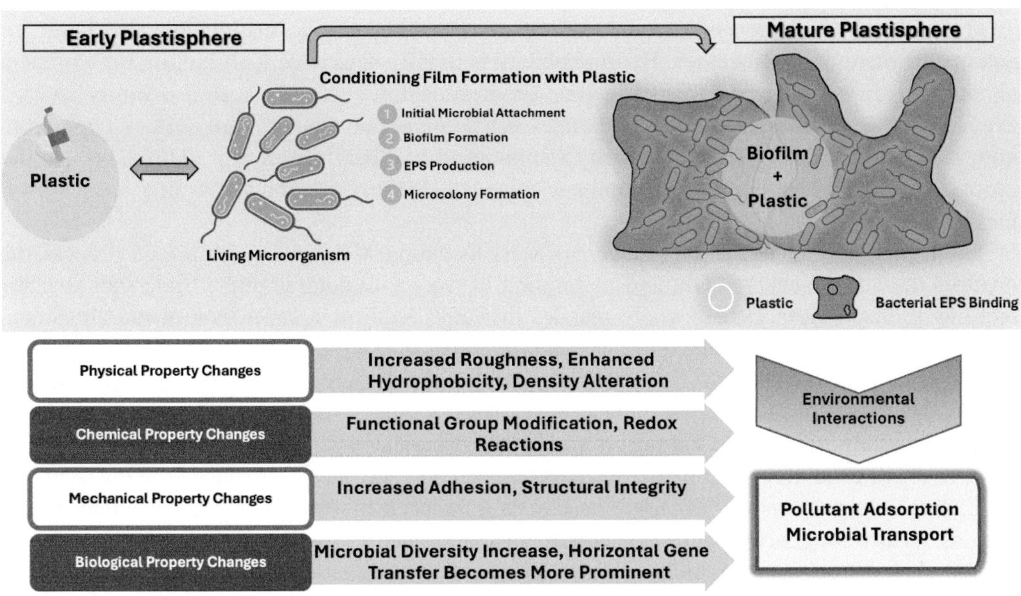

FIGURE 7.1 Progression of biofilm formation on plastic surfaces from early plastisphere stages to mature biofilms.

of exposure to water, altering the surface properties of the plastic by reducing its hydrophobicity and providing attachment sites for microbial colonizers. It acts as a crucial mediator between the inert plastic surface and the microbial community, facilitating the initial adhesion and subsequent biofilm development (Oberbeckmann et al., 2018). As the biofilm matures, microorganisms produce EPSs), creating a complex matrix that binds tightly to the plastic, forming microcolonies (Dussud et al., 2018; Rummel et al., 2021).

The interaction between biofilms and plastic results in significant alterations in the plastic's properties. Physically, the surface roughness and hydrophobicity of the plastic increase, along with changes in its density. Chemically, the biofilm induces modifications such as the introduction of functional groups and facilitation of redox reactions. Mechanically, the biofilm enhances the adhesion and structural integrity of the plastic. Biologically, the plastic surface becomes a niche for a diverse array of microbial communities. As the biofilm matures, microbial diversity increases, creating microenvironments that support different species with distinct metabolic activities. Moreover, this increased diversity fosters horizontal gene transfer (HGT), a critical process where microbes exchange genetic material, such as ARGs, through mechanisms like conjugation, transformation, and transduction. HGT becomes more prominent as microbial communities grow denser and more diverse, allowing the spread of adaptive traits that enhance the resilience and adaptability of the biofilm (Unnikrishnan et al., 2024).

Biofilm formation on the surfaces of MPs enhances their adhesiveness and density, largely due to the production of EPS. This process promotes the creation of aggregates and dense complexes (Sooriyakumar et al., 2022). In aquatic environments, these changes in MPs' buoyancy result in vertical transport, causing the complexes to sink to deeper water layers and concentrate in the sediment. This phenomenon could explain the higher concentrations of MPs observed in sediments compared to the water column (Ghanadi et al., 2024; Sri Hari et al., 2023).

In soil environments, similar patterns of accumulation occur as those observed in aquatic systems, where MPs aggregate in deeper layers. In soils, however, the process is driven by factors such as water infiltration from rainfall or irrigation and the presence of plant roots, which facilitate the movement of MPs into deeper soil layers. In both cases, biofilm formation alters the physical properties of MPs, such as their surface roughness and density, promoting vertical transport and accumulation in either the sediment or deeper soil strata (Dussud et al., 2018; Moyal et al., 2023; Tu et al., 2020). The increased specific surface area due to biofilm formation enhances biodegradation and weathering processes. Hydrophobicity generally decreases with biofilm development, although it can be affected by other factors. Environmental conditions such as salinity and UV exposure, as well as biofilm-induced changes in chemical composition and surface charge, are known to influence MPs' hydrophobicity (Rummel et al., 2017; Tu et al., 2020). This parallel process highlights how biofilm development influences MP distribution across different environmental media.

The formation of biofilms on plastic surfaces, including MPs, is a multifaceted process that involves the attachment, growth, and maturation of microbial communities. Understanding how biofilms form on surfaces, especially plastics in water, requires a close look at the attachment process. It starts with the creation of a conditioning film within seconds. When a plastic item enters the aquatic environment, it quickly acquires a coating of inorganic and organic substances (Jachimowicz et al., 2022; Oberbeckmann et al., 2018; Renner & Weibel, 2011). The formation of a biofilm on its surface follows almost immediately, occurring within minutes to hours. This film, made of organic and inorganic matter from the water, coats the plastic's surface. The initial attachment of microorganisms to the plastic surface is influenced by various physicochemical factors, such as surface roughness, hydrophobicity, and the presence of specific functional groups. Once the initial attachment occurs, microorganisms begin to form microcolonies. After adhering to the plastic surface, they start producing EPS, which form a protective matrix that encapsulates the microbial community. This matrix can enhance the adhesion of the biofilm to the plastic surface as well as provide protection from environmental stressors and facilitate the acquisition of nutrients.

Living in a biofilm offers several advantages for competition and survival, such as forming stable consortia, horizontal gene exchange, nutrient accumulation, and protection against toxic substances and desiccation (Jachimowicz et al., 2022; Oberbeckmann et al., 2018).

As the biofilm matures, it can undergo structural changes, including the formation of microcolonies, channels, and pores, which allow for the efficient transport of nutrients and the removal of waste products (Schulze et al., 2022). This complex architecture also contributes to the increased density and cohesion of the biofilm, which can subsequently impact the behavior and fate of the plastic substrate. Rummel et al. (2021) highlighted the importance of conditioning film in shaping biofilm development on plastic surfaces. This film is formed within seconds of contact with water and plays a crucial role in microbial attachment and shaping subsequent development of biofilm growth. This film can alter the electrostatic properties of the plastic, influencing the attachment of microbial cells based on their own surface charges. Additionally, the conditioning film can impact the hydrophobicity of the plastic, favoring the attachment of microorganisms with a matching hydrophobic nature. Furthermore, the conditioning film modifies the overall surface-free energy of the plastic, which encompasses the energy required for microorganisms to overcome repulsive forces and successfully attach to the surface. By lowering this energy barrier, the conditioning film can promote more favorable conditions for microbial attachment and biofilm formation. This section will discuss the various factors that influence the development of biofilms on MPs.

7.2.1 Factors Influencing Biofilm Development on Microplastics

The formation of biofilms on MPs is a complex and dynamic process influenced by a myriad of interconnected factors. This intricate interplay between the plastic substrate, environmental conditions, and microbial communities shapes the development, structure, and function of biofilms in aquatic ecosystems. Understanding these factors is crucial for predicting the fate and ecological impact of MPs in the environment.

7.2.1.1 Microplastic Properties and Their Influence on Biofilm Formation

The physical and chemical characteristics of MPs play a fundamental role in biofilm formation, serving as the initial substrate for microbial colonization. These properties not only influence the rate and extent of biofilm development but also shape the composition of the microbial communities that colonize the plastic surface.

7.2.1.1.1 Surface Properties

Surface roughness, hydrophobicity, and charge can influence microbial attachment. These interactions are primarily influenced by interfacial electrostatic and van der Waals forces, as well as other non-specific forces like hydration, hydrophobic interactions, and steric forces. The properties of the MP substrate, such as its stiffness, mechanical stability, elasticity, and topography, can regulate the physiology of bacteria and their growth into biofilms on the MP surface (Tu et al., 2020). In response to these surface properties, bacterial cells secrete various biomolecules, including DNA, proteins, lipids, and lipopolysaccharides, which accumulate to form the EPS (Chen et al., 2019; Meng et al., 2024). This EPS can influence the stiffness and elasticity of the biofilm on the MP, posing a challenge for its eradication. While individual physical properties of the MP surface may have a significant impact on bacterial attachment, the detailed mechanisms underlying cell-surface interactions on MPs are not fully understood. Furthermore, hydrophobic interaction also plays a role in the initial attachment of bacteria to the MP surfaces. These binding events can trigger the genetic regulation, expression, and secretion of chemical factors, such as quorum-sensing molecules, which can induce biofilm formation and the production of EPS on the MPs. The design of specific MP topography can also influence the initial attachment of bacterial cells and regulate biofilm development on these surfaces. A study by Pompilio et al. (2008) investigated the factors influencing biofilm formation of *Stenotrophomonas maltophilia*, a common bacterium, on polystyrene (PS), a ubiquitous plastic in the environment. The researchers found that the hydrophobicity of the bacterial cell surface positively correlated with the strength of biofilm formation. This suggests that hydrophobic plastics may be more prone to colonization by a diverse range of bacteria. Additionally, the study revealed that motile strains of *S. maltophilia* formed stronger biofilms, indicating that bacterial motility is a crucial factor in the ability of microorganisms to colonize plastic surfaces in aquatic ecosystems (Chen et al., 2019; Moyal et al., 2023; Renner & Weibel, 2011).

7.2.1.1.2 Polymer Type

Different types of plastics, such as polyethylene (PE), polypropylene (PP), and PS, have distinct characteristics that can selectively favor certain microbial communities (Sutkar et al., 2023). In scenarios where the solid surface also acts as the substrate, such as PE exposed to microbial communities, the carbon availability within a biofilm is significantly higher. However, PE's extreme hydrophobic nature can initially support microbial attachment but may pose challenges to sustained colonization and biofilm development. While hydrophobic interactions facilitate the early attachment of certain bacterial cells, PE's low surface energy and lack of polar groups may inhibit the progression from attachment to stable biofilm formation. As a result, despite the initial bacterial adherence, the conditions required for the development of a mature biofilm, such as EPS production and stable microbial growth, may be less favorable on PE compared to other plastics like PS. This suggests that while hydrophobicity generally promotes bacterial attachment, the specific surface properties of each polymer type, including roughness and topography, ultimately regulate the success of biofilm formation (Rummel et al., 2021; Sooriyakumar et al., 2022). To overcome this barrier, Orr et al. (2004) integrated a nonionic surfactant into the culture medium of *Pseudomonas aeruginosa*. This surfactant appeared to increase the hydrophilicity of the PE surface, thereby enhancing bacterial adhesion to the polymer. In another study, a two-step enrichment method was used to isolate bacteria capable of degrading PE (Gong et al., 2018). The initial enrichment was conducted in soil supplemented with PE, followed by a second enrichment in a synthetic medium (SM) where PE served as the sole carbon source. This approach successfully isolated a *Rhodococcus* strain that could form biofilms on PE. Throughout the study, they assessed the cell-surface hydrophobicity of *Rhodococcus* and evaluated its biofilm formation potential (Gong et al., 2018; Vimala & Mathew, 2016).

The rapid initial colonization of PE is attributed to the microorganisms' use of mineral oil on the PE surface as a carbon source. Once the mineral oil is depleted, most of the biofilm, unable to access the PE surface, disperses into the medium. This suggests that PE biodegradation is primarily carried out by a low-density population with a slow growth rate that can utilize PE as a carbon source. One of the earliest studies, Albertsson and Karlsson (1993), also noted that high-density polyethylene (HDPE) biodegradation is hindered by the presence of mineral oil, which is preferentially used as a carbon source. Using a pure culture obtained through enrichment and selection, rather than nonselected microbial populations, results in microorganisms better suited for the biodegradation of PE (Orr et al., 2004; Xu et al., 2019).

7.2.1.2 Environmental Conditions and Biofilm Dynamics

Environmental factors play a critical role in shaping both the rate of biofilm formation and the composition of the microbial community on plastic surfaces. Warmer temperatures tend to accelerate microbial metabolic rates and growth, resulting in faster biofilm development. For example, isolated bacteria from mangrove sediments have been shown to degrade plastics such as PP, PE, and PS. The high temperature, high salinity, pH, organic matter concentration, and low ventilation of mangrove sediments create favorable conditions for diverse microbial communities capable of degrading various plastics (Renner & Weibel, 2011). Similarly, in ocean environments, the unique physicochemical properties, including nutrient availability, pH, and oxygen levels, select for specialized microbial communities that can colonize and degrade MPs (Sri Hari et al., 2023; Ventura et al., 2024).

The biofilms that form in the ecological niches of these MPs are shaped by the prevailing environmental conditions. Additionally, higher concentrations of essential nutrients like nitrogen and phosphorus support the formation of larger, more diverse biofilms. A comparative study found that biofilm growth was greatest on nutrient-rich polylactic acid (PLA) substrates compared to other biodegradable polyesters (Cifuentes et al., 2023; Dodhia et al., 2023). Salinity also influences microbial osmotic balance, impacting which species can thrive. Other environmental factors, including pH, light exposure, and water flow, similarly influence biofilm formation and growth on MPs (Oberbeckmann et al., 2018; Wang et al., 2020).

Hydraulic conditions, such as water flow speed, also significantly affect biofilm formation on MPs. Higher-speed flows can physically disrupt MPs, leading to their fragmentation and releasing nutrients that enhance microbial growth (Mathieu et al., 2014; Zhang et al., 2019). Smaller MPs, acting as carbon sources, are often associated with increased nutrient levels and higher rates of microbial proliferation (Ateia et al., 2020; Chen et al., 2023). Biofilm development is a dynamic process, continuously subjected to shear forces that influence bacterial attachment, detachment, and transport (Chen et al., 2019, 2023; Zhang et al., 2019). While most studies in the field of drinking water distribution systems (DWDS) have focused on biofilms within pipes, the biofilm formation on plastic surfaces in natural environments has been relatively overlooked. MPs of different material compositions uniquely affect microbial community structures, including opportunistic pathogens (Ogonowski et al., 2018; Zhang et al., 2019). The microbial communities within the plastisphere differ significantly from those on other materials or in water environments, with Alpha-proteobacteria and Beta-proteobacteria being the dominant species (Ogonowski et al., 2018).

However, integrated studies on the effects of hydraulic conditions on the plastisphere are limited. Previous studies involving biofilm–MP interactions often covered one or two aspects of hydraulic conditions, MPs, and biofilms, with less comprehensive studies on shear stress distribution and biofilm development in the presence of MPs in DWDS. Moreover, analyzing shear stresses could also theoretically help assess potential mechanical damage to blood vessels when plastic particles are ingested, but such studies are currently scarce. To fully understand the influence of hydraulic conditions on biofilm formation and behavior, more research is needed to investigate the full range of hydraulic variables, including flow rate, shear stress, and duration of exposure, and how they interact with MPs to shape biofilm dynamics.

7.2.2 Biodegradation and Biofilm–Plastic Interactions

Biodegradation, or biotic degradation, refers to the breakdown of MPs into smaller molecules by living organisms without producing toxic byproducts. This process involves enzymatic decomposition, where organic matter is converted into methane, biomass, water, and inorganic compounds. Conventional tests can monitor this decomposition over time, illustrating disposal options (Bacha et al., 2021; Paço et al., 2017; Rajcoomar et al., 2024). For example, a study investigating the biofilm formation and degradation of commercially available biodegradable plastic films by bacteria found that specific bacterial isolates, such as *Bacillus cereus*, were able to utilize the plastic as a carbon source leading to a significant reduction in the plastic's molecular weight over time. MPs provide a unique ecological niche for microorganisms by offering a carbon source and supporting their growth (Sutkar et al., 2023; Yuan et al., 2020). Microbial degradation can be influenced by metabolic products or the microbes themselves. Mixed cultures tend to be more effective at degradation than pure strains (Miloloža et al., 2022; Skariyachan et al., 2018; Sutkar et al., 2023).

Microbial degradation, in conjunction with physicochemical processes, may be an important pathway in breaking down plastic litter in the marine environment. Yet, little is known about potential microorganisms degrading marine plastic litter because most quantitative techniques do not allow resolving degradation in the subpercent range, nor do the experimental setups provide unambiguous proof of microbial utilization of plastics. Vaksmaa et al. (2023) presented data from short (<1 week) experimental assays that allowed to quantitatively measure plastic mineralization rates. In addition, they traced and visualized the assimilation of plastic-derived carbon into individual cells of *R. mucilaginosa*. Central to this approach is the utilization of isotopically labeled PE, containing a high degree of 13°C instead of 12°C, and tracing the PE-derived 13°C in the degradation product CO_2 and biomass.

Plastics' basic chemistry determines their utility and environmental impact. Plastic degradation involves three main steps: colonization by microorganisms, fragmentation, and mineralization. Various microbes, including bacteria, fungi, algae, and protozoa, participate in biodegradation (Kettner et al., 2017; Masó et al., 2003). Hydrophilic transformations facilitated by microbe-derived enzymes are essential for further biodegradation and microbial adhesion (Pinto et al., 2019; Rüthi et al., 2023). Research indicates that certain bacterial genera like *Pseudomonas*, *Bacillus*, and *Brevibacillus* are effective in MP degradation (Bacha et al., 2021; Matjašič et al., 2021; Sutkar et al., 2023).

To date, most research on microbial plastic degradation has been conducted under controlled laboratory conditions, focusing on single-strain isolates from both terrestrial and marine environments. Many of these microorganisms were sourced from biofilms found on plastic waste, such as a marine fungus isolated from a coastal dumping site (Sangeetha Devi et al., 2015), and *Ideonella sakaiensis*, a bacterium identified at a PET recycling plant that demonstrated polymer degradation through scanning electron microscopy (SEM) imaging and weight loss analysis (Roager & Sonnenschein, 2019; Yoshida et al., 2016). Several studies have explored the biodegradation potential of different bacterial genera. Kyaw et al. examined the biodegradation of low-density PE by various *Pseudomonas* strains, reporting weight losses of up to 20% after 120 days. In another study, Vimala and Mathew (2016) enhanced PE degradation using *Bacillus subtilis*. Similarly, Auta et al. (2018) investigated the degradation of PE, PS, polyethylene terephthalate, and PP by *Bacillus* strains, noting significant weight loss. Furthermore, Auta et al. explored PP degradation by *Rhodococcus sp.* and *Bacillus* sp., confirming biodegradation through SEM and Fourier-transform infrared spectroscopy analysis. These studies demonstrate the capacity of various bacteria to degrade multiple plastic types, offering insights into the mechanisms and environmental factors that influence the biodegradation process (Roager & Sonnenschein, 2019; Shabbir et al., 2020; Shilpa et al., 2022).

Recent studies highlight the importance of fungal strains in plastic degradation. Fungal species like *Aspergillus*, *Penicillium*, *Aureobasidium*, *Fusarium*, and *Rhizopus* have shown promising results in the biodegradation of various plastics, including polyolefins, polyesters, and PAs. Enzymes, such as hydrolases, oxidases, and depolymerases, produced by these fungi can effectively break down the polymer chains, contributing to the degradation of plastic materials. For instance, Rüthi et al. (2023) found significant weight loss in commercial plastic films degraded by *Pseudogymnoascus pannorum* and *Thelebolus globosus*. Gong et al. (2018) isolated *Comamona*s testosterone F4 from wastewater and examined its polyethylene terephthalate degradation potential. The study revealed enhanced polyethylene terephthalate decomposition in alkaline conditions due to increased bacterial growth and biomass. Chen et al. (2020) investigated MP degradation in sewage sludge using hyperthermophilic composting (hTC). The study found significant weight loss of MPs, particularly PS-MPs, under hTC conditions compared to conventional composting.

Overall, the literature highlights the diverse range of microorganisms capable of colonizing and degrading plastic materials, providing insights into the complex interactions between microbes and plastics in the environment (Shilpa et al., 2022; Wright et al., 2020). These studies collectively emphasize the potential of biodegradation methods, including microbial-mediated processes, for addressing MP pollution.

7.3 MICROPLASTICS AS SINKS AND SOURCES OF POLLUTANTS THROUGH BIOFILM INTERACTIONS

Biofilms can act as sinks, trapping and accumulating various pollutants like heavy metals, persistent organic pollutants, and even MPs. The plastisphere can facilitate the adsorption and concentration of pollutants, making plastics a potential vector for the transport of contaminants in aquatic environments. Multiple studies have investigated the ability of biofilms to accumulate pollutants on plastic surfaces. In the sections that follow, the mechanisms behind the accumulation of heavy metals and organic contaminants within biofilms on plastic surfaces will be discussed in detail.

7.3.1 Biofilm-Colonized Microplastics as Sinks for Heavy Metals

Studies have explored the behavior of heavy metal adsorption on biofilm-colonized MPs. Ahamed et al. (2020) found that biofilm presence on low-density polyethylene (LDPE) surfaces led to a 13-fold rise in Pb adsorption capacity. Li et al. (2022a) and Lin et al. (2021) explored the behavior of lead (Pb) adsorption on biodegradable poly(butylene succinate) (PBS)-MPs during biodegradation. Their research indicated that biofilm-colonized, degraded PBS-MPs adsorbed around ten times more Pb than pristine PBS-MPs (647.09 vs. 64.13 μg/g). This significant increase was due to biofilm colonization and PBS degradation. Biofilm presence on LDPE surfaces led to a 13-fold rise in Pb adsorption capacity, reaching 1,602 g/m^2, compared to 124 g/m^2 without biofilm (Ahamed et al., 2020). Similarly, PP MPs had shown higher adsorption capacities for Pb and Cu than for Cd and Zn, with physisorption identified as the primary mechanism (Lin et al., 2021). The concentrations and fractions of contaminants in MP biofilms are comparable to those in natural components of the marine environment, such as suspended particulate matter (SPM) and sediments. The season is another factor that influences the concentration of Hg and its fractions in MP biofilms. Graca et al. (2024) found that the concentration of total mercury (HgT) in MP biofilms ranged from 78 to 234 ng/g dry weight, with a median of 138 ng/g dry weight. These concentrations were comparable to those in sediments (136 ng/g dry weight) and significantly higher than those in SPM (63 ng/g dry weight). The results further showed that both the concentration of total Hg and the distribution of its fractions (organic, inorganic, and elemental Hg) varied seasonally, being higher in the warmer months. This indicates that biofilm colonization on MPs can act as an important vector for transporting and accumulating mercury and other pollutants in aquatic ecosystems. Another study by Rummel et al. (2021) examined the adsorption of hydrophobic organic contaminants to MP–biofilm complexes. The researchers found that the presence of biofilms significantly enhanced the adsorption of contaminants, such as polycyclic aromatic hydrocarbons (PAHs) and polychlorinated biphenyls (PCBs), compared to clean MPs. Several other studies also confirmed metal accumulation acts as important vectors for transporting and accumulating a wide range of pollutants, including heavy metals, persistent organic pollutants, and even MPs themselves (Ashraf, 2017; Fan et al., 2021; Nguyen et al., 2023). The adsorption and concentration of these contaminants onto biofilm-coated plastic surfaces can lead to their increased bioavailability and potential transfer through aquatic food webs, posing significant environmental and health risks. Conversely, biofilms can also serve as sources of pollutants, potentially releasing the accumulated contaminants back into the environment under changing conditions, such as shifts in water chemistry or physical disruption of the biofilm. This dynamic interplay between biofilms as sinks and sources of pollutants highlights the complex and far-reaching implications for the transport and fate of contaminants in aquatic ecosystems (Meng et al., 2020; Moyal et al., 2023; Shabbir et al., 2020). This enhanced metal adsorption is attributed to several factors:

1. **Surface Property Alterations:** Biofilms modify the surface characteristics of MPs, making them more hydrophilic and increasing the presence of functional groups such as carboxyl, hydroxyl, and sulfhydryl. These functional groups have a high affinity for metal ions, thereby enhancing the adsorption of metals onto biofilm-covered microplastics (BMPs).
2. **Particle Size Influence:** Smaller MPs are more easily colonized by biofilms compared to larger MPs. The increased biofilm coverage on smaller MPs results in a higher metal adsorption capacity, making them more effective metal carriers.
3. **Redox Reactions:** BMPs can facilitate redox reactions, altering the oxidation states of adsorbed metals. These reactions can impact the chemical behavior and mobility of the metals, influencing their transport and potential environmental effects (Chen et al., 2022).
4. Redox reactions are crucial in shaping the interactions between biofilms and the metals adsorbed on MPs. In aquatic ecosystems, the biogeochemical cycling of elements such as carbon, nitrogen, and metals is closely tied to the redox environment, which dictates the availability of electron acceptors like oxygen, nitrate, sulfate, and ferric iron.

Smaller MPs inherently have a higher surface area-to-volume ratio than larger particles, which makes them more effective at supporting biofilm colonization and subsequent metal adsorption. A larger surface area allows more space for biofilm development, increasing the density of microbial biomass that can adhere to the plastic surface. Studies have shown that biofilm formation is proportional to the available surface area, meaning that the colonization potential increases as the particle size decreases (An et al., 2024; Gong et al., 2019). This increased biofilm coverage enhances the adsorption capacity of smaller MPs because more functional groups (such as carboxyl, hydroxyl, and sulfhydryl groups) are available to bind metal ions. Furthermore, the larger biofilm mass on smaller MPs increases the potential for redox reactions and metal transformations, contributing to the overall effectiveness of smaller MPs as carriers of metals in aquatic environments (An et al., 2024; Tang & Li, 2024).

When biofilms colonize MP surfaces, they can create microenvironments where redox reactions take place, altering the oxidation states of metals adsorbed onto the MPs (Dodhia et al., 2023; Rogers et al., 2020). For example, BMPs have been shown to reduce hexavalent chromium (Cr^{6+}), a highly toxic form of the metal, to trivalent chromium (Cr^{3+}), which is less toxic and more stable. This suggests that BMPs not only act as carriers of metals but also influence the chemical forms and toxicity of the metals they transport (Chen et al., 2022). These transformations are driven by microbial metabolic processes within the biofilm, which utilize available electron acceptors, and are further influenced by the surrounding biogeochemical conditions, such as those found in sediment or water column layers. Understanding how BMPs mediate redox reactions is crucial for assessing their role in the transport, fate, and potential detoxification of metals in aquatic ecosystems. The study by Graca et al. (2024) revealed that while original MPs could adsorb metals, their capacity was relatively limited. The order of metal adhesion to original MPs (PP, PS, PLA) was Cr (VI)>Ni (II)>Cu (II)>Zn (II)>Co (II)>Cd (II), with Cr exhibiting the highest capacity at 17.1 μg/g. Following biofilm formation, the adsorption capacity of biofilm-coated MPs (BMPs) significantly increased. BMP75 showed the greatest enhancement, with metal adsorption capacity rising from 205.8% to 701.8% Cu. BMP4000-1 and BMP4000-2 experienced moderate improvements of 114.1% and 49.3% on average, respectively. Notably, BMP75 exhibited unique behavior in Cr (VI) adsorption, where the capacity continued to increase over time without reaching equilibrium, indicative of an ongoing redox reaction. This reaction facilitated the reduction of Cr (VI) to Cr (III), thereby enhancing adsorption. Overall, the research underscored the enhanced metal adsorption capacity of biofilm MP, especially BMP75, due to biofilm formation and associated redox reactions. The accumulation of pollutants on biofilm-colonized MPs can have significant implications for the transport, fate, and bioavailability of these contaminants in aquatic environments (Sarma et al., 2022). Fan et al. (2021) showed that PP-MPs had higher adsorption capacities for Pb and Cu than for Cd and Zn, with physisorption identified as the primary mechanism. Graca et al. (2024) aimed to determine whether the concentrations and fractions of mercury (Hg) in MP biofilms are comparable to those in natural components of the marine environment, such as SPM and sediments. Another study by Nguyen et al. (2023) demonstrated the ability of biofilms to accumulate heavy metals like Cu, Cd, and Pb on MP surfaces. The biofilm matrix and the diverse microbial community within can act as effective sinks for various pollutants, potentially leading to their concentration and transfer through the food web (Rogers et al., 2020). Similarly, Li et al. (2022a) investigated the bioaccumulation of heavy metals (e.g., Cd, Cu, Pb, and Zn) by biofilms on MPs in a freshwater lake. Wu et al. (2022) demonstrated that biofilm-colonized MPs have much higher adsorption capacities for heavy metals like Cu and Cr compared to larger, pristine MPs. This enhanced metal adsorption is attributed to several factors, including surface property alterations, particle size influence, and redox reactions.

7.3.2 Organic Pollutants and Biofilm Dynamics

Organic pollutants like PCBs, PAHs, and others can adsorb to MPs facilitated by biofilms. Biofilms have been shown to enhance the adsorption of these pollutants to MP surfaces, leading to increased bioavailability and potential for trophic transfer (Ashraf, 2017; Hirai et al., 2011). Conversely, biofilms can also degrade some organic pollutants, potentially reducing their environmental persistence. The dual role of biofilms on MPs as both sinks and sources of pollutants in aquatic environments is influenced by various factors. He et al. (2023) explored the adsorption behavior of norfloxacin, a commonly used antibiotic, on both original and biologically aged MPs like polyvinyl chloride (PVC), polyamide (PA), and HDPE. The study showed how environmental factors, such as pH and salinity, affect the adsorption capacity of these MPs. Multiple studies have investigated the ability of biofilms to accumulate pollutants on plastic surfaces. For instance, Rummel et al. (2021) also examined the adsorption of hydrophobic organic contaminants to MP–biofilm complexes. The presence of biofilms significantly enhanced the adsorption of contaminants, such as PAHs and PCBs, compared to clean MPs. The adsorption behavior of PAHs on MPs is strongly influenced by their hydrophobicity and polarity. For example, benzo(a)pyrene (BaP) exhibits higher adsorption onto MPs compared to pyrene, primarily due to the latter's greater hydrophilicity (José & Jordao, 2022). The surface functional groups of MPs play a crucial role in their adsorption capacity. For instance, oxygen-containing functional groups on the surface of degradable MPs can form intermolecular hydrogen bonds with the polar portions of pollutants like fipronil. Various microorganisms, including bacteria, fungi, and algae, have demonstrated potential in degrading and transforming PAHs (He et al., 2023; Nguyen et al., 2023; Yan et al., 2024).

The dynamic nature of biofilm–MP interactions and their role in the transport and fate of pollutants emphasize the need for further research and the development of effective strategies to mitigate the environmental and health risks posed by MP pollution. For instance, studies have investigated the behavior of BMPs in various environments. One such study by Su et al. (2022) examined BMP formation and dynamics in constructed wetlands, using microcosm systems with different MP types. The study found that PLA MPs exhibited higher microbial diversity, connectivity, and stability than LDPE, likely due to PLA's lower hydrophobicity. This highlights the significant impact of MPs on microbial community assembly, potentially contributing to environmental risks in constructed wetland systems.

In addition, PAHs, recognized for their mutagenic and carcinogenic properties, can also interact with MPs, influencing their environmental transport. The adsorption of PAHs on MPs is primarily driven by hydrophobic interactions, with the affinity of different PAHs varying according to their chemical structure. For example, BaP exhibits stronger adsorption on MPs compared to pyrene due to its greater hydrophobicity (Bao et al., 2021). Similarly, PS demonstrates higher adsorption capacity for phenanthrene than PVC, due to both hydrophobic and π-π interactions. The surface functional groups of MPs, such as oxygen-containing groups on degradable plastics, play a critical role in determining the adsorption capacity for pollutants like bisphenols and fipronil. In this regard, biofilms colonizing MPs can enhance the adsorption, concentration, and even degradation of pollutants such as PAHs (José & Jordao, 2022).

Microbial communities adhering to MPs can metabolize both the MP substrate and adsorbed pollutants (Ghosal et al., 2016). For example, various microorganisms, including bacteria and fungi, have demonstrated the ability to degrade PAHs, with low molecular weight PAHs being preferentially utilized by microbes in some studies. The selective utilization of PAHs by microorganisms like *Klebsiella pneumoniae* highlights the potential of biofilms on MPs to influence the biodegradation of these harmful compounds. However, BMPs can also act as sources of pollutants, potentially releasing them back into the environment under changing conditions, such as shifts in water chemistry or physical disturbances (José & Jordao, 2022). The pathways for PAH biodegradation and bioconversion vary depending on the microbial type (Sakshi & Haritash, 2020). Wu and colleagues found that when PAH-loaded LDPE sheets were incubated in water, the residual fractions of low molecular weight PAHs (e.g., naphthalene-d8, acenaphthene-d10, and fluorene-d10) were below 20%, whereas high molecular weight PAHs (e.g., benzo[a]pyrene-d12 and benzo[ghi]perylene-d12) exceeded 30%, indicating a selective microbial utilization favoring low molecular weight PAHs (Bao et al., 2021; Ghosal et al., 2016; Su et al., 2022; Wu et al., 2019a).

Therefore, understanding the interactions between biofilms, MPs, and pollutants is critical for assessing the environmental risks posed by BMPs and for developing pollution mitigation strategies. Given the complexity of environmental conditions and pollutant diversity, further research is needed to explore the specific mechanisms driving pollutant adsorption, retention, and release in aquatic ecosystems.

7.4 ANTIMICROBIAL AND DISINFECTANT RESISTANCE

Antibiotic resistance is a natural phenomenon where bacteria evolve to survive antibiotics, a process accelerated by the misuse and overuse of antibiotics in medicine, agriculture, and industry. This creates selective pressure, favoring resistant strains that can spread widely. Additionally, disinfection processes, such as chlorine use, can induce oxidative stress in bacteria, enhancing their resistance to antibiotics (Abe et al., 2020; Zhang et al., 2024). This section explores the development of antibiotic and disinfectant resistance in plastic-associated biofilms, mechanisms of resistance gene transfer within biofilms, and the impact of MPs on the spread of disinfectant resistance genes and environmental implications.

7.4.1 Development of Antibiotic and Disinfectant Resistance in Plastic-Associated Biofilms

The overuse of antibiotics puts selective pressure on bacteria, favoring the survival and proliferation of resistant strains, which can then spread to humans, animals, and the environment. HGT is a key mechanism by which ARGs can spread rapidly among different bacterial species and strains. Bacteria can share resistance genes through various processes, such as conjugation, transformation, and transduction. The presence of antimicrobial compounds, including antibiotics and heavy metals, can further promote HGT, as bacteria are incentivized to acquire resistance mechanisms to survive in these challenging environments (Figure 7.2) (Hegarty et al., 2022; Malayil et al., 2022). This results in the widespread dissemination of antibiotic resistance, making it a growing global public health concern.

The colonization of MPs by biofilms containing diverse microbial communities has raised concerns about the potential role of these plastic-associated microbes in the proliferation and dissemination of antimicrobial resistance genes. Biofilms can harbor and concentrate antimicrobial

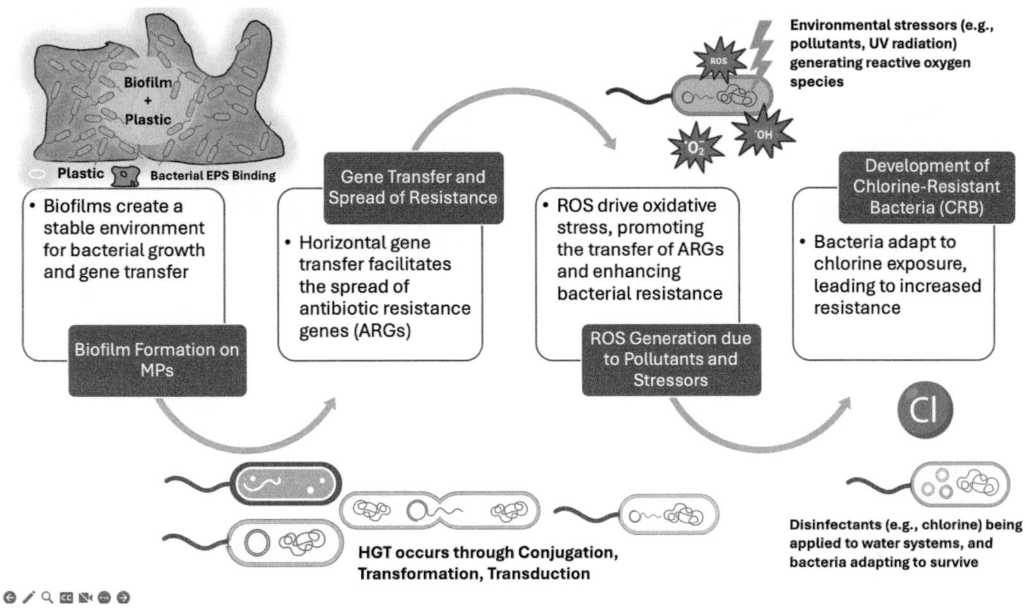

FIGURE 7.2 Key processes in generating resistance in plastic-associated biofilms.

resistance genes, which can then be transferred to other microorganisms in the environment. This can lead to the emergence and spread of multidrug-resistant pathogens, posing a significant public health threat (Moyal et al., 2023). These biofilms can harbor pathogenic bacteria, including pathogens like *Aeromonas* spp. and *Vibrio* spp., as well as opportunistic human pathogens such as *E. coli* (Kirstein et al., 2016; Marathe & Bank, 2021; Viršek et al., 2017). Notably, biofilms on MPs can also harbor ARGs, which are genetic elements that enable bacteria to survive exposure to antibiotics. MPs can selectively accumulate antibiotics and antibiotic-resistant bacteria on their surfaces in various environments, including landfill leachates, freshwater, and seawater (Su et al., 2021; Wu et al., 2019b; Zhang et al., 2018). Thus, MPs may serve as vectors for the transport of ARGs and pathogens in aquatic environments, potentially facilitating the spread of antibiotic resistance.

Various techniques have been utilized to detect and quantify ARGs associated with marine plastics. These include selective isolation of resistant bacteria, phenotypic antibiotic sensitivity assays, whole genome sequencing, shotgun metagenomics, and quantitative polymerase chain reaction (qPCR) (Wu et al., 2019b). Traditional culture-based techniques involve isolating bacteria on culture media followed by antibiotic susceptibility testing (Khan et al., 2019). Zhang et al. (2020) isolated and characterized antibiotic-resistant marine bacteria from MP particles collected at marine aquaculture sites using a combination of seven antibiotics and nonselective media. Their study identified multiple multidrug-resistant marine bacteria, including pathogenic *Vibrio* species on MPs. In contrast, other studies have focused on the selective isolation of pathogens like *Vibrio* spp. (Laverty et al., 2020) and *E. coli* (Håkonsholm et al., 2023), demonstrating the presence of multidrug-resistant pathogens on marine MPs. Recently, a study reported the whole genome sequences of antibiotic-resistant fish pathogens isolated from marine plastics (Radisic et al., 2024; Su et al., 2021).

7.4.2 Mechanisms of Resistance Gene Transfer within Biofilms

Mechanisms of resistance gene transfer within biofilms are critical to understanding the spread of antibiotic resistance. Imran et al. (2019) concluded that the co-contamination of MPs and heavy metals fosters the development and dissemination of multidrug-resistant human pathogens through coselection mechanisms. Studies have demonstrated that even minimal concentrations of antibiotics and biocides can select for antibiotic resistance and promote the horizontal transfer of ARGs (Gullberg et al., 2011; Jutkina et al., 2018; Marathe & Bank, 2021; Zhang et al., 2018). Additionally, bacteria within biofilms exhibit greater efficiency in HGT compared to planktonic bacteria. Research has shown that the presence of MPs enhances HGT via conjugation (Arias-Andres et al., 2018; Marathe & Bank, 2021). Although significant research has been conducted on the selection of resistance and the promotion of HGT by antimicrobial compounds, there is limited understanding of the effects of adsorbed chemicals on plastisphere bacteria, particularly regarding the selection and transfer of ARGs on MP particles (Abe et al., 2020; Arias-Andres et al., 2018).

Overall, the intricate relationship between biofilms, MPs, and antimicrobial resistance underscores the need for comprehensive studies to understand the complex environmental and public health implications (Attrassi et al., 1993; Castro et al., 2014; Le et al., 2018; Marathe & Bank, 2021). The close proximity of bacteria within biofilms facilitates HGT, including the transfer of ARGs, through mechanisms such as conjugation, transformation, and transduction. As a result, biofilms act as reservoirs for resistant bacteria, supporting the persistence and spread of antimicrobial resistance in various environments. Thus, the intersection of MPs, biofilm formation, and antibiotic resistance represents a critical area of research with significant implications for environmental and human health (Marathe & Bank, 2021; Mofijur et al., 2021).

7.4.3 Impact of Microplastics on the Spread of Disinfectant Resistance Genes

Disinfectant resistance is another critical concern, particularly in environments where disinfection processes, such as chlorine treatment, are used. Chlorine-induced oxidative stress in bacteria triggers pathways that enhance the expression of both antibiotic and disinfectant resistance genes, creating a dual threat. The correlation between chlorine resistance and antibiotic resistance is particularly troubling, as antibiotic-resistant bacteria can survive longer in the presence of chlorine, which is commonly used in water treatment facilities (Liu et al., 2016). Studies have shown that antibiotic-resistant bacteria can survive longer when exposed to free chlorine compared to nonresistant bacteria, indicating a correlation between chlorine resistance and antibiotic resistance (Zhang et al., 2024).

MPs further exacerbate this issue. Studies have shown that common antibiotics, like tetracycline and ciprofloxacin, can adsorb onto MPs, increasing their bioavailability and contributing to the persistence of disinfectant-resistant bacteria in water systems. Effluents from water treatment plants can introduce MPs and adsorbed antibiotics into DWDSs, influencing the composition of chlorine-resistant bacteria (CRB) and posing a serious threat to water quality (Kirstein et al., 2016; Zhang et al., 2024). These findings underscore the need for optimizing water treatment processes to mitigate the risks associated with MP contamination and the proliferation of disinfectant-resistant bacteria.

Biofilms on MPs serve as reservoirs for disinfectant-resistant bacteria, with oxidative stress—generated by reactive oxygen species (ROS)—playing a central role. ROS drive oxidative stress responses in bacteria, facilitating the transfer of resistance genes under pollutant exposure. Moreover, pollutants carried by MPs inhibit antioxidant enzymes like superoxide dismutase (SOD), leading to excessive ROS accumulation. This oxidative stress not only promotes the spread of ARGs but also increases bacterial resistance to chlorine, compounding the challenges of effective disinfection (Figure 7.2). ROS are generated as byproducts of normal cellular metabolism, particularly from the mitochondrial electron transport chain during aerobic respiration. They can also be produced in response to environmental stressors, such as exposure to pollutants, UV radiation, and pathogens. ROS drive oxidative stress in ARGs, facilitating the transfer of these genes among bacteria under pollutant exposure (Li et al., 2022b; Ma et al., 2021; Marathe & Bank, 2021; Yu et al., 2021). Intracellular oxidative stress is also linked to increased bacterial chlorine resistance. Enzymes like

SOD, catalase (CAT), and glutathione peroxidase (GSH-PX) play roles in mitigating ROS-induced damage. However, pollutants carried by MPs can inhibit SOD activity, likely due to excessive ROS accumulation, which surpasses the antioxidant system's capacity (Cheng et al., 2020). Unlike SOD, CAT activity increases postdomestication, while GSH-PX activity varies depending on the pollutant, suggesting a complex interplay between oxidative stress and enzyme activity (Li et al., 2022b, c). The term "domestication" in this context refers to the adaptation of bacterial populations through prolonged exposure to pollutants and MPs, resulting in the selection of strains that are more resilient to stressors like oxidative stress and disinfection processes. For example, while SOD activity is inhibited by pollutant exposure, CAT activity increases after this domestication period, reflecting the bacteria's adaptive response. GSH-PX activity varies depending on the pollutant, suggesting a complex interaction between oxidative stress and enzyme regulation. Additionally, ATPase activity, which reflects environmental stress on organisms, changes under different pollutant conditions (Zhang et al., 2024). Domestication with MPs and pollutants like sulfamethoxazole (SMX) can increase ATPase activity, indicating altered bacterial metabolism and physiological activity (Jing et al., 2020). These changes highlight the differential effects of MPs on bacterial stress responses. ROS are relevant to CRB because they drive oxidative stress, which facilitates the transfer of ARGs among bacteria under pollutant exposure. This oxidative stress is linked to increased bacterial chlorine resistance. Pollutants carried by MPs can exacerbate this by inhibiting antioxidant enzymes like SOD, leading to excessive ROS accumulation. The challenges of disinfection are compounded by the presence of biofilms, which can harbor disinfectant-resistant bacteria. The accumulation of ROS under pollutant exposure increases oxidative stress, facilitating the transfer of ARGs among bacteria. This oxidative stress is linked to enhanced bacterial resistance to chlorine, a common disinfectant. Pollutants carried by MPs inhibit antioxidant enzymes such as SOD, leading to excessive ROS accumulation and further promoting resistance. Additionally, studies have highlighted changes in ATPase activity, a marker of environmental stress, under different pollutant conditions. Such alterations in bacterial metabolism and physiological activity contribute to the persistence of CRB in polluted environments, further underlining the complexity of disinfectant resistance in BMPs.

These changes highlight the differential effects of MPs on bacterial stress responses. ROS accumulation plays a critical role in CRB as it facilitates the transfer of ARGs among bacteria exposed to pollutants. This oxidative stress is also linked to enhanced bacterial resistance to chlorine, a common disinfectant. The presence of biofilms, which can harbor disinfectant-resistant bacteria, compounds the challenges of effective disinfection (Zhang et al., 2020, 2024). Thus, understanding the complex interactions between MPs, biofilms, oxidative stress, and antimicrobial resistance is crucial for developing effective strategies to mitigate these environmental and public health threats.

7.5 CONCLUSION

The complex interactions between MPs, biofilm formation, and antimicrobial resistance represent a critical area of research with far-reaching implications for environmental and public health. MPs can serve as substrates for biofilm formation, harboring diverse microbial communities, including antibiotic-resistant bacteria. The accumulation of antibiotics and other pollutants on MPs can increase the bioavailability of these compounds, further exacerbating the problem. In summary, biofilm formation on MPs significantly influences their biodegradation, pollutant dynamics, microbial transport, and antimicrobial resistance. Biofilms enhance MPs' surface area, promoting biodegradation and weathering processes, with specific microorganisms like bacteria and fungi playing key roles in breaking down various plastics. These biofilms act as both sinks and sources of pollutants; they accumulate contaminants such as heavy metals but can also release these pollutants back into the environment under certain conditions, posing substantial environmental and health risks. Furthermore, biofilms facilitate the spread of microorganisms, including pathogens, across aquatic environments, thereby impacting microbial diversity and local ecosystems. Additionally, plastic-associated biofilms contribute to the development and dissemination of antibiotic and disinfectant resistance genes, exacerbating public health challenges. Understanding these multifaceted interactions is crucial for developing effective strategies to mitigate the environmental impacts of plastic pollution.

REFERENCES

Abe, K., Nomura, N., & Suzuki, S. (2020). Biofilms: Hot spots of horizontal gene transfer (HGT) in aquatic environments, with a focus on a new HGT mechanism. *FEMS Microbiology Ecology*, *96*(5), fiaa031. https://doi.org/10.1093/femsec/fiaa031

Ahamed, T., Brown, S. P., & Salehi, M. (2020). Investigate the role of biofilm and water chemistry on lead deposition onto and release from polyethylene: An implication for potable water pipes. *Journal of Hazardous Materials*, *400*, 123253. https://doi.org/10.1016/j.jhazmat.2020.123253

Albertsson, A.-C., & Karlsson, S. (1993). Aspects of biodeterioration of inert and degradable polymers. *International Biodeterioration & Biodegradation*, *31*(3), 161–170. https://doi.org/10.1016/0964-8305(93)90002-J

Allemann, M. N., Tessman, M., Reindel, J., Scofield, G. B., Evans, P., Pomeroy, R. S., Burkart, M. D., Mayfield, S. P., & Simkovsky, R. (2024). Rapid biodegradation of microplastics generated from bio-based thermoplastic polyurethane. *Nature Portfolio*, *14*(1). https://doi.org/10.1038/s41598-024-56492-6

An, Q., Zheng, N., Pan, J., Ji, Y., Wang, S., Li, X., Chen, C., Peng, L., & Wang, B. (2024). Association between plant microbiota and cadmium uptake under the influence of microplastics with different particle sizes. *Environment International*, *190*, 108938. https://doi.org/10.1016/j.envint.2024.108938

Arias-Andres, M., Klümper, U., Rojas-Jimenez, K., & Grossart, H.-P. (2018). Microplastic pollution increases gene exchange in aquatic ecosystems. *Environmental Pollution*, *237*, 253–261.

Ashraf, M. A. (2017). Persistent organic pollutants (POPs): A global issue, a global challenge. In Philippe Garrigues (Ed.), *Environmental Science and Pollution Research* (Vol. 24, pp. 4223–4227). Springer.

Ateia, M., Kanan, A., & Karanfil, T. (2020). Microplastics release precursors of chlorinated and brominated disinfection byproducts in water. *Chemosphere*, *251*, 126452. https://doi.org/10.1016/j.chemosphere.2020.126452

Attrassi, B., Saghi, M., & Flatau, G. (1993). Multirésistance bactérienne aux antibiotiques en milieu marin (côte atlantique, Maroc). *Environmental Technology*, *14*(12), 1179–1186. https://doi.org/10.1080/09593339309385396

Auta, H. S., Emenike, C. U., Jayanthi, B., & Fauziah, S. H. (2018). Growth kinetics and biodeterioration of polypropylene microplastics by *Bacillus* sp. and *Rhodococcus* sp. isolated from mangrove sediment. *Marine Pollution Bulletin*, *127*, 15–21. https://doi.org/10.1016/j.marpolbul.2017.11.036

Bacha, A.-U.-R., Nabi, I., & Zhang, L. (2021). Mechanisms and the engineering approaches for the degradation of microplastics. *ACS ES&T Engineering*, *1*(11), 1481–1501. https://doi.org/10.1021/acsestengg.1c00216

Bao, Z.-Z., Chen, Z.-F., Zhong, Y., Wang, G., Qi, Z., & Cai, Z. (2021). Adsorption of phenanthrene and its monohydroxy derivatives on polyvinyl chloride microplastics in aqueous solution: Model fitting and mechanism analysis. *Science of the Total Environment*, *764*, 142889. https://doi.org/10.1016/j.scitotenv.2020.142889

Castro, A. P. de, Fernandes, G. da R., & Franco, O. L. (2014, September). Insights into novel antimicrobial compounds and antibiotic resistance genes from soil metagenomes. *Frontiers Media*, *5*, 489. https://doi.org/10.3389/fmicb.2014.00489

Chen, W., Tanaka, K., Tani, Y., Bi, X., Liu, J., & Yu, Q. (2022). Effect of particle size on the colonization of biofilms and the potential of biofilm-covered microplastics as metal carriers. *Science of the Total Environment*, *821*, 153265.

Chen, X., Lian, X., Wang, Y., Chen, S., Sun, Y., Tao, G., Tan, Q., & Feng, J. (2023). Impacts of hydraulic conditions on microplastics biofilm development, shear stresses distribution, and microbial community structures in drinking water distribution pipes. *Journal of Environmental Management*, *325*, 116510. https://doi.org/10.1016/j.jenvman.2022.116510

Chen, X., Xiong, X., Jiang, X., Shi, H., & Wu, C. (2019). Sinking of floating plastic debris caused by biofilm development in a freshwater lake. *Chemosphere*, *222*, 856–864. https://doi.org/10.1016/j.chemosphere.2019.02.015

Chen, Z., Zhao, W., Xing, R., Xie, S., Yang, X., Cui, P., Lü, J., Liao, H., Yu, Z., Wang, S., & Zhou, S. (2020). Enhanced *in situ* biodegradation of microplastics in sewage sludge using hyperthermophilic composting technology. *Journal of Hazardous Materials*, *384*, 121271. https://doi.org/10.1016/j.jhazmat.2019.121271

Cheng, Y., Zhu, L., Song, W., Jiang, C., Li, B., Du, Z., Wang, J., Wang, J., Li, D., & Zhang, K. (2020). Combined effects of mulch film-derived microplastics and atrazine on oxidative stress and gene expression in earthworm (*Eisenia fetida*). *Science of the Total Environment*, *746*, 141280. https://doi.org/10.1016/j.scitotenv.2020.141280

Cifuentes, I. E. M., Degenhardt, J., Neumann-Schaal, M., Jehmlich, N., Ngugi, D. K., & Öztürk, B. (2023). Comparative biodegradation analysis of three compostable polyesters by a marine microbial community. *American Society for Microbiology*, *89*(12). https://doi.org/10.1128/aem.01060-23

De-la-Torre, G. E. (2019, October). Microplastics: An emerging threat to food security and human health. *Journal of Food Science & Technology*, *57*(5), 1601–1608. https://doi.org/10.1007/s13197-019-04138-1

Dodhia, M. S., Rogers, K. L., Fernández-Juárez, V., Carreres-Calabuig, J. A., Löscher, C. R., Tisserand, A. A., Keulen, N., Riemann, L., Shashoua, Y., & Posth, N. R. (2023). Microbe-mineral interactions in the Plastisphere: Coastal biogeochemistry and consequences for degradation of plastics. *Frontiers in Marine Science*, *10*. https://doi.org/10.3389/fmars.2023.1134815

Donlan, R. M. (2002, September). Biofilms: Microbial life on surfaces. *Emerging Infectious Diseases*, *8*(9), 881–890. https://doi.org/10.3201/eid0809.020063

Dussud, C., Meistertzheim, A. L., Conan, P., Pujo-Pay, M., George, M., Fabre, P., Coudane, J., Higgs, P., Elineau, A., Pedrotti, M. L., Gorsky, G., & Ghiglione, J. F. (2018). Evidence of niche partitioning among bacteria living on plastics, organic particles and surrounding seawaters. *Environmental Pollution*, *236*, 807–816. https://doi.org/10.1016/j.envpol.2017.12.027

Fan, T., Zhao, J., Chen, Y., Wang, M., Wang, X., Wang, S., Chen, X., Lu, A., & Zha, S. (2021). Coexistence and adsorption properties of heavy metals by polypropylene microplastics. *Adsorption Science & Technology*, 2021, 4938749.

Ghanadi, M., Joshi, I., Dharmasiri, N., Jaeger, J. E., Burke, M., Bebelman, C., Symons, B., & Padhye, L. P. (2024). Quantification and characterization of microplastics in coastal environments: Insights from laser direct infrared imaging. *Science of the Total Environment*, *912*, 168835. https://doi.org/10.1016/j.scitotenv.2023.168835

Ghosal, D., Ghosh, S., Dutta, T. K., & Ahn, Y. (2016). Current state of knowledge in microbial degradation of polycyclic aromatic hydrocarbons (PAHs): A review. *Frontiers in Microbiology*, *7*. https://doi.org/10.3389/fmicb.2016.01369

Gong, J., Kong, T., Li, Y., Li, Q., Li, Z., & Zhang, J. (2018). Biodegradation of microplastic derived from poly(ethylene terephthalate) with bacterial whole-cell biocatalysts. *Polymers*, *10*(12), Article 12. https://doi.org/10.3390/polym10121326

Gong, W., Jiang, M., Han, P., Liang, G., Zhang, T., & Liu, G. (2019). Comparative analysis on the sorption kinetics and isotherms of fipronil on nondegradable and biodegradable microplastics. *Environmental Pollution*, *254*, 112927. https://doi.org/10.1016/j.envpol.2019.07.095

Graca, B., Rychter, A., Bełdowska, M., & Wojdasiewicz, A. (2024). Seasonality of mercury and its fractions in microplastics biofilms-comparison to natural biofilms, suspended particulate matter and bottom sediment. *Science of the Total Environment*, *948*, 174814. https://doi.org/10.1016/j.scitotenv.2024.174814

Gullberg, E., Cao, S., Berg, O. G., Ilbäck, C., Sandegren, L., Hughes, D., & Andersson, D. I. (2011). Selection of resistant bacteria at very low antibiotic concentrations. *PLOS Pathogens*, *7*(7), e1002158. https://doi.org/10.1371/journal.ppat.1002158

Håkonsholm, F., Hetland, M. A. K., Löhr, I. H., Lunestad, B. T., & Marathe, N. P. (2023). Co-localization of clinically relevant antibiotic- and heavy metal resistance genes on plasmids in Klebsiella pneumoniae from marine bivalves. *MicrobiologyOpen*, *12*(4), e1368. https://doi.org/10.1002/mbo3.1368

He, S., Tong, J., Xiong, W., Xiang, Y., Peng, H., Wang, W., Yang, Y., Ye, Y., Hu, M., Yang, Z., & Zeng, G. (2023). Microplastics influence the fate of antibiotics in freshwater environments: Biofilm formation and its effect on adsorption behavior. *Journal of Hazardous Materials*, *442*, 130078. https://doi.org/10.1016/j.jhazmat.2022.130078

Hegarty, B., Dai, Z., Raskin, L., Pinto, A., Wigginton, K., & Duhaime, M. (2022). A snapshot of the global drinking water virome: Diversity and metabolic potential vary with residual disinfectant use. *Water Research*, *218*, 118484. https://doi.org/10.1016/j.watres.2022.118484

Hirai, H., Takada, H., Ogata, Y., Yamashita, R., Mizukawa, K., Saha, M., Kwan, C., Moore, C., Gray, H., Laursen, D., & others. (2011). Organic micropollutants in marine plastics debris from the open ocean and remote and urban beaches. *Marine Pollution Bulletin*, *62*(8), 1683–1692.

Imran, M., Das, K. R., & Naik, M. M. (2019). Co-selection of multi-antibiotic resistance in bacterial pathogens in metal and microplastic contaminated environments: An emerging health threat. *Chemosphere*, *215*, 846–857. https://doi.org/10.1016/j.chemosphere.2018.10.114

Jachimowicz, P., Nosek, D., & Cydzik-Kwiatkowska, A. (2022). Chemical and microbiological changes on the surface of microplastic after long term exposition to different concentrations of ammonium in the environment. *Science of the Total Environment*, *830*, 154784. https://doi.org/10.1016/j.scitotenv.2022.154784

Jing, M., Han, G., Wan, J., Zhang, S., Yang, J., Zong, W., Niu, Q., & Liu, R. (2020). Catalase and superoxide dismutase response and the underlying molecular mechanism for naphthalene. *Science of the Total Environment*, *736*, 139567. https://doi.org/10.1016/j.scitotenv.2020.139567

José, S., & Jordao, L. (2022). Exploring the interaction between microplastics, polycyclic aromatic hydrocarbons and biofilms in freshwater. *Polycyclic Aromatic Compounds*, *42*(5), 2210–2221. https://doi.org/10.1080/10406638.2020.1830809

Jutkina, J., Marathe, N. P., Flach, C.-F., & Larsson, D. G. J. (2018). Antibiotics and common antibacterial biocides stimulate horizontal transfer of resistance at low concentrations. *Science of the Total Environment*, *616–617*, 172–178. https://doi.org/10.1016/j.scitotenv.2017.10.312

Kettner, M. T., Rojas-Jimenez, K., Oberbeckmann, S., Labrenz, M., & Grossart, H.-P. (2017). Microplastics alter composition of fungal communities in aquatic ecosystems. *Environmental Microbiology*, *19*(11), 4447–4459. https://doi.org/10.1111/1462-2920.13891

Khan, Z. A., Siddiqui, M. F., & Park, S. (2019). Current and emerging methods of antibiotic susceptibility testing. *Diagnostics*, *9*(2), Article 2. https://doi.org/10.3390/diagnostics9020049

Kirstein, I. V., Kirmizi, S., Wichels, A., Garin-Fernandez, A., Erler, R., Löder, M., & Gerdts, G. (2016). Dangerous hitchhikers? Evidence for potentially pathogenic *Vibrio* spp. on microplastic particles. *Marine Environmental Research*, *120*, 1–8. https://doi.org/10.1016/j.marenvres.2016.07.004

Kumar, P., Lee, J., Brown, R. J. C., & Kim, K.-H. (2021). Micro- and nano-plastic pollution: Behavior, microbial ecology, and remediation technologies. *Journal of Cleaner Production*, *291*, 125240–125240. https://doi.org/10.1016/j.jclepro.2020.125240

Kvale, K., Hunt, C., James, A., & Koeve, W. (2023). Regionally disparate ecological responses to microplastic slowing of faecal pellets yields coherent carbon cycle response. *Frontiers Media*, *10*. https://doi.org/10.3389/fmars.2023.1111838

Laverty, A. L., Primpke, S., Lorenz, C., Gerdts, G., & Dobbs, F. C. (2020). Bacterial biofilms colonizing plastics in estuarine waters, with an emphasis on Vibrio spp. and their antibacterial resistance. *PLoS One*, *15*(8), e0237704. https://doi.org/10.1371/journal.pone.0237704

Le, T.-H., Ng, C., Trần, N. H., Chen, H., & Gin, K. Y.-H. (2018). Removal of antibiotic residues, antibiotic resistant bacteria and antibiotic resistance genes in municipal wastewater by membrane bioreactor systems. *Water Research*, *145*, 498–508. https://doi.org/10.1016/j.watres.2018.08.060

Li, W., Zu, B., Yang, Q., Huang, Y., & Li, J. (2022a). Adsorption of lead and cadmium by microplastics and their desorption behavior as vectors in the gastrointestinal environment. *Journal of Environmental Chemical Engineering*, *10*(3), 107379. https://doi.org/10.1016/j.jece.2022.107379

Li, Z., Chang, X., Hu, M., Fang, J. K.-H., Sokolova, I. M., Huang, W., Xu, E. G., & Wang, Y. (2022b). Is microplastic an oxidative stressor? Evidence from a meta-analysis on bivalves. *Journal of Hazardous Materials*, *423*, 127211. https://doi.org/10.1016/j.jhazmat.2021.127211

Li, Z., Gao, J., Guo, Y., Cui, Y., Wang, Y., Duan, W., & Wu, Z. (2022c). Enhancement of antibiotic resistance dissemination by artificial sweetener acesulfame potassium: Insights from cell membrane, enzyme, energy supply and transcriptomics. *Journal of Hazardous Materials*, *422*, 126942. https://doi.org/10.1016/j.jhazmat.2021.126942

Lin, Z., Hu, Y., Yuan, Y., Hu, B., & Wang, B. (2021). Comparative analysis of kinetics and mechanisms for Pb(II) sorption onto three kinds of microplastics. *Ecotoxicology and Environmental Safety*, *208*, 111451. https://doi.org/10.1016/j.ecoenv.2020.111451

Liu, S., Gunawan, C., Barraud, N., Rice, S. A., Harry, E. J., & Amal, R. (2016). Understanding, monitoring, and controlling biofilm growth in drinking water distribution systems. *Environmental Science & Technology*, *50*(17), 8954–8976. https://doi.org/10.1021/acs.est.6b00835

Ma, X., Zhang, X., Xia, J., Sun, H., Zhang, X., & Ye, L. (2021). Phenolic compounds promote the horizontal transfer of antibiotic resistance genes in activated sludge. *Science of the Total Environment*, *800*, 149549. https://doi.org/10.1016/j.scitotenv.2021.149549

Malayil, L., Ramachandran, P., Chattopadhyay, S., Allard, S. M., Bui, A., Butron, J., Callahan, M. T., Craddock, H. A., Murray, R., East, C., Sharma, M., Kniel, K., Micallef, S., Hashem, F., Gerba, C. P., Ravishankar, S., Parveen, S., May, E., Handy, E., … Sapkota, A. R. (2022). Variations in bacterial communities and antibiotic resistance genes across diverse recycled and surface water irrigation sources in the mid-Atlantic and Southwest United States: A CONSERVE two-year field study. *Environmental Science & Technology*, *56*(21), 15019–15033. https://doi.org/10.1021/acs.est.2c02281

Marathe, N. P., & Bank, M. S. (2021). The microplastic-antibiotic resistance connection. In M. S. Bank (Ed.), *Microplastic in the Environment: Pattern and Process* (pp. 311–322). Springer International Publishing. https://doi.org/10.1007/978-3-030-78627-4_9

Masó, M., Garcés, E., Pagès, F., & Camp, J. (2003). Drifting plastic debris as a potential vector for dispersing Harmful Algal Bloom (HAB) species. *Scientia Marina, 67*(1). https://doi.org/10.3989/scimar.2003.67n1107

Mathieu, L., Bertrand, I., Abe, Y., Angel, E., Block, J. C., Skali-Lami, S., & Francius, G. (2014). Drinking water biofilm cohesiveness changes under chlorination or hydrodynamic stress. *Water Research*, *55*, 175–184. https://doi.org/10.1016/j.watres.2014.01.054

Matjašič, T., Simčič, T., Medveščeк, N., Bajt, O., Dreo, T., & Mori, N. (2021). Critical evaluation of biodegradation studies on synthetic plastics through a systematic literature review. *Science of the Total Environment*, *752*, 141959. https://doi.org/10.1016/j.scitotenv.2020.141959

Meng, J., Zhang, Y., Yao, Z., Tao, P., Li, G., Yu, X., Zhao, J.-L., & Peng, J. (2020). How biofilms affect the uptake and fate of hydrophobic organic compounds (HOCs) in microplastic: Insights from an In situ study of Xiangshan Bay, China. *Water Research*, *184*, 116118–116118. https://doi.org/10.1016/j.watres.2020.116118

Meng, L., Liang, L., Shi, Y., Yin, H., Li, L., Xiao, J., Huang, N., Zhao, A., Xia, Y., & Hou, J. (2024). Biofilms in plastisphere from freshwater wetlands: Biofilm formation, bacterial community assembly, and biogeochemical cycles. *Journal of Hazardous Materials*, *476*, 134930. https://doi.org/10.1016/j.jhazmat.2024.134930

Miloloža, M., Cvetnić, M., Kučić Grgić, D., Ocelić Bulatović, V., Ukić, Š., Rogošić, M., Dionysiou, D. D., Kušić, H., & Bolanča, T. (2022). Biotreatment strategies for the removal of microplastics from freshwater systems. A review. *Environmental Chemistry Letters*, *20*(2), 1377–1402. https://doi.org/10.1007/s10311-021-01370-0

Mofijur, M., Ahmed, S. F., Mofijur, M., Siddiki, S. Y. A., Islam, A. B. M. S., Shahabuddin, M., Ong, H. C., Mahlia, T. M. I., Djavanroodi, F., & Show, P. L. (2021). Source, distribution and emerging threat of micro- and nanoplastics to marine organism and human health: Socio-economic impact and management strategies. *Environmental Research*, *195*, 110857–110857. https://doi.org/10.1016/j.envres.2021.110857

Moyal, J., Dave, P. H., Wu, M., Karimpour, S., Brar, S. K., Zhong, H., & Kwong, R. W. M. (2023). Impacts of biofilm formation on the physicochemical properties and toxicity of microplastics: A concise review. *Reviews of Environmental Contamination and Toxicology*, *261*(1), 8. https://doi.org/10.1007/s44169-023-00035-z

Munhoz, D. R., Harkes, P., Bériot, N., Larreta, J., & Basurko, O. C. (2022, December). Microplastics: A review of policies and responses. *Microplastics*, *2*(1), 1–26. https://doi.org/10.3390/microplastics2010001

Nguyen, H. T., Lee, Y. K., Kwon, J.-H., & Hur, J. (2023). Microplastic biofilms in water treatment systems: Fate and risks of pathogenic bacteria, antibiotic-resistant bacteria, and antibiotic resistance genes. *Science of the Total Environment*, *892*, 164523. https://doi.org/10.1016/j.scitotenv.2023.164523

Oberbeckmann, S., Kreikemeyer, B., & Labrenz, M. (2018). Environmental factors support the formation of specific bacterial assemblages on microplastics. *Frontiers in Microbiology*, *8*. https://doi.org/10.3389/fmicb.2017.02709

Oberbeckmann, S., Loeder, M. G. J., Gerdts, G., & Osborn, A. M. (2014). Spatial and seasonal variation in diversity and structure of microbial biofilms on marine plastics in Northern European waters. *FEMS Microbiology Ecology*, *90*(2), 478–492. https://doi.org/10.1111/1574-6941.12409

Ogonowski, M., Motiei, A., Ininbergs, K., Hell, E., Gerdes, Z., Udekwu, K. I., Bacsik, Z., & Gorokhova, E. (2018). Evidence for selective bacterial community structuring on microplastics. *Environmental Microbiology*, *20*(8), 2796–2808. https://doi.org/10.1111/1462-2920.14120

Orr, I. G., Hadar, Y., & Sivan, A. (2004). Colonization, biofilm formation and biodegradation of polyethylene by a strain of Rhodococcus ruber. *Applied Microbiology and Biotechnology*, *65*(1), 97–104. https://doi.org/10.1007/s00253-004-1584-8

Paço, A., Duarte, K., da Costa, J. P., Santos, P. S. M., Pereira, R., Pereira, M. E., Freitas, A. C., Duarte, A. C., & Rocha-Santos, T. A. P. (2017). Biodegradation of polyethylene microplastics by the marine fungus *Zalerion maritimum. Science of the Total Environment*, *586*, 10–15. https://doi.org/10.1016/j.scitotenv.2017.02.017

Pinto, M., Langer, T. M., Hüffer, T., Hofmann, T., & Herndl, G. J. (2019). The composition of bacterial communities associated with plastic biofilms differs between different polymers and stages of biofilm succession. *PLoS One*, *14*(6), e0217165. https://doi.org/10.1371/journal.pone.0217165
Pompilio, A., Piccolomini, R., Picciani, C., D'Antonio, D., Savini, V., & Di Bonaventura, G. (2008). Factors associated with adherence to and biofilm formation on polystyrene by Stenotrophomonas maltophilia: The role of cell surface hydrophobicity and motility. *FEMS Microbiology Letters*, *287*(1), 41–47. https://doi.org/10.1111/j.1574-6968.2008.01292.x
Radisic, V., Grevskott, D. H., Junghardt, N., Øvreås, L., & Marathe, N. P. (2024). Multidrug-resistant Enterococcus faecium strains enter the Norwegian marine environment through treated sewage. *MicrobiologyOpen*, *13*(2), e1397. https://doi.org/10.1002/mbo3.1397
Rajcoomar, S., Amoah, I. D., Abunama, T., Mohlomi, N., Bux, F., & Kumari, S. (2024). Biofilm formation on microplastics in wastewater: Insights into factors, diversity and inactivation strategies. *International Journal of Environmental Science and Technology*, *21*(4), 4429–4444. https://doi.org/10.1007/s13762-023-05266-0
Ramasamy, M., & Lee, J. (2016, January). Recent nanotechnology approaches for prevention and treatment of biofilm-associated infections on medical devices. *Biomed Research International*, 2016, 1–17. https://doi.org/10.1155/2016/1851242
Renner, L. D., & Weibel, D. B. (2011). Physicochemical regulation of biofilm formation. *MRS Bulletin*, *36*(5), 347–355. https://doi.org/10.1557/mrs.2011.65
Rogers, K. L., Carreres-Calabuig, J. A., Gorokhova, E., & Posth, N. R. (2020). Micro-by-micro interactions: How microorganisms influence the fate of marine microplastics. *Limnology and Oceanography Letters*, *5*(1), 18–36. https://doi.org/10.1002/lol2.10136
Roager, L., & Sonnenschein, E. C. (2019). Bacterial candidates for colonization and degradation of marine plastic debris. *Environmental Science & Technology*, *53*(20), 11636–11643. https://doi.org/10.1021/acs.est.9b02212
Rummel, C. D., Jahnke, A., Gorokhova, E., Kühnel, D., & Schmitt-Jansen, M. (2017). Impacts of biofilm formation on the fate and potential effects of microplastic in the aquatic environment. *Environmental Science & Technology Letters*, *4*(7), 258–267. https://doi.org/10.1021/acs.estlett.7b00164
Rummel, C. D., Lechtenfeld, O. J., Kallies, R., Benke, A., Herzsprung, P., Rynek, R., Wagner, S., Potthoff, A., Jahnke, A., & Schmitt-Jansen, M. (2021). Conditioning film and early biofilm succession on plastic surfaces. *Environmental Science & Technology*, *55*(16), 11006–11018. https://doi.org/10.1021/acs.est.0c07875
Rüthi, J., Cerri, M., Brunner, I., Stierli, B., Sander, M., & Frey, B. (2023). Discovery of plastic-degrading microbial strains isolated from the alpine and Arctic terrestrial plastisphere. *Frontiers in Microbiology*, *14*. https://doi.org/10.3389/fmicb.2023.1178474
Sakshi, & Haritash, A. K. (2020). A comprehensive review of metabolic and genomic aspects of PAH-degradation. *Archives of Microbiology*, *202*(8), 2033–2058. https://doi.org/10.1007/s00203-020-01929-5
Sangeetha Devi, R., Rajesh Kannan, V., Nivas, D., Kannan, K., Chandru, S., & Robert Antony, A. (2015). Biodegradation of HDPE by *Aspergillus* spp. From marine ecosystem of Gulf of Mannar, India. *Marine Pollution Bulletin*, *96*(1), 32–40. https://doi.org/10.1016/j.marpolbul.2015.05.050
Sarma, H., Hazarika, R. P., Kumar, V., Roy, A., Pandit, S., & Prasad, R. (2022, March). Microplastics in marine and aquatic habitats: Sources, impact, and sustainable remediation approaches. *Environmental Sustainability (Singapore)*, *5*(1), 39–49. https://doi.org/10.1007/s42398-022-00219-8
Schulze, S., Schiller, H., Solomonic, J., Telhan, O., Costa, K. C., & Pohlschröder, M. (2022). Advanced understanding of prokaryotic biofilm formation through use of a cost-effective and versatile multipanel adhesion (mPAD) mount. *American Society for Microbiology*, *88*(4). https://doi.org/10.1128/aem.02283-21
Shabbir, S., Faheem, M., Ali, N., Kerr, P. G., Wang, L., Sathishkumar, K., & Li, Y. (2020). Periphytic biofilm: An innovative approach for biodegradation of microplastics. *The Science of the Total Environment*, *717*, 137064–137064. https://doi.org/10.1016/j.scitotenv.2020.137064
Shilpa, S., Basak, N., & Meena, S. S. (2022, July). Microbial biodegradation of plastics: Challenges, opportunities, and a critical perspective. *Frontiers of Environmental Science and Engineering*, *16*(12), 161. https://doi.org/10.1007/s11783-022-1596-6
Skariyachan, S., Patil, A. A., Shankar, A., Manjunath, M., Bachappanavar, N., & Kiran, S. (2018). Enhanced polymer degradation of polyethylene and polypropylene by novel thermophilic consortia of *Brevibacillus* sps. and *Aneurinibacillus* sp. screened from waste management landfills and sewage treatment plants. *Polymer Degradation and Stability*, *149*, 52–68. https://doi.org/10.1016/j.polymdegradstab.2018.01.018
Sooriyakumar, P., Bolan, N., Kumar, M., Singh, L., Yu, Y., Li, Y., Weralupitiya, C., Vithanage, M., Ramanayaka, S., Sarkar, B., Wang, F., Gleeson, D. B., Zhang, D., Kirkham, M. B., Rinklebe, J., & M Siddique, K. H. (2022). Biofilm formation and its implications on the properties and fate of microplastics in aquatic environments: A review. *Journal of Hazardous Materials Advances*, *6*, 100077. https://doi.org/10.1016/j.hazadv.2022.100077

Sri Hari, S. Subramani., T., Prapanchan, V. N., & Li, P. (2023). Human health risk perspective study on characterization, quantification and spatial distribution of microplastics in surface water, groundwater and coastal sediments of thickly populated Chennai coast of South India. *Human and Ecological Risk Assessment: An International Journal*, *29*(1), 222–244. https://doi.org/10.1080/10807039.2022.2154635

Su, X., Yuan, J., Lu, Z., Xu, J., & He, Y. (2022). An enlarging ecological risk: Review on co-occurrence and migration of microplastics and microplastic-carrying organic pollutants in natural and constructed wetlands. *Science of the Total Environment*, *837*, 155772. https://doi.org/10.1016/j.scitotenv.2022.155772

Su, Y., Zhang, Z., Zhu, J., Shi, J., Wei, H., Xie, B., & Shi, H. (2021). Microplastics act as vectors for antibiotic resistance genes in landfill leachate: The enhanced roles of the long-term aging process. *Environmental Pollution*, *270*, 116278. https://doi.org/10.1016/j.envpol.2020.116278

Sutkar, P. R., Gadewar, R. D., & Dhulap, V. P. (2023). Recent trends in degradation of microplastics in the environment: A state-of-the-art review. *Journal of Hazardous Materials Advances*, *11*, 100343. https://doi.org/10.1016/j.hazadv.2023.100343

Tang, K. H. D., & Li, R. (2024). The effects of plastisphere on the physicochemical properties of microplastics. *Bioprocess and Biosystems Engineering, 48*, 1–15. https://doi.org/10.1007/s00449-024-03059-4

Tu, C., Zhou, Q., Zhang, C., Liu, Y., & Luo, Y. (2020). Biofilms of microplastics. In D. He & Y. Luo (Eds.), *Microplastics in Terrestrial Environments: Emerging Contaminants and Major Challenges* (pp. 299–317). Springer International Publishing. https://doi.org/10.1007/698_2020_461

Unnikrishnan, V., Anusree, S., Shaikh, I., D'Costa, P. M., Chandran, T., Valsan, G., Vandana, T. U., Tamrakar, A., Paul, M. M., Rangel-Buitrago, N., & Warrier, A. K. (2024). Insights into the seasonal distribution of microplastics and their associated biofilms in the water column of two tropical estuaries. *Marine Pollution Bulletin*, *206*, 116750. https://doi.org/10.1016/j.marpolbul.2024.116750

Vaksmaa, A., Polerecky, L., Dombrowski, N., Kienhuis, M. V. M., Posthuma, I., Gerritse, J., Boekhout, T., & Niemann, H. (2023). Polyethylene degradation and assimilation by the marine yeast Rhodotorula mucilaginosa. *ISME Communications*, *3*(1), 68. https://doi.org/10.1038/s43705-023-00267-z

Ventura, E., Marín, A., Gámez-Pérez, J., & Cabedo, L. (2024). Recent advances in the relationships between biofilms and microplastics in natural environments. *World Journal of Microbiology and Biotechnology*, *40*(7), 220. https://doi.org/10.1007/s11274-024-04021-y

Vimala, P. P., & Mathew, L. (2016). Biodegradation of polyethylene using *Bacillus Subtilis*. *Procedia Technology*, *24*, 232–239. https://doi.org/10.1016/j.protcy.2016.05.031

Viršek, M. K., Lovšin, M. N., Koren, Š., Kržan, A., & Peterlin, M. (2017). Microplastics as a vector for the transport of the bacterial fish pathogen species *Aeromonas salmonicida*. *Marine Pollution Bulletin*, *125*(1), 301–309. https://doi.org/10.1016/j.marpolbul.2017.08.024

Wang, S., Xue, N., Li, W., Zhang, D., Pan, X., & Luo, Y. (2020). Selectively enrichment of antibiotics and ARGs by microplastics in river, estuary and marine waters. *Science of the Total Environment*, *708*, 134594. https://doi.org/10.1016/j.scitotenv.2019.134594

Wright, R., Erni-Cassola, G., Zadjelovic, V., Latva, M., & Christie-Oleza, J. A. (2020, September). Marine plastic debris: A new surface for microbial colonization. *Environmental Science & Technology*, *54*(19), 11657–11672. https://doi.org/10.1021/acs.est.0c02305

Wu, C., Tanaka, K., Tani, Y., Bi, X., Liu, J., & Yu, Q. (2022). Effect of particle size on the colonization of biofilms and the potential of biofilm-covered microplastics as metal carriers. *Science of the Total Environment*, *821*, 153265. https://doi.org/10.1016/j.scitotenv.2022.153265

Wu, P., Cai, Z., Jin, H., & Tang, Y. (2019a). Adsorption mechanisms of five bisphenol analogues on PVC microplastics. *Science of the Total Environment*, *650*, 671–678. https://doi.org/10.1016/j.scitotenv.2018.09.049

Wu, X., Pan, J., Li, M., Li, Y., Bartlam, M., & Wang, Y. (2019b). Selective enrichment of bacterial pathogens by microplastic biofilm. *Water Research*, *165*, 114979. https://doi.org/10.1016/j.watres.2019.114979

Xu, X., Wang, S., Gao, F., Li, J., Zheng, L., Sun, C., He, C., Wang, Z., & Qu, L. (2019). Marine microplastic-associated bacterial community succession in response to geography, exposure time, and plastic type in China's coastal seawaters. *Marine Pollution Bulletin*, *145*, 278–286. https://doi.org/10.1016/j.marpolbul.2019.05.036

Yan, X., Chio, C., Li, H., Zhu, Y., Chen, X., & Qin, W. (2024). Colonization characteristics and surface effects of microplastic biofilms: Implications for environmental behavior of typical pollutants. *Science of the Total Environment*, *937*, 173141. https://doi.org/10.1016/j.scitotenv.2024.173141

Yoshida, S., Hiraga, K., Takehana, T., Taniguchi, I., Yamaji, H., Maeda, Y., Toyohara, K., Miyamoto, K., Kimura, Y., & Oda, K. (2016). A bacterium that degrades and assimilates poly(ethylene terephthalate). *Science*, *351*(6278), 1196–1199. https://doi.org/10.1126/science.aad6359

Yu, Z., Wang, Y., Lu, J., Bond, P. L., & Guo, J. (2021). Nonnutritive sweeteners can promote the dissemination of antibiotic resistance through conjugative gene transfer. *The ISME Journal*, *15*(7), 2117–2130. https://doi.org/10.1038/s41396-021-00909-x

Yuan, J., Ma, J., Sun, Y., Zhou, T., Zhao, Y., & Yu, F. (2020). Microbial degradation and other environmental aspects of microplastics/plastics. *Science of the Total Environment*, *715*, 136968. https://doi.org/10.1016/j.scitotenv.2020.136968

Zettler, E. R., Mincer, T. J., & Amaral-Zettler, L. A. (2013). Life in the "plastisphere": microbial communities on plastic marine debris. *Environmental Science & Technology*, *47*(13), 7137–7146. doi: 10.1021/es401288x.

Zhang, J., Li, W., Chen, J., Wang, F., Qi, W., Li, Y., & Xie, B. (2019). Effect of hydraulic conditions on the prevalence of antibiotic resistance in water supply systems. *Chemosphere*, *235*, 354–364. https://doi.org/10.1016/j.chemosphere.2019.06.157

Zhang, J., Ma, W., Li, Y., Zhong, D., Zhou, Z., & Ma, J. (2024). The resistance change and stress response mechanisms of chlorine-resistant bacteria under microplastic stress in drinking water distribution system. *Environmental Pollution*, *356*, 124331. https://doi.org/10.1016/j.envpol.2024.124331

Zhang, Y., Gu, A. Z., Cen, T., Li, X., He, M., Li, D., & Chen, J. (2018). Sub-inhibitory concentrations of heavy metals facilitate the horizontal transfer of plasmid-mediated antibiotic resistance genes in water environment. *Environmental Pollution*, *237*, 74–82. https://doi.org/10.1016/j.envpol.2018.01.032

Zhang, Y., Lu, J., Wu, J., Wang, J., & Luo, Y. (2020). Potential risks of microplastics combined with superbugs: Enrichment of antibiotic resistant bacteria on the surface of microplastics in mariculture system. *Ecotoxicology and Environmental Safety*, *187*, 109852. https://doi.org/10.1016/j.ecoenv.2019.109852

8 How Do Microplastics Distribute Through Freshwater Ecosystems? Which Biota, Feeding Groups, and Trophic Levels Are Most at Risk?

Heinrich T.J. Dahms, Shako Senyolo, Simone Dahms-Verster, and Richard Greenfield

8.1 INTRODUCTION

The discovery of plastic pollution in oceans that was made in 1971 by Carpenter and Smith (1972) was a warning to the threat of plastic pollution in the environment. Similarly, small plastic beads and particles were discovered worldwide such as those reported by Gregory (1977) around beaches in New Zealand and by Ryan (1990) on South African beaches termed "microplastics". Arguably, the publication by Thompson et al. (2004) created a surge of research into plastic and microplastics (MPs) in marine environments and biota in the last 20 years (Blettler et al., 2018; Eerkes-Medrano and Thompson, 2018; He et al., 2018; Li et al., 2018). This, however, led to a scenario where research into MPs in freshwater environments was lagging far behind marine environments (Blettler et al., 2018; Eerkes-Medrano and Thompson, 2018; He et al., 2018; Li et al., 2018). Researchers have tried to rectify this over the last several years.

MPs have now been detected in freshwater ecosystems worldwide, in multiple biota, and various environments including pristine and urban rivers, streams, dams, lakes, ponds, and reservoirs (Li et al., 2018; Lv et al., 2019; Watkins et al., 2019; Xiong et al., 2018; Dahms et al., 2020; Dalu et al., 2023; Dahms and Greenfield, 2024). However, how MPs are distributed in freshwater environments is uncertain, with various anthropogenic and environmental factors influencing it. Unfortunately, the distribution of MPs is not limited to the environment, with various freshwater biota having been found to ingest significant amounts of various MP shapes, colours, and polymers.

MPs have been found in simple freshwater organisms such as benthic macroinvertebrates to complex fish, birds, and mammal species (Li et al., 2018). MPs can enter biota either directly by animals mistaking it as a food source or secondarily through bioaccumulation (de Sá et al., 2018). When these MPs are ingested, they can have severe negative health impacts. This includes gut blockages in smaller organisms, reduced movement and reproduction, cancer, inhibited ability to expel toxins, and the death of the animal (Guo and Wang, 2019).

Understanding how and where MPs are distributed in ecosystems is important to determine how they impact various organisms. The aim of this chapter is to discuss how MPs distribute in freshwater ecosystems and which biota, functional feeding groups, and trophic levels are most at risk. This will aid researchers in assessing which organisms and environments require more research, and where greater MP mitigation is required to protect freshwater ecosystems.

DOI: 10.1201/9781032717814-8

8.2 HOW DO MICROPLASTICS DISTRIBUTE IN FRESHWATER ENVIRONMENTS?

Freshwater ecosystems are not homogenous environments and may have significant differences in their characteristics. This is evident in the River Continuum Concept (RCC), which describes how river habitats change along a river from source to sea. The RCC explains that along a river system, there are changes in flow, velocity, geology, plant life, and alterations in riverine biota due to topographical changes in the surrounding environment (Vannote et al., 1980; Dodds and Maasri, 2022). These changes along the river system lead to different river habitats such as rapids, riffles, runs, cascades, pools, and habitats like wetlands, lakes, and dams (DWAF, 2009). Each of these environments has different physical and chemical properties that impact plant life and biota (Vannote et al., 1980; Dobbs and Maasri, 2022). How MPs are distributed according to the RCC must similarly be considered.

Once the widespread presence of MPs in freshwater was determined, researchers started to investigate which abiotic components may impact the distribution of MPs. Studies by Nel et al. (2018), Dahms et al. (2020), Park et al. (2020), Huang et al. (2020), Tien et al. (2020), and Lu et al. (2022) highlighted how environmental variables impacted MP abundances. These aspects included water quality, organic content, water velocity, sediment grain sizes, plant life, and depth (Nel et al. 2018; Dahms et al., 2020; Huang et al., 2020; Park et al., 2020; Tien et al., 2020; Lu et al., 2022). Environments with finer sediment grain sizes were correlated to higher MP concentrations in sediment, primarily due to decreased river velocity, which led to the deposition of MPs (Nel et al., 2018; Dahms et al., 2020; Huang et al., 2020; Tien et al., 2020). This was again related to increased MP concentrations in water in environments with low depth, increased flow, and larger sediment grain sizes (Dahms et al., 2020; Huang et al., 2020; Park et al., 2020)

The findings in these studies highlight that without proper contextualizing information, a concentration of MPs cannot be correctly understood, and high or low concentrations may provide a pseudo-indication of the extent of MP pollution at a site. Water velocity and most notably water discharge at a site can completely alter how MPs may be distributed in freshwater environments, and critically, how they may impact biota (Nel et al., 2018; Dahms et al., 2020; Park et al., 2020; Tien et al., 2020; Lu et al., 2022). These points highlight the importance of collecting environmental or contextual data of not only the site and season, but also the spots in the water body where MP samples are collected.

Rivers were initially characterized as highways for plastic pollution where plastic and MPs from inland sources could be transported downstream into the oceans (Schmidt et al., 2017; Lebreton et al., 2017; Meijer et al., 2021). This definition became less frequent with increased MP research in river systems and the impact that abiotic characteristics have on MP distribution (Weideman et al., 2019; Watkins et al., 2019; Dahms et al., 2020; Lu et al., 2022). Due to the heterogeneity of rivers, sections of river habitats may have different characteristics where MPs can be distributed accordingly (DWAF, 2009).

Typical river habitats like cascades, rapids, and riffles are environments with increased velocity (DWAF, 2009). These habitats create an environment that may be more inclined to increased resuspension of MPs into the water column (Dahms et al., 2020; Owowenu et al., 2023). Similarly, the higher velocity and decreased depths correlate to large sediment grain sizes, where MPs may become more easily resuspended after periods of decreased flow (Dahms et al., 2020; Owowenu et al., 2023). River habitats such as runs and pools where rivers lose their energy may then provide environments more conducive to the deposition of MPs. These environments could allow for biofilm growth on MPs, increasing the density of MPs, and becoming trapped with the fine sediment that deposits in these environments (Guo and Wang, 2019). In these environments, it is critical to determine the MP concentrations in the sediment, rather than only the surface water.

Dams and lakes within river systems are environments similar to pools where MPs may become deposited from reduced flow (Watkins et al., 2019; Dahms et al., 2022). These environments have been identified as points that might act as sinks for MPs, with low concentrations of MPs found in water (Weideman et al., 2020; Watkins et al., 2019; Dahms et al., 2022). Studies that compared the sediment and water MP content found that sediment in dams may be more contaminated, highlighting how, in areas of reduced flow, many MPs may deposit into the sediment (Watkins et al., 2019; Dahms et al., 2022). Dams could therefore be environments where benthic organisms may be impacted to a greater extent by MPs than pelagic species. Results indicating lower numbers of MPs in water than in sediments in dams highlight the importance of assessing the entire environment to gain a more accurate indication of MP distribution (Watkins et al., 2019; Weideman et al., 2019, 2020; Dahms et al., 2022; Dahms and Greenfield, 2024). Furthermore, it must be noted that many dam walls release a consistent volume of water downstream, potentially releasing large quantities of MPs downstream of the dam as detected in the studies by Watkins et al. (2019) and Dahms et al. (2022).

Wetlands are known to naturally act as biofilters, with increased plant life filtering suspended particles, improving water quality, and reducing the presence of toxicants downstream (Dordio et al., 2008). These environments lead to reduced water flow creating an environment where MPs may deposit into the sediment at greater concentrations (Lu et al., 2022). This was found in a study by Lu et al., 2022, where an urban natural wetland showed a significant ability of filtering out MPs from water and depositing it in the sediment. These findings highlight the importance of continued research on the role of wetlands in MP distribution, which is an under-researched environment (Lu et al., 2022; Owowenu et al., 2023; Dahms and Greenfield, 2024). The role of how plant life and algae could contribute to MPs becoming trapped in freshwater habitats outside of wetlands needs to be explored.

Increased MP content can be correlated with poor water quality, particularly in urban freshwater environments close to Wastewater Treatment Plants (WWTPs) (Dahms et al., 2020; Tien et al., 2020). This correlation is possibly due to the release of MPs from WWTP that could release billions of MPs each day (Li et al., 2018). These stretches within a freshwater system can be regarded as areas of concern where large abundances of MPs are entering freshwater habitats which are then transported downstream (Li et al., 2018; Sun et al., 2019).

MP, however, is a broad term for a variety of plastic polymers, each with unique characteristics, and different densities (Jansen, 2016; Coppock et al., 2017; Geyer, 2020). In freshwater environments, where the freshwater is less dense than the saltwater of marine environments, a question can be raised whether polymers distribute according to density. The mean density of freshwater correlates to ≈ 1 g/cm^3, which is less dense than polymers such as polyvinyl chloride which would not float, but denser than polymers such as polypropylene and polyethylene which could remain in the water column (Mark, 2009). However, the density of polymers can change in the environment by biofilm growth (Guo and Wang, 2019). How the densities of different polymers impact their distribution requires much-needed research. Organisms that live in the sediment or benthic fish and invertebrates could therefore be exposed to different polymers of plastic than pelagic species. MPs and rivers are equally heterogeneous, and this creates a variety of challenges to describe and compare concentrations of MPs accurately. It is therefore important to continue assessing and researching MPs and how they interact with freshwater environments.

Several abiotic components impact how MPs distribute through freshwater environments. It is, however, necessary to stress that freshwater ecosystems are extremely complex and heterogeneous in nature. An example might be pools found in river sections with riffles or cascades that may have large quantities of plant life. The overall goal of this description of how MPs distribute through ecosystems is to highlight which environments in a freshwater environment may have higher or lower MP abundances. It also highlights the critical importance of contextualizing the environment where sampling of MPs occurs and to sample various habitats along freshwater systems.

8.3 MICROPLASTICS IN FRESHWATER BIOTA

When MPs enter aquatic environments, they can become ingested by organisms (de Sá et al., 2018). This can range from small filter-feeding Zooplankton to large predatory organisms, such as sperm whales (de Sá et al., 2018). The ingestion of MPs by biota occurs through two pathways in aquatic environments (Li et al., 2018). MPs are either ingested directly through mistaken identity or when drinking or by secondary uptake where MPs bioaccumulate to larger predatory organisms (Ory et al., 2017; Lehtiniemi et al., 2018; Li et al., 2018; de Sá et al., 2018). Many filter-feeding organisms such as mussels, clams, and other detritivores mistakenly ingest MPs in the water column with no ability to prevent their ingestion (Berglund et al., 2019; Baldwin et al., 2020). Other organisms, such as certain fish species, have mistaken MPs for prey, such as marine fish mistaking blue MPs that look similar in size, colour, and shape to the blue copepods they feed on (Ory et al., 2017). In laboratory experiments, to prove bioaccumulation of MPs, species such as the three-fin stickleback (*Gasterosteus aculeatus*) ingested MPs when feeding on littoral mysids (*Praunus* sp) that were exposed to MPs (Lehtiniemi et al., 2018). MPs have been found in all classes of vertebrates including multiple invertebrate groups. Fish remain the most studied group of organisms for MP ingestion in freshwater environments followed by invertebrates, birds, amphibians, reptiles, and mammals being the least studied group (Figure 8.1).

8.3.1 Microplastics in Invertebrates

The consumption of MPs by different types of freshwater invertebrates has not been widely reported (Table 8.1) (Akindele et al., 2019). Freshwater invertebrates are important consumers in the freshwater ecosystem and serve as potential entry points for contaminants, such as MPs, into the food chain. When these organisms ingest MPs, the particles can accumulate in their bodies and be transferred through different levels of the food chain, as these organisms are eaten in numbers by various organisms (both vertebrates and invertebrates). The ingestion of MPs by freshwater invertebrates has been observed in all continents, except Antarctica, and includes various phyla such as Arthropoda, Mollusca, and Annelida (Table 8.1) (Li et al., 2018). MPs do not only need to be in the water column to become available for ingestion by biota. The MPs that become trapped between sediment can also become bioavailable when being ingested by benthic macroinvertebrates (Nel et al., 2018). These organisms form a large percentage of the bio load, are a food source to many predators, and therefore can allow MPs to distribute from the sediment into the food chain.

Some of these arthropods that are known to ingest MPs include (Table 8.1) *Chironomus* spp. (Nel et al., 2018; Akindele et al., 2020; Dahms et al., 2020), *Chalcolestes viridis* (formerly *Lestes viridis*), *Siphlonurus* sp. (Akindele et al., 2020), *Cryptorchestia garbinii* (Iannilli et al., 2020), and *Paratya australiensis* (Nan et al., 2019). *Chironomus* spp. larvae are deposit feeders and therefore feed directly from the river sediment (Nel et al., 2018). The riverbed and sediment aquatic systems have a higher retention capacity and are considered the most significant sink for pollutants in freshwater environments (Akindele et al., 2020). The diet of the larvae includes detritus, associated microorganisms, algae, macrophytes, and other macroinvertebrates (Fry, 2021) and so their broad food selection habits can lead them to mistake MPs for dead plant debris or food.

The research conducted on Chironomid larvae by Nel et al. (2018) highlighted interesting findings regarding seasonal differences in the maximum MP burden in organisms. They found that the MP burden was significantly higher in the winter months than in the summer months. This is important as the Bloukrans River sampled in the study is located in a winter rainfall area, and one would therefore assume increased water velocities would reduce the concentration of MPs in the sediment. Season and associated precipitation patterns therefore have an impact on the digestion of MPs by benthic macroinvertebrates. This suggests that chironomids can effectively expel MP particles from their bodies. Additionally, it is proposed that the transition of chironomid larvae into the pupa stage of development, characterized by changes in their cephalothorax as noted by

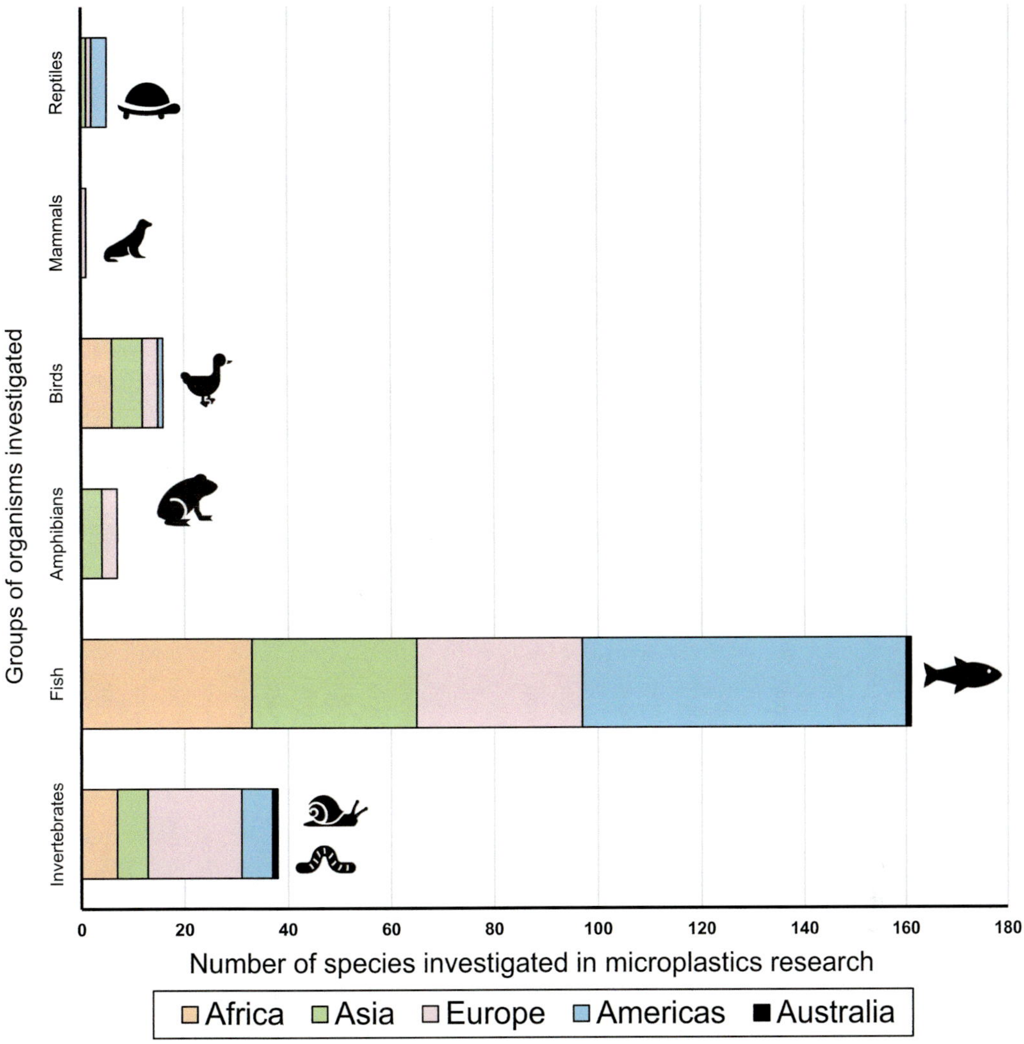

FIGURE 8.1 Number of species found to ingest microplastics in freshwater systems across several continents and which major group the species belong to (see references in Table 8.1).

Karima (2021), leads to a shift in their diet. As a result, they ingest fewer MPs during the summer compared to the winter season.

Nel et al. (2018) reported that deposit feeders, especially Chironomus species, are suitable bioindicators for MPs in a river system in South Africa. This suggests that deposit feeders could be effective indicators of MP pollution in fluvial freshwater environments, as they are specific to their habitat and can provide insights into long-term environmental impacts. Additionally, Akindele et al. (2020) found that deposit feeders encounter a wider range of polymers compared to predatory insects.

Various species of molluscs have been reported to ingest MPs. These species include *Lanistes varicus*, *Theodoxus fluviatilis*, and *Melanoides tuberculata* (Akindele et al., 2019); *Corbicula fluminea* (Su et al., 2016; Su et al., 2018; Baldwin et al., 2020); *Nerita articulata*, *Nerita polita*, and *Chicoreus capucinus* (Zaki et al., 2021); *Dreissena bugensis* (Baldwin et al., 2020); *Lasmigona costata* (Wardlaw and Prosser, 2020); and *Anodonta anatina* (Berglund et al., 2019). In various studies, the soft tissue (whole organism excluding the shell) of molluscs has been analysed for MP

TABLE 8.1
A Summary of Freshwater Species Found to Ingest Microplastics, Where They Were Collected, Which Class of Organisms They Belong to, Their Feeding Strategy, and the Abundance of Microplastics Found in Them

Continent	Study Location	Class	Organism	Feeding Strategy	Microplastics Concentrations	Reference
Africa	South Africa, Braamfontein Spruit	Invertebrates	*Chironomidae* sp.	Detritivore	53.4 particles g^{-1} ww	Dahms et al. (2020)
	South Africa, Bloukrans River		*Chironomidae* sp.	Detritivore	0.37 ± 0.44 (summer) and 1.12 ± 1.19 (winter) particles mg^{-1} ww	Nel et al. (2018)
	Nigeria, Osun River		*Lestes virdis*	Carnivore	43.29 ± 43.29 particles per gram ww	Akindele et al. (2020)
			Siphlonurus sp.	Omnivore	62.36 ± 3.53 particles per gram ww	
			Chironomus sp.	Detritivore	291.76 ± 26.55 Particles per gram ww	
	Nigeria, Osun River		*Lanistes varicus*	Herbivore	3.8 ± 0.83 particles per mg	Akindele et al. (2019)
			Melanoides tuberculata	Herbivore	1.7 ± 0.42 particles per mg	
	South Africa, Livuhu, and Crocodile Rivers	Fish			Ranges per species of:	Dalu et al. (2023)
			Labeobarbus marequensis	Omnivore	10–87	
			Labeo rosae	Omnivore	15–47	
			Opsaridium peringueyi	Carnivore	20–71	
			Chiloglanis Paratus	Omnivore	42–54	
			Enteromius trimaculatus	Carnivore	21–21	
			Enteromius eutaenia	Carnivore	23–47	
			Micralestes acutidens	Carnivore	20–119	
			Oreochromis mossambicus	Omnivore	16–73	
			Serranochromis meridianus	Carnivore	22–22	
			Talipia sparrmanii	Omnivore	24–59	
			Enteromius paludinosus	Omnivore	24–43	
			Oreochromis niloticus	Omnivore	37–61	
			Schilbe Intermedius	Carnivore	54–54	

(Continued)

TABLE 8.1 (*Continued*)
A Summary of Freshwater Species Found to Ingest Microplastics, Where They Were Collected, Which Class of Organisms They Belong to, Their Feeding Strategy, and the Abundance of Microplastics Found in Them

Continent	Study Location	Class	Organism	Feeding Strategy	Microplastics Concentrations	Reference
	South Africa, Vaal River		*Clarias gariepinus*	Omnivore	7.47 particles per individual	Dahms et al. (2022)
	South Africa, Vaal River		*Cyprinus carpio*	Omnivore	26.23 ± 12.57 particles per individual	Saad et al. (2022)
	Nigeria, Eleyele Lake				Range per species of:	Adeogun et al. (2020)
			Hepsetus odie	Carnivore	1–2	
			Lates niloticus	Carnivore	1–4	
			Parachanna obscura	Omnivore	1–5	
			Chrysicthys nigrodigitatus	Carnivore	1–3	
			Sarotherodon melanotheron	Herbivore	1–34	
			Coptodon zillii	Herbivore	1–6	
			Oreochromis niloticus	Omnivore	1–4	
	Ethiopia, Lake Ziway		*Oreochromis niloticus*	Omnivore	2.06	Merga et al. (2020)
			Clarias gariepinus	Omnivore	21.2	
			Cyprinus carpio	Omnivore	56.3	
			Carassius carassius	Omnivore	4.8	
	Tanzania, Lake Victoria		*Oreochromis niloticus*	Omnivore	35% prevalence	Biginagwa et al. (2016)
			Lates niloticus	Carnivore	55% prevalence	Adu-Boahen et al. (2020)
	Ghana, Akora River		*Oreochromis niloticus*	Omnivore	0.17 particles per individual	
			Oreochromis aureus	Omnivore	2 particles per individual	
			Oreochromis mosssambicus	Omnivore	0.16 particles per individual	
			Sarotherodon melanotheron	Herbivore	0.1 particles per individual	
			Clarias anguillaris	Omnivore	1 particle per individual	

(*Continued*)

TABLE 8.1 (*Continued*)
A Summary of Freshwater Species Found to Ingest Microplastics, Where They Were Collected, Which Class of Organisms They Belong to, Their Feeding Strategy, and the Abundance of Microplastics Found in Them

Continent	Study Location	Class	Organism	Feeding Strategy	Microplastics Concentrations	Reference
	South Africa, two wetlands	Birds	*Anas erythrorhyncha*	Herbivore	2% prevalence	Reynolds and Ryan (2018)
			Anas smithii	Omnivore	17% prevalence	
			Anas undulata	Omnivores	3% prevalence	
			Dendrocygna viduata	Omnivore	0%	
			Alopochen aegyptiaca	Omnivores	1% prevalence	
			Plectropterus gambensis	Herbivore	3%	
Asia	China, Lake Taihu	Invertebrates	*Corbicula fluminea*	Filter feeder	0.2–12.5 items g^{-1} ww	Su et al. (2016)
	China, Yangtze River basin		*Corbicula fluminea*	Filter feeder	0.3–4.9 items g^{-1} ww; 0.4–5.0 items per individual	Su et al. (2018)
	Malaysia, Klang River		*Nerita articulate*	Herbivore	0.5–1.75 items g^{-1} ww (average for all species)	Zaki et al. (2021)
			Nerita polita	Herbivore		
			Chicoreus capucinus	Carnivore		
	China, Rice-fish cultures		*Procambarus clarkii*	Omnivore	2.5 particles per individual	Lv et al. (2019)
		Fish	*Monopterus albus*	Omnivore	3.3 particles per individual	
			Misgurnus anguillicaudatus	Omnivore	1.8 particles per individual	
	Thailand, Chi River				Prevalence	Kasamesiri and Thaimuangphol (2020)
			Labiobarbus siamensis	Detrivore	50%	
			Puntioplites proctozyson	Omnivore	13.30%	
			Cyclochelichthy repasson	Omnivore	29.60%	
			Henicorhynchus siamensis	Omnivore	28.60%	
			Labeo chrysophekadion	Detrivore	25%	
			Mystus bocourti	Carnivore	26.70%	
			Hemibagrus spilopterus	Carnivore	30%	
			Laides longibarbis	Detrivore	16.70%	

(*Continued*)

TABLE 8.1 (*Continued*)
A Summary of Freshwater Species Found to Ingest Microplastics, Where They Were Collected, Which Class of Organisms They Belong to, Their Feeding Strategy, and the Abundance of Microplastics Found in Them

Continent	Study Location	Class	Organism	Feeding Strategy	Microplastics Concentrations	Reference
	China, Xiangxi Bay		*Culter alburnus*	Carnivore	1.5 ± 1.38 particles per individual	Zhang et al. (2017)
			Culter dabryi	Carnivore	0.5 ± 0.71 particles per individual	
			Culter mongolicus	Carnivore	Not detected	
			Culter oxycephaloides	Carnivore	Not detected	
			Cyprinus carpio	Carnivore	Not detected	
			Pelteobagrus nitidus	Carnivore	Not detected	
			Pelteobagrus fulvidraco	Carnivore	0.33 ± 0.58 particles per individual	
			Pelteobaggrus vachelli	Carnivore	1 ± 1.41 particles per individual	
			Pseudobagrus ussuriensis	Carnivore	1 particle per individual	
			Siluriformes	Carnivore	Not detected	
			Siniperca chuatsi	Carnivore	Not detected	
			Sinibrama wui	Omnivore	Not detected	
			Squalidus argentatus	Omnivore	Not detected	
	Philippines, Lawaan River		*Kuhlia rupestris*	Omnivore	70 particles in total	Cabansag et al. (2021)
	Indonesia, Citarum River		*Chanos chanos*	Omnivore	1.111 ± 0.838 tissue items/fish	Sembiring et al. (2020)
	Indonesia, Citarum River		*Chanos chanos*	Omnivore	2.2 items/fish (gut and gill)	Fareza and Sembiring (2020)
	Thailand, Chi River				Prevalence of:	Kasamesiri and Thaimuang phol (2020)
			Mystus bocourti	Detritivore	73%	
			Puntioplites proctozysron	Omnivore	86.70%	
			Henicorhynchus siamensis	Omnivore	71.40%	
			Laides longibarbis	Omnivore	83.30%	
			Labiobarbus siamensis	Detritivore	50%	
			Cyclocheilichthys repasson	Carnivore	70.40%	

(*Continued*)

TABLE 8.1 (*Continued*)
A Summary of Freshwater Species Found to Ingest Microplastics, Where They Were Collected, Which Class of Organisms They Belong to, Their Feeding Strategy, and the Abundance of Microplastics Found in Them

Continent	Study Location	Class	Organism	Feeding Strategy	Microplastics Concentrations	Reference
			Hemibagrus spilopterus	Carnivore	70%	
			Labeo chrysophekadion	Detritivore	75%	
	South Korea, Han River				One fish per species	Park et al. (2020)
			Cyrinus carpio	Omnivore	48	
			Carassius cuvieri	Omnivore	4	
			Lepomis marcochirus	Carnivore	10	
			Micropterus salmoides	Carnivore	16	
			Silurus asotus	Carnivore	22	
			Channa argus	Carnivore	32	
	Taiwan, Fengshan River		*Oreochromis niloticus*	Omnivore	42 particles per individual	Tien et al. (2020)
			Pterygoplichthys pardalis	Omnivore	51 particles per individual	
			Carassius auratus	Omnivore	55 particles per individual	
			Leiognathus equulus	Carnivore	62 particles per individual	
			Pomadasys argenteus	Carnivore	45 particles per individual	
	China, Qinghai Lake		*Gymnocypris przewalskii*	Omnivore	5.4 ± 3.6 particles per individual	Xiong et al. (2018)
	China, Poyang Lake		*Carassius auratus*	Omnivore	Range of 0–18 particles per individual	Yuang et al. (2019)
	China, Taihu Lake		*Cyprinus carpio*	Omnivore	2.5 ± 1.3 particles per individual	Jabeen et al. (2017)
			Carassius auratus	Omnivore	1.9 ± 1.0 particles per individual	
			Hypophthalmichthys molitrix	Planktivore	3.8 ± 2.0 particles per individual	
			Pseudorasbora parva	Omnivore	2.5 ± 1.8 particles per individual	
			Megalobrama amblycephala	Herbivore	1.8 ± 1.7 particles per individual	
			Hemiculter bleekeri	Planktivore	2.1 ± 1.1 particles per individual	
	China, Lijang River		*Cyprinus carpio*	Omnivore	0.4 ± 0.7 particles per individual	Zhang et al. (2021)

(*Continued*)

TABLE 8.1 (*Continued*)
A Summary of Freshwater Species Found to Ingest Microplastics, Where They Were Collected, Which Class of Organisms They Belong to, Their Feeding Strategy, and the Abundance of Microplastics Found in Them

Continent	Study Location	Class	Organism	Feeding Strategy	Microplastics Concentrations	Reference
			Pelteobagrus fulvidraco	Carnivore	0.3 ± 0.1 particles per individual	
			Mystus macropterus	Carnivore	1.0 ± 0.5 particles per individual	
			Pelteobagrus vachelli	Carnivore	0.4 ± 0.4 particles per individual	
	China, Yangtze River	Amphibians	*Bufo gargarizans*	Omnivore	2.44–56.88 items g^{-1}	Hu et al. (2018)
			Microhyla ornate	Filter feeder	0.53–2.6 items g^{-1}	
			Rana limnochari	Omnivore	2.73 items g^{-1}	
			Pelophylax nigromaculatus	Omnivore	1.27–1.8 items g^{-1}	
	China	Reptiles	*Trachemys scripta*	Carnivore	N/A	Wang et al. (2010)
	China, Yancheng wetland	Birds	*Egretta garzetta*	Carnivore	N/A_refer to Jiang et al. (2024)	Jiang et al. (2024)
			Ardea cinerea	Carnivore	N/A_refer to Jiang et al. (2024)	
			Larus saundersi	Carnivore	N/A_refer to Jiang et al. (2024)	
			Anser cygnoides	Herbivore	43.43 ± 61.98 items g^{-1} faeces	
			Phalacrocorax sp.	Carnivore	N/A_refer to Jiang et al. (2024)	
			Alcedo atthis	Carnivore	N/A_refer to Jiang et al. (2024)	
Europa	Italy, Lakes Vico, and Bracciano	Invertebrates	*Cryptorchestia garbinii*	Detritivore	2.2; 1.8; 5; 4.6 Particles per individual (four sites)	Iannilli et al. (2020)
	United Kingdom, England, Irwell River		*Tubifex tubifex*	Detritivore	129 ± 65.4 particles g^{-1} tissue	Hurley et al. (2017)
	Wales, Taff, Usk, and Wye Rivers		Heptageniidae (Family)		50% prevalence (refer to Windsor et al., 2019)	Windsor et al. (2019)
			Baetidae (Family)			
			Hydropsychidae (Family)			
	Sweden, Hgije River		*Anodonta anatina*	Filter feeder	12.2 ± 9.1 particles per individual	Berglund et al. (2019)

(*Continued*)

TABLE 8.1 (*Continued*)
A Summary of Freshwater Species Found to Ingest Microplastics, Where They Were Collected, Which Class of Organisms They Belong to, Their Feeding Strategy, and the Abundance of Microplastics Found in Them

Continent	Study Location	Class	Organism	Feeding Strategy	Microplastics Concentrations	Reference
					3.2 ± 3.5 particles per individual	
	United Kingdom, England		Amphipoda		0.07 particles per individual	Parker et al. (2022)
			Annelida		0.56 particles per individual	
			Coleoptera		0.29 particles per individual	
			Diptera		0.06 particles per individual	
			Ephemeroptera		0.74 particles per individual	
			Gastropoda		0.29 particles per individual	
			Hemiptera (herbivorous)	Herbivores	0.12 particles per individual	
			Hemiptera (predatory)	Carnivores	0.12 particles per individual	
			Isopoda		0.17 particles per individual	
			Megaloptera		0.25 particles per individual	
			Odonata		0.24 particles per individual	
			Trichoptera		0.62 particles per individual	
		Fish	*Alburnus alburnus*	Carnivore	0.45 particles per individual	Parker et al. (2022) (continued)
			Barbatula barbatula	Carnivore	0.89 particles per individual	
			Cottus gobio	Carnivore	0.71 particles per individual	
			Gasterosteus aculeatus	Carnivore	0.56 particles per individual	
			Leuciscus leuciscus	Carnivore	0.57 particles per individual	
			Perca fluviatilis	Carnivore	0.35 particles per individual	
			Phoxinus phoxinus	Omnivore	0.76 particles per individual	
			Rutilus rutilus	Omnivore	0.57 particles per individual	
			Squalius cephalus	Omnivore	0.69 particles per individual	

(*Continued*)

TABLE 8.1 (*Continued*)

A Summary of Freshwater Species Found to Ingest Microplastics, Where They Were Collected, Which Class of Organisms They Belong to, Their Feeding Strategy, and the Abundance of Microplastics Found in Them

Continent	Study Location	Class	Organism	Feeding Strategy	Microplastics Concentrations	Reference
	Belgium, Flemish Rivers		*Gobio gobio*	Carnivore	9% prevalence	Slootmaekers et al. (2019)
	Germany, Baden-Württemberg		*Leuciscus leuciscus*	Carnivore	Mean of 0.2 ± 0.5 particles per individual of all fish collected	Roch et al. (2019)
			Barbus barbus	Carnivore		
			Squalius cephalus	Omnivore		
			Barbatula barbatula	Carnivore		
			Gobio gobio	Carnivore		
			Phoxinus phoxinus	Omnivore		
			Alburnus alburnus,	Carnivore		
			Neogobius melanostomus	Carnivore		
			Cobitis taenia,	Detritivore		
			Rutilus rutilus,	Omnivore		
			Scardinius erythrophthalmus	Omnivore		
			Coregonus wartmanni	Carnivore		
			Tinca tinca	Omnivore		
			Perca fluviatilis,	Carnivore		
			Blicca bjoerkna	Carnivore		
			Gasterosteus aculeatus	Carnivore		
			Lota lota	Carnivore		
			Gymnocephalus cernua	Carnivore		
			Esox Lucius	Carnivore		
			Abramis Brama	Omnivore		
			Silurus glanis	Omnivore		
			Sander lucioperca	Carnivore		

(Continued)

TABLE 8.1 (*Continued*)

A Summary of Freshwater Species Found to Ingest Microplastics, Where They Were Collected, Which Class of Organisms They Belong to, Their Feeding Strategy, and the Abundance of Microplastics Found in Them

Continent	Study Location	Class	Organism	Feeding Strategy	Microplastics Concentrations	Reference
	Turkey, Rize waterbodies	Amphibians	*Pelophylax ridibundus*	Omnivore	302.62 ± 159.05 items g^{-1}	Karaoglu and Gul (2020)
			Rana macrocnemis	Omnivore	306.69 ± 182.72 items g^{-1}	
	Italy, Karst Ponds, Central Apennines		*Triturus carnifex*	Carnivore	N/A_refer to Ianella et al. (2020)	Iannella et al. (2020)
	France, Sausset park	Reptiles	*Trachemys scripta*	Carnivore	N/A	Prevot-Julliard et al. (2024)
	Italy, Ticino River	Birds	*Alcedo atthis*	Carnivore	7.5% prevalence	Winkler et al. (2020)
	United Kingdom, Wales		*Cinclus cinclus*	Carnivore	0.9 ± 0.1 (per faecal sample); (11.3 ± 1.6) particles g^{-1} dw	Souza et al. (2020)_
	Italy, Ticino River	Mammal	*Lutra lutra*	Carnivore	Undefined: Presence noted	Smiroldo et al. (2019)
Americas	Canada, Grand River watershed	Invertebrates	*Lasmigona costata*	Filter feeder	71% prevalence	Wardlaw and Prosser (2020)
	United States of America, Lake Mead		*Corbicula fluminea*	Filter feeder	51.7 items per individual	Baldwin et al. (2020)
			Dreissena bugensis	Filter feeder	13; 2.7 items per individual	
	United States of America, Wisconsin, Kinnickinnic River		*Heptageniidae*	Herbivore	Range for all species: 1.2–33.2 particles g^{-1} dw	Simmerman and Wasik (2019)
			Hydropsychidae	Omnivore		
			Gammaridae	Detritivore		
		Fish	*Salmon trutta*	Carnivore	Range for all species 0–1242 items per individual	
			Salvelinus fontinalis	Carnivore		

(*Continued*)

TABLE 8.1 (*Continued*)

A Summary of Freshwater Species Found to Ingest Microplastics, Where They Were Collected, Which Class of Organisms They Belong to, Their Feeding Strategy, and the Abundance of Microplastics Found in Them

Continent	Study Location	Class	Organism	Feeding Strategy	Microplastics Concentrations	Reference
	United States of America, Lake Mead		*Morone saxatilis*	Carnivore	4.2 items per individual	Baldwin et al. (2020) (continued)
			Cyprinus carpio	Omnivore	11 items per individual	
	United States of America, reservoirs		*Micropterus salmoides*	Carnivore	24.7 ± 2.5 items per individual	Hurt et al. (2020)
			Dorosoma cepedianum	Omnivore	5.2 ± 0.4 items per individual	
	Canada, Wascana creek				Prevalence of:	Campbell et al. (2017)
			Esox lucius	Carnivore	83%	
			Catostomus commersoni	Omnivore	70%	
			Notropis atherinoides	Omnivore	70%	
			Pimephales promelas	Omnivore	50%	
			Eucalia inconstans	Omnivore	70%	
	Mexico, Gulf of Mexico watershed				Microplastics present (Yes or No)	Phillips and Bonner (2015)
			Dorosoma cepedianum	Omnivore	N/A	
			Dorosoma petenense	Filter feeder	N/A	
			Campostoma anomalum	Omnivore	N/A	
			Cyprinella lutrensis	Omnivore	N/A	
			Cyprinella venusta	Carnivore	N/A	
			Notropis amabilis	Omnivore	N/A	
			Notropis volucellus	Carnivore	N/A	
			Pimephales vigilax	Carnivore	N/A	
			Notropis sabinae	Carnivore	N/A	
			Notropis stramineus	Omnivore	N/A	

(Continued)

TABLE 8.1 (*Continued*)

A Summary of Freshwater Species Found to Ingest Microplastics, Where They Were Collected, Which Class of Organisms They Belong to, Their Feeding Strategy, and the Abundance of Microplastics Found in Them

Continent	Study Location	Class	Organism	Feeding Strategy	Microplastics Concentrations	Reference
			Astyanax mexicanus	Carnivore	N/A	
			Ameiurus natalis	Carnivore	N/A	
			Ictalurus punctatus	Carnivore	N/A	
			Noturus gyrinus	Carnivore	N/A	
			Fundulus notatus	Carnivore	N/A	
			Gambusia affinis	Omnivore	N/A	
			Lepomis auritus	Carnivore	N/A	
			Lepomis cyanellus	Carnivore	N/A	
			Lepomis humilis	Carnivore	N/A	
			Lepomis macrochirus	Carnivore	N/A	
			Lepomis megalotis	Carnivore	N/A	
			Lepomis microlophus	Carnivore	N/A	
			Micropterus salmoides	Carnivore	N/A	
	Brazil, Pajeu River		*Hoplosternum littorale*	Omnivore	3.6 particles per individual	Silva-Cavalcanti et al. (2017)
	Brazil, Amazon River		*Bagre bagrea*	Carnivore	12.8 particles per individual	Souza et al. (2018)
			Bagre marinusa	Carnivore	7.8 particles per individual	
			Caranx hipposa	Carnivore	30.7 particles per individual	
			Selene setapinnis	Carnivore	2 particles per individual	
			Lutjanus analisa	Carnivore	1 particle per individual	
			Lutjanus synagrisa	Carnivore	1 particle per individual	
			Narcine brasiliensisa	Carnivore	3 particles per individual	
			Polydactylus oligodona	Carnivore	3 particles per individual	
			Cynoscion leiarchusa	Carnivore	2 particles per individual	

(*Continued*)

TABLE 8.1 (*Continued*)
A Summary of Freshwater Species Found to Ingest Microplastics, Where They Were Collected, Which Class of Organisms They Belong to, Their Feeding Strategy, and the Abundance of Microplastics Found in Them

Continent	Study Location	Class	Organism	Feeding Strategy	Microplastics Concentrations	Reference
			Cynoscion microlepidotusa	Carnivore	1.3 particles per individual	
			Cynoscion virescensa	Carnivore	3 particles per individual	
			Macrodon ancylodona	Carnivore	2 particles per individual	
			Sphyrnidae Sphyrna tiburoa	Carnivore	9 particles per individual	
			Trichiuridae lepturus	Carnivore	2 particles per individual	
	Canada, Lakes Ontario and Superior		*Ameiurus nebulosus*	Omnivore	160 particles per individual	Munno et al. (2022)
			Catostomus commersonii	Carnivore	120 particles per individual	
			Neogobius melanostomus	Carnivore	24 particles per individual	
			Pimephales promelas	Omnivore	17 particles per individual	
			Perca flavescens	Carnivore	30 particles per individual	
			Luxilus cornutus	Omnivore	11 particles per individual	
			Notropis atherinoides	Omnivore	20 particles per individual	
			Notropis hudsonius	Omnivore	26 particles per individual	
			Coregonus spp	N/A	17 particles per individual	
			Salvelinus namaycush	Carnivore	27 particles per individual	
			Coregonus clupeaformis	Carnivore	17 particles per individual	
			Catostomus catostomus	Carnivore	11 particles per individual	
			Prosopium cylindraceum	Carnivore	14 particles per individual	
	United States of America, Texas, Brazos River Basin		*Lepomis auritus*	Carnivore	45% prevalence	Peters and Bratton (2016)
	United States of America, Massachusetts lake	Reptiles	*Chrysemys Picta Picta*	Herbivore	N/A	Padgett et al. (2010)

(*Continued*)

TABLE 8.1 (*Continued*)
A Summary of Freshwater Species Found to Ingest Microplastics, Where They Were Collected, Which Class of Organisms They Belong to, Their Feeding Strategy, and the Abundance of Microplastics Found in Them

Continent	Study Location	Class	Organism	Feeding Strategy	Microplastics Concentrations	Reference
	Brazil, Rio Grande		Emydidae	Carnivore	N/A	Silveira et al. (2019)
	United States of America, Mississippi		*Graptemys flavimaculata*	Omnivore	N/A	Selman et al. (2018)
	United States of America, Great Lakes	Birds	*Phalacrocorax auritus*	Carnivore	86% prevalence	Brookson et al. (2019_
Australia	Victoria	Invertebrates	*Paratya australiensis*	Herbivore	36.2% prevalence	Nan et al. (2020)
	Melbourne, several wetlands	Fish	Gambusia holbrooki	Carnivore	0.6 items per individual	Su et al. (2019)

contamination. The reported MP loads are typically measured in units of particles/organism or items/gram (wet weight).

Macrozoobenthic animals, such as snails (e.g., *Theodoxus fluviatilis*), are opportunistic feeders, which means they eat whatever comes their way. This makes them vulnerable to a high risk of MP contamination (Akindele et al., 2019). Carnivorous gastropods, like *Chicoreus capucinus*, are active predators that can unintentionally ingest prey contaminated with MPs. This indicates the potential for biomagnification and trophic transfer of MPs among organisms in the food web, which could ultimately transfer to humans through the consumption of seafood (Zaki et al., 2021). This potential pathway of MPs into humans from macroinvertebrates again highlights the importance of continued research of MPs in freshwater biota. Herbivorous gastropods, such as *Nerita* spp., can encounter MPs while grazing on algae. Feeding on algae in the presence of MPs can result in MP ingestion (Windsor et al., 2019; Zaki et al., 2021).

Various invertebrate bioindicators have been proposed to understand MP pollution in freshwater environments. These include the Asian clam (*Corbicula fluminea*) (Su et al., 2016, 2018) and macrozoobenthos in flowing water systems, as their long life span allows for effective assessment of pollution over time (Akindele et al., 2019). Species of annelids known to ingest MPs and used as bioindicators include *Tubifex tubifex*, which are common in freshwater systems (Hurley et al., 2017). They live in the top layers of freshwater sediment and are deposit feeders. The *Tubifex* sp. entire life cycle takes place within sediments. They can survive in highly polluted environments and are one of the last species to remain in deteriorating environmental conditions. *Tubifex tubifex* serves as a food source for many macroinvertebrates, including leeches and small benthivorous fish (Hurley et al., 2017). Additionally, they are consumed by salmon and trout, creating a potential link to the human food chain and increasing the risk of bioaccumulation. This is due to the worms' ability to withstand high concentrations of plastic additives and desorbed contaminants.

8.3.2 Microplastics in Fish

Fish are an important economic and environmental resource, being a source of food for many organisms including humans (Ali and Jauncey, 2004; Vitule et al., 2006). They can range throughout various trophic levels from primary to tertiary consumers, with multiple strategies to survive in different habitats (Bleckmann, 2024). Fish are similarly an important resource for environmental science and can be used as a bioindicator to assess anthropogenic impacts on biota in both laboratory and field studies (Barboza et al., 2018; Su et al., 2019). Due to their economic and ecological importance, various fish species in marine and freshwater environments have been investigated and found to ingest MPs (Table 8.1). In freshwater environments, fish collected in the Nile River and Lake Victoria in Africa (Biginagwa et al., 2016; Khan et al., 2020), the Amazon River in South America (Pegado et al., 2018), the Yangtze River in Eurasia (Hu et al., 2018; Su et al., 2018), and in the Great Lakes in North America (Munno et al., 2022) have all been found to ingest MPs (Table 8.1). The distribution of MPs in fish is concerning, being found in benthic and pelagic fish in each trophic system (Jabeen et al., 2017; Merga et al., 2020; Dahms et al., 2022; Dalu et al., 2023).

Species of fish across Africa (Table 8.1) have been found to ingest MPs (Alimi et al., 2021). This includes herbivorous species such as *Coptodon zillii*, which ingested a range of 1–6 plastics per fish in a study by Adeogun et al. (2020), as well as large predatory fish such as *Clarias gariepinus*, which ingested a mean of 7.47 particles per fish (Dahms et al., 2022). An invasive species, *Cyprinus carpio*, was investigated in Lake Ziway in Ethiopia and in the Vaal River in South Africa, with both studies noting much higher concentrations of MPs in the fish compared to other species (Merga et al., 2020; Saad et al., 2022). In both studies, *Cyprinus carpio* ingested a mean of 26.23 ± 12.57 particles per fish in the Vaal River and 56.3 mg/kg ww of plastic in fish (Merga et al., 2020; Saad et al., 2022). This species is an omnivorous fish that feeds by foraging on the sediment when feeding on detritus and macroinvertebrates (Dadedo et al., 2015). The feeding habit of this species and high abundances of MPs found in them suggests that they could possibly be ingesting MPs directly and through bioaccumulation. This is concerning considering *Cyprinus carpio* is a food source to many people across Africa and Asia, who might be impacted by toxicants that may be bioaccumulated by the fish from the plastic they ingest (Saad et al., 2022). Research of MPs in Africa has been described as lacking behind other continents, and with fish being such a key source of nutrition in many African countries, it is critical to continue to research the presence and impacts of MPs in freshwater fish of Africa.

A similar trend has been noticed in Eurasian and American freshwater bodies with multiple fish from many trophic levels and feeding strategies ingesting MPs. Concentrations of fish in Eurasia ranged from as low as 0.3 ± 0.1 items per fish in the benthic catfish species *Pelteobagrus fulvidraco* to as high as 58 particles per fish in *Oreochromis niloticus* (Tien et al., 2020; Zhang et al., 2021). *Cyprinus carpio* was one of the most investigated species highlighting its distribution and importance, although had lower concentrations than those investigated in Africa (Jabeen et al., 2017; Zhang et al., 2017; Merga et al., 2020; Park et al., 2020; Zhang et al., 2021; Saad et al., 2022).

Currently, no fish species have been selected as a universal bioindicator for MP pollution, although *Cyprinus carpio* and *Oreochromis niloticus* have been widely investigated and may be a good proposed fish bioindicator for MPs (Jabeen et al., 2017; Zhang et al., 2017; Adeogun et al., 2019; Adu Boahen et al., 2022; Baldwin et al., 2020; Khan et al., 2020; Merga et al., 2020; Park et al., 2020; Tien et al., 2020; Zhang et al., 2021; Saad et al., 2022; Dalu et al., 2023). Concentrations found in *Cyprinus carpio* range from as little as 0.4 ± 0.7 particles per fish to as high as 48 particles per fish (Park et al., 2020; Zhang et al., 2021). Similarly, concentrations of MPs in *Oreochromis niloticus* have ranged from 1 to 4 particles per fish at the lowest concentration and 58 particles per fish at the highest average found at a single site (Adeogun et al., 2019; Tien et al., 2020). Both *Cyprinus carpio* and *Oreochromis niloticus* are omnivores, feeding on a variety of material including plankton, detritus, and invertebrates. Similarly, both species are important as a food source to humans, with *Cyprinus carpio* having a wider distribution compared to *Oreochromis niloticus*, with both species being classified as invasive in some regions (Biginagwa et al., 2016; Saad et al., 2022).

A universal bioindicator would require a singular species that is found globally, with a standard method to be sampled, and must be known to ingest MPs from multiple pathways (directly and indirectly). The species cannot be sensitive to MPs and must be able to survive in organic-rich, highly polluted aquatic environments (Holt and Miller, 2010). This is challenging as fish are not site specific and can travel over long distances; therefore, the MP found within them could be a representation of a much larger area than the site where they were collected (Holt and Miller, 2010). A multi-layered index that would incorporate multiple trophic layers, classes, and species, used to produce an MP-biota score, would provide a more representative evaluation.

The wide variety of methods used between sampling MPs in water and sediment prevents accurate comparisons between studies. Fish as a metric can easily be used to compare MP concentrations between studies, as sampling methods are largely standardized (Weis, 2020; Dahms and Greenfield, 2024). However, no standard unit for reporting MPs in fish exists. Researchers have used units such as particles/items/MPs per fish, range of plastics per species, mg of MPs per kg of fish ww, percentage prevalence, and particles per kg of fish ww (Lv et al., 2019; Adeogun et al., 2020; Merga et al., 2020; Cabansag et al., 2021). Notably, researchers also collect fish for MPs through various methods such as gill nets, electronarcosis, netting, angling, or buying fish from local fishermen (Xiong et al., 2018; Baldwin et al., 2020; Dahms et al., 2022; Saad et al., 2022). Fish are then dissected and the whole gut is extracted for MP analysis through digestion with KOH or H_2O_2 (Xiong et al., 2018; Baldwin et al., 2020; Dahms et al., 2022; Saad et al., 2022). Other studies have also removed the gills and muscle tissue for MP analysis (Hurt et al., 2020; Park et al., 2020).

MPs are distributed widely in freshwater environments and therefore in many freshwater fish species. How these MPs could be impacting the health of fish in different habitats are unknown, as accurately contextualizing the size, number, ratio of polymers, and pollutants on MPs at a site is challenging (Weis, 2020; Owowenu et al., 2023). This leads to many toxicological studies using inaccurate concentrations of MPs significantly higher than environmentally relevant concentrations (Weis, 2020).

8.3.3 Microplastics in Amphibians

Amphibian species worldwide are facing a threat of extinction, largely due to their sensitivity to changes in climate, habitat, and contaminants (IUCN, 2019). MPs may be one of the contaminants affecting these amphibians. The issue of MP pollution in freshwater systems is receiving increasing attention, but there has been little investigation into small water bodies like streams, wetlands, vernal ponds, and farm ponds (Iannella et al., 2020).

There have been reports of amphibian species ingesting MPs, including *Bufo gargarizans*, *Microhyla ornate*, *Rana limnochari*, *Pelophylax nigromaculatus* (Hu et al., 2018), *Pelophylax ridibundus*, *Rana macrocnemis* (Karaoglu and Gul, 2020), and *Triturus carnifex* (Iannella et al., 2020). It is worth noting that all of these assessments were done on the tadpole stage, except for *Triturus carnifex*, also known as the Italian newt.

Research has shown that amphibians can be directly exposed to MPs as early as the tadpole stage, and there is a correlation between increasing tadpole size and the amount of MPs in their tissues (Hu et al., 2018). Tadpoles are primarily omnivorous and feed on a variety of nutritional sources, including biofilm, periphytic algae, invertebrates, and plants (Dodd, 2009). These nutritional sources could all lead to the intake and bioaccumulation of MPs. One study recorded the highest average abundance of MPs in *Rana macrocnemis* in the northeast of Turkey (Karaoglu and Gul, 2020), suggesting that tadpoles may transport MP contamination through the food chain to other trophic levels (both aquatic and terrestrial). The dominant contaminants found ingested by amphibians across studies were micropolymers such as polyester, polyethylene, polyethylene terephthalate, nylon, and acrylic (Hu et al., 2018; Karaoglu and Gul, 2020).

8.3.4 Microplastics in Waterfowl

Birds form part of the vertebrate class Aves, which consists of approximately 11,208 different species listed on the IUCN database in 2024 (IUCN, 2024). These organisms have a global distribution inhabiting habitats from the equator to the polar regions including oceans and freshwater environments (Ducatez and Lefebvre, 2014; Wang et al., 2021). The interaction of birds and plastic pollution was inevitable as they each share a wide and overlapping distribution. Initially, it was suggested that 78% of seabird species ingest MPs, which will increase to 99% in 2050 as plastic pollution increases globally (Wilcox et al., 2015; Basto et al., 2019; Wang et al., 2021).

The ingestion of MPs in birds inhabiting freshwater environments, however, remains less clear with little research conducted on these species. Six species of wetland birds were investigated across two wetlands in South Africa by Reynolds and Ryan (2018). MPs in faecal samples and feather brushes of the birds were collected from two sites (Reynolds and Ryan, 2018). MPs were then divided and reported into percentage of plastics found in each species (Reynolds and Ryan, 2018). The birds had a mean of 1.53 ± 0.64 particles per faecal sample and 1.44 ± 0.72 particles per feather brushing (Reynolds and Ryan, 2018). MPs were then also detected in the pellets of common kingfisher (*Alcedo atthis*) along the Ticino River in Northern Italy (Winkler et al., 2020). Only 12 MPs were collected from the 133 pellets that were collected in the study (Winkler et al., 2020). The MPs collected were investigated through Fourier Transform-Infrared Spectroscopy (FT-IR) analysis and three predominant polymers were found (polyethylene, polyurethane, and polypropylene) (Winkler et al., 2020).

When a study in China compared the presence of MPs in bird faeces between terrestrial and aquatic birds, Jiang et al. (2024), no significant differences were found between the two groups. It was, however, noted that the aquatic bird species Swan goose (*Anser cygnoides*) ingested significantly more MPs than any of the other aquatic and terrestrial species (Jiang et al., 2024). It was further noted that the distribution of the MPs was due to what the specific bird species feed on, with terrestrial birds having a diet made of fruits and seed, but the aquatic species ingesting aquatic invertebrates and fish (Jiang et al., 2024). With MPs being more prevalent in fish and invertebrates, it may lead to bioaccumulation of MPs in the aquatic birds (Jiang et al., 2024).

The concerns raised by Jiang et al. (2024) have also previously been raised when Brookson et al. (2019) investigated and found that double-crested cormorants (*Phalacrocorax auratus*) ingest MPs in the Great Lakes in North America. The study found a prevalence of MPs in the gut of 86% in the birds investigated (Brookson et al., 2019). In South Wales, another riverine predatory bird, the Eurasian dipper (*Cinclus cinclus*), was investigated and MPs were found in 45% of the faecal samples that were collected (Souza et al., 2020). MPs were collected in faecal samples with a mean of 0.9 ± 0.1 particles per faecal sample or 11.3 ± 1.6 particles per g dry weight (Souza et al., 2020).

The ever-increasing presence of MPs in rivers and other freshwater ecosystems will lead to an increased accumulation of MPs in aquatic bird species. These MPs that can be polluted with harmful toxicants, such as persistent organic pollutants (POPs), could impact bird species on a global scale (Guo and Wang, 2019). More research on MPs in not only aquatic but terrestrial bird species is needed, as research tends to be aimed at fish and macroinvertebrates (de Sá et al., 2018).

8.3.5 Microplastics in Freshwater Reptiles

MPs have been detected in all seven of the marine turtle species, however, of the freshwater species, only 5 of 352 species of freshwater turtles have ever been found to ingest MPs (Clause et al., 2021). These studies did not aim to quantify the concentration of MPs ingested by the biota, but only noted their presence in the gut of these species (Prévot-Julliard et al., 2007; Padgett et al., 2010; Wang et al., 2010; Selman and Lindeman, 2018). These studies include turtles from the Americas and Eurasia with no reptiles investigated on the other continents. Although only five species in total were noted to ingest MPs, they include herbivorous, carnivorous, and omnivorous species highlighting MPs penetration into multiple trophic levels. No other freshwater-inhabiting reptiles have been investigated. It would be recommended that apex species like crocodilians and alligators are investigated for MPs using the non-lethal technique published by Gonzalez-Jauregui et al. (2019). These organisms would be ingesting a variety of fish, birds, and other aquatic animals that could lead to bioaccumulation of MPs in them.

8.3.6 Microplastics in Freshwater Mammals

The presence of MPs in terrestrial and aquatic mammals remains low (Cheng et al., 2022). Apart from some marine mammals, almost no freshwater or terrestrial mammals other than rats, chickens, pigs, and humans have been investigated for MPs (Cheng et al., 2022). One study that investigated the feeding habits of otters (*Lutra lutra*) detected MPs in the gut of these animals (Smiroldo et al., 2019). Round amber-coloured MP beads were detected, which the authors suggested may look similar to fish eggs that the otters feed on (Smiroldo et al., 2019). This could be an indication that the animals had rather mistaken the beads as a food source rather than only bioaccumulating it through the food chain, as some fish species have (Ory et al., 2017). The limited knowledge on the presence and abundance of MPs in mammals that live in freshwater environments is concerning. Mammals play an important role in the ecosystem and feed on a variety of organisms through the food chain. It could be safe to assume that they would similarly bioaccumulate MPs as seen in the lower trophic levels. Similarly, mammals that drink water from freshwater environments potentially contaminated with MPs could be transporting it from the rivers into terrestrial habitats (Chang et al., 2022). As noted in previous sections, MPs are distributed in multiple species from small benthic macroinvertebrates to large predatory birds and mammals. It must be considered whether these MPs have an impact on freshwater mammals and terrestrial mammals, and how they could be impacted by them.

8.4 ECOTOXICOLOGICAL EFFECTS OF MICROPLASTICS ON FRESHWATER BIOTA

MPs can have a variety of toxicological impacts on biota, not only due to the size of the MPs but also due to their ability to act as a vector for other toxicants (Guo and Wang, 2019). MPs around the world have been found to carry a variety of pollutants, such as metals, and POPs, such as DDT and other pesticides (Guo and Wang, 2019). Although toxicological studies use concentrations of MPs that cannot be related to environmental abundances, they highlight how the presence of MPs can increase the toxicological impact of certain metals such as mercury (Barboza et al., 2018).

The impact of various MP polymers on different species in different regions is a crucial area of research. Studies in Table 8.2 have shown that polymers, such as polyethylene, polystyrene, and polypropylene, can have diverse effects on freshwater species. These effects range from mortality and decrease reproductive rates to physiological changes, such as decreased lipid concentration, increased oxidative stress, and altered gene expression. Invertebrates like *Potamoryrgus antipodarum*, *Gammarus fossarum*, and *Hydra attenuata*, as well as fish like *Danio rerio*, *Oreochromis niloticus*, *Barbodes gonionotus*, *Acanthurus dussumieri*, and *Cyprinus carpio* can suffer mortality when exposed to certain micropolymers (Blarer and Burkhardt-Holm, 2016; Imhof and Laforsch, 2016; Murphy and Quinn, 2017; Romano et al., 2017; Naidoo and Glassom, 2019; Xia et al., 2020; Ahmadifar et al., 2020). Amphibians like *Alytes obstetricans* can also be threatened by exposure to MP polymers (Bosch et al., 2021). The current body of research suggests a positive correlation between higher concentrations of MP particles and adverse effects (Murphy and Quinn, 2017) and emphasizes that the toxicity of MPs is dependent on exposure duration. Moreover, existing studies indicate that the type and shape of MP polymers can influence their toxicity (Imhof and Laforsch, 2016; Murphy and Quinn, 2017). It is, however, important to note that further research is required to comprehensively understand these relationships, as a limited number of polymers have been examined in exposure studies that can be directly extrapolated to real-time freshwater environments. This discrepancy between field and laboratory conditions is notable, as most field studies provide data on a mixture of polymers. Hence, there is a clear need for more in-depth investigations to bridge the gap between laboratory-controlled and real-world settings.

It is important to note that while the impact of MP polymers on invertebrates, fish, and amphibians has been demonstrated, no data has been provided for birds and mammals to our knowledge. This highlights a gap in our understanding of how these MP polymers may be affecting these animal groups. Further research is needed to assess the potential impact on birds and mammals to gain a comprehensive understanding of the overall ecotoxicological implications.

According to Table 8.2, Europe leads in this research field with a greater number of studies conducted on the effects of different polymer types on various species. The data indicates that polystyrene and polyethylene are the most concerning polymers due to their consistent negative effects on species across different continents. These effects include increased mortality rates, decreased reproductive rates, increased oxidative stress, and various physiological and biochemical changes in the exposed organisms. The findings underline the necessity for further research and environmental considerations regarding the use and disposal of these polymers to minimize their impact on freshwater ecosystems and species.

TABLE 8.2
Biological Effects of Microplastic Polymer Exposure on Freshwater Organisms from Different Continents

Species	Continent	Exposure/Period	Polymer Type	Biological Effect	Reference
			Invertebrates		
Potamoryrgus antipodarum	Europe	Added to food – 8 weeks	Polyamide Polyethylene terephthalate Polycarbonate Polystyrene Polyvinyl chloride	Mortality	Imhof and Laforsch (2016)
	Europe	Added to sediment – 31 days	Polystyrene	Mortality Decreased reproductive rates	Romero-Blanco et al. (2021)
Gammarus fossarum	Europe	Added to food – 0.5 hours to 28 days	Polyamide, polystyrene	Mortality	Blarer and Burkhardt-Holm (2016)
Palaemonetes pugio	North America	3 hours Added to water	Polyethylene, polypropylene, polystyrene	Mortality	Gray and Weinstein (2017)
Dreissena polymorpha	Europe	Added to water – 6 days	Polystyrene	Increased dopamine concentration	Magni et al. (2018)
Gammarus pulex	Europe	Added to sediment – 28 days	Polystyrene	Significant growth reduction	Redondo-Hasselerharm et al. (2018)
Hydra attenuata	North America	Added to water – 3, 24, 48, and 96 hours	Polyethylene	Decreased feeding rates, morphological changes	Murphy and Quinn (2017)
Eriocheir sinensis	Asia	Added to water – 7 to 21 days	Polystyrene	Decreased specific growth rate (SGR), Decreased hepatosomatic index level	Yu et al. (2018)
	Asia	Added to water – 7–14–21 days	Polystyrene	Decreased lysozyme activity, Increased phenoloxidase activity increased	Liu et al. (2019)

(Continued)

TABLE 8.2 (*Continued*)
Biological Effects of Microplastic Polymer Exposure on Freshwater Organisms from Different Continents

Species	Continent	Exposure/Period	Polymer Type	Biological Effect	Reference
Caenorhabditis elegans	Asia	Added to water – 72 hours	Polystyrene	Increased reactive oxygen species production, lipofuscin accumulation, and the expression of oxidative stress-related genes	Liu et al. (2019)
Chironomus tepperi	Australia	Added to water – 5–10 days	Polyethylene	Decreased number of juveniles, Decreased larval growth rates, Reduced antenna length, Decreased, survival rates	Ziajahromi et al. (2018)
Chironomus riparius	Europe	Added to water – 28 days	Co-polyamide	Increased number of larvae, Altered egg shape	Khosrovyan and Kahru (2019)
	Europe	Added to water – 36 hours	Polystyrene and tyre rubber	Over-expression of heat shock proteins	Carrasco-Navarro et al. (2021)
Lumbriculus variegatus	Europe	Added to water – 48 hours	Polyethylene	Decreased lipid concentration, Increased lipid peroxidation rates, Decreased catalase and Glutathione S activities	Silva et al. (2020)
Caenorhabditis elegans	Asia	Added to water – 72 hours	Polystyrene	Decreased mobility, Reduced growth rate, Increased reactive oxygen species concentration, Increased expression of daclk-1, ctk-1, ctk-2, ctk-3, glutathione S-transferase-4, isp-1, skn-4, sod-1, sod-2, sod-3, sod-4, and sod-5	Yu et al. (2020)
	Asia	Added to water – 24 hours	Polystyrene	Increased reactive oxygen species, concentration Increased expression of oxidative stress-related genes clk-1, ctl-1, sod-3, sod- 4, and sod-5	Chen et al. (2021)

(*Continued*)

TABLE 8.2 (*Continued*)
Biological Effects of Microplastic Polymer Exposure on Freshwater Organisms from Different Continents

Species	Continent	Exposure/Period	Polymer Type	Biological Effect	Reference
Dreissena bugensis	North America	Added to water – 25 days	Polyethylene	Decreased oxygen consumption rate, Mortality.	Pedersen et al. (2020)
Dreissena polymorpha	Europe	Added to water – 14 days	Polystyrene	Mortality.	Weber et al. (2020)
Neocaridina palmata	Europe	Added to food – 4–24 hours	Polyethylene, polyvinyl chloride, polystyrene	Decreased reproductive rates	Klein et al. (2021)
Fish					
Danio rerio (larvae)	Europe	Injected into embryos – 1–2–3–4 days	Polystyrene	Increased oxidative stress and lipid metabolism rate	Veneman et al. (2017)
	Europe	Added to water – 2–7–14 days	Polyethylene	Decreased oxidative stress, steroid synthesis, and muscle development	LeMoine et al. (2018)
	Asia	Added to water – 7 days	Polystyrene	Decreased transcription of genes related to glycolysis and lipid metabolism Decreased glutathione concentration and catalase activity	Wan et al. (2018)
Danio rerio	Asia	Added to water – 120 hours	Polystyrene	Decreased free swimming performance of larvae Increased inflammation and oxidative stress	Qiang and Cheng (2019)
	Asia	Added to water – 21 days	Polystyrene	Increased reactive oxygen species concentration Increased cell apoptosis rates in the testicle	Qiang and Cheng (2021)
	Asia	Added to water – 7–14–21–28–35 days	Polystyrene	Increased reactive oxygen species, production and lipid peroxidation Increased lactate dehydrogenase, alanine aminotransferase, and aspartate aminotransferase activities Decreased glutathione peroxidase, catalase, and superoxide dismutase activities	Umamaheswari et al. (2020)

(*Continued*)

TABLE 8.2 (*Continued*)
Biological Effects of Microplastic Polymer Exposure on Freshwater Organisms from Different Continents

Species	Continent	Exposure/Period	Polymer Type	Biological Effect	Reference
	Asia	Added to water – 21 days	Polypropylene	Physical damage to the intestines Oxidative stress Inflammation Decreased lipid concentration	Zhao et al. (2021)
	Europe	Added to water – 21 days	Polyethylene Polypropylene Polystyrene Polyvinyl chloride	Decreased circulation rate Decreased acetylcholinesterase activity	Hanslik et al. (2021)
	Europe	Added to water – 24–48–72–96 hours	Polypropylene Polystyrene	Mortality Decreased heart rate Reduced body length of larvae	Prata et al. (2021)
	South America	Added to water – 5 days	Polyethylene	Increased glutathione, catalase, superoxide dismutase, and acetylcholinesterase activities	Guimarães et al. (2020)
	South America	Added to water – 10–20 days	Polystyrene	Increased catalase, glutathione S-transferase, Na+/K+ ATPase activities and lipid peroxidation concentration	Rangasamy et al. (2021)
Symphysodon aequifosciotus	Asia	Added to water – 30 days	Polyethylene	Decreased trypsin, alkaline phosphatase and acetylcholinesterase activities Increased amylase, citrate synthase, cytochrome oxidase, and lactate dehydrogenase activities	Wen et al. (2017)
Oreochromis niloticus	Asia	Added to water – 0–1–3–6–10–14 days	Polystyrene	Mortality Decreased acetylcholinesterase activity	Ding et al. (2018)
	Asia	Added to water – 28 days	Polystyrene	Decreased superoxide dismutase and glutathione peroxidase activities Decreased malondialdehyde concentration	Ahmadifar et al. (2020)

(*Continued*)

TABLE 8.2 (*Continued*)
Biological Effects of Microplastic Polymer Exposure on Freshwater Organisms from Different Continents

Species	Continent	Exposure/Period	Polymer Type	Biological Effect	Reference
Carassius auratus	Asia	Added to food – 6 weeks	Ethylene-vinyl acetate copolymer Polystyrene Polyamide	Decreased weight Damaged oral cavity Microgranuloma and inflammation	Jabeen et al. (2018)
	Asia	Added to water – 1–3–7 days	Polystyrene	Retarded growth Decreased catalase and glutathione peroxidase activities Increased superoxide dismutase activity Increased heart rate Enlarged intestinal cavity	Yang et al. (2020)
Barbodes gonionotus	Asia	Added to water – 96 hours	Polyvinyl chloride	Mortality Increased trypsin and chymotrypsin activities	Romano et al. (2018)
Acanthurus dussumieri	Africa	Added to food – 95 days	Polyethylene Polyvinyl chloride Polystyrene	Mortality Decreased body length and weight	Naidoo and Glassom (2019)
Prochilodus lineatus	South America	Added to water – 24–96 hours	Polyethylene	DNA damage in liver cells Decreased glutathione conc. and acetylcholinesterase activities	Roda et al. (2020)
Cyprinus carpio	Asia	30–60 days Added to food	Polyvinyl chloride	Mortality Decreased malondialdehyde concentration Decreased superoxide dismutase, catalase, and glutathione S-transferase activities	Xia et al. (2020)
	Asia	Added to water – 10 days	Polyvinyl chloride	Mortality Abnormal swimming behaviour Increased mucus secretions Structural abnormalities	Darabi et al. (2022)

(*Continued*)

TABLE 8.2 (*Continued*)
Biological Effects of Microplastic Polymer Exposure on Freshwater Organisms from Different Continents

Species	Continent	Exposure/Period	Polymer Type	Biological Effect	Reference
Clarias gariepinus	Africa	Add to food – 45 days	Polyvinyl chloride	Increased Haemoglobin, mean corpuscular haemoglobin, red blood cell count, and lipid peroxidation level Decreased mean corpuscular volume and white blood cell counts Decreased glutathione peroxidase, superoxide dismutase, acetylcholinesterase, and catalase activities	Iheanacho and Odo (2020)
Amphibians					
Physalaemus cuvieri	South America	Added to water – 7 days	Polyethylene	Decreased locomotor activity No social aggregation in the presence of a potential predator Histopathological damage to the liver Mutagenic effects Cytotoxicity	da Costa Araújo and Malafaia (2020), da Costa Araújo et al. (2020°), da Costa Araújo et al. (2020b), and da Costa Araújo and Malafaia (2020)
Alytes obstetricans	Europe	Added to water – 14 days	Polystyrene	Mortality	Boyero et al. (2019)
	Europe	Added to water – 6 days	Polystyrene	Increased Batrachochytrium dendrobatidis (Bd) infection susceptibility	Bosch et al. (2021)
Birds, reptiles, and mammals (N/A)					

8.5 CONCLUSION

MPs are an emerging contaminant with global implications and can be linked to 12 of the United Nations Sustainable Development Goals (Walker, 2021). These contaminants were initially described widely in marine environments, but recent research highlights their global distribution in freshwater environments. The MPs behave uniquely according to their freshwater environment, distributing and accumulating in specific river habitats that could lead to interactions with various different species of biota from simple macroinvertebrates to specialized mammals and birds. It is no surprise that MPs have therefore been detected in over 250 different species of organisms across the globe. Although the toxicological impact is still being determined, initial studies indicate how MPs are not only harmful themselves but also increases toxicity of already known harmful pollutants. Whether known sublethal concentrations may become lethal in the presence of MPs due to increased toxicity must be investigated in future studies. Microplastic research has rapidly increased over the last decade and a more standardized approach, with standardized methods and reporting units, is necessary to accurately describe MPs in freshwater environments.

REFERENCES

Adeogun, A.O., Ibor, O.R., Khan, E.A., Chukwuka, A.V., Omogbemi, E.D., Arukwe, A., 2020. Detection and occurrence of microplastics in the stomach of commercial fish species from a municipal water supply lake in southwestern Nigeria. *Environ. Sci. Pollut. Res.* 27, 31035–31045. https://doi.org/10.1007/s11356-020-09031-5

Adu-Boahen, K., Dadson, I.Y., Mensah, D.K.D., Kyermeh, S., 2022. Mapping ecological impact of microplastics on freshwater habitat in the central region of Ghana: a case study of River Akora. *Geo. J.* 87, 621–639. https://doi.org/10.1007/s10708-020-10273-6

Ahmadifar, E., Kalhor, N., Dawood, M.A.O., Ahmadifar, M., Moghadam, M.S., Abarghouei, S., Hedayati, A., 2020. Effects of polystyrene microparticles on inflammation, antioxidant enzyme activities, and related gene expression in Nile tilapia (*Oreochromis niloticus*). *Environ. Sci. Pollut.* 28, 14909–14916. https://doi.org/10.1007/s11356-020-11731-x

Akindele, E.O., Ehlers, S.M., Koop, J.H.E. 2019. First empirical study of freshwater microplastics in West Africa using gastropods from Nigeria as bioindicators. *Limnologica.* 78, 125708. https://doi.org/10.1016/j.limno.2019.125708

Akindele, E.O., Ehlers, S.M., Koop, J.H.E. 2020. Freshwater insects of different feeding guilds ingest microplastics in two Gulf of Guinea tributaries in Nigeria. *Environ. Sci. Pollut. Res.* 27, 33373–33379. https://doi.org/10.1007/s11356-020-08763-8

Ali, M.Z., Jauncey, K., 2004. Optimal dietary carbohydrate to lipid ratio in African catfish *Clarias gariepinus* (Burchell 1822). *Aquac. Int.* 12, 169–180. https://doi.org/10.1023/B:AQUI.0000032065.28059.5b

Alimi, O.S., Fadare, O.O., Okoffo, E.D., 2021. Microplastics in African ecosystems: current knowledge, abundance, associated contaminants, techniques, and research needs. *Sci. Tot. Environ.* 755, 1442422. https://doi.org/10.1016/j.scitotenv.2020.142422

Baldwin, A.K., Spanjer, A.R., Rosen, M.R., Thom, T., 2020. Microplastics in Lake Mead National Recreation Area, USA: occurrence and biological uptake. *Plos One*. https://doi.org/10.1371/journal.pone.0228896

Barboza, L.G.A., Vieira, L.R., Branco, V., Carvalho, C., Guilhermino, L., 2018. Microplastics increase mercury bioconcentration in gills and bioaccumulation in the liver, and cause oxidative stress and damage in *Dicentrarchus labrax* juveniles. *Sci. Rep.* 8, 1–9. https://doi.org/10.1038/s41598-018-34125-z

Basto, M.N., Nicastro, K.R., Tavares, A.I., McQuaid, C.D., Casero, M., Azevedo, F., Zardi, G.I., 2019. Plastic ingestion in aquatic birds in Portugal. *Mar. Pollut. Bull.* 138, 19–24. https://doi.org/10.1016/j.marpolbul.2018.11.024

Berglund, E., Fogelberg, V., Nilsson, P.A., Hollander, J., 2019. Microplastics in a freshwater mussel (*Anodonta anatina*) in Northern Europe. *Sci. Tot. Environ.* 697, 134192. https://doi.org/10.1016/j.scitotenv.2019.134192

Biginagwa, F.J., Mayoma, B.S., Shashoua, Y., Syberg, K., Khan, F.R., 2016. First evidence of microplastics in the African Great Lakes: Recovery from Lake Victoria Nile perch and Nile tilapia. *J. Great Lakes Res.* 42, 146–149. https://doi.org/10.1016/j.jglr.2015.10.012

Blarer, P., Burkhardt-Holm, P., 2016. Microplastics affect assimilation efficiency in the freshwater amphipod *Gammarus fossarum*. *Environ. Sci. Pollut. Res*. 23, 23522–23532. https://doi.org/10.1007/s11356-016-7584-2

Bleckmann, H., 2024. What is a fish? In *Stupid as a Fish?* Springer. https://doi.org/10.1007/978-3-662-68376-7_2

Blettler, M.C.M., Abrial, E., Khan, F.R., Sivri, N., Espinola, L.A., 2018. Freshwater plastic pollution: recognizing research biases and identifying knowledge gaps. *Water Res*. 143, 416–424. https://doi.org/10.1016/j.watres.2018.06.015

Bosch, J., Thumsová, B., López-Rojo, N., Pérez, J., Alonso, A., Fisher, M. C., Boyero, L., 2021. Microplastics increase susceptibility of amphibian larvae to the chytrid fungus *Batrachochytrium dendrobatidis. Nature*. 11, 22438. https://doi.org/10.1038/s41598-021-01973-1

Boyero, L., López-Rojo, N., Bosch, J., Alonso, A., Correa-Araneda, F., Pérez, J., 2019. Microplastics impair amphibian survival, body condition and function. *Chemosphere*. 244, 125500. https://doi.org/10.1016/j.chemosphere.2019.125500

Brookson, C.B., de Solla, S.R., Fernie, K.J., Cepeda, M., Rochman, C.M., 2019. Microplastics in the diet of nestling double-crested cormorants (*Phalacrocorax auritus*), an obligate piscivore in a freshwater ecosystem. *Can. Sci. Pub*. 76, 11. https://doi.org/10.1139/cjfas-2018-0388

Cabansag, J.B.P., Olimberio, R.B., Villanobos, Z.M.T., 2021. Microplastics in some fish species and their environs in Eastern Visayas, Philippines. *Mar. Pol. Bull*. 167, 112312. https://doi.org/10.1016/j.marpolbul.2021.112312

Campbell, S.H., Williamson, P.R., Hall, B.D., 2017. Microplastics in the gastrointestinal tracts of fish and the water from an urban prairie creek. *Facets*. 2, 395–409. https://doi.org/10.1139/facets-2017-0008

Carpenter, E.J., Smith, K.L., 1972. Plastics on the Sargasso sea surface. *Science*. 175, 1240–1241. https://doi.org/10.1126/science.175.4027.1240

Carrasco-Navarro, V., Muñiz-González, A., Sorvari, J., Martínez-Guitarte, J., 2021. Altered gene expression in *Chironomus riparius* (Insecta) in response to tire rubber and polystyrene microplastics. *Environ. Pollut*. 15, 117462. https://doi.org/10.1016/j.envpol.2021.117462

Chang, X., Fang, Y., Wang, Y., Wang, F., Shang, L., Zhong, R., 2022. Microplastic pollution in soils, plants, and animals: a review of distributions, effects and potential mechanisms. *Sci. Tot. Environ*. 850, 157857. https://doi.org/10.1016/j.scitotenv.2022.157857

Chen, H., Hua, X., Li, H., Wang, C., Dang, Y., Ding, P., Yu, Y., 2021. Transgenerational neurotoxicity of polystyrene microplastics induced by oxidative stress in *Caenorhabditis elegans*. *Chemosphere*. 272, 129642. https://doi.org/10.1016/j.chemosphere.2021.129642

Chen, M., Yue, Y., Bao, X., Yu, H., Tan, Y., Tong, B., Kumkhong, S., Yu, Y., 2022. Microplastics as contaminants in water bodies and their threat to the aquatic animals: a mini-review. *Animals*, 12(20), p. 2864. https://doi.org/10.3390/ani12202864.

Clause, A.G., Celestian, A.J., Pauly, G.B., 2021. Plastic ingestion by freshwater turtles: a review and call to action. *Sci. Rep*. 11, 5672. https://doi.org/10.1038/s41598-021-84846-x

Coppock, R.L., Cole, M., Lindeque, P.K., Queirós, A.M., Galloway, T.S., 2017. A small-scale, portable method for extracting microplastics from marine sediments. *Environ. Pollut*. 230, 829–837. https://doi.org/10.1016/j.envpol.2017.07.017

da Costa Araújo, A.P.D.C., Malafaia, G., 2020. Can short exposure to polyethylene microplastics change tadpoles' behavior? A study conducted with neotropical tadpole species belonging to order Anura (*Physalaemus cuvieri*). *J. Haz. Mat*. 391, 122214. https://doi.org/10.1016/j.jhazmat.2020.122214

da Costa Araújo, A.P., De Melo, N.F.S., De Oliveira Junior, A.G., Rodrigues, F.P., Fernandes, T., De Andrade Vieira, J.E., Rocha, T.L., Malafaia, G., 2020a. How much are microplastics harmful to the health of amphibians? A study with pristine polyethylene microplastics and *Physalaemus cuvieri*. *J. Haz. Mat*. 382, 121066. https://doi.org/10.1016/j.jhazmat.2019.121066

da Costa Araújo, A.P.D.C., Gomes, A.R., Malafaia, G., 2020b. Hepatotoxicity of pristine polyethylene microplastics in neotropical *Physalaemus cuvieri* tadpoles (Fitzinger, 1826). *J. Haz. Mat*. 386, 121992. https://doi.org/10.1016/j.jhazmat.2019.121992

Dadedo, E., Eyayu, A., Sorsa, S., Tilahun, G., 2015. Food and feeding habits of the Common carp (*Cyprinus carpio* L. 1758) (Pisces: Cyprinidae) in Lake Koka, Ethiopia. *MEJS*. 7, 16–31. https://doi.org/10.4314/mejs.v7i1.117233

Dahms, H.T.J., Greenfield, R., 2024. A review of the environments, biota, and methods used in microplastics research in South Africa. *S. Afr. J. Sci*. 120, 16669. https://doi.org/10.17159/sajs.2024/16669

Dahms, H.T.J., Tweddle, G.P., Greenfield, R., 2022. Gastric microplastics in *Clarias gariepinus* of the upper Vaal River, South Africa. *Front. Environ. Sci*. 10, 931073. https://doi.org/10.3389/fenvs.2022.931073

Dahms, H.T.J., van Rensburg, G.J., Greenfield, R., 2020. The microplastic profile of an urban African stream. *Sci. Total Environ.* 731, 138893. https://doi.org/10.1016/j.scitotenv.2020.138893

Dalu, T., Themba, N.N., Dondofema, F., Cuthbert, R.N., 2023. Nowhere to go! Microplastic abundances in freshwater fishes living near wastewater plants. *Environ. Toxicol. Pharmacol.* 104210. https://doi.org/10.1016/j.etap.2023.104210

Darabi, H., Baradaran, A., Ebrahimpour, K., 2022. Subacute toxic effects of polyvinyl chloride microplastics (PVC-MPs) in juvenile common carp, *Cyprinus carpio* (Pisces: Cyprinidae). *Casp. J. Environ. Sci.* 20, 233–242. https://doi.org/10.22124/CJES.2022.5551

de Sá, L.C., Oliveira, M., Ribeiro, F., Lopes, T., Futter, M., 2018. Studies of the effects of microplastics on aquatic organisms: what do we know and where should we focus our efforts in the future? Sci. *Total Environ.* 645, 1029–1039. https://doi.org/10.1016/j.scitotenv.2018.07.207

Ding, J., Zhang, S., Razanajatovo, R.M., Zou, H., Zhu, W., 2018. Accumulation, tissue distribution, and biochemical effects of polystyrene microplastics in the freshwater fish red tilapia (*Oreochromis niloticus*). *Environ. Pollut.* 238, 1–9. https://doi.org/10.1016/j.envpol.2018.03.001

Dodd, C.K., 2009. *Amphibian Ecology and Conservation: A Handbook of Techniques*. Oxford University Press.

Dodds, W.K., Maasri, A., 2022. The river continuum concept. *Inland Waters.* 2, 237–248. https://doi.org/10.1016/B978-0-12-819166-8.00105-5

Dordio, A.V., Palace Carvalho, A.J., Honrado Pinto, A.P., 2008. Chapter 3: Wetlands: water "living filters". In Raymundo E. Russo (ed.) *Wetlands: Ecology, Conservation and Restoration*, ISBN: 978-1-60456-995-7, Nova Science Publishers, Inc., 16–51.

Ducatez, S., Lefebvre, L., 2014. Patterns of research effort in birds. *Plos One*. 9, e89955. https://doi.org/10.1371/journal.pone.0089955

DWAF, 2009. Department of Water Affairs. Rapid Habitat Assessment Model Manual. Report no RDM/ Nat/00/CON/0707. Authors: D Louw & CJ Kleynhans Submitted by Water for Africa.

Eerkes-Medrano, D., Thompson, R., 2018. Occurrence, fate, and effect of microplastics in freshwater systems. In E.Y. Zeng (Ed.) *Microplastic Contamination in Aquatic Environments*, 95–132. Elsevier. https://doi.org/10.1016/B978-0-12-813747-5.00004-7

Fareza, A.A., Sembiring, E., 2020. Occurrence of microplastics in water, sediment and milkfish (*Chanos chanos*) in Citarum River Downstream (Case Study: Muara Gembong). *Web. Conf.* 148, 07005. https://doi.org/10.1051/e3sconf/202014807005

Fry, C., 2021. A field guide to freshwater macroinvertebrates of Southern Africa. *Jacana Media (Pty) Ltd.* 10 Orange Street, Sunnyside, Auckland Park, 2092, South Africa.

Geyer, R., 2020. *A Brief History of Plastics. Mare Plasticum-The Plastic Sea: Combatting Plastic Pollution through Science and Art*. Springer. https://doi.org/10.1007/978-3-030-38945-1

Gonzalez-Jauregui, M., Borges-Ramirez, M., Barao-Nobrega, J.A.L., Escamilla, A., Dzul-Caamal, R., von Osten, J.R., 2019. Stomach flushing technique applied to quantify microplastics in crocodilians. *MethodsX*. 6, 2677–2685. https://doi.org/10.1016/j.mex.2019.11.013

Gray, A.D., Weinstein, J.E., 2017. Size- and shape-dependent effects of microplastic particles on adult daggerblade grass shrimp (*Palaemonetes pugio*). *Environ. Tox. Chem.* 36, 3074–3080. https://doi.org/10.1002/etc.3881

Gregory, M.R., 1977. Plastic pellets on New Zealand beaches. *Mar. Pollut. Bull.* 8, 82–84. https://doi.org/10.1016/0025-326X(77)90193-X

Guimarães, A.T.B., Charlie-Silva, I., Malafaia, G., 2020. Toxic effects of naturally-aged microplastics on zebrafish juveniles: a more realistic approach to plastic pollution in freshwater ecosystems. *J. Haz. Mat.* 407, 124833. https://doi.org/10.1016/j.jhazmat.2020.124833

Guo, X., Wang, J., 2019. The chemical behaviours of microplastics in marine environment: a review. *Mar. Pollut. Bull.* 142, 1–14. https://doi.org/10.1016/j.marpolbul.2019.03.019

Hanslik, L., Huppertsberg, S., Kämmer, N., Knepper, T.P., Braunbeck, T., 2021. Rethinking the relevance of microplastics as vector for anthropogenic contaminants: adsorption of toxicants to microplastics during exposure in a highly polluted stream – analytical quantification and assessment of toxic effects in zebrafish (*Danio rerio*). *Sci. Tot. Environ.* 816, 151640. https://doi.org/10.1016/j.scitotenv.2021.151640

He, D., Luo, Y., Lu, S., Liu, M., Song, Y., Lei, L., 2018. Microplastics in soils: analytical methods, pollution characteristics and ecological risks. *Trends Anal. Chem.* 109, 163–172. https://doi.org/10.1016/j.trac.2018.10.006

Holt, E.A., Miller, S.W., 2010. Bioindicators: using organisms to measure environmental impacts. *Nature Education Knowledge*. 3, 8. https://www.nature.com (Accessed 5 August 2024).

Hu, L., Chernick, M., Hinton, D.E., Shi, H., 2018. Microplastics in small waterbodies and tadpoles from Yangtze River Delta, China. *Environ. Sci. Technol.* 52, 8885–8893. https://doi.org/10.1021/acs.est.8b02279

Huang, D., Tao, J., Cheng, M., Deng, R., Chen, S., Yin, L., Li, R., 2020. Microplastics and nanoplastics in the environment: Macroscopic transport and effect on creatures. *J. Hazard. Mater.* 407, 124399. https://doi.org/10.1016/j.jhazmat.2020.124399

Hurley, R.R., Woodward, J.C., Rothwell, J.J., 2017. Ingestion of microplastics by freshwater tubifex worms. *J. Am. Chem. Soc.* 51, 12844–12851. https://doi.org/10.1021/acs.est.7b03567

Hurt, R., O'Reilly, C.M., Perry, W.L., 2020. Microplastic prevalence in two fish species in two U.S. reservoirs. *Limnol. Oceanogr. Lett.* 5, 147–153. https://doi.org/10.1002/lol2.10140

Iannella, M., Console, G., D'Alessandro, P., Cerasoli, F., Mantoni, C., Ruggieri, F., Di Donato, F., Biondi, M., 2020. Preliminary analysis of the diet of *Triturus carnifex* and pollution in Mountain Karst Ponds in Central Apennines. *Water.* 12, 44. https://doi.org/10.3390/w12010044

Iannilli, V., Corami, F., Grasso, P., Lecce, F., Buttinelli, M., Setini, A., 2020. Plastic abundance and seasonal variation on the shorelines of three volcanic lakes in Central Italy: can amphipods help detect contamination? *Environ. Sci. Pollut. Res.* 27, 14711–14722. https://doi.org/10.1007/s11356-020-07954-7

Iheanacho, S.C., Odo, G.E., 2020. Neurotoxicity, oxidative stress biomarkers and haematological responses in African catfish (*Clarias gariepinus*) exposed to polyvinyl chloride microparticles. *Tox. Pharm.* 232, 108741. https://doi.org/10.1016/j.cbpc.2020.108741

Imhof, H.K., Laforsch, C., 2016. Hazardous or not – are adult and juvenile individuals of *Potamopyrgus antipodarum* affected by non-buoyant microplastic particles? *Environ. Pollut.* 218, 383–391. https://doi.org/10.1016/j.envpol.2016.07.017

IUCN, 2019https://iucn.org/our-union/commissions/group/iucn-ssc-amphibian-specialist-group

IUCN, 2024. The IUCN red list of species. *Red list.* https://www.iucnredlist.org/ (Accessed 11 June 2024).

Jabeen, K., Li, B., Chen, Q., Su, L., Wu, C., Hollert, H., Shi, H., 2018. Effects of virgin microplastics on goldfish (*Carassius auratus*). *Chemosphere.* 213, 323–332. https://doi.org/10.1016/j.chemosphere.2018.09.031

Jabeen, K., Su, L., Li, J., Yang, D., Tong, C., Mu, J., Shi, H., 2017. Microplastics and mesoplastics in fish from coastal and fresh waters of China. *Environ. Pollut.* 221, 141–149. https://doi.org/10.1016/j.envpol.2016.11.055

Jansen, J.A., 2016. [WWW Document]. Ascend. Plastics-it's all about molecular structure. *Plastic Engineering.* https://read.nxtbook.com/wiley/plasticsengineering/september2016/consultantscorner_plastics.html. (Accessed 24 June 2023).

Jiang, H., Cheng, H., Wu, S., Li, H., Chen, H., Li, Z., Yao, X., Zhang, Y., Chen, Y., Chen, S., Chen, S., Zheng, L., Sui, Y., Shao, R., 2024. Microplastics footprint in nature reserves-a case study on the microplastics in the guano from Yancheng Wetland Rare Birds National Nature Reserve, China. *Environ. Res.* 265, 119252. https://doi.org/10.1016/j.envres.2024.119252

Karaoglu, K., Gul, S., 2020. Characterization of microplastic pollution in tadpoles living in small water-bodies from Rize, the northeast of Turkey. *Chemosphere.* 255, 126915. https://doi.org/10.1016/j.chemosphere.2020.126915

Karima, Z., 2021. Chironomidae: biology, ecology and systematics. In F.K. Perveen (Ed.) *The Wonders of Diptera - Characteristics, Diversity, and Significance for the World's Ecosystems*, 1–25. IntechOpen. https://dx.doi.org/10.5772/intechopen.95577

Kasamesiri, P., Thaimuangphol, W., 2020. Microplastics ingestion by freshwater fish in the Chi river, Thailand. *Int. J. Geomate.* 18, 114–119. https://doi.org/10.21660/2020.67.9110

Khan, F.R., Shashoua, Y., Crawford, A., Drury, A., Sheppard, K., Stewart, K., Sculthorp, T., 2020. 'The plastic nile': First evidence of microplastic contamination in fish from the nile river (Cairo, Egypt). *Toxics*, 8(2), p. 22. https://doi.org/10.3390/toxics8020022.

Khosrovyan, A., Kahru, A., 2019. Evaluation of the hazard of irregularly-shaped co-polyamide microplastics on the freshwater non-biting midge *Chironomus riparius* through its life cycle. *Chemosphere.* 244, 125487. https://doi.org/10.1016/j.chemosphere.2019.125487

Klein, K., Heß, S., Nungeß, S., Schulte-Oehlmann, U., Oehlmann, J., 2021. Particle shape does not affect ingestion and egestion of microplastics by the freshwater shrimp *Neocaridina palmata. Environ. Sci. Pollut. Res.* 28, 62246–62254. https://doi.org/10.1007/s11356-021-15068-x

Lebreton, L.C.M., Van Der Zwet, J., Damsteeg, J.W., Slat, B., Andrady, A., Reisser, B. 2017. River plastic emissions to the world's oceans. *Nat. Commun.* 8, 15611. https://doi.org/10.1038/ncomms15611

Lehtiniemi, M., Hartikainen, S., Näkki, P., Engström-Öst, J., Koistinen, A., Setälä, O., 2018. Size matters more than shape: ingestion of primary and secondary microplastics by small predators. *Food Webs* 17, e00097. https://doi.org/10.1016/j.fooweb.2018.e00097

LeMoine, C.M.R., Kelleher, B.M., Lagarde, R., Northam, C., Elebute, O.O., Cassone, B.J., 2018. Transcriptional effects of polyethylene microplastics ingestion in developing zebrafish (*Danio rerio*). *Environ. Pollut.* 243, 591–600. https://doi.org/10.1016/j.envpol.2018.08.084

Li, J., Liu, H., Paul Chen, J., 2018. Microplastics in freshwater systems: a review on occurrence, environmental effects, and methods for microplastics detection. *Water Res.* 137, 362–374. https://doi.org/10.1016/j.watres.2017.12.056

Liu, Z., Yu, P., Cai, M., Wu, D., Zhang, M., Chen, M., Zhao, Y., 2019. Effects of microplastics on the innate immunity and intestinal microflora of juvenile *Eriocheir sinensis*. *Sci. Tot. Environ.* 685, 836–846. https://doi.org/10.1016/j.scitotenv.2019.06.265

Lu, H.C., Ziajahromi, S., Locke, A., Neale, P.A., Leusch, F.D.L., 2022. Microplastics profile in constructed wetlands: distribution, retention and implications. *Environ. Pollut.* 313, 120079. https://doi.org/10.1016/j.envpol.2022.120079

Lv, W., Zhou, W., Lu, S., Huang, W., Yuan, Q., Tian, M., Lv, W., He, D., 2019. Microplastic pollution in rice-fish co-culture system: a report of three farmland stations in Shanghai, China. *Sci. Tot. Environ.* 652, 1209–1218. https://doi.org/10.1016/j.scitotenv.2018.10.321

Magni, S., Gagné, F., André, C., Della Torre, C., Auclair, J., Hanana, H., Parenti, C.C., Bonasoro, F., Binelli, A., 2018. Evaluation of uptake and chronic toxicity of virgin polystyrene microbeads in freshwater zebra mussel *Dreissena polymorpha* (mollusca: Bivalvia). *Sci. Tot. Environ.* 631–632, 778–788. https://doi.org/10.1016/j.scitotenv.2018.03.075

Mark, J.E. (Ed.), 2009. *Polymer Data Handbook*. Oxford University Press.

Meijer, L.J.J., van Emmerik, T., van der Ent, R., Schmidt, C., Lebreton, L., 2021. More than 1000 rivers account for 80% of global riverine plastic emissions into the ocean. *Sci. Adv.* 7, 1–13. https://doi.org/10.1126/sciadv.aaz5803

Merga, L.B., Redondo-Hasselerharm, P.E., Van den Brink, P.J., Koelmans, A.A., 2020. Distribution of microplastic and small macroplastic particles across four fish species and sediment in an African lake. *Sci. Tot. Environ.* 741, 140527. https://doi.org/10.1016/j.scitotenv.2020.140527

Munno, K., Helm, P.A., Rochman, C., George, T., Jackson, D.A., 2022. Microplastic contamination in Great Lakes fish. *Con. Bio.* 36, e13794. https://doi.org/10.1111/cobi.13794

Murphy, F., Quinn, B., 2017. The effects of microplastic on freshwater *Hydra attenuata* feeding, morphology & reproduction. *Environ. Pollut.* 234, 487–494. https://doi.org/10.1016/j.envpol.2017.11.029

Naidoo, T., Glassom, D., 2019. Decreased growth and survival in small juvenile fish, after chronic exposure to environmentally relevant concentrations of microplastic. *Mar. Pollut. Bull.* 145, 254–259. https://doi.org/10.1016/j.marpolbul.2019.02.037

Nan, B, Su, L., Kellar, C., Craig, N.J., Keough, M.J., Pettigrove, V., 2019. Identification of microplastics in surface water and Australian freshwater shrimp *Paratya australiensis* in Victoria, Australia. *Environ. Pollut.* 259, 113865. https://doi.org/10.1016/j.envpol.2019.113865

Nel, H.A., Dalu, T., Wasserman, R.J., 2018. Sinks and sources: assessing microplastic abundance in river sediment and deposit feeders in an austral temperate urban river system. *Sci. Tot. Environ.* 612, 950–956. https://doi.org/10.1016/j.scitotenv.2017.08.298

Ory, N.C., Sobral, P., Ferreira, J.L., Thiel, M., 2017. Amberstripe scad *Decapterus muroadsi* (Carangidae) fish ingest blue microplastics resembling their copepod prey along the coast of Rapa Nui (Easter Island) in the South Pacific subtropical gyre. *Sci. Tot. Environ.* 586, 430–437. https://doi.org/10.1016/j.scitotenv.2017.01.175

Owowenu, E., Nnadozie, C., Akamagwuna, F.C., Noundou, X.S., Uku, J.E., Odume, O.N., 2023. A critical review of environmental factors influencing the transport dynamics of microplastics in riverine systems: implications for ecological studies. *Aquat. Ecol.* 57, 557–570. https://doi.org/10.1007/s10452-023-10029-7

Padgett, D.J., Carboni, J.J., Schepis, D.J., 2010. The dietary composition of *Chrysemys picta picta* (Eastern Painted Turtles) with special reference to the seeds of aquatic macrophytes. *Northeast. Sci. Rep.* 17, 305–312. https://doi.org/10.1656/045.017.0212

Park, T.-J., Lee, S.-H., Lee, M.-S., Lee, J.-K., Lee, S.-H., Zoh, K.-D., 2020. Occurrence of microplastics in the Han River and riverine fish in South Korea. *Sci. Tot. Environ.* 708, 134535. https://doi.org/10.1016/j.scitotenv.2019.134535

Parker, B., Andreou, D., Pabortsava, K., Barrow, M., Green, I.D., Britton, J.R., 2022. Microplastic loads within riverine fishes and macroinvertebrates are not predictable from ecological or morphological characteristics. *Sci. Tot. Environ.* 839, 156321. https://doi.org/10.1016/j.scitotenv.2022.156321

Pedersen, A.F., Gopalakrishnan, K., Boegehold, A.G., Peraino, N.J., Westrick, J.A., Kashian, D.R., 2020. Microplastic ingestion by quagga mussels, *Dreissena bugensis*, and its effects on physiological processes. *Environ. Pollut.* 260, 113964. https://doi.org/10.1016/j.envpol.2020.113964

Pegado, T.S., Schmid, K., Winemiller, K.O., Chelazzi, D., Cincinelli, A., Dei, L., Giarrizzo, T., 2018. First evidence of microplastic ingestion by fishes from the Amazon River estuary. *Mar. Pollut. Bull.* 133, 814–821. https://doi.org/10.1016/j.marpolbul.2018.06.035

Peters, C.A., Bratton, S.P., 2016. Urbanization is a major influence on microplastic ingestion by sunfish in the Brazos River Basin, Central Texas, USA. *Environ. Pollut.* 210, 380–387. https://doi.org/10.1016/j.envpol.2016.01.018

Phillips, M.B., Bonner, T.H., 2015. Occurrence and amount of microplastic ingested by fishes in watersheds of the Gulf of Mexico. *Mar. Pol. Bull.* 100, 264–269. https://doi.org/10.1016/j.marpolbul.2015.08.041

Prata, J.C., Venâncio, C., Girão, A.V., Da Costa, J.P., Lopes, I., Duarte, A.C., Rocha-Santos, T., 2021. Effects of virgin and weathered polystyrene and polypropylene microplastics on *Raphidocelis subcapitata* and embryos of *Danio rerio* under environmental concentrations. *Sci. Total Environ.* 816, 151642. https://doi.org/10.1016/j.scitotenv.2021.151642

Prévot-Julliard, A.-C., Gousset, E., Archinard, C., Cadi, A., Girondot, M., 2007. Pets and invasion risks: is the slider turtle strictly carnivorous? *Amphibia-Reptilia.* 28, 139–143. https://doi.org/10.1163/156853807779799036

Qiang, L., Cheng, J., 2019. Exposure to microplastics decreases swimming competence in larval zebrafish (*Danio rerio*). *Ecotox. Environ. Saf.* 176, 226–233. https://doi.org/10.1016/j.ecoenv.2019.03.088

Qiang, L., Cheng, J., 2021. Exposure to polystyrene microplastics impairs gonads of zebrafish (*Danio rerio*). *Chemosphere.* 263, 128161. https://doi.org/10.1016/j.chemosphere.2020.128161

Rangasamy, B., Malafaia, G., Maheswaran, R., 2021. Evaluation of antioxidant response and na+-K+-ATPase activity in zebrafish exposed to polyethylene microplastics: shedding light on a physiological adaptation. *J. Haz. Mat.* 426, 127789. https://doi.org/10.1016/j.jhazmat.2021.127789

Redondo-Hasselerharm, P.E., Falahudin, D., Peeters, E.T.H.M., Koelmans, A.A., 2018. Microplastic effect thresholds for freshwater benthic macroinvertebrates. *Environ. Sci. Technol.* 52, 2278–2286. https://doi.org/10.1021/acs.est.7b05367

Reynolds, C., Ryan, P.G., 2018. Micro-plastic ingestion by waterbirds from contaminated wetlands in South Africa. *Mar. Pollut. Bull.* 126, 330–333. https://doi.org/10.1016/j.marpolbul.2017.11.021

Roch, S., Walter, T., Ittner, L.D., Friedrich, C., Brinker, A., 2019. A systematic study of the microplastic burden in freshwater fishes of south-western Germany - are we searching at the right scale? *Sci. Tot. Environ.* 689, 1001–1011. https://doi.org/10.1016/j.scitotenv.2019.06.404

Roda, J.F.B., Lauer, M.M., Risso, W.E., Bueno Dos Reis Martinez, C., 2020. Microplastics and copper effects on the neotropical teleost *Prochilodus lineatus*: is there any interaction? *Mol. Integ. Phys.* 242, 110659. https://doi.org/10.1016/j.cbpa.2020.110659

Romano, N., Ashikin, M., Teh, J.C., Syukri, F., Karami, A., 2017. Effects of pristine polyvinyl chloride fragments on whole body histology and protease activity in silver barb *Barbodes gonionotus* fry. *Environ. Pollut.* 237, 1106–1111. https://doi.org/10.1016/j.envpol.2017.11.040

Romero-Blanco, A., Remón-Elola, A., Alonso, Á., 2021. Assessment of the effects of environmental concentrations of microplastics on the aquatic snail *Potamopyrgus antipodarum.* Water Air. *Soil Pollut.* 232, 438. https://doi.org/10.1007/s11270-021-05379-7

Ryan, P.G., 1990. Plastic and other artefacts on South African beaches: temporal trends in abundance and composition. *S. Afr. J. Sci.* 86, 450–452.

Saad, D., Chauke, P., Cukrowska, E., Richards, H., Nikiema, J., Chimuka, L., Tutu, H., 2022. First biomonitoring of microplastic pollution in the Vaal River using Carp fish (*Cyprinus carpio*) "as a bio-indicator". *Sci. Tot. Environ.* 836, 155623. https://dx.doi.org/10.1016/j.scitotenv.2022.155623

Schmidt, C., Krauth, T., Wagner, S., 2017. Export of plastic debris by rivers into the sea. *Environ. Sci. Technol.* 51, 12246–12253. https://doi.org/10.1021/acs.est.7b02368

Selman, W., Lindeman, P.V., 2018. Spatial, seasonal, and sexual variation in the diet of *Graptemys flavimaculata*, a threatened turtle of the Pascagoula River system, Mississippi, USA. *Copeia.* 106, 247–254. https://doi.org/10.1643/CH-17-644

Sembiring, E., Fareza, A.A., Suendo, V., Reza, M., 2020. The presence of microplastics in water, sediment, and milkfish (*Chanos chanos*) at the downstream area of Citarum River, Indonesia. *Water Air Soil Pollut.* 231, 355. https://doi.org/10.1007/s11270-020-04710-y

Silva, C.J., Patrício Silva, A.L., Campos, D., Soares, A.M., Pestana, J.L., Gravato, C., 2020. *Lumbriculus variegatus* (oligochaeta) exposed to polyethylene microplastics: biochemical, physiological and reproductive responses. *Ecotox. Environ. Saf.* 207, 111375. https://doi.org/10.1016/j.ecoenv.2020.111375

Silva-Cavalcanti, J.S., Silva, J.D.B., de Franca, E.J., de Araujo, M.C.B., Gusmao, F., 2017. Microplastics ingestion by a common tropical freshwater fishing resource. *Environ. Pollut.* 221, 218–226. https://doi.org/10.1016/j.envpol.2016.11.068

Simmerman, C., Wasik, L.K.C., 2019. The effect of urban point source contamination on microplastic levels in water and organisms in a cold-water stream. *Limnol. Oceanogr. Lett.* 5, 137–146. https://doi.org/10.1002/lol2.10138

Smiroldo, G., Balestrieri, A., Pini, E., Tremolada, P., 2019. Anthropogenically altered trophic webs: alien catfish and microplastics in the diet of Eurasian otters. *Mamm. Res.* 64, 165–174. https://doi.org/10.1007/s13364-018-00412-3

Souza, J.M., Windsor, F., Santillo, D., Ormerod, S.J., 2020. Food web transfer of plastics to an apex riverine predator. *Glob. Change Biol.* 26, 3846–3857. https://doi.org/10.1111/gcb.15139

Su, L., Cai, H., Kolandhasamy, P., Wu, C., Rochman, C.R., Shi, H., 2018. Using the Asian clam as an indicator of microplastic pollution in freshwater ecosystems. *Environ. Pollut.* 234, 347–355. https://doi.org/10.1016/j.envpol.2017.11.075

Su, L., Nan, B., Hassell, K., Craig, N.J., Pettigrove, V., 2019. Microplastics biomonitoring in Australian urban wetlands using a common noxious fish (*Gambusia holbrooki*). *Chemosphere.* 228, 65.74. https://doi.org/10.1016/j.chemosphere.2019.04.114

Su, L., Xue, Y., Li, L., Yang, D., Kolandhasamy, P., Li, D., Shi, H., 2016. Microplastics in Taihu Lake, China. *Environ. Pollut.* 216, 711–719. https://doi.org/10.1016/j.envpol.2016.06.036

Sun, J., Dai, X., Wang, Q., van Loosdrecht, M.C.M., Ni, B.J., 2019. Microplastics in wastewater treatment plants: detection, occurrence and removal. *Water Res.* 152, 21–37. https://doi.org/10.1016/j.watres.2018.12.050

Tien, C-J., Wang, Z-X., Chen, C.S., 2020. Microplastics in water, sediment and fish from the Fengshan River system: relationship to aquatic factors and accumulation of polycyclic aromatic hydrocarbons by fish. *Environ. Pollut.* 265, 114962. https://doi.org/10.1016/j.envpol.2020.114962

Thompson, R. C., Olsen, Y., Mitchell, R. P., Davis, A., Rowland, S. J., John, A. W., McGonigle, D., & Russell, A. E. (2004). Lost at sea: where is all the plastic? *Science*, *304*(5672), 838–838. doi: 10.1126/science.1094559.

Umamaheswari, S., Priyadarshinee, S., Bhattacharjee, M., Kadirvelu, K., Ramesh, M., 2020. Exposure to polystyrene microplastics induced gene modulated biological responses in zebrafish (*Danio rerio*). *Chemosphere.* 281, 128592. https://doi.org/10.1016/j.chemosphere.2020.128592

Vannote, R.L., Minshall, G.W., Cummins, K.W., Sedell, J.R., Cushing, C.E., 1980. The river continuum concept. *Can. J. Fish. Aquat. Sci.* 37, 130–137. https://doi.org/10.1139/f80-017

Veneman, W. J., Spaink, H.P., Brun, N.R., Bosker, T., Vijver, M.G., 2017. Pathway analysis of systemic transcriptome responses to injected polystyrene particles in zebrafish larvae. *Aquat. Toxicol.* 190, 112–120. https://doi.org/10.1016/j.aquatox.2017.06.014

Vitule, J.R.S., Umbria, S.C., Aranha, J.M.R., 2006. Introduction of the African catfish *Clarias gariepinus* (Burchell, 1822) into Southern Brazil. *Biol. Invasions.* 8, 677–681. https://doi.org/10.1007/s10530-005-2535-8

Walker, T.R., 2021. (Micro)plastics and the UN sustainable development goals. *Curr. Opin. Green. Sustain. Chem.* 30, 100497. https://doi.org/10.1016/j.cogsc.2021.100497

Wan, Z., Wang, C., Zhou, J., Shen, M., Wang, X., Fu, Z., Jin, Y., 2018. Effects of polystyrene microplastics on the composition of the microbiome and metabolism in larval zebrafish. *Chemosphere.* 217, 646–658. https://doi.org/10.1016/j.chemosphere.2018.11.070

Wang, J., Shi, H., Hu, S., Ma, K., Li, C., 2010. Interspecific differences in diet between introduced red-eared sliders and native turtles in China. *Asian Herpetol. Res.* 4, 190–196. https://doi.org/10.3724/SP.J.1245.2013.000190

Wang, L., Nabi, G., Yin, L., Wang, Y., Li, S., Hao, Z., Li, D., 2021. Birds and plastic pollution: recent advances. *Avian Res.* 12, 59. https://doi.org/10.1186/s40657-021-00293-2

Wardlaw, C., Prosser, R.S., 2020. Investigation of microplastics in freshwater mussels (*Lasmigona costata*) from the Grand River Watershed in Ontario, Canada. *Water Air Soil Pollut.* 231, 405. https://doi.org/10.1007/s11270-020-04741-5

Watkins, L., Mcgrattan, S., Sullivan, P.J., Walter, M.T., 2019. The effect of dams on river transport of microplastic pollution. *Sci. Total Environ.* 664, 834–840. https://doi.org/10.1016/j.scitotenv.2019.02.028

Weber, A., Jeckel, N., Wagner, M., 2020. Combined effects of polystyrene microplastics and thermal stress on the freshwater mussel *Dreissena polymorpha. Sci. Tot. Environ.* 718, 137253. https://doi.org/10.1016/j.scitotenv.2020.137253

Weideman, E.A., Perold, V., Ryan, P.G., 2019. Little evidence that dams in the Orange–Vaal River system trap floating microplastics or microfibres. *Mar. Pollut. Bull.* 149, 110664. https://doi.org/10.1016/j.marpolbul.2019.110664

Weideman, E.A., Perold, V., Ryan, P.G., 2020. Limited long-distance transport of plastic pollution by the Orange-Vaal River system, South Africa. *Sci. Tot. Environ.* 727, 138653. https://doi.org/10.1016/j.scitotenv.2020.138653

Weis, J.A., 2020. Aquatic microplastic research—a critique and suggestions for the future. *Water.* 12, 1475. https://doi.org/10.3390/w12051475

Wen, B., Zhang, N., Jin, S., Chen, Z., Gao, J., Liu, Y., Liu, H., Xu, Z., 2017. Microplastics have a more profound impact than elevated temperatures on the predatory performance, digestion and energy metabolism of an Amazonian cichlid. *Aquat. Toxicol.* 195, 67–76. https://doi.org/10.1016/j.aquatox.2017.12.010

Wilcox, C., Sebille, E.V., Hardesty, B.D., 2015. Threat of plastic pollution to seabirds is global, pervasive, and increasing. *PNAS.* 112, 38. https://doi.org/10.1073/pnas.1502108112

Windsor, F.M., Tilley, R.M., Tyler, C.R., Ormerod, S.J., 2019. Microplastic ingestion by riverine macroinvertebrates. *Sci. Tot. Environ.* 646, 68–74. https://doi.org/10.1016/j.scitotenv.2018.07.271

Winkler, A., Nessi, A., Antoniolo, D., Laus, M., Santo, N., Parolini, M., Tremolada, P., 2020. Occurrence of microplastics in pellets from the common kingfisher (*Alcedo atthis*) along the Ticino River, North Italy. *Environ. Sci. Pollut. Res.* 27, 41731–41739. https://doi.org/10.1007/s11356-020-10163-x

Xia, X., Sun, M., Zhou, M., Chang, Z., Li, L., 2020. Polyvinyl chloride microplastics induce growth inhibition and oxidative stress in *Cyprinus carpio* var. larvae. *Sci. Tot. Environ.* 716, 136479. https://doi.org/10.1016/j.scitotenv.2019.136479

Xiong, X., Zhang, K., Chen, X., Shi, H., Luo, Z., Wu, C., 2018. Sources and distribution of microplastics in China's largest inland lake – Qinghai Lake. *Environ. Pollut.* 235, 889–906. https://doi.org/10.1016/j.envpol.2017.12.081

Yang, H., Xiong, H., Mi, K., Xue, W., Wei, W., Zhang, Y., 2020. Toxicity comparison of nano-sized and micron-sized microplastics to goldfish *Carassius auratus* larvae. *J. Haz. Mat.* 388, 122058. https://doi.org/10.1016/j.jhazmat.2020.122058

Yu, P., Liu, Z., Wu, D., Chen, M., Lv, W., Zhao, Y., 2018. Accumulation of polystyrene microplastics in juvenile *Eriocheir sinensis* and oxidative stress effects in the liver. *Aquat. Toxicol.* 200, 28–36. https://doi.org/10.1016/j.aquatox.2018.04.015

Yu, Y., Chen, H., Hua, X., Dang, Y., Han, Y., Yu, Z., Chen, X., Ding, P., Li, H., 2020. Polystyrene microplastics (PS-MPs) toxicity induced oxidative stress and intestinal injury in nematode *Caenorhabditis elegans*. *Sci. Tot. Environ.* 726, 138679. https://doi.org/10.1016/j.scitotenv.2020.138679

Zaki, M.R.M., Yaid, S.H.M., Zainuddin, A.H., Ariy, A.Z., 2021. Microplastic pollution in tropical estuary gastropods: abundance, distribution and potential sources of Klang River estuary, Malaysia. *Mar. Pol. Bull.* 162, 111866. https://doi.org/10.1016/j.marpolbul.2020.111866

Zhang, K., Xiong, X., Hu, H., Wu, C., Bi, Y., Wu, Y., Zhou, B., Lam, P.S., Liu, J., 2017. Occurrence and characteristics of microplastic pollution in Xiangxi Bay of Three Gorges Reservoir, China. *Environ. Sci. Technol.* 51, 3794–3801. https://doi.org/10.1021/acs.est.7b00369

Zhang, L., Xie, Y., Zhong, S., Liu, J., Qin, Y., Gao, P., 2021. Microplastics in freshwater and wild fishes from Lijiang River in Guangxi, Southwest China. *Sci. Tot. Environ.* 755, 142428. https://doi.org/10.1016/j.scitotenv.2020.142428

Zhao, Y., Qiao, R., Zhang, S., Wang, G., 2021. Metabolomic profiling reveals the intestinal toxicity of different length of microplastic fibers on zebrafish (*Danio rerio*). *J. Haz. Mat.* 403, 123663. https://doi.org/10.1016/j.jhazmat.2020.123663

Ziajahromi, S., Kumar, A., Neale, P.A., Leusch, F.D.L., 2018. Environmentally relevant concentrations of polyethylene microplastics negatively impact the survival, growth and emergence of sediment-dwelling invertebrates. *Environ. Pollut.* 236, 425–431. https://doi.org/10.1016/j.envpol.2018.01.094

9 Plastisphere in Aquatic Food Production Systems and Seafood Safety

Idil Can-Tuncelli, Nuray Erkan, Gokhan Tuncelli, and Devrim Memis

9.1 INTRODUCTION

Microplastics (MPs) and nanoplastics (NPs) are now found in a wide range of ecosystems, from the depths of the oceans to freshwater bodies and terrestrial environments (Jiang et al., 2020). Exposed to abiotic and biotic factors, plastics break down into MPs and NPs, which can persist in the environment for hundreds or even thousands of years (Udovicki et al., 2022). MPs have emerged as environmental pollutants of concern due to the limited understanding of their impact on human health, even though their presence in food and air has been documented (Wright & Kelly, 2017). Due to their small particle size and low density, plastics act as carriers for other environmental pollutants, such as pathogenic microorganisms, phytoestrogens, mycotoxins, organohalogens, pesticides, phthalates, heavy metals, and pharmaceutical residues, posing significant risks to humans, the top consumers in the food pyramid. In the food safety approach, which aims to deliver a healthy product to the consumer by eliminating factors harmful to human health throughout the production-to-consumption process, plastic and its derivatives constitute a group of hazards that must be considered, examined, and mitigated, alongside known dangers (Andrady, 2017; De-la-Torre, 2020).

The plastisphere is characterized by diverse microbial communities that colonize plastic surfaces in aquatic environments. These microbial biofilms are composed of bacteria, fungi, algae, and other microorganisms, forming complex and dynamic ecosystems (Battin et al., 2016). While the plastic surfaces provide a durable substrate that supports the growth and persistence of these biofilms, it is important to note that plastics are not chemically stable over time. They can degrade under environmental conditions, which may influence the composition and function of these biofilms. The presence of these microbial communities can alter the chemical properties of the plastic particles, making them more attractive to other microorganisms and potentially facilitating the spread of pathogens and invasive species. These interactions within the microbial communities can affect nutrient cycles and ecological balance in aquatic systems (Rogers et al., 2020). The pervasive presence of MPs in the environment and their complex interactions with various pollutants also impact the urgent need for comprehensive research and effective mitigation strategies. Further, understanding the dynamics of the plastisphere and its impacts on aquatic food production systems and seafood safety is crucial for addressing the challenges posed by plastic pollution. This is particularly relevant in the context of international food safety directives, such as those outlined by the Codex Alimentarius Commission, which set global standards for food safety to protect consumer health and ensure fair practices in food trade (FAO & WHO, 2023).

In response to these challenges, innovative aquatic food production systems are being developed to mitigate MP pollution. These systems aim to promote sustainable seafood production while

DOI: 10.1201/9781032717814-9

contributing to the ecological remediation of contaminated water bodies. By integrating multiple species from different trophic levels, these systems can improve water quality, enhance economic viability, and reduce the environmental footprint of aquaculture practices (Okeke et al., 2022).

9.2 PLASTISPHERE AND MICROPLASTIC CONTAMINATION

The increasing prevalence of plastic pollution in aquatic ecosystems has led to the formation of a unique ecological habitat known as the "plastisphere" (Figure 9.1). This term, first introduced by Zettler et al. (2013), refers to the diverse microbial communities that colonize plastic debris in marine and freshwater environments, creating new ecological niches or ecosystems distinct from the surrounding water or sediment. It is important to note that the plastisphere encompasses not only microscopic organisms but also larger life forms that may attach to or associate with plastic debris. The formation of the plastisphere begins when plastic waste, through physical and chemical degradation processes, such as fragmentation and photo-oxidation, breaks down into MPs. These particles provide a stable and persistent substrate for microbial colonization, creating new ecological niches distinct from those found in the surrounding water or sediment. This process has been well-documented by Amaral-Zettler et al. (2020), who first described the plastisphere, as well as by other researchers such as Wright et al. (2021) and Oberbeckmann et al. (2018), who have explored the complex interactions within microbial communities on plastic debris (Amaral-Zettler et al., 2020; Oberbeckmann et al., 2018; Jacquin et al., 2019; Wright et al., 2021; Yang et al., 2020). The microbial assemblages on these plastic surfaces include a variety of autotrophs, heterotrophs, predators, and pathogens, significantly impacting the biogeochemical cycles of carbon, nitrogen, and sulfur (Ogonowski et al., 2018; Rogers et al., 2020; Dodhia et al., 2023). The plastisphere not only supports diverse microbial life but also plays a role in the transport of invasive species and pathogens, raising concerns about its ecological and health implications (Bryant et al., 2016; Mishra et al., 2024).

MPs, defined as plastic particles less than 5 mm in diameter, originate from the degradation of larger plastic items or are intentionally manufactured at small sizes for use in products like cosmetics and cleaning agents. They are pervasive in marine environments, where they are ingested by

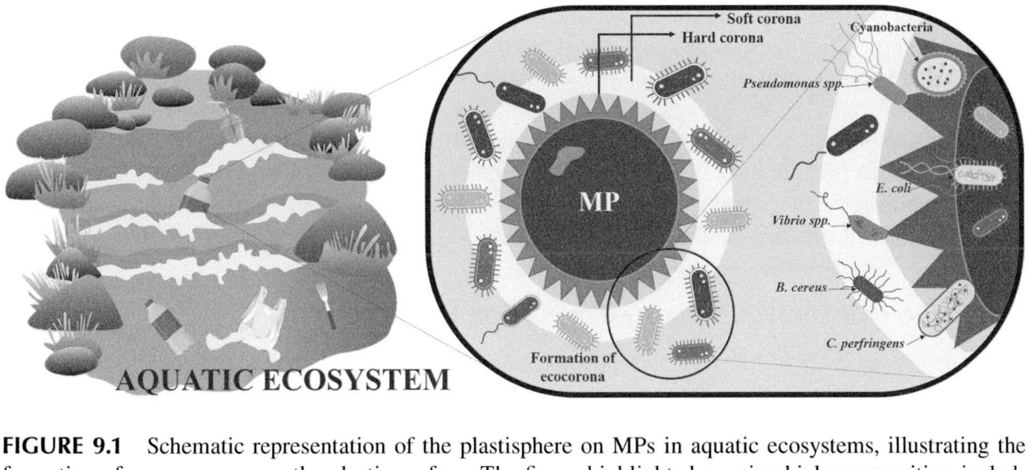

FIGURE 9.1 Schematic representation of the plastisphere on MPs in aquatic ecosystems, illustrating the formation of an ecocorona on the plastic surface. The figure highlights key microbial communities, including various bacteria (e.g., *Pseudomonas* spp., *Vibrio* spp., *Bacillus cereus*, *Escherichia coli*, and *Clostridium perfringens*) and cyanobacteria, which colonize the MPs. The ecocorona is depicted with a differentiation between the soft and hard corona layers, showing how these microorganisms attach and potentially alter the biogeochemical properties of the MPs. (This figure was adapted and modified from Tavelli et al. (2022) to fit the specific scope of this study.)

various aquatic organisms, leading to potential bioaccumulation and biomagnification up the food chains. The presence of MPs in these environments is a critical issue due to their potential to act as vectors for various chemical contaminants, including heavy metals, persistent organic pollutants (POPs), and endocrine-disrupting chemicals like Bisphenol A (BPA). These contaminants pose significant risks to both aquatic organisms and human health.

MPs in aquatic environments quickly become covered with an organic layer known as an ecocorona, which consists of proteins and other biomolecules. This rapid colonization process has been documented in several studies, including those by Lobelle and Cunliffe (2011), Galloway et al. (2017), and Dussud et al. (2018), which demonstrate how MPs rapidly acquire an ecocorona upon entering aquatic environments. This layer can be divided into an inner "hard corona" and an outer "soft corona." The 'hard corona' refers to the tightly bound layer of biomolecules, such as proteins and lipids, that are adsorbed directly onto the surface of the MP. This layer is relatively stable and forms quickly when MPs enter the aquatic environment. In contrast, the 'soft corona' consists of loosely bound biomolecules that can more readily exchange with the surrounding environment. These distinctions are crucial because they influence how MPs interact with microorganisms and other substances in the water. As shown in Figure 9.1, the formation of the hard and soft corona layers plays a key role in shaping the microenvironment around MPs (Milani et al., 2012; Galloway et al., 2017).

The ecocorona attracts bacteria, leading to the formation of biofilms, which are stratified structures composed primarily of water, extracellular polymeric substances (EPS), and microorganisms. The microbial composition of the plastisphere—the microbial habitat on plastic debris—varies depending on location, season, and polymer type. However, core biofilm-forming taxa such as *Acidovorax*, *Aquabacterium*, *Rhodobacter*, *Sphingorhabdus*, and *Sphingomonas* are commonly found. Other frequently identified genera include *Pseudomonas* and *Vibrio*, which may play roles in both degradation processes and pathogenic interactions (Parrish & Fahrenfeld, 2019; Qin et al., 2022; Tang, 2024; Zhu et al., 2020). The development of these biofilms follows four stages: initial attachment, irreversible attachment, biofilm maturation (including cluster and microcolony formation), and dispersion. Pathogenic genera such as *Vibrio* are also early colonizers, initiating MP colonization and aiding the attachment of other microbes (Oberbeckmann et al., 2014). Plastisphere microorganisms also play a direct role in polymer biodegradation (Du et al., 2022; Mishra et al., 2024; Yuan et al., 2022). As plastics travel through aquatic environments, abiotic factors alter their mechanical and physicochemical properties. Plastisphere microbes can further modify MP surfaces by degrading additives and secreting MP-degrading enzymes. This biodegradation process involves two main steps: bio-fragmentation, where extracellular enzymes break down polymers, and assimilation, where the resulting small, water-soluble molecules are absorbed by the microbes (Luo et al., 2022).

The plastisphere exacerbates the impacts of MP pollution by providing a habitat for pathogenic microorganisms (Zettler et al., 2013; Oberbeckmann et al., 2015; Keswani et al., 2016), which may adhere to plastic surfaces and spread to new environments. Additionally, the chemical additives in plastics can leach into the water, posing further risks to marine life and human health.

9.2.1 Accumulation of Microplastics in Food Ecosystems

Humans are exposed to MPs and NPs through multiple pathways, including the consumption of contaminated foods from agriculture, livestock, fisheries, and aquaculture. In addition to dietary intake, inhalation of airborne MPs and NPs, or contact with these particles through textiles and dust, further contributes to human exposure (Prata et al., 2020) (Figure 9.2).

MPs pose a significant threat to food systems, which are critical for human health and nutrition. These particles are introduced into terrestrial and aquatic ecosystems through domestic and industrial waste, synthetic textiles, and the breakdown of macroplastics via processes such as erosion, runoff, and degradation. MPs accumulate in the environment, particularly in agricultural and aquatic ecosystems, leading to contamination of food sources.

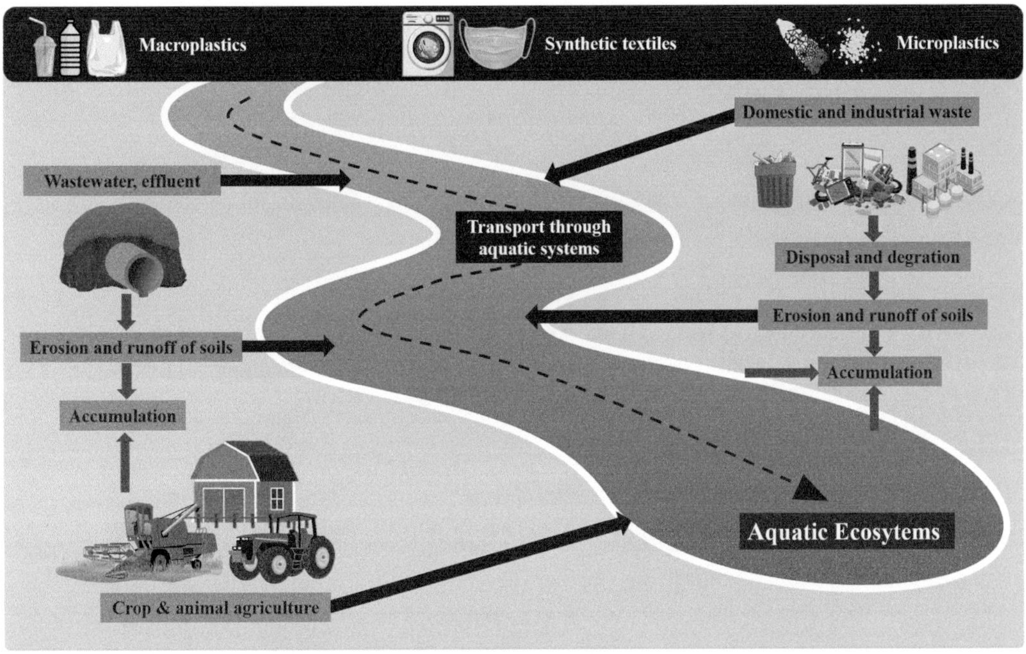

FIGURE 9.2 Sources of MPs in terrestrial and aquatic ecosystems. (Adapted from Horton et al., 2017.)

In terrestrial environments, MPs enter agricultural fields through mulching films, fertilizers, irrigation, atmospheric deposition, and wastewater (Figure 9.3). The breakdown of these plastic materials contaminates the soil, which can enter the food chain through crops and livestock. MPs are then consumed by humans through contaminated food or can be inhaled, with some studies suggesting that inhalation poses a greater health risk than dietary intake (Wright & Kelly, 2017; Vethaak & Legler, 2021).

Aquatic ecosystems are also heavily impacted by MPs, which enter through the degradation of fishing gear, nets, and other equipment. Severe weather events, such as flash floods or hurricanes, exacerbate the movement of plastic waste from terrestrial to aquatic environments, where MPs are ingested by organisms ranging from plankton to larger marine animals. This bioaccumulation poses risks to humans consuming seafood (Laskar & Kumar, 2019; De-la-Torre, 2020).

MP and NP contamination can also arise from environmental sources like feed and food packaging. While MPs are often visibly detected in packaging, NPs, due to their smaller size, can penetrate deeper into food products. Additionally, water used in food processing can become contaminated with MPs, exacerbating the issue. Consumers unknowingly ingest MPs through drinking water and contaminated food. Reducing plastic pollution and adopting sustainable waste management solutions across the entire food production chain are crucial to mitigate the risks posed by MPs and NPs (Yuan et al., 2022).

The accumulation of MPs in food ecosystems is a multifaceted challenge, impacting both terrestrial and aquatic food chains. When plastics are released into the environment through anthropogenic activities, they contaminate land and aquatic sources, eventually entering oceans, lakes, and other bodies of water via waterways and winds. In aquatic environments, MPs are ingested by organisms, which can lead to bioaccumulation in the food web and pose risks to human health (De-la-Torre, 2020) (Figure 9.4).

Overall, addressing the interconnectedness of pollution in both terrestrial and aquatic ecosystems is essential for protecting food safety and human health. Strategies to reduce MP pollution include limiting plastic use, improving waste management practices, and exploring sustainable alternatives.

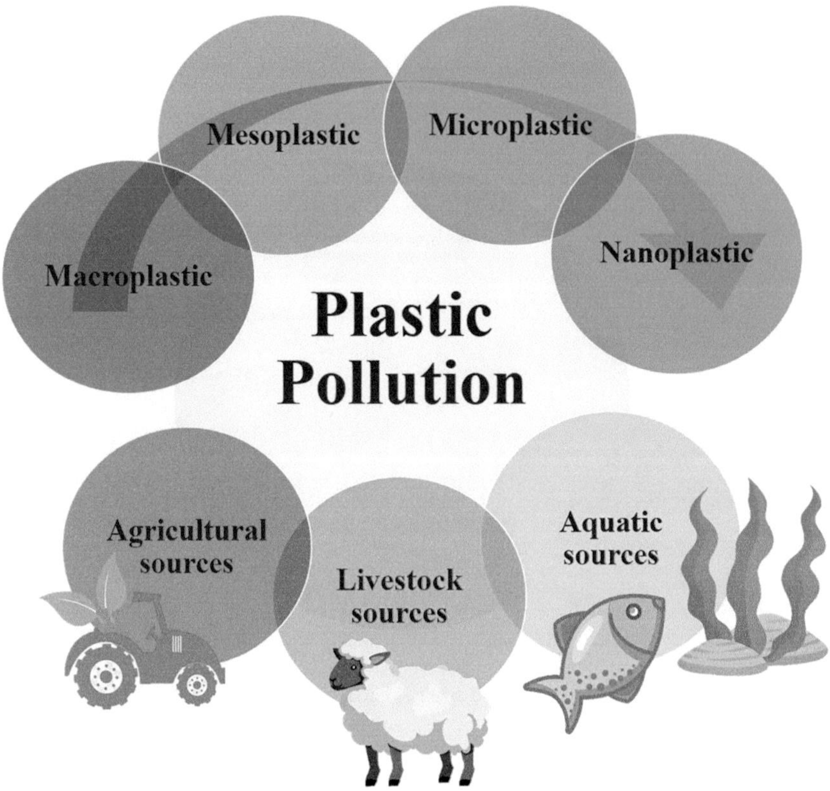

FIGURE 9.3 Plastic pollution pathways: the contribution of macro-, meso-, micro-, and nanoplastics from agricultural, livestock, and aquatic sources. (Adapted from Erkan and Can Tunçelli, 2024.)

9.2.2 Dynamics of Plastisphere in Aquaculture Systems

In aquaculture systems, the presence of MPs and their associated microbial communities, termed the plastisphere, raises concern about potential impacts on water quality, fish health, and overall aquaculture productivity. The plastisphere consists of diverse microbial taxa, including bacteria, fungi, and algae, that attach to plastic surfaces and form biofilms. Among these, certain strains such as *Vibrio* spp., *Aeromonas* spp., and *Pseudomonas* spp. are particularly common and concerning in aquaculture environments due to their pathogenic potential and ability to cause diseases in fish and shellfish, thereby threatening food security (Mishra et al., 2024; Oberbeckmann et al., 2018). These biofilms provide a habitat for various microbial interactions, including symbiosis, competition, and predation.

Biofilm formation on plastic surfaces in aquaculture systems can influence microbial community dynamics. This formation is influenced by factors such as plastic type, surface roughness, and environmental conditions like temperature, pH, and nutrient availability. In aquaculture settings, where nutrient levels are often elevated, the plastisphere can thrive and potentially alter the microbial balance in the system (Mishra et al., 2024).

The presence of the plastisphere in aquaculture systems can have several ecological and health implications. For one, the plastisphere can act as a vector for pathogenic microorganisms. Evidence from aquaculture systems shows that pathogens such as *Vibrio* spp., which are known to cause diseases in fish and shellfish, can colonize plastic debris and spread within these environments. This can lead to increased disease outbreaks, affecting aquaculture productivity and posing risks to food safety (Kirstein et al., 2016; Vezzulli et al., 2016; Oberbeckmann et al., 2018). Additionally, the

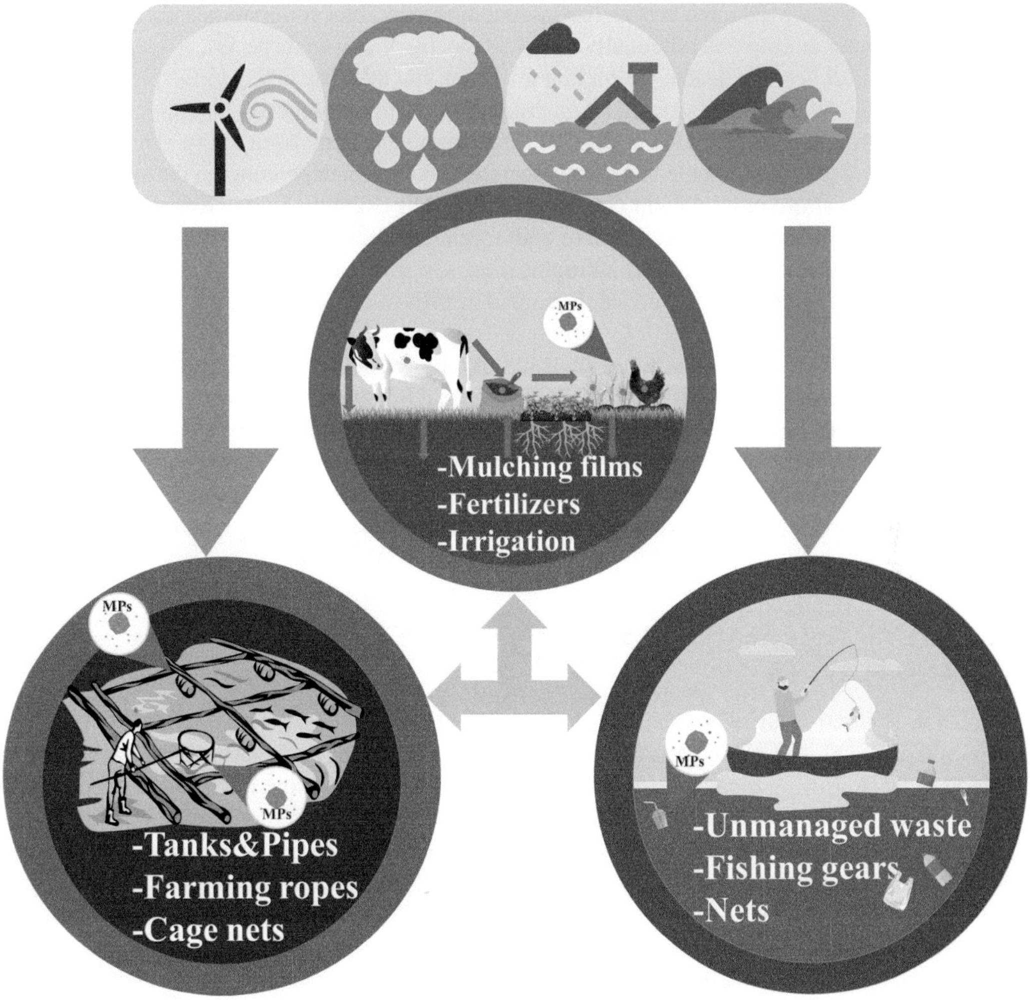

FIGURE 9.4 MP risks in aquatic and terrestrial food systems.

plastisphere can influence nutrient dynamics in aquaculture systems. Microbial degradation of organic matter on plastic surfaces can release nutrients into the water, potentially contributing to eutrophication and harmful algal blooms. These blooms can deplete oxygen levels, create dead zones, and further stress aquatic organisms (Oberbeckmann et al., 2014).

Microorganisms within the plastisphere have the potential to degrade plastics, albeit slowly. Certain bacteria and fungi produce enzymes capable of breaking down plastic polymers into smaller molecules, which can then be further degraded. However, the efficiency of plastic degradation in natural environments is limited, and complete mineralization is rare. Instead, plastics often fragment into smaller pieces, contributing to the persistence of MPs in aquatic systems.

In the context of aquaculture, whether in situ at sea, in freshwater systems like hatcheries, or in land-based tanks, these dynamics take on additional significance. For in situ aquaculture systems, MPs can originate from the degradation of aquaculture equipment, such as nets and ropes, directly contributing to the local plastisphere. In land-based aquaculture tanks, while the controlled environment might limit exposure to external MP sources, plastic infrastructure and equipment can still introduce MPs. These MPs interact with the surrounding microbial communities, forming biofilms that can alter water quality and potentially affect the health of cultured species.

The interactions between plastic-degrading microbes and other members of the plastisphere are complex. In aquaculture systems, degradation products from these processes can serve as substrates for other microorganisms, creating a dynamic and interdependent community. This interaction could either mitigate or accentuate the risks associated with MPs, depending on the specific conditions of the aquaculture environment (Du et al., 2022). For example, in nutrient-rich aquaculture settings, microbial communities may thrive and accelerate the degradation of organic matter on plastics, possibly leading to localized eutrophication. On the other hand, the contained environment of land-based systems might reduce the spread of MPs but could also concentrate their impact within the system.

Addressing plastisphere challenges in aquaculture requires a multifaceted approach. Key strategies include reducing plastic waste by using biodegradable materials, improving waste management, and raising awareness. Regular monitoring of plastic pollution and plastisphere dynamics can help manage risks. Research into microbial interactions and their impacts on aquaculture health is crucial. Exploring plastic-degrading microorganisms for bioremediation could offer sustainable solutions, though ecological impacts must be considered. The plastisphere's dynamics highlight the complex interplay between plastic pollution and microbial communities.

9.3 SUSTAINABLE SEAFOOD SECURITY, SEAFOOD SAFETY, AND PLASTICS

Sustainable seafood security and seafood safety are vital components of global food systems, especially in the face of increasing plastic pollution. These concepts not only ensure the availability of seafood to meet growing global demand but also protect consumers from the health risks posed by contaminated marine food products. With the rise of plastic pollution in marine environments, safeguarding seafood security and safety has become more challenging, necessitating rigorous monitoring and management practices to protect both the environment and human health (FAO, 2020; Thilsted et al., 2016).

Plastics, particularly MPs, can absorb and concentrate harmful pollutants, such as POPs and heavy metals, which can then be transferred through the food web. For example, research conducted in the North Pacific Central Gyre revealed that MPs could absorb polychlorinated biphenyls (PCBs) at concentrations up to 1 million times higher than the surrounding seawater. These contaminated MPs were then ingested by fish species, leading to bioaccumulation and potential biomagnification of these toxic substances as they move up the food chain (Rochman et al., 2013). Aquatic organisms, including commercially important species such as tuna and mackerel, can ingest these MPs, resulting in bioaccumulation and potential biomagnification of pollutants (Fauser et al., 2022; Rochman et al., 2013).

The presence of MPs in marine environments raises significant concerns for the health of aquatic organisms and human consumers. Marine ecosystems, particularly fish and shellfish, are especially vulnerable to the detrimental effects of MPs. Research indicates that MPs not only permeate the water column but also accumulate in marine sediments, forming long-term pollution reservoirs (Ghosh et al., 2023). These particles can disrupt feeding behaviors and digestive processes in marine organisms, leading to reduced growth and reproduction rates. Studies like those by Thompson et al. (2016) and Wright and Kelly (2017) highlight the ecological damage caused by MPs, which impact species at various trophic levels, from plankton to top predators (Parrish & Fahrenfeld, 2019). The trophic transfer of MPs through food webs presents risks to entire ecosystems, potentially leading to biodiversity loss and ecological imbalance (Ebrahimi et al., 2022).

Furthermore, MP exposure threatens the biochemical systems of aquatic organisms, potentially causing oxidative stress, endocrine disruption, and inflammation. Ingested MPs can interfere with enzymatic activities, immune responses, and hormone regulation, which are crucial for the survival of marine species (Ghosh et al., 2023). Wright and Kelly (2017) demonstrated that such biochemical disruptions lead to reduced fitness, making organisms more susceptible to disease and less capable of adapting to environmental changes. The continuous ingestion of MPs by filter feeders, such as mussels and oysters, introduces pollutants into human food chains, raising concerns for food security. Many of these species are primary food sources in coastal communities, and the threat posed by MPs extends to human health (Tang, 2024).

In addition to direct biological impacts, MPs in food-producing ecosystems foster the formation of the plastisphere, a microbial community that colonizes plastic debris (Ebrahimi et al., 2022). These biofilms alter the surface properties of MPs, enhancing their ability to adsorb toxic pollutants like POPs and heavy metals. Zettler et al. (2013) and Oberbeckmann et al. (2015) show how these plastispheres degrade the nutritional quality of aquatic organisms by accumulating hazardous chemicals. Moreover, the plastisphere can make MPs resemble food, leading to increased ingestion by marine organisms, further compromising the health of aquatic animals and posing risks for humans consuming seafood contaminated by MPs.

Emerging evidence also suggests that MPs act as vectors for toxic chemicals and pathogens. Research by Rochman et al. (2013) reveals that MPs have a high affinity for adsorbing pollutants like PCBs and dioxins. Once ingested, these pollutants bioaccumulate in fish and shellfish, raising significant food safety concerns. Additionally, MPs leach toxic additives such as BPA and phthalates, which have been linked to reproductive toxicity and carcinogenesis (Fauser et al., 2022). The European Food Safety Authority (EFSA) has emphasized the importance of banning harmful chemicals, setting strict limits on pollutants, and conducting comprehensive risk assessments to prevent chemical hazards from entering the food chain (Qin et al., 2022).

To address the microbial safety of food products, effective packaging and maintaining a healthy cold chain are critical, as recommended by the Food and Agriculture Organization (FAO) and the World Health Organization (WHO). These measures help minimize microbial contamination during food transport and storage, reducing the risk of microbial growth on MP-contaminated seafood (Qin et al., 2022). Maintaining proper cold storage can also limit the leaching of chemicals from plastics into food products, providing additional protection for consumers.

As the issue of MPs in marine environments escalates, comprehensive research and mitigation strategies are urgently needed. Studies by Udovicki et al. (2022) and Ozden (2023) call for policy interventions aimed at reducing plastic waste, improving waste management, and promoting biodegradable alternatives. Collaboration between governments, industries, and research institutions is essential to develop effective solutions that protect both ecosystems and human health. The growing concern over MPs in food-producing ecosystems underscores the urgent need for global action to address this emerging threat. Food sources provided through fishing and aquaculture in aquatic ecosystems are guarantees of sustainable food security. Aquaculture alone contributed approximately 46% of the total global fish production in 2018, playing a significant role in meeting the growing demand for protein and other essential nutrients, particularly in developing countries (FAO, 2020). By 2030, it is projected that aquaculture will account for more than 60% of fish consumed worldwide, highlighting its crucial role in global food security (FAO, 2020). Sustainable food security is not only about providing enough food but also ensuring that the food is safe and nutritious. One of the main factors ensuring food security is food safety, a comprehensive management system that prevents biological, chemical, and physical hazards that threaten consumer health from production to consumption. These hazards naturally exist in food sourced from agriculture, livestock, and fisheries (FAO/WHO, 2009). For example, livestock can be exposed to mycotoxins in feed, while fish in aquaculture may accumulate contaminants from their environment, both of which can impact food safety (Bhat et al., 2010; Love et al., 2020; Erkan, 2023).

Health is a fundamental human need, and adequate and balanced nutrition is essential for maintaining health. The sustainability of healthy and adequate nutrition is closely linked to a country's economy and development, as malnutrition can lead to decreased productivity and increased healthcare costs, impeding economic growth (FAO, 2013; WHO, 2018). The reliable and high-quality delivery of strategic products like food has been recognized throughout history as the "right to access sufficient and healthy food," enshrined in international frameworks such as the Universal Declaration of Human Rights and the International Covenant on Economic, Social and Cultural Rights (UN, 1948, 1966). Ensuring the sustainability of available resources, particularly in agriculture, livestock, and fisheries, is essential for maintaining secure food systems that support this fundamental right (FAO, 2020).

Plastic polymers and their additives are integral to modern industrial production and found in many everyday products, especially in the food packaging sector. However, the potential impacts of these materials on human health and food safety have sparked significant debate in society and the scientific community. MP contamination can occur during food processing and preservation (Can Tunçelli & Erkan, 2020). Traditional methods like refrigeration, freezing, cooking, drying, smoking, salting, fermentation, and canning, as well as newer techniques like high hydrostatic pressure, ozonation, irradiation, pulsed light, plasma treatments, bioprotection, and alternative packaging, all involve direct contact with plastic-containing products (Kontominas et al., 2021). The presence of MPs has been reported in various processed seafoods (Silva et al., 2024; Venkatesh et al., 2024).

Research on MPs in the context of food safety is crucial for understanding their sources, distribution, and potential impacts on human health (De-la-Torre, 2020). The potential health effects of MPs include adverse impacts on the gastrointestinal system, carcinogenic and toxic effects, and endocrine system disruptions (Wright & Kelly, 2017; Rochman et al., 2013; EFSA Panel on Contaminants in the Food Chain, 2016). Additionally, the toxic substances that MPs carry or adsorb can accumulate in the human body, leading to long-term health issues such as inflammation, oxidative stress, immune system disruption, and potential carcinogenic effects (Prata et al., 2020; Fauser et al., 2022). Research focused on understanding these health impacts is crucial for developing effective risk assessment and management strategies. Identifying the pathways through which MPs contaminate food products can help develop methods to prevent contamination throughout the food chain, thereby ensuring safer food consumption in the future.

Food safety is critical to ensure our food is free from contaminants that pose health risks. In the seafood industry, the presence of MPs is an emerging concern. MPs can enter the marine food chain, accumulate in seafood, and carry harmful chemicals and pathogens, posing direct health risks to consumers and reflecting broader environmental issues. Mitigating these risks involves monitoring aquatic environments, improving food production processes, and raising consumer awareness. Sustainable food production systems are crucial for reducing the impacts of MPs and related hazards. Continuous monitoring of plastic contamination in seafood and rigorous safety assessments are essential to protect public health and maintain consumer confidence. Ensuring sustainable seafood security requires managing plastic pollution through reducing waste inputs, improving waste management, and promoting biodegradable alternatives. Addressing the intersection of plastic pollution and seafood safety is vital for developing more sustainable and resilient aquatic food production systems (Wright et al., 2021).

9.4 METHOD AND APPROACHES TO ELIMINATE MP AND MP-RELATED POLLUTION IN AQUATIC FOOD PRODUCTION SYSTEMS

Amid the dilemma of dwindling food sources and an increasing global population, aquaculture has become a critical socioeconomic parameter for ensuring sustainable food security. Aquaculture has also made significant advancements in the sustainable use of natural resources and the adoption of eco-friendly technologies (Dewali et al., 2023; Estim et al., 2024). Aquatic food production systems, particularly innovative approaches such as Recirculating Aquaculture Systems (RASs), Integrated Multi-Trophic Aquaculture (IMTA), and aquaponics, hold significant potential in mitigating MP pollution. These systems minimize the discharge of pollutants into natural environments, thereby indirectly reducing the release of MPs (Espinal & Matulić, 2019; Azhar & Memiş, 2023; Memiş et al., 2024; Tunçelli et al., 2024b). These systems not only promote sustainable seafood production but also contribute to the ecological remediation of water bodies contaminated with MPs. By incorporating filtration mechanisms, such as biofilters and sedimentation units, RAS and IMTA systems can actively reduce the presence of MPs in water. Additionally, the integration of various trophic levels in IMTA allows for the absorption and biodegradation of certain pollutants, potentially including MPs, by lower trophic organisms such as mollusks or seaweed. Studies have shown that seaweed can trap MPs, thus functioning as bioremediators within the system (Nguyen et al., 2023; Lusher & Covernton, 2022). However, further research is necessary to understand the full extent of MP capture and degradation in these systems. Today, aquaculture production surpasses the total seafood obtained through wild fishing, ranking first in terms of healthy and sustainable ecosystem-based food production (Bjørndal et al., 2024; Kadac-Czapska et al., 2024).

The most critical factor in making aquaculture sustainable is the availability of clean water sources. Contaminants, including MPs, in the water used for aquaculture can directly impact the health and growth rates of farmed species (Wu et al., 2023). Therefore, effective water management practices are essential to ensure aquaculture remains viable and environmentally sustainable. MP pollution, one of the major environmental pollution challenges, poses significant risks to aquatic ecosystems and aquaculture operations. MPs in water negatively affect the habitats, food intake, and reproduction of living organisms in aquatic ecosystems. In recent years, studies have highlighted the adverse effects of MP pollution, which increase as particle size decreases, leading to greater uptake by aquatic organisms across different trophic levels (Skirtun et al., 2022). MPs are transferred along the food chain, potentially causing negative impacts on the life activities of aquatic species. This issue complicates the production of healthy seafood in aquaculture settings.

Industrial activities and agricultural practices exacerbate MP pollution in both freshwater and marine environments. Wastewater, especially untreated or insufficiently treated effluent, serves as a significant pathway for MPs to enter aquatic systems, as particles can pass through treatment plants and reach natural water sources (Kadac-Czapska et al., 2024).

While the best approach to sustainable seafood production is to cultivate shellfish in areas free from external pollution sources, truly unpolluted waters are rare. Therefore, sourcing shellfish from regions with relatively low pollution levels and employing depuration techniques can help ensure food safety. For example, in Italy and France, bivalve depuration centers are commonly used in shellfish farming. These facilities use controlled, clean water to purge contaminants from bivalves before they enter the market, significantly reducing the risks associated with MPs and other pollutants (De-la-Torre et al. 2024; Fraissinet et al., 2024b). Depuration removes microbial contaminants from lightly or moderately polluted bivalves, enhancing the availability and supply of safe, nutritious shellfish. Additionally, depuration has been shown to be effective in mitigating MP pollution, further contributing to the sustainability and safety of seafood production. This process also aligns with legal requirements in many countries, where depuration is mandated under specific conditions (Fraissinet et al., 2024a). Effective depuration depends on systems that maximize the biological activity of bivalves while preventing the reuptake of excreted contaminants. Adhering to recognized food hygiene standards in depuration centers is crucial to avoid increasing contamination levels or causing cross-contamination. Operators must also understand the limitations of depuration, as it does not consistently remove all types of contaminants. By implementing these measures, the seafood industry can contribute to sustainable and safe production, aligning with HACCP recommendations (Fraissinet et al., 2024a).

Aquaculture food production systems are as directly affected by ecosystem pollution as natural habitats. Aquaculture, especially in developing countries, plays a significant role in global food production and nutrition supply. However, the widespread detection of MPs in aquaculture systems has raised concerns regarding the safety of the food products produced (Wu et al., 2023). MPs detected in farmed species often originate from feed ingredients such as fishmeal and fish feeds (Egea-Corbacho et al., 2023). With the large-scale production and use of plastics today, MPs have become a pervasive environmental pollutant with increasingly adverse effects (Skirtun et al., 2022). Therefore, adopting sustainable and environmentally friendly practices to reduce MP pollution in food production systems is essential. Regular monitoring and management practices should ensure that aquaculture facilities do not become sources of MP pollution.

9.4.1 Impacts of MP Presence on Aquatic Food Production Systems

Aquatic food production systems, comprising both fisheries and aquaculture, are essential to global food security, providing high-quality protein and essential nutrients with relatively low environmental footprints. In 2022, these systems produced approximately 223.2 million tonnes of seafood and algae, of which 185.4 million tonnes were aquatic animals and 37.8 million tonnes were algae (FAO, 2024). Despite their vital role, both fisheries and aquaculture face significant challenges from MP pollution, which threatens marine biodiversity, food safety, and human health.

Fisheries involve the harvesting of aquatic species from natural water bodies, including wild capture fisheries and recreational fisheries. While wild capture fisheries contribute substantially to global seafood production, they are highly vulnerable to environmental threats such as plastic pollution. Overfishing and unsustainable fishing practices, such as trawling, longlining, and purse seining, not only threaten marine biodiversity but also exacerbate MP pollution. These methods contribute to the degradation of marine ecosystems through the breakdown of synthetic fishing gear into MPs (FAO, 2020).

A significant portion of this plastic pollution comes from ghost gear—discarded or lost fishing nets, ropes, and buoys—that degrade into MPs. These MPs are then ingested by fish, shellfish, and other marine organisms, leading to bioaccumulation of toxic chemicals and eventual transfer into the human food chain (Lusher et al., 2017). This issue is particularly concerning in industrial fisheries, where large-scale mechanized operations produce substantial amounts of plastic debris, further contaminating marine ecosystems. Even artisanal fisheries, which rely on smaller, traditional boats, contribute to MP pollution on a smaller scale (Wright et al., 2021).

Aquaculture, the controlled farming of aquatic organisms, plays an increasingly important role in global food security as wild fish stocks are depleted due to overfishing, climate change, and pollution (Naylor et al., 2021). Aquaculture systems include marine aquaculture (mariculture) in coastal waters and freshwater aquaculture in ponds, lakes, and rivers. Innovative approaches like IMTA and RAS have been developed to improve sustainability and minimize environmental impact (Ebeling & Timmons; 2010; Chopin, 2013; Badiola et al., 2012).

However, aquaculture systems face substantial risks from MP contamination, which can enter through various pathways:

- **Fishmeal and Fish Feeds**: MPs have been detected in fishmeal (Gündoğdu et al., 2021; Wang et al., 2022), a primary feed for farmed species like sea bass (Matias et al., 2023) and bivalves (Zhang et al., 2020; Tunçelli & Memis, 2024; Tunçelli et al., 2024a). This leads to the contamination of the aquaculture food web, as MPs accumulate in the tissues of farmed species, posing direct health risks to human consumers (Egea-Corbacho et al., 2023).
- **Plastic Equipment**: Aquaculture operations rely heavily on plastic-based materials, such as nets, cages, pipes, and tanks, which degrade over time and release MPs into the aquatic environment (Tidwell, 2012). Studies from regions like China and Norway have shown

that aquaculture environments have high levels of MP contamination, which threatens the health of farmed species and the surrounding ecosystems (Lusher & Pettersen, 2021; Wu et al., 2023).

- **Environmental and Health Concerns**: The use of antifouling agents, pesticides, and heavy metals in aquaculture compounds the MP pollution problem. These chemicals can leach into the environment and interact with MPs, forming a toxic mixture that leads to antimicrobial resistance, environmental degradation, and food safety risks for consumers (Elgarahy et al., 2021; Li et al., 2022)

MP contamination in aquatic food production systems presents numerous health risks for both marine organisms and human consumers:

- **Physical Damage**: MPs cause physical damage to the digestive systems of marine organisms, reducing their feeding efficiency, growth, and reproductive success. This is particularly harmful in aquaculture settings, where the health and growth of farmed species directly impact economic viability (Anbumani & Kakkar, 2018; Franzellitti et al., 2019, Yin et al., 2021)
- **Chemical Contamination**: MPs act as vectors for toxic chemicals such as POPs and heavy metals (Sheng et al., 2021). These pollutants are resistant to environmental degradation and can bioaccumulate in the tissues of marine organisms, eventually being transferred up the food chain to humans (Wright & Kelly, 2017). Additionally, plastic additives like plasticizers and flame retardants, commonly used in plastic products, can leach from MPs, adding further chemical hazards (Andrady, 2017; Cheng et al., 2020; Fauser et al., 2022).
- **Microbial Contamination**: MPs also provide surfaces for the growth of harmful microbial communities, including pathogens, which further compromise the health of farmed species and the safety of seafood products (Luo et al., 2022). This "plastisphere" of microbial life has raised concerns about its role in spreading diseases within aquatic food systems (Amobonye et al., 2021).

In addition to impacting live aquatic species, MPs have been found in processed seafood products. During seafood processing, plastic-containing tools, additives, and packaging come into contact with food. Studies report that MPs can be found in various seafood products after processing (Can Tunçelli & Erkan, 2020; Kadac-Czapska et al., 2024). Additionally, plastic cutting boards used in food preparation are a significant source of MP contamination (Habib et al., 2022). Research shows that cutting styles and board materials, such as polypropylene and polyethylene, influence the amount of MPs released into food (Yadav et al., 2023). Furthermore, commonly used food items like table salt and sugar have also been found to contain MPs, further highlighting the pervasiveness of plastic pollution (Habib et al., 2022).

MPs not only affect food safety but also threaten marine biodiversity. Ingested MPs cause physical damage to marine organisms, impair their growth and reproductive capabilities, and disrupt ecosystem functions. For example, studies have shown that MPs accumulate in the tissues of fish and shellfish, leading to ecosystem imbalances as contaminants move up the food chain (Ahmad et al., 2020). Additionally, MPs degrade habitats by altering water quality and sediment composition, further threatening biodiversity (Wright et al., 2021).

Given the growing threat posed by MP pollution in aquatic food production systems, several strategies have been proposed:

1. **Sustainable Fishing and Aquaculture Practices**: Encouraging the use of sustainable fishing techniques and eco-friendly aquaculture practices is essential for reducing MP release into the environment. This includes replacing plastic-based equipment with

biodegradable alternatives and improving waste management systems to prevent plastic debris from entering marine ecosystems (Skirtun et al., 2022).

2. **Regulation of Plastic Use**: Stricter regulations on the use of plastic in aquaculture, particularly for fishing gear and packaging, can help curb MP pollution. For example, the European Union's Single-Use Plastics Directive aims to reduce plastic waste through recovery and recycling initiatives (Elliott et al., 2020).
3. **Advanced Filtration Systems**: In aquaculture systems like RASs, advanced filtration technologies can be used to remove MPs from water supplies, reducing exposure to plastic contaminants (Schuhen & Sturm, 2022; Kılıç et al., 2023). This is especially important in water-scarce regions where water must be reused within the system.
4. **Public Awareness and Research**: Increasing public awareness about the dangers of MP pollution is crucial for reducing plastic consumption. Ongoing research into the effects of MPs on marine organisms and human health will help to develop more effective mitigation strategies (Ghosh, 2023).

MPs represent a significant and growing threat to the sustainability of aquatic food production systems. Their presence in both fisheries and aquaculture undermines the health of marine ecosystems, poses direct risks to human health through contaminated seafood, and threatens global biodiversity. To safeguard the future of aquatic food systems, it is crucial to implement sustainable practices, enhance waste management, adopt advanced filtration technologies, and raise public awareness about the impact of plastic pollution.

9.4.2 Ecosystem-Based Production Models in Aquatic Environment

Ecosystem-based food production integrates ecological principles into food systems to optimize resource use, reduce environmental impacts, and promote sustainability. In the aquatic environment, this approach has led to the development of several production models, such as IMTA, RAS, and aquaponics. These models offer more sustainable alternatives to traditional aquaculture methods, which often face challenges related to MP and NP pollution, resource inefficiency, and environmental degradation. A summary of these ecosystem-based models, their benefits, and their mechanisms for mitigating MP pollution is provided in Table 9.1.

Aquatic food production faces increasing contamination from MPs and NPs, which can enter the food chain through water sources, packaging, and production processes (Okeke et al., 2022). These pollutants present significant risks to food safety and threaten the sustainability of aquatic

TABLE 9.1
Overview of Ecosystem-Based Aquaculture Models and Their Benefits in Potential Mitigating MP Pollution

Ecosystem-Based Model	Description	Benefits	Potential MP Mitigation Mechanism
IMTA	Integration of species from different trophic levels, such as fish, shellfish, and seaweed.	Natural nutrient recycling, improved water quality.	Seaweeds and mussels filter and sequester MPs from water (Fraissinet et al., 2024b).
RAS	Closed-loop system that recirculates water through advanced filtration.	Conserves water, reduces discharge of pollutants.	Filters MPs, controls waste within the system (Can Tunçelli & Erkan, 2020).
Aquaponics	Combines aquaculture and hydroponics; fish waste fertilizes plants.	Reduces water use, eliminates chemical fertilizers.	Microbial communities and plants capture and degrade MPs (Tunçelli et al., 2023).

environments. To address these issues, ecosystem-based production models have emerged as a solution, offering both environmental and economic benefits by integrating natural ecosystem functions into aquaculture practices (Tunçelli et al., 2023).

9.4.2.1 Integrated Multitrophic Aquaculture

IMTA operates by cultivating species from different trophic levels within the same system (Figure 9.5). This setup mirrors natural ecosystems, where waste from one species is utilized by another, creating a self-sustaining, resource-efficient cycle. For example, fish excretions provide nutrients for seaweeds, which improve water quality by absorbing excess nutrients and filtering out pollutants like MPs. Filter-feeders such as mussels can also capture MPs from the water, reducing their presence in the ecosystem (Fraissinet et al., 2024b).

IMTA has emerged as a critical model for promoting sustainability in aquaculture. By integrating various species such as fish, shellfish, and seaweed, IMTA minimizes waste and recycles nutrients within the system, resulting in improved water quality and reduced pollution. In contrast to traditional monoculture systems, which often lead to nutrient buildup and environmental degradation, IMTA creates a balanced and ecologically harmonious production environment (Granada et al., 2015).

The potential for IMTA to mitigate MP contamination is particularly notable. Filter feeders like mussels are known for their ability to trap and sequester MPs, while seaweeds can absorb these pollutants. Studies have shown that mussels filter out MPs from fish feces, which are then retained within the mussels' tissues. This MP retention process, followed by depuration, significantly reduces the environmental MP load (Stollberg et al., 2021). Therefore, IMTA not only enhances production efficiency but also contributes to MP pollution control.

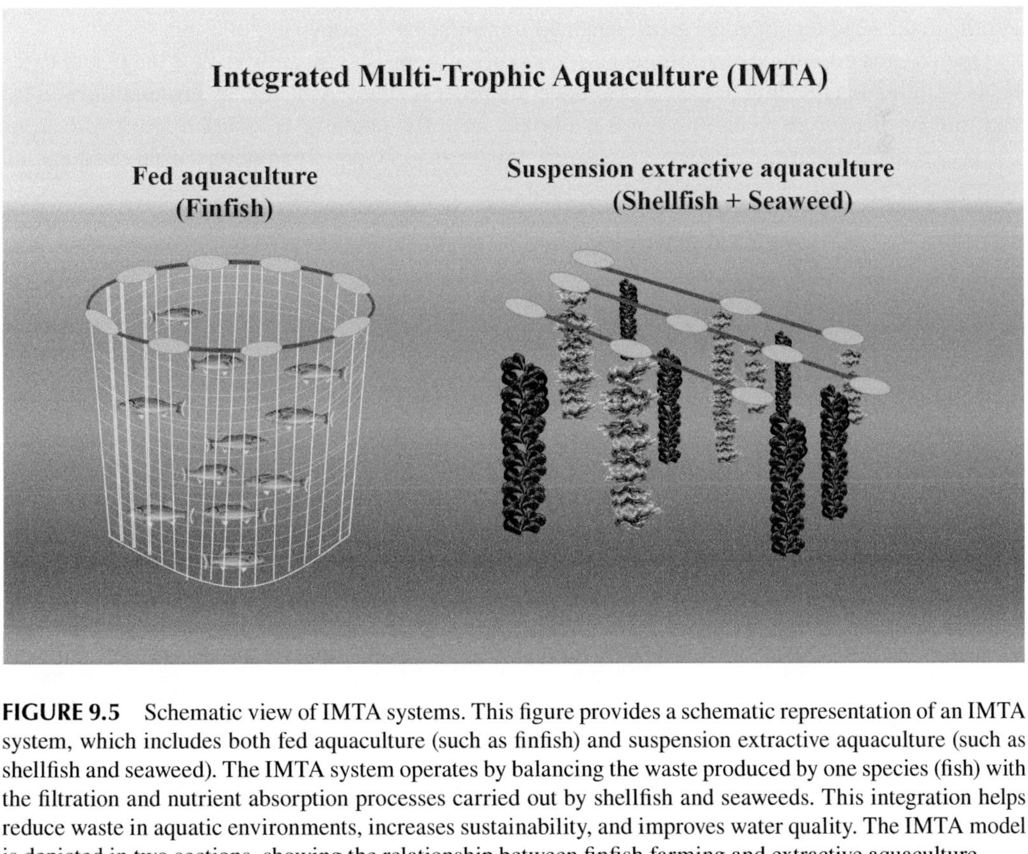

FIGURE 9.5 Schematic view of IMTA systems. This figure provides a schematic representation of an IMTA system, which includes both fed aquaculture (such as finfish) and suspension extractive aquaculture (such as shellfish and seaweed). The IMTA system operates by balancing the waste produced by one species (fish) with the filtration and nutrient absorption processes carried out by shellfish and seaweeds. This integration helps reduce waste in aquatic environments, increases sustainability, and improves water quality. The IMTA model is depicted in two sections, showing the relationship between finfish farming and extractive aquaculture.

9.4.2.2 Recirculating Aquaculture Systems

RASs are designed to recycle water within an aquaculture facility, significantly reducing environmental impacts while providing more controlled production conditions. By filtering and reusing water, RASs conserve valuable water resources and minimize the discharge of pollutants into natural ecosystems. Advanced filtration technologies in RAS help remove organic waste and MPs, thus preventing the accumulation of these pollutants. This system is particularly advantageous in regions with water scarcity or strict environmental regulations, as it reduces water consumption and allows precise control over key factors like water quality, temperature, and oxygen levels (Badiola et al., 2012).

However, recent research has raised concerns that RAS may also serve as a source of MP and NP pollution. Studies by Wei et al. (2024) and Blonç et al. (2023) have detected MPs and NPs within RASs and even found evidence of their translocation into the organs of farmed fish. This suggests that RAS, despite its sustainability advantages, may contribute to plastic contamination if the materials used within the system degrade into MPs over time.

To mitigate this risk, it is essential that materials less prone to fragmentation and MP release are selected for system components. Such an approach would help maintain the environmental benefits of RAS while reducing the potential for secondary plastic pollution. Additionally, regular monitoring and maintenance of filtration systems are necessary to prevent MPs from entering or being released into the system.

9.4.2.3 Aquaponics

Aquaponics is a closed-loop system that integrates aquaculture (raising fish) with hydroponics (growing plants in water) (Figure 9.6). In this system, fish waste serves as a nutrient source for plants, which, in turn, purify the water for the fish. Aquaponics systems reduce water consumption, eliminate the need for chemical fertilizers, and minimize environmental pollution.

Operating in controlled environments, aquaponics systems significantly reduce the risk of external contamination, leading to a cleaner production process. These systems, by eliminating soil use and minimizing the reliance on plastic materials, have the potential to contain fewer MPs compared to traditional farming. However, plastic components like pipes and pumps used in aquaponic

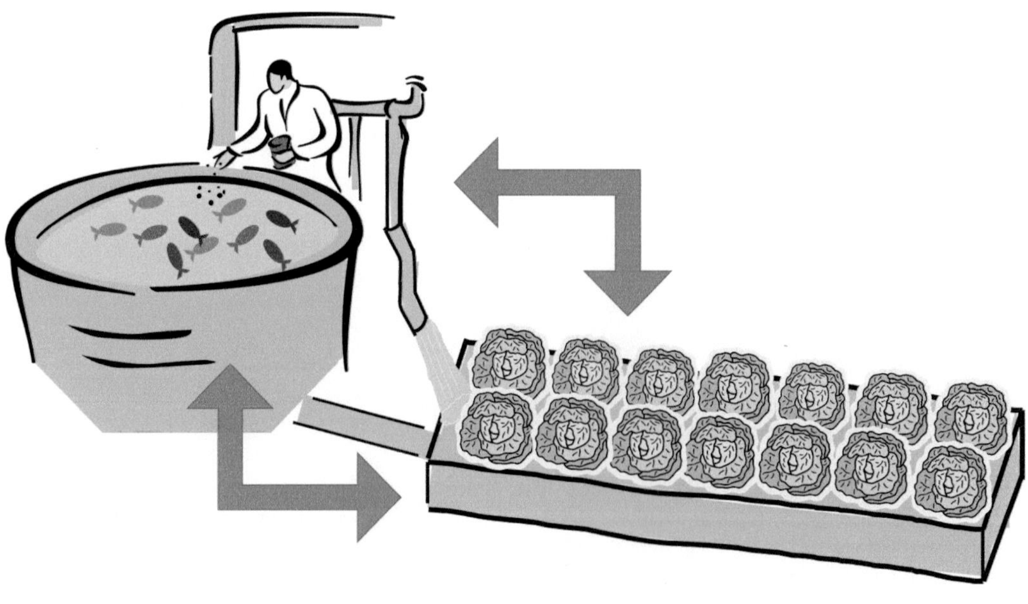

FIGURE 9.6 Schematic view of aquaponics systems.

systems may degrade over time, contributing to MP contamination. While aquaponic systems may not fully eliminate MPs, regular maintenance of water and equipment can minimize their formation and accumulation. Moreover, the use of more durable and sustainable materials could further reduce MP pollution.

Research suggests that the combination of aquatic organisms and plants creates a system that not only recycles nutrients but also offers the potential for trapping MPs, preventing their accumulation in the environment. Although more empirical studies are needed to fully understand the role of aquaponics in mitigating MPs, the controlled nature of these systems shows promise for future research (Okeke et al., 2022).

Additionally, similar to bioremediation studies in aquaculture, processes like aggregation and flocculation observed in microalgae-based systems could enhance MP removal efficiency in aquaponics (Sillanpää et al., 2022), further highlighting their potential as a bioremediation tool.

Aquaponics is especially well-suited for urban environments where space is limited, and water resources are scarce. This model allows for fresh food production with a minimal ecological footprint, reducing carbon emissions typically associated with traditional farming and long food supply chains (Somerville et al., 2014). Along with environmental benefits, aquaponics provides economic advantages by offering dual revenue streams from both fish and plant production.

9.4.2.4 Challenges and Future Directions

Despite the numerous benefits of ecosystem-based models like IMTA, RAS, and aquaponics, there are challenges that need to be addressed to fully realize their potential. These systems require significant investment in technology and knowledge transfer, particularly in the context of managing high biomass production and waste filtration. Additionally, while IMTA and RAS can reduce MP concentrations, complete removal of MPs and NPs from aquatic environments remains a challenge.

Further research and technological innovation are essential to improving the efficiency of these systems, particularly regarding their ability to filter out and sequester MPs. Ongoing studies on the interaction between MPs, aquatic species, and ecosystem dynamics will be key to advancing sustainable aquaculture practices.

Greater public awareness and stricter regulations on plastic use, especially in aquaculture equipment and packaging, are crucial to minimizing plastic pollution. Encouraging eco-friendly materials and biodegradable alternatives can further reduce the environmental footprint of these systems. The development of more effective depuration techniques for species like mussels will also be critical to controlling MP pollution in IMTA systems (Stollberg et al., 2021).

Ecosystem-based production models such as IMTA, RAS, and aquaponics represent innovative and sustainable solutions to the growing problem of MP and NP pollution in aquatic food systems. These models not only help mitigate plastic pollution but also enhance resource efficiency, reduce environmental impact, and promote sustainable seafood production. By leveraging natural processes and employing closed-loop systems, ecosystem-based models can significantly contribute to ensuring long-term food security and protecting marine ecosystems from the harmful effects of plastic contamination.

9.5 CONCLUSION

MP pollution represents a growing challenge to aquatic ecosystems, food production systems, and public health. Within aquaculture, MPs not only affect the health of farmed species but also risk contamination in the food chain, making it essential to explore solutions for reducing this contamination. One critical aspect of MP pollution is the plastisphere, a microbial community that colonizes plastic surfaces in water, which complicates the management of MP pollution in aquaculture systems.

To address these risks, sustainable aquaculture practices like IMTA, RAS, and aquaponics offer promising pathways. These ecosystem-based models optimize resource use, recycle nutrients, and reduce pollution, potentially capturing and even degrading MPs in the process. While there is evidence suggesting that microbial communities within these systems can mitigate MP accumulation, the precise mechanisms of this interaction remain unclear and require further investigation (Okeke et al., 2022). More research is needed to determine how these systems can be optimized to prevent MPs from entering the broader environment and food chain.

The plastisphere adds complexity to the issue. While microbial biofilms on plastic can aid in nutrient cycling and waste degradation, they may also fragment plastics into smaller particles, making them more bioavailable to aquatic species. The interaction between the plastisphere and MP pollution requires further exploration, particularly in terms of plastic breakdown, chemical adsorption, and the potential for pathogens to proliferate (Zettler et al., 2013). It is not yet fully understood how these dynamics impact food safety and aquaculture productivity, and this should be a priority for future research.

Moving forward, scaling up ecosystem-based production models like IMTA and RAS will be crucial to minimizing MP pollution, but these systems also need regular monitoring and technological improvements. For example, the degradation of plastic components in aquaculture equipment can introduce MPs, though the extent to which beneficial microbial communities in these systems can mitigate this impact is still uncertain and demands further empirical studies (Fraissinet et al., 2024a).

Equally important are stricter regulations on plastic use in aquaculture equipment and packaging. Encouraging the adoption of biodegradable materials and eco-friendly practices, supported by public awareness campaigns, can further help reduce the environmental impact of plastic contamination in aquaculture systems (FAO, 2020).

Addressing MP pollution in aquaculture requires a multidisciplinary approach that includes scientific innovation, policy reform, and community engagement. By continuing to develop and refine sustainable, ecosystem-based models, we can reduce the risks posed by MP contamination while ensuring resilient food production systems. Finally, a deeper understanding of the plastisphere and its role in plastic degradation and ecosystem health will be essential to achieving long-term sustainability in aquatic food systems.

REFERENCES

Ahmad, M., Li, J. L., Wang, P. D., Hozzein, W. N., & Li, W. J. (2020). Environmental perspectives of microplastic pollution in the aquatic environment: a review. *Marine Life Science & Technology*, *2*, 414–430.

Amaral-Zettler, L. A., Zettler, E. R., & Mincer, T. J. (2020). Ecology of the plastisphere. *Nature Reviews Microbiology*, *18*(3), 139–151.

Amobonye, A., Bhagwat, P., Singh, S., & Pillai, S. (2021). Plastic biodegradation: frontline microbes and their enzymes. *Science of the Total Environment*, *759*, 143536.

Anbumani, S., & Kakkar, P. (2018). Ecotoxicological effects of microplastics on biota: a review. *Environmental Science and Pollution Research, 25*, 14373–14396.

Andrady, A. L. (2017). The plastic in microplastics: a review. *Marine Pollution Bulletin*, *119*(1), 12–22.

Azhar, M., & Memiş, D. (2023). Application of the IMTA (integrated multi-trophic aquaculture) system in freshwater, brackish and marine aquaculture. *Aquatic Sciences and Engineering*, *38*(2), 101–121.

Badiola, M., Mendiola, D., & Bostock, J. (2012). Recirculating aquaculture systems (RAS) analysis: main issues on management and future challenges. *Aquacultural Engineering, 53*, 1–26.

Battin, T. J., Besemer, K., Bengtsson, M. M., Romani, A. M., & Packmann, A. I. (2016). The ecology and biogeochemistry of stream biofilms. *Nature Reviews Microbiology, 14*(4), 251–263.

Bhat, R., Rai, R. V., & Karim, A. A. (2010). Mycotoxins in food and feed: present status and future concerns. *Comprehensive Reviews in Food Science and Food Safety*, *9*(1), 57–81. https://doi.org/10.1111/j.1541-4337.2009.00094.x.

Bjørndal, T., Dey, M., & Tusvik, A. (2024). Economic analysis of the contributions of aquaculture to future food security. *Aquaculture, 578*, 740071.

Blonç, M., Husson, F., Llorca, M., Farré, M., Tort, L., Brandts, I., & Teles, M. (2023). Occurrence of micro-nanoplastics in a commercial recirculated aquaculture system and their translocation to cultured fish organs: a baseline study. *Journal of Hazardous Materials Advances, 12*, 100381.

Bryant, J. A., Clemente, T. M., Viviani, D. A., Fong, A. A., Thomas, K. A., Karl, P. K., … & DeLong, E. F. (2016). Diversity and activity of communities inhabiting plastic debris in the North Pacific Gyre. *mSystems, 1*(3), e00024-16.

Can Tunçelli, İ., & Erkan, N. (2020). Gıda güvenliği açısından su ürünlerinde mikroplastik riski ve araştırma yöntemleri. *Aquatic Research, 4*(1), 73–87. (In Turkish)

Can Tunçelli, İ., & Erkan, N. (2024). Microplastic pollution in wild and aquacultured Mediterranean mussels from the Sea of Marmara: abundance, characteristics, and health risk estimations. *Environmental Research, 242*, 117787.

Cheng, H., Luo, H., Hu, Y., & Tao, S. (2020). Release kinetics as a key linkage between the occurrence of flame retardants in microplastics and their risk to the environment and ecosystem: a critical review. *Water Research, 185*, 116253.

Chopin, T. (2013). Integrated multi-trophic aquaculture: anticipating the future, devising the pathway. *Aquaculture, 43*(3), 9–13.

De-la-Torre, G. E. (2020). Microplastics: an emerging threat to food security and human health. *Journal of Food Science and Technology, 57*(5), 1601–1608.

De-la-Torre, G.E., Dioses-Salinas, D.C., Fernández Severini, M.D., Forero López, A.D. (2024). Microplastics and food safety. In: A. K. Mishra, P. Raizada, T. Helmy, E., S. Arockiasamy, R. Selvasembian (Eds.), *Microplastics*. Singapore: Springer. https://doi.org/10.1007/978-981-97-6461-7_8.

Dewali, S., Sharma, N., Melkani, D., Arya, M., Kathayat, N., Panda, A. K., & Bisht, S. S. (2023). Aquaculture: contributions to global food security. In: S. Ghosh, A. K. Panda, C. Jung, & S. S. Bisht (Eds.), *Emerging Solutions in Sustainable Food and Nutrition Security* (pp. 123–139). Cham: Springer International Publishing.

Dodhia, M. S., Rogers, K. L., Fernández-Juárez, V., Carreres-Calabuig, J. A., Löscher, C. R., Tisserand, A. A., Keulen, N., Riemann, L., Shashoua, Y., & Posth, N. R., (2023). Microbe-mineral interactions in the plastisphere: Coastal biogeochemistry and consequences for degradation of plastics. *Frontiers in Marine Science, 10*, p.1134815. https://doi.org/10.3389/fmars.2023.1134815.

Du, Y., Liu, X., Dong, X., & Yin, Z. (2022). A review on marine plastisphere: biodiversity, formation, and role in degradation. *Computational and Structural Biotechnology Journal, 20*, 975–988.

Dussud, C., Meistertzheim, A. L., Conan, P., Pujo-Pay, M., George, M., Fabre, P., ... & Ghiglione, J. F. (2018). Evidence of niche partitioning among bacteria living on plastics, organic particles and surrounding seawaters. *Environmental Pollution, 236*, 807–816, https://doi.org/10.1016/j.envpol.2017.12.027.

Ebeling, J. M., & Timmons, M. B. (2010). *Recirculating Aquaculture*. Ithaca, NY: Cayuga Aqua Ventures.

Ebrahimi, P., Abbasi, S., Pashaei, R., Bogusz, A., & Oleszczuk, P. (2022). Investigating impact of physicochemical properties of microplastics on human health: A short bibliometric analysis and review. *Chemosphere, 289*, 133146, https://doi.org/10.1016/j.chemosphere.2021.133146.

EFSA Panel on Contaminants in the Food Chain. (2016). Presence of microplastics and nanoplastics in food, with particular focus on seafood. *EFSA Journal, 14*(6), 4501.

Egea-Corbacho, A., Martín-García, A. P., Franco, A. A., Albendín, G., Arellano, J. M., Rodríguez-Barroso, R., … & Malta, E. J. (2023). Microplastic in industrial aquaculture: occurrence in the aquatic environment, feed and organisms (*Dicentrarchus labrax*). *Science of the Total Environment, 904*, 166774.

Elgarahy, A. M., Akhdhar, A., & Elwakeel, K. Z. (2021). Microplastics prevalence, interactions, and remediation in the aquatic environment: a critical review. *Journal of Environmental Chemical Engineering, 9*(5), 106224.

Elliott, T., Gillie, H., & Thomson, A. (2020). European Union's plastic strategy and an impact assessment of the proposed directive on tackling single-use plastics items. In: T. M. Letcher (Ed.), *Plastic Waste and Recycling* (pp. 601–633). Amsterdam: Academic Press.

Erkan, N. (2023). Evaluation of marine mucilage in terms of effects on food security and food safety. In: M. Albay (Ed.), *Mucilage Problem in the Sea of Marmara*. Istanbul: Istanbul University Press, pp. 253–297. https://doi.org/10.26650/B/LS32.2023.003.12

Erkan, N., & Can Tunçelli, İ. (2024). Gıdalarda plastik problemi [The plastic problem in foods]. In: M. E. Aydın, & M. Tayanç (Eds.), *Mikroplastikler, endokrin bozucular ve* çevresel *etkileri* [*Microplastics, Endocrine Disruptors and Their Environmental Effects*] (pp. 451–486). Türkiye Bilimler Akademisi. https://doi.org/10.53478/TUBA.978-625-8352-95-5

Espinal, C. A., & Matulić, D. (2019). Recirculating aquaculture technologies. *Aquaponics Food Production Systems*, *1*, 35–76.
Estim, A., Shapawi, R., Shaleh, S. R. M., Fui-Fui, C., & Mustafa, S. (2024). Transformative research in aquaculture for sustainable seafood security. In: W. L. Filho, T. F. Ng, U. Iyer-Raniga, A. Ng, & A. Sharifi (Eds.), *SDGs in the Asia and Pacific Region* (pp. 415–444). Cham: Springer International Publishing.
FAO (Food and Agriculture Organization) (2013). *Dietary Protein Quality Evaluation in Human Nutrition: Report of an FAO Expert Consultation*. FAO, Rome: Food and nutrition paper, 92. https://www.fao.org/ag/humannutrition/35978-02317b979a686a57aa4593304ffc17f06.pdf
FAO. (2020). *The State of World Fisheries and Aquaculture 2020: Sustainability in Action*. Rome: Food and Agriculture Organization of the United Nations.
FAO and WHO. 2023. *Codex Alimentarius Commission Procedural Manual*. Twenty-eighth edition, revised. Rome. https://doi.org/10.4060/cc5042en.
FAO/WHO. (2009). *Principles and guidelines for the conduct of microbiological risk assessment*. Food and Agriculture Organization of the United Nations/World Health Organization. https://www.fao.org/4/i0680e/i0680e.pdf
Fauser, P., Vorkamp, K., & Strand, J. (2022). Residual additives in marine microplastics and their risk assessment–a critical review. *Marine Pollution Bulletin*, *177*, 113467.
Fraissinet, S., Arduini, D., Martines, A., De Benedetto, G. E., Malitesta, C., Giangrande, A., & Rossi, S. (2024b). Microplastics uptake by four filter feeders. *Journal of Marine Science and Engineering*, *12*(6), 1000.
Franzellitti, S., Canesi, L., Auguste, M., Wathsala, R. H., & Fabbri, E. (2019). Microplastic exposure and effects in aquatic organisms: a physiological perspective. *Environmental Toxicology and Pharmacology*, *68*, 37–51.
Fraissinet, S., De Benedetto, G. E., Malitesta, C., Holzinger, R., & Materić, D. (2024a). Microplastics and nanoplastics size distribution in farmed mussel tissues. *Communications Earth & Environment*, *5*(1), 128.
Galloway, T. S., Cole, M., & Lewis, C. (2017). Interactions of microplastic debris throughout the marine ecosystem. *Nature Ecology & Evolution*, *1*(5), 0116.
Ghosh, S., Sinha, J. K., Ghosh, S., Vashisth, K., Han, S., & Bhaskar, R. (2023). Microplastics as an emerging threat to the global environment and human health. *Sustainability*, *15*(14), 10821.
Granada, L., Sousa, N., Lopes, S., & Lemos, M. F. L. (2015). Is integrated multitrophic aquaculture the solution to the sector's major challenges? A review. *Reviews in Aquaculture*, *6*, 1–18. https://doi.org/10.1111/raq.12093
Gündoğdu, S. E. D. A. T., Eroldoğan, O. T., Evliyaoğlu, E., Turchini, G. M., & Wu, X. G. (2021). Fish out, plastic in: global pattern of plastics in commercial fishmeal. *Aquaculture*, *534*, 736316.
Habib, R. Z., Poulose, V., Alsaidi, R., Al Kendi, R., Iftikhar, S. H., Mourad, A. H. I., ... & Thiemann, T. (2022). Plastic cutting boards as a source of microplastics in meat. *Food Additives & Contaminants: Part A*, *39*(3), 609–619.
Horton, A. A., Walton, A., Spurgeon, D. J., Lahive, E., & Svendsen, C. (2017). Microplastics in freshwater and terrestrial environments: evaluating the current understanding to identify the knowledge gaps and future research priorities. *Science of the Total Environment*, *586*, 127–141.
Jacquin, J., Cheng, J., Odobel, C., Pandin, C., Conan, P., Pujo-Pay, M., ... & Ghiglione, J.-F. (2019). Microbial ecotoxicology of marine plastic debris: a review on colonization and biodegradation by the 'plastisphere'. *Frontiers in Microbiology, 10*, 865.
Jiang, B., Kauffman, A. E., Li, L., McFee, W., Cai, B., Weinstein, J., ... & Xiao, S. (2020). Health impacts of environmental contamination of micro-and nanoplastics: a review. *Environmental Health and Preventive Medicine, 25*, 1–15.
Kadac-Czapska, K., Knez, E., & Grembecka, M. (2024). Food and human safety: the impact of microplastics. *Critical Reviews in Food Science and Nutrition*, *64*(11), 3502–3521.
Keswani, A., Oliver, D. M., Gutierrez, T., & Quilliam, R. S. (2016). Microbial hitchhikers on marine plastic debris: human exposure risks at bathing waters and beach environments. *Marine Environmental Research, 118*, 10–19.
Kılıç, E., Yücel, N., & Şahutoğlu, S. M. (2023). Microplastic composition, load and removal efficiency from wastewater treatment plants discharging into Orontes River. *International Journal of Environmental Research, 17*(2), 25.
Kirstein, I. V., Kirmizi, S., Wichels, A., Garin-Fernandez, A., Erler, R., Löder, M., & Gerdts, G. (2016). Dangerous hitchhikers? Evidence for potentially pathogenic Vibrio spp. on microplastic particles. *Marine Environmental Research*, *120*, 1–8. https://doi.org/10.1016/j.marenvres.2016.07.004.

Kontominas, M. G., Badeka, A. V., Kosma, I. S., & Nathanailides, C. I. (2021). Innovative seafood preservation technologies: recent developments. *Animals*, *11*(1), 92.

Laskar, N., & Kumar, U. (2019). Plastics and microplastics: a threat to environment. *Environmental Technology & Innovation*, *14*, 100352.

Li, Z., Junaid, M., Chen, G., & Wang, J. (2022). Interactions and associated resistance development mechanisms between microplastics, antibiotics and heavy metals in the aquaculture environment. *Reviews in Aquaculture, 14*(2), 1028–1045.

Lobelle, D., & Cunliffe, M. (2011). Early microbial biofilm formation on marine plastic debris. *Marine pollution bulletin*, *62*(1), 197–200, https://doi.org/10.1016/j.marpolbul.2010.10.013.

Love, D. C., Fry, J. P., Cabello, F., Good, C. M., & Lunestad, B. T. (2020). Veterinary drug use in United States net pen Salmon aquaculture: Implications for drug use policy. *Aquaculture*, *518*, 734820, https://doi.org/10.1016/j.aquaculture.2019.734820.

Luo, H., Liu, C., He, D., Xu, J., Sun, J., Li, J., & Pan, X. (2022). Environmental behaviors of microplastics in aquatic systems: a systematic review on degradation, adsorption, toxicity and biofilm under aging conditions. *Journal of Hazardous Materials*, *423*, 126915.

Lusher, A. L., & Covernton, G. A. (2022). Microplastics in fish and seafood species. *Plastics and the Ocean: Origin, Characterization, Fate, and Impacts*, 367–388.

Lusher, A., & Pettersen, R. (2021). Sea-based sources of microplastics to the Norwegian marine environment. *NIVA-rapport*, ISSN 1894-7948.

Lusher, A. L., Hollman, P., & Mendoza-Hill, J. (2017). Microplastics in fisheries and aquaculture: status of knowledge on their occurrence and implications for aquatic organisms and food safety. *FAO Fisheries and Aquaculture Technical Paper No. 615*.

Matias, R. S., Gomes, S., Barboza, L. G. A., Salazar-Gutierrez, D., Guilhermino, L., & Valente, L. M. (2023). Microplastics in water, feed and tissues of European seabass reared in a recirculation aquaculture system (RAS). *Chemosphere, 335*, 139055.

Memiş, D., Azhar, M. H., Tunçelli, G., & Erkan, N. (2024). Güvenli gıda üretiminde entegre multi-trofik akuakültür (IMTA) sistem modeli. In: N. Erkan & S. Oğur (Eds.), *Gıda güvencesinde akuatik kaynaklar ve sürdürülebilir gıda güvenliği* (pp. 117–144). İstanbul Üniversitesi Yayınları (In Turkish with English abstract). https://doi.org/10.26650/B/LSB23LSB24.2024.026.004

Milani, S., Baldelli Bombelli, F., Pitek, A. S., Dawson, K. A., & Radler, J. (2012). Reversible versus irreversible binding of transferrin to polystyrene nanoparticles: soft and hard corona. *ACS Nano*, *6*(3), 2532–2541.

Mishra, S., Dash, D., & Das, A. P. (2024). Aquatic microbial diversity on plastisphere: colonization and potential role in microplastic biodegradation. *Geomicrobiology Journal*, *41*(4), 312–323.

Naylor, R. L., Hardy, R. W., Buschmann, A. H., Bush, S. R., Cao, L., Klinger, D. H., … & Troell, M. (2021). A 20-year retrospective review of global aquaculture. *Nature, 591*(7851), 551–563.

Nguyen, D., Ovadia, O., & Guttman, L. (2023). Temporal force governs the microbial assembly associated with *Ulva fasciata* (Chlorophyta) from an integrated multi-trophic aquaculture system. *Frontiers in Microbiology*, *14*, 1223204, https://doi.org/10.3389/fmicb.2023.1223204.

Oberbeckmann, S., Kreikemeyer, B., & Labrenz, M. (2018). Environmental factors support the formation of specific bacterial assemblages on microplastics. *Frontiers in Microbiology, 8*, 2709.

Oberbeckmann, S., Löder, M. G., & Labrenz, M. (2015). Marine microplastic-associated biofilms–a review. *Environmental chemistry*, *12*(5), 551–562, https://doi.org/10.1071/EN15069

Oberbeckmann, S., Loeder, M. G., Gerdts, G., & Osborn, A. M. (2014). Spatial and seasonal variation in diversity and structure of microbial biofilms on marine plastics in Northern European waters. *FEMS Microbiology Ecology*, *90*(2), 478–492.

Ogonowski, M., Motiei, A., Ininbergs, K., Hell, E., Gerdes, Z., Udekwu, K. I., … & Gorokhova E. (2018). Evidence for selective bacterial community structuring on microplastics. *Environmental Microbiology, 20*(8), 2796–2808.

Okeke, E. S., Okoye, C. O., Atakpa, E. O., Ita, R. E., Nyaruaba, R., Mgbechidinma, C. L., & Akan, O. D. (2022). Microplastics in agroecosystems-impacts on ecosystem functions and food chain. *Resources, Conservation and Recycling*, *177*, 105961.

Ozden, O. (2023). *Su Ürünlerinde Gıda Güvenliği ve Kalite Kontrol*. Istanbul: Istanbul University Press. (In Turkish).

Parrish, K., & Fahrenfeld, N. L. (2019). Microplastic biofilm in fresh- and wastewater as a function of microparticle type and size class. *Environmental Science: Water Research & Technology*, *5*(3), 495–505.

Prata, J. C., da Costa, J. P., Lopes, I., Duarte, A. C., & Rocha-Santos, T. (2020). Environmental exposure to microplastics: an overview on possible human health effects. *Science of the Total Environment*, *702*, 134455.

Qin, Z., Zhao, Z., Xia, L., & Ohore, O. E. (2022). Research trends and hotspots of aquatic biofilms in freshwater environment during the last three decades: a critical review and bibliometric analysis. *Environmental Science and Pollution Research, 29*(32), 47915–47930.

Rochman, C. M., Hoh, E., Kurobe, T., & Teh, S. J. (2013). Ingested plastic transfers hazardous chemicals to fish and induces hepatic stress. *Scientific Reports, 3*, 3263.

Rogers, K. L., Carreres-Calabuig, J. A., Gorokhova, E., & Posth, N. R., (2020). Micro-by-micro interactions: How microorganisms influence the fate of marine microplastics. Limnology and Oceanography Letters. In: E. Granek, S. Brander, E. Holland (Eds.), *Special Issue: Microplastics in Marine and Freshwater Organisms: Presence and Potential Effects*, doi: 10.1002/lol2.10136.

Schuhen, K., & Sturm, M. T. (2022). Microplastic pollution and reduction strategies. In: T. Rocha-Santos, M. F. Costa, & C. Mouneyrac (Eds.), *Handbook of Microplastics in the Environment* (pp. 1097–1128). Cham: Springer International Publishing.

Sheng, Y., Ye, X., Zhou, Y., & Li, R. (2021). Microplastics (MPs) Act as sources and vector of pollutants-impact hazards and preventive measures. *Bulletin of Environmental Contamination and Toxicology, 107*(4), 722–729.

Sillanpää, M., Khadir, A., & Muthu, S. S. (Eds.). (2022). *Microplastics Pollution in Aquatic Media: Occurrence, Detection, and Removal*. Singapore: Springer.

Silva, D. M., Almeida, C. M. R., Guardiola, F. A., Pereira, R., Rodrigues, S. M., & Ramos, S. (2024). Uncovering microplastics contamination in canned seafood. *Food Chemistry, 448*, 139049.

Skirtun, M., Sandra, M., Strietman, W. J., van den Burg, S. W., De Raedemaecker, F., & Devriese, L. I. (2022). Plastic pollution pathways from marine aquaculture practices and potential solutions for the North-East Atlantic region. *Marine Pollution Bulletin, 174*, 113178.

Somerville, C., Cohen, M., Pantanella, E., Stankus, A., & Lovatelli, A. (2014). *Small-Scale Aquaponic Food Production: Integrated Fish and Plant Farming*. Rome: FAO. Retrieved from https://www.fao.org/3/i4021e/i4021e.pdf

Stollberg, N., Kröger, S. D., Reininghaus, M., Forberger, J., Witt, G., & Brenner, M. (2021). Uptake and absorption of fluoranthene from spiked microplastics into the digestive gland tissues of blue mussels, Mytilus edulis L. *Chemosphere, 279*, 130480.

Tang, K. H. D. (2024). Terrestrial and aquatic plastisphere: formation, characteristics, and influencing factors. *Sustainability, 16*(5), 2163.

Tavelli, R., Callens, M., Grootaert, C., Abdallah, M. F., & Rajkovic, A. (2022). Foodborne pathogens in the plastisphere: can microplastics in the food chain threaten microbial food safety? *Trends in Food Science & Technology, 129*, 1–10.

Thilsted, S. H., Thorne-Lyman, A., Webb, P., Bogard, J. R., Subasinghe, R., Phillips, M. J., & Allison, E. H. (2016). Sustaining healthy diets: the role of capture fisheries and aquaculture for improving nutrition in the post-2015 era. *Food Policy, 61*, 126–131.

Thompson, R. C., Swan, S. H., Moore, C. J., & vom Saal, F. S. (2016). Our plastic age. *Philosophical Transactions of the Royal Society B: Biological Sciences, 364*(1526), 1973–1980.

Tidwell, J. H. (2012). *Aquaculture Production Systems* (Vol. 434). Oxford: Wiley-Blackwell.

Tunçelli, G., & Memiş, D. (2024). The effect of swimming activity and feed restriction of rainbow trout (*Oncorhynchus mykiss*) on water quality and fish-plant growth performance in aquaponics. *Journal of Fish Biology, 104*, 1493–1502.

Tunçelli, G., Memiş, D., & Erkan, N. (2024b). Akuaponik sistemler: Sürdürülebilir gıda üretimi için umut veren bir çözüm. In: N. Erkan & S. Oğur (Eds.), *Gıda güvencesinde akuatik kaynaklar ve sürdürülebilir gıda güvenliği* (pp. 85–116). İstanbul: İstanbul Üniversitesi Yayınları (In Turkish with English abstract). https://doi.org/10.26650/B/LSB23LSB24.2024.026.003

Tunçelli, G., Tunçelli, İ., & Memiş, D. (2023). Evaluation of lettuce (Lactuca sativa L.) in aquaponic system in terms of food safety. *Ege Journal of Fisheries and Aquatic Sciences, 40*(1), 27–34.

Tunçelli, G., Tuncelli, I. C., Dagsuyu, E., Turkyilmaz, I. B., Yanardag, R., & Erkan, N. (2024a). The effect of different types of microplastic and acute cadmium exposure on the *Mytilus galloprovincialis* (Lamarck, 1819). *Science of the Total Environment, 936*, 173505.

UN (United Nations) (1948). Universal Declaration of Human Rights (Article 25). United Nations General Assembly. https://www.un.org/en/about-us/universal-declaration-of-human-rights.

UN (United Nations) (1966). International Covenant on Economic, Social and Cultural Rights (Article 12). United Nations General Assembly. https://www.ohchr.org/en/instruments-mechanisms/instruments/international-covenant-economic-social-and-cultural-rights.

Udovicki, B., Andjelkovic, M., Cirkovic-Velickovic, T., & Rajkovic, A. (2022). Microplastics in food: scoping review on health effects, occurrence, and human exposure. *International Journal of Food Contamination*, *9*(1), 7.

Venkatesh, S., Naidu, B. C., Palanisamy, S., Ngasotter, S., Shaju, S. S., Porayil, L., … & Xavier, K. M. (2024). Microplastic accumulation dynamics and risk assessment in dried fish processed with sea salt at different salting ratios. *Journal of Hazardous Materials Advances*, *14*, 100415.

Vethaak, A. D., & Legler, J. (2021). Microplastics and human health. *Science, 371*(6530), 672–674.

Vezzulli, L., Grande, C., Reid, P. C., Hélaouët, P., Edwards, M., Höfle, M. G., ... & Pruzzo, C. (2016). Climate influence on Vibrio and associated human diseases during the past half-century in the coastal North Atlantic. *Proceedings of the National Academy of Sciences*, *113*(34), E5062–E5071, https://doi.org/10.1073/pnas.1609157113.

Wang, Q., Li, J., Zhu, X., Sun, C., Teng, J., Chen, L., … & Zhao, J. (2022). Microplastics in fish meals: an exposure route for aquaculture animals. *Science of the Total Environment*, *807*, 151049.

Wei, L., Su, Z., Yue, Q., Huang, X., Wei, M., & Wang, J. (2024). Microplastics, heavy metals, antibiotics, and antibiotic resistance genes in recirculating aquaculture systems. *TrAC Trends in Analytical Chemistry*, *172*, 117564.

WHO (World Health Organisation) (2018). Global Nutrition Policy Review 2016–2017: Country Progress in Creating Enabling Policy Environments for Promoting Healthy Diets and Nutrition. Geneva: World Health Organization. https://www.who.int/publications/i/item/9789241514873.

Wright, R. J., Langille, M. G., & Walker, T. R. (2021). Food or just a free ride? A meta-analysis reveals the global diversity of the Plastisphere. *The ISME Journal*, *15*(3), 789–806.

Wright, S. L., & Kelly, F. J. (2017). Plastic and human health: a micro issue? *Environmental Science & Technology*, *51*(12), 6634–6647.

Wu, H., Hou, J., & Wang, X. (2023). A review of microplastic pollution in aquaculture: sources, effects, removal strategies and prospects. *Ecotoxicology and Environmental Safety*, *252*, 114567.

Yadav, H., Khan, M. R. H., Quadir, M., Rusch, K. A., Mondal, P. P., Orr, M., … & Iskander, S. M. (2023). Cutting boards: an overlooked source of microplastics in human food? *Environmental Science & Technology*, *57*(22), 8225–8235.

Yang, Y., Liu, W., Zhang, Z., Grossart, H. P., & Gadd, G. M. (2020). Microplastics provide new microbial niches in aquatic environments. *Applied Microbiology and Biotechnology, 104*(14), 6501–6511.

Yin, K., Wang, Y., Zhao, H., Wang, D., Guo, M., Mu, M., … & Xing, M. (2021). A comparative review of microplastics and nanoplastics: toxicity hazards on digestive, reproductive and nervous system. *Science of the Total Environment, 774*, 145758.

Yuan, Z., Nag, R., & Cummins, E. (2022). Human health concerns regarding microplastics in the aquatic environment-from marine to food systems. *Science of the Total Environment*, *823*, 153730.

Zettler, E. R., Mincer, T. J., & Amaral-Zettler, L. A. (2013). Life in the "plastisphere": microbial communities on plastic marine debris. *Environmental Science & Technology, 47*(13), 7137–7146.

Zhang, F., Man, Y. B., Mo, W. Y., Man, K. Y., & Wong, M. H. (2020). Direct and indirect effects of microplastics on bivalves, with a focus on edible species: a mini-review. *Critical Reviews in Environmental Science and Technology, 50*(20), 2109–2143.

Zhu, Z., Shan, L., Hu, F., Li, Z., Zhong, D., Yuan, Y., & Zhang, J. (2020). Biofilm formation potential and chlorine resistance of typical bacteria isolated from drinking water distribution systems. *RSC Advances, 10*(52), 31295–31304.

10 Community Ecology and Gene Flow Within Biofilms Formed on Plastic Surfaces

Atif Aziz Chowdhury, Nilendu Basak, Neelakanta Sarvashiva Kiran, Chandrashekar Yashaswini, Dinesh Karthiban, and Ankita Chatterjee

10.1 INTRODUCTION

The global issue of plastic pollution has spread widely and has serious effects on the terrestrial and marine ecosystems (Li et al., 2024). The surface of the plastic polymer offers a robust substrate that microbes may adhere to and spread across vast distances (Amaral-Zettler et al., 2015). Microorganisms can form extracellular polymeric matrix surrounding them and develop biofilm on such surfaces. As compared to their natural habitats, microorganisms in such communities also known as "plastisphere" display different ecological and genetic dynamics (Sooriyakumar et al., 2022). In recent years, because of their implications for environmental health and microbial ecology, the plastisphere has become an important area of research. Microbes in the plastisphere form biofilms. The biofilms are communities of bacteria, algae, fungi, and protozoa, attached to each other and to a surface through the production of extracellular polymeric substances (EPSs) (Böhning et al., 2024; Flemming et al., 2023, 2016; Sharma and Rajpurohit, 2024). This EPS matrix protects the biofilm cells from environmental and chemical stresses. The main components of this EPS matrix are lipids, proteins, nucleic acids, and polysaccharides (Flemming and Wingender, 2010). This matrix delivers mechanical stability, regulates surface adhesion, and creates a unified three-dimensional network that renders biofilm cells immobile (Limoli et al., 2015).

The type and structure of plastics and environmental factors have a considerable impact on the microbial community within the plastisphere (Shi et al., 2023). This community often includes genera such as *Azotobacter*, *Bacillus*, *Rhodococcus*, and *Pseudomonas* (Du et al., 2022). Because of the plastisphere's special qualities, some bacteria may predominate and affect the biodiversity of the area (Shen et al., 2021). Plastisphere has been found to have a significant role in biogeochemical cycles (Miao et al., 2021, 2020). However, there is a chance that these microbial communities might have altered metabolic potentials, which could impact biogeochemical cycles. For example, Miao et al. (2020) reported altered carbon cycling in plastisphere as compared to microbial communities on natural compounds. Furthermore, the plastisphere can act as a hub for the transfer of resistance genes, including those conferring resistance to antibiotics and metals (Yang et al., 2019). It can also accumulate and concentrate heavy metals and other pollutants, facilitating their transfer from one location to another and thus spreading pollution (Bhagwat et al., 2021). Furthermore, the plastisphere may encourage the development of harmful strains, endangering human health (Singh et al., 2022). The comprehension of microbial interactions, survival, and community organization in the plastisphere bears significant consequences for human, ecological, and environmental health (Li et al., 2024). However, significant gaps remain in our knowledge regarding the detailed mechanisms of microbial interactions in the plastisphere, its impact on biogeochemical cycles, the extent of resistance gene transfer, and its effects on both human and environmental health, as well as the

 DOI: 10.1201/9781032717814-10

long-term impacts on ecological functioning (Shen et al., 2021). Additionally, the study of community ecology within these biofilms is essential to understand the interactions among different microbial species and their collective impact on the environment. Understanding the microbial interactions in plastisphere can provide a path for the management of plastic pollution, thereby contributing to the "One Health" goal (Li et al., 2024).

This chapter aims to summarize the current understanding of the plastisphere and its microbial communities. Significant emphasis is placed on the microbial dynamics within plastisphere communities, preceded by a brief overview of biofilm formation. Key aspects discussed include horizontal gene transfer (HGT) and genetic diversity within the plastisphere. Furthermore, this chapter explores the ecological implications associated with plastic-associated biofilms.

10.2 FORMATION AND COMPOSITION OF PLASTIC-ASSOCIATED BIOFILMS

The term "plastisphere" is used to define different complex microbial biofilm communities that colonize plastic debris and microplastics in aquatic environments (Zettler et al., 2013). The composition of microbial communities depends on many variables, including the type of plastic, the environment of the surrounding, and the biofilm's maturity stage. Diatoms, Gammaproteobacteria, and Alphaproteobacteria are some of the first classes of microbes to form biofilms on the plastic surface (Zhai et al., 2023). Both the physical properties and chemical composition of the plastic substrates impact this colonization. Various plastic types (such as foam, soft, and hard plastics) support different microbial biofilm formation (Chaudhary et al., 2022). Environmental factors significantly influence biofilm development on plastics. Factors such as temperature variations, salinity, nutrient availability, and hydrodynamic conditions affect microbial colonization patterns (Sooriyakumar et al., 2022). Its development proceeds in four stages, as in other biofilms, i.e., initial attachment, irreversible attachment, and biofilm maturation and dispersion (Figure 10.1).

Additionally, HGT plays an important role in the initial attachment phase. The dense microbial communities within biofilms make them capable of higher HGT rates than planktonic states (Wang et al., 2024). This increase in the rate enhances genetic diversity and adaptability among microorganisms colonizing plastic surfaces. The mobile genetic elements (MGEs) within biofilms increase plasmid stability and can promote the rapid propagation of traits, such as antibiotic resistance,

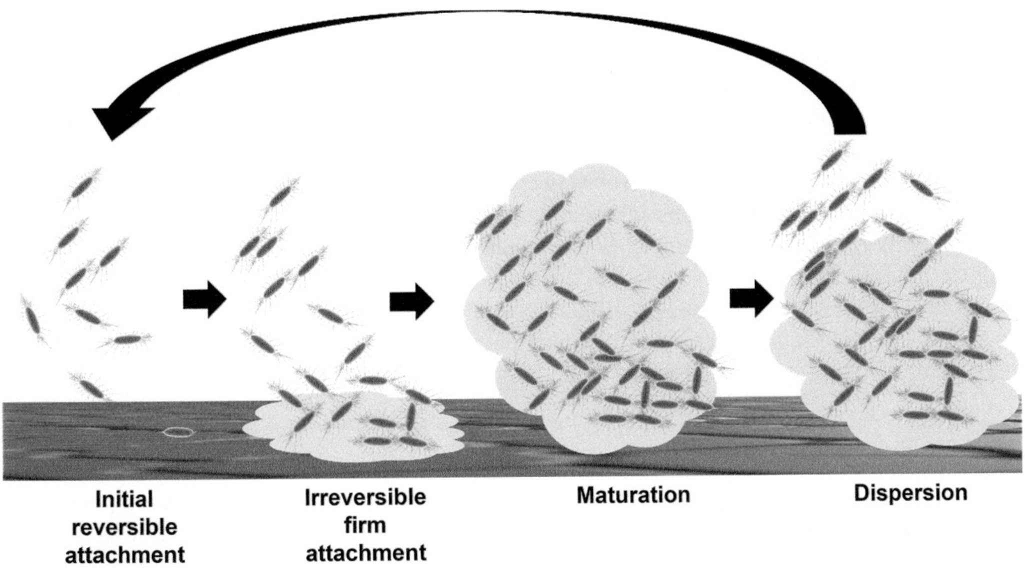

FIGURE 10.1 Overview of the four-staged biofilm formation on plastic surfaces by microorganisms.

within the community (Jiang et al., 2022). Besides that, HGT mechanisms such as conjugation are more effective in biofilm environments due to the close proximity of cells allowing active genetic exchange. Consequently, the initial attachment sets the stage not only for the development of a biofilm community but also for critical genetic interactions influencing microbial community dynamics.

10.2.1 Initial Attachment

Following the interaction of the environment and MPs, organic and/or inorganic molecules are absorbed by the MPs, producing a surface coating or film. Microbes attached to the surfaces can introduce new genetic traits through HGT that enhance the survival and colonization capabilities of other microbes on plastic surfaces. This makes a favourable environment for microorganisms to survive and reproduce. However, several factors including surface hydrophobicity, type, shape, and surface roughness influence the interaction of microbes with the MP surface (Shi et al., 2023). Furthermore, the formation of biofilms depends on the effectiveness of conjugation and that may be dependent on interactions among multiple plasmids (Gama et al., 2020). Many plastics have hydrophobic properties, which facilitate interactions with hydrophobic areas on the surfaces of microorganisms, enhancing early microbial adhesion (Messer et al., 2024). Additionally, microorganisms have specific structures such as flagella, pili, and fimbriae. These appendages enable microorganisms to securely adhere to the plastic surface, thereby starting the process of forming microcolonies (Kirstein et al., 2019). Structural stability conferred by the F-pilus improves DNA delivery during conjugation and promotes biofilm formation in harsh conditions (Patkowski et al., 2023). Notably, bacteria can revert to their planktonic state, indicating the reversible nature of this initial attachment (Shi et al., 2023). Electrostatic, van der Waals, hydrophobic, steric, and protein adhesion forces can regulate the degree of attachment (Sharma et al., 2023). Bos et al. (2023) noticed that Alteromonadaceae is the key player in the early colonization of plastic surfaces containing genes associated with adhesion, biofilm formation, and chemotaxis. The mannose-sensitive haemagglutinin (MSHA) operon was described as the most important factor for the early hydrophobic plastic surface colonization.

10.2.2 Irreversible Attachment

After initial attachment, microorganisms start to produce an EPS, which binds the cells to the surface and each other. In this second stage, molecular coordinated binding takes place between certain adhesins and the outermost layer. Additionally, in the absence of physical and chemical intervention, the microbes keep on aggregating on the surface (Sharma et al., 2023). Therefore, a durable biofilm matrix is formed that aids in the development of biofilms and defence against environmental stresses. Additionally, microbes become more adept at adhering to plastic surfaces by adjusting to their chemical and physical characteristics (Ganesan et al., 2022). These adhesins vary among microorganisms. For example, *Staphylococcus epidermidis* produces polysaccharide intercellular adhesin that helps in binding and forming biofilm (Sharma et al., 2023). Moreover, initial colonization is typically mediated by primary colonizers such as bacteria belonging to genera like *Pseudomonas*, *Vibrio*, and *Alteromonas*, which possess the ability to adhere to surfaces and form microcolonies (Caruso, 2020; Elumalai et al., 2024).

10.2.3 Biofilm Maturation

Following firm attachment, microbes divide rapidly and form microcolonies. Various chemical signalling within the EPS matrix and microcolonies facilitate further development and produced EPS confers the structural integrity of the biofilm (Sharma et al., 2023). This EPS structure helps maintain the biofilm from environmental stresses. With maturity, the biofilm gets into its three-dimensional structure. The channels and pores within the microcolonies help maintain the nutrient flow (Mirghani et al., 2022). With time, the biofilm continues to grow and becomes thicker and more robust. The maturation might involve a quorum-sensing mechanism where surfactants and modulins are reported to play an important part in Staphylococcal biofilm (Otto, 2013). Finally, a three-layered mature biofilm is formed consisting of a regulating inner layer, a basement layer with microbes, and an outermost layer with ready to disperse planktonic microorganisms (Rather et al., 2021). The physiological changes due to biofilm growth could influence the dynamics of the plasmids such as their copy number control and the efficiency of conjugative transfer, thereby further shaping the genetic landscape of the microbial community (Cook and Dunny, 2014; Jones et al., 2021).

10.2.4 Detachment

The detachment stage of biofilm is a critical phase in the biofilm life cycle, involving the release of cells from the biofilm matrix into the surrounding environment. This allows the biofilm to spread and colonize new surfaces. Due to its ability to synchronize gene expression in response to population density, a quorum sensing system is essential in initiating biofilm dispersal. This regulatory system makes sure that when the environment is suitable for survival outside the biofilm matrix, detachment takes place (Solano et al., 2014). Additionally, enzymatic pathways can help in dispersal of biofilm. In *Staphylococcus aureus*, proteases and nuclease enzymes degrade the extracellular matrix and release cells from biofilm (Lister and Horswill, 2014). Molecular signalling, for example, cyclic di-GMP (c-di-GMP) can regulate biofilm development and dispersal of biofilm of *Vibrio chloreae*. A high concentration of c-di-GMP favours the growth of biofilms, while a decrease causes motility and decreases EPS synthesis, which in turn causes dispersal (Teschler et al., 2022). Moreover, mechanical forces and environmental conditions can further facilitate biofilm dispersal.

10.3 DYNAMICS OF MICROBIAL INTERACTIONS WITHIN PLASTIC-ASSOCIATED BIOFILMS

Understanding the dynamics of microbial communities in plastic-associated biofilms is crucial for comprehending their role in plastic degradation, nutrient cycling, and potentially harmful effects on aquatic ecosystems (Koh et al., 2023). "Dynamics" in the context of microbial interactions within plastic-associated biofilms refers to the continuous and evolving processes of competition, cooperation, and adaptation between microbial species as they interact with each other and their environment, influencing plastic degradation over time (Caroppo et al., 2022). Plastic-associated biofilms serve as dynamic microenvironments where diverse microbial communities thrive.

Once established, plastic-associated biofilms undergo a series of dynamic interactions among microbial inhabitants. These interactions are governed by factors such as nutrient availability, interspecies communication, and competition for space and resources (Yu et al., 2023). Cooperative behaviours, such as the exchange of metabolites and the formation of consortia, contribute to the resilience and functionality of these biofilms (Joshi et al., 2021). For example, some bacteria produce EPSs that form a matrix, providing structural support and facilitating nutrient exchange within the biofilm community (Kiran et al., 2024).

Competition within plastic-associated biofilms is also a significant driving force shaping microbial dynamics (Yu et al., 2023). Microorganisms compete for nutrients, space, and metabolic substrates, leading to niche differentiation and the selection of specific microbial taxa (Kiran et al., 2021). This competitive environment can drive the evolution of specialized metabolic capabilities, including the ability to degrade plastic polymers as a carbon and energy source. Studies have identified microbial species capable of producing plastic-degrading enzymes such as lipases, esterases, and polyethylene terephthalate (PET) hydrolases (Figure 10.2), which play a crucial role in plastic degradation within biofilms (Oberbeckmann et al., 2016; Priya et al., 2022; Yoshida et al., 2021).

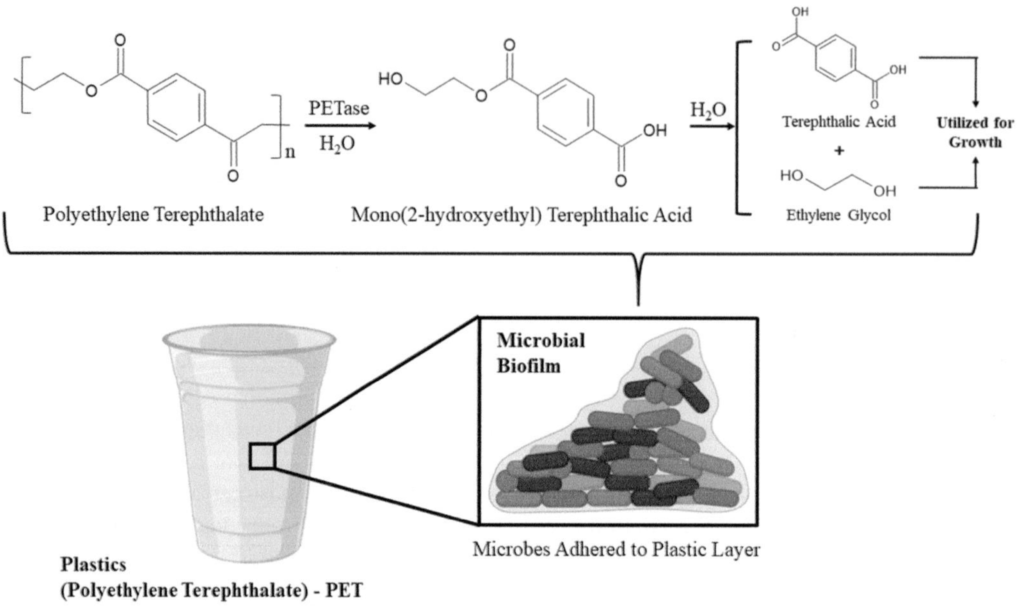

FIGURE 10.2 Mechanism of degradation of PET by microbial biofilm.

Furthermore, microbial interactions within plastic-associated biofilms can have broader ecological implications. These biofilms serve as hotspots of microbial activity, influencing nutrient cycling and biogeochemical processes in aquatic environments. For example, the degradation of plastic polymers by microbial enzymes releases organic carbon and other nutrients into the surrounding ecosystem, potentially impacting nutrient dynamics and microbial community composition (Salam et al., 2023; Tagg et al., 2019; Yang et al., 2020).

10.4 METABOLIC INTERACTIONS AND NUTRIENT EXCHANGE IN THE PLASTISPHERE

Metabolic interactions and nutrient exchange are fundamental processes driving the dynamics of microbial communities within plastic-associated biofilms. These interactions play a crucial role in shaping the structure, function, and resilience of biofilm ecosystems, ultimately influencing plastic degradation and ecosystem health (Bocci et al., 2024; Sooriyakumar et al., 2022). Understanding the mechanisms underlying metabolic interactions and nutrient exchange within these biofilms is essential for elucidating their role in plastic degradation and nutrient cycling in aquatic environments (Figure 10.3)

Within plastic-associated biofilms, metabolic interactions occur through a variety of mechanisms, including the exchange of metabolites, cooperative metabolic pathways, and cross-feeding interactions among microbial species (Bos et al., 2023). Cooperative behaviours, such as syntrophy and cross-feeding, enable microbial communities to exploit complex substrates and optimize resource utilization. For example, some bacteria within biofilms may produce extracellular enzymes capable of degrading complex organic compounds present in plastic polymers, releasing simpler metabolites that can be utilized by other microbial taxa within the community (Mishra et al., 2023, 2022).

Nutrient exchange within plastic-associated biofilms is facilitated by the production and secretion of EPSs, which form a matrix that surrounds microbial cells and provides a scaffold for nutrient transport and exchange. EPS serves as a reservoir for organic carbon, nitrogen, and other essential nutrients, enabling microbial communities to thrive in nutrient-limited environments. Additionally, EPS promotes adhesion to plastic surfaces, stabilizing biofilm structure and facilitating the

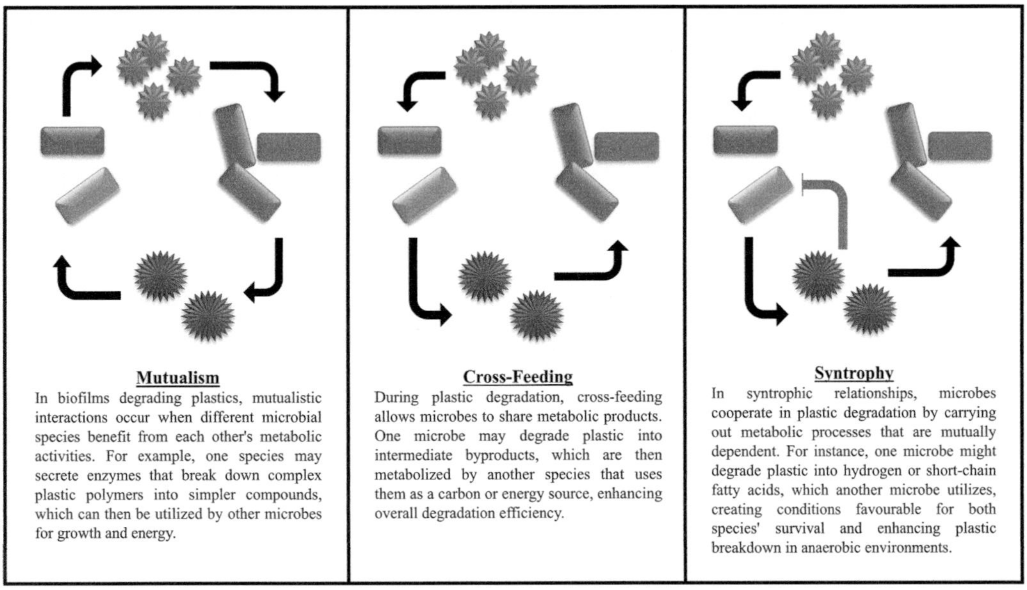

FIGURE 10.3 Diverse mechanisms and pathways of metabolic interactions and nutrient exchange by microbes in biofilm.

colonization of new microorganisms (Flemming et al., 2023; Flemming and Wingender, 2010; Kiran et al., 2024).

One key mechanism driving metabolic interactions and nutrient exchange within plastic-associated biofilms is quorum sensing, a cell-to-cell communication process mediated by signalling molecules called autoinducers (Wright et al., 2020). Quorum sensing allows microbial communities to coordinate gene expression and behaviour in response to changes in population density, environmental conditions, and nutrient availability. In plastic-associated biofilms, quorum sensing regulates the expression of genes involved in biofilm formation, nutrient acquisition, and the production of extracellular enzymes, thereby influencing metabolic interactions and nutrient exchange among microbial species (Caruso, 2020; Caruso et al., 2023). Another important mechanism facilitating metabolic interactions within plastic-associated biofilms is HGT, the process by which genetic material is transferred between different microbial species (Urbanek et al., 2018). HGT enables the acquisition of novel metabolic capabilities, such as the ability to degrade plastic polymers, through the transfer of genes encoding plastic-degrading enzymes or metabolic pathways. This process contributes to the functional diversity and adaptability of microbial communities within biofilms, enhancing their capacity to degrade plastic and exploit complex environmental substrates (Hu et al., 2022; Purohit et al., 2020).

Furthermore, metabolic interactions and nutrient exchange within plastic-associated biofilms are influenced by environmental factors such as temperature, pH, and nutrient availability. For example, variations in temperature can affect the rate of enzymatic plastic degradation and the metabolic activity of microbial communities within biofilms (Oberbeckmann et al., 2021). Similarly, fluctuations in nutrient concentrations can alter the composition and function of biofilm communities, leading to shifts in metabolic interactions and nutrient exchange dynamics (Weissbrodt et al., 2014).

10.5 HORIZONTAL GENE TRANSFER (HGT) AND GENETIC DIVERSITY

The biofilm formation helps in the survival of bacteria from some harsh environmental stress like increases in adhesion to the surface during heavy flow, drought, and low nutrient availability (Madsen et al., 2012). Biofilms lead to communication between microbes through cell signalling and by chemical means which helps the microbial community share metabolites and gain resistance to antibiotics under the safety of an exopolysaccharide matrix when compared with its planktonic forms (Rabin et al., 2015). Biofilm formation on plastic is common due to various biotic and abiotic factors but is not always evidence of plastic degradation (Bos et al., 2023). The presence of conjugative plasmids that just codes for antibiotic resistance like RP4 is also proven to be the reason for biofilm formation in plastics (Zhang et al., 2022b). However, the formation of biofilms on plastic alters the physical and chemical nature of the surface, and most of the time it causes fragmentation of plastics through biofouling. In some cases, plastic-degrading microbes were also reported in biofilm, considering it an effective method in plastic degradation (Delacuvellerie et al., 2021; Howard et al., 2023; Misic and Covazzi Harriague, 2019).

HGT is an important phenomenon in the evolution of microbes in which genes of one organism is transferred to another organism without the reproduction means. The genes can be exchanged even between unrelated individuals through HGT but the gene exchange is very easy when the organism is closely related (Abe et al., 2020; Arnold et al., 2022). The HGT within the community of biofilm is commonly due to conjugation, nanotube formation, transformation, and viral-mediated transduction. Because the organisms are more intact with each other, the conjugation tube formation is stable for a long time under the protection of the exopolysaccharide matrix. The transfer plasmid, or extrachromosomal additional DNA, through conjugation or the uptake of naked DNA by a competent cell, i.e., transformation, increases in rate and occurs very frequently inside the biofilm. However, the transduction, i.e., the transfer of DNA through vector phage, is not well studied in the biofilm state (Abe et al., 2020; Zhang et al., 2022a). Cytochromes P450, alkB1, alkB2, almA, H39 lipase, cut 1, ISF6_4831 and many other bacterial genes are responsible for degradation of plastics (Priya et al., 2022). In appropriate conditions, any of these genes are able to undergo transformation (Molin and Tolker-Nielsen, 2003), which can increase the efficiency of plastic degradation in biofilm.

The horizontal transferring of plastic-degrading genes under the biofilm is mostly theoretical; still, it has to be studied in detail and more practically. Biofilm always has a consortium of cultivatable and non-cultivatable organisms protected by the matrix, which doesn't allow the entry of phage infection to cause a decrease in the rate of transduction and also make it hard to study. It has major drawbacks that the consortium of each biofilm is unique and cannot be recreated in lab conditions easily.

10.6 EXTRACELLULAR POLYMERIC SUBSTANCES AND GENETIC MATERIAL ADSORPTION

EPSs hold a crucial role in microbial communities and formation of biofilms on plastic surfaces in aquatic environment. EPS is a complex matrix containing polysaccharides, proteins, lipids, and nucleic acids, which are secreted by microorganisms to form a protective and adhesive environment. EPS maintains the structural integrity and stability of biofilms, thereby helping in adhesion to the plastic surfaces and protection against environmental stresses. A primary function of EPS is their ability to adsorb genetic material, such as DNA and RNA fragments, from their surrounding environment (Karygianni et al., 2020). Genetic material adsorption to EPS allows microorganisms within biofilms to pick up new genetic material and exchange genetic information with other microbial species. This process can contribute to the adaptability and evolution of microbial communities by promoting genetic diversity and facilitating the spread of advantageous traits, such as antibiotic-resistance genes (ARGs) (Li et al., 2022). Overall, EPS and their role in genetic material adsorption highlight the importance of biofilms in microbial ecology and their potential implications for various environmental and industrial processes, including wastewater treatment, bioremediation, and medical infections. Understanding the interplay between EPS and genetic material adsorption in plastic-associated biofilms is important for several reasons. It throws light on the persistence and adaptation mechanisms of microbial communities on plastic debris in marine and freshwater environments (Wani et al., 2024). Moreover, it underscores the potential environmental implications of biofilm formation on plastic, such as altering plastic degradation rates or influencing the fate of associated pollutants. In summary, EPS and their role in genetic material adsorption are integral to the formation, stability, and function of plastic-associated biofilms, highlighting their significance in microbial ecology and environmental biotechnology (Okeke et al., 2022).

10.7 ECOLOGICAL IMPLICATIONS OF PLASTIC-ASSOCIATED BIOFILMS

In recent years, the study of the plastisphere has gained significant interest because of its ecological implications. The altered microbial communities, enhanced carbon metabolism, and repositories for pollutants and resistance genes in the plastisphere have far-reaching consequences, especially in aquatic ecosystems. Additionally, they can have the potential to degrade the plastic polymers. Therefore, understanding the ecological effects of plastic pollution is essential to assessing the risks it poses to the ecosystem and developing strategies to mitigate those effects. Moreover, in the plastisphere, new patterns of microbial cohabitation develop that can endanger the health of natural ecosystems (Li et al., 2024). Similarly, the "One Biofilm" idea raises the possibility of a complex connection between environmental health, animal health, and human health due to common biofilm production processes in various contexts (Jacques and Malouin, 2022).

10.7.1 Diversity and Composition of Microbial Communities

When compared to natural substrates, biofilms on plastic substrates show different microbial community compositions, with reduced diversity and species richness (Moyal et al., 2023). The chemical characteristics of the plastic have an impact on this variance by favouring distinct microbial species. For example, in an experimental design, after 1 week of incubation, the PVC surface showed to be dominated by microbes from families of Gammaproteobacteria, whereas the other plastic surface is dominated by the microbes of *Flavobacteriaceae* and *Oleiphilaceae* families (Pinto et al., 2019). Additionally, the plastisphere reported to harbour both traditional heterotrophic bacteria like *Pseudomonas*, *Azotobacter*, *Bacillus*, *Rhodococcus*, *Pirellulaceae*, *Phycisphaerales*, *Cyclobacteriaceae*, *Roseococcus*, *Vibrionaceae*, and *Mycobacteriacea*, as well as photoheterotrophic bacteria like *Erythrobacter* and *Roseobacter* (Du et al., 2022). However, the plastisphere often harbours less diverse and less rich microbial communities than natural substrates, which has been explained by a lack of naturally occurring organic nutrients in the plastisphere (Galli et al., 2022; Shen et al., 2021). As a consequence of a reduction in biodiversity, the ecosystem might become less resilient to environmental changes including pollution and global warming (Shen et al., 2021).

10.7.2 Microbial Metabolic Diversity

Plastisphere microbes harbour microbial communities with different metabolic potential than that of the natural substrate. The BIOLOG EcoPlate analysis of biofilms from different plastic surfaces showed alteration in carbon utilization of PVC and PE as compared to natural substrate in the freshwater ecosystems (Miao et al., 2020). Another study also showed lower carbon metabolism in PET as compared to PVC (Miao et al., 2021). These variations may result from variations in temperature, nutrition availability, plastic structure, biofilm architecture, and microbial community composition in the plastisphere (Miao et al., 2020).

10.7.3 Spread of Antibiotic-Resistance Genes (ARGs) and Metal-Resistance Genes (MRGs)

The surface structure of plastics often favours the colonization and proliferation of resistant bacteria and causes the accumulation of ARGs and MRGs (Guo et al., 2020; Yang et al., 2019). Furthermore, plastispheres can efficiently accumulate and concentrate these resistance genes. As a consequence, these genes persist in the plastisphere for a long time and possibly spread in the environment (Di Cesare et al., 2021; Yang et al., 2019). Studies suggest that the biofilm formation on polyethylene surface harbours Firmicutes, which has a significant role in ARGs (*tet*A, *sul*1, *aac*(6′)-Ib, and *tet*W) abundance. Moreover, MRGs (*cop*A and *znt*B) and ARGs have shared a common correlation in the biofilm of plastic surfaces (Guo et al., 2020). Liang et al. (2023) found the potential pathogenic bacteria such as *Pseudomonas* and *Vibrio* spp. on the surface of floating plastic around the sea, along with several ARGs increasing the threat by many folds. Nguyen et al. (2023) reported the ARGs and resistance bacteria on the surface of MP, sometimes escaped from the conventional wastewater treatment plants. The pass-through of these MPs may be due to poor management in the treatment plants, which possess a significant threat to human life and environment. One of the most abundant ARGs found on the plastic surface is identified as *sul*1 which is responsible for the sulphonamide resistance; moreover, *Erm*B is responsible for the resistance against macrolide-type antibiotics (Nguyen et al., 2023; Parrish and Fahrenfeld, 2019). The abundance and proliferation of ARGs on the plastic surface is co-related with several factors such as environment, incubation time, and types of plastic materials along with the EPS of the residential microbiota (Syranidou and Kalogerakis, 2022; Zheng et al., 2023). Furthermore, the accumulation, adsorption, and concentration of antibiotics on the plastispheric biofilm enhance the diversity, abundance, and enrichment of ARGs (Zheng et al., 2023). The co-occurrence of ARGs and MRGs might be the coherence of the same mobile gene and influence the microbial community of resistome, although the abundance of ARGs and MRGs does not rely on the microplastic or macroplastic biota (Yang et al., 2019).

10.7.4 Accumulation of Pollutants

Biofilm matrices often attract and trap pollutants (Chowdhury et al., 2023). The plastisphere can adsorb more organic and heavy metal pollutants than virgin plastics (Bhagwat et al., 2021; Liu et al., 2021). More surface area and changes in the hydrophobic/hydrophilic properties of plastics due to the presence of biofilms allow for increased adsorption of contaminants (Bhagwat et al., 2021). Additionally, the spread of MP-associated biofilms in diverse environments can trap and concentrate pollutants. These have a far-reaching consequence in agricultural soils and marine environments and can impact soil health, crop productivity, and marine biodiversity (Yu et al., 2022).

10.7.5 Microbiological Degradation of Plastics

Bacteria, fungi, viruses, archaea, algae, and protozoans can all colonize in the plastisphere. Microorganisms in the plastisphere can break down plastics more efficiently than marine planktonic strains due to MPs' hydrophobicity. Plastisphere communities may partially or totally break down and rupture polymeric polymers or hydrocarbons. Additionally, the microbial communities in the plastisphere may break down the adsorbed organic pollutants (Du et al., 2022). It's speculated that quorum sensing is essential for the synthesis of plastispheres and the degradation of microplastics. Proteobacteria and Bacteriodetes are the most dominant bacterial phyla in the plastisphere (Zhai et al., 2023). Nevertheless, several environmental variables, including temperature, pH, and nutrition availability, can affect how quickly plastics degrade (Yuan et al., 2020).

10.7.6 Colonization of Pathogens

Organic pollutants adsorbed in the plastisphere can be used as nutrients by various microbes, serving as a source of pathogens. Additionally, the exchange of the resistance genes can be acquired within the plastisphere and can cause severe challenges to healthcare facilities. Bacteria that cause human disease, like the cholera bacterium *Vibrio cholerae*, can be abundant in plastispheres (Laverty et al., 2020). Sequence analysis further revealed the plastisphere's association with toxoplasmosis and cancer pathways (Zhang et al., 2022a). Moreover, the co-relation between the ARGs and the pathogenic bacteria on the plastisphere has been addressed by many researchers (Liang et al., 2023; Nguyen et al., 2023). Martínez-Campos et al. (2021) reported the total 51 bacterial phyla present in the plastisphere, whereas predominance was observed in the case of potentially pathogenic species of *Mycobacterium*, *Pseudomonas*, and *Vibrio* while the abundance of other bacterial taxa is also noteworthy. The presence of *Escherichia coli* (Boni et al., 2021) and other species of *Actinobacteria*, *Legionella*, *Mycobacterium*, *Neisseria*, *Proteobacteria*, and *Verrucomicrobia* (Wang et al., 2021) on the surface of MPs found in the wastewater treatment plants has been reported. Silva et al. (2023) displayed the virulence characteristics, carrying ARGs and antibiotic resistance pathogenic strains such as *Citrobacter* sp., *E. coli*, *Klebsiella pneumoniae*, and *Shigella* sp. indicating the role of MPs (found in River) in the spread of these infectious pathogens. Another recent report by Garcés-Ordóñez et al. (2024) specified the 19 bacterial species among 1,760 genera identified as pathogenic strains found throughout different seasons in the lagoon of Ciénaga Grande de Santa Marta. Meanwhile, another pathogenic strain *Ralstonia pickettii* is found on the surface of microfibres from face masks in the coastal sea (Cheng et al., 2024).

10.8 CONCLUSION

Globally, increased plastic usage and disposal have led to the accumulation of plastics in various environments, including aquatic, terrestrial, and marine ecosystems. These plastics often break down into microplastics, complicating pollution management. The plastisphere—the surface of plastics and microplastics—acts as a microbial hotspot, facilitating the distribution of antibiotic and metal-resistant genes and concentrating environmental pollutants. The microbial selection pressure from plastic types and environmental conditions promotes the proliferation of specific microbes with altered metabolic functions. This can change microbial diversity, increasing dominance indices and disrupting vulnerable biogeochemical cycles. Consequently, ecosystems may become less resilient to environmental changes. Thus, future studies should focus on the metabolic capacity and microbiological diversity of plastisphere microorganisms in a variety of plastic varieties and contexts. In addition, assessing the plastisphere's long-term effects will improve our comprehension of the effects of plastic pollution. Furthermore, since these microbial communities have the ability to break down plastics, research should look into ways to use their metabolic capability to provide long-term solutions for environmental plastic and microplastic degradation.

ACKNOWLEDGEMENT

The authors would like to acknowledge the support of their respective departments and the universities. The corresponding author would like to acknowledge REVA University for extensive support.

REFERENCES

Abe, K., Nomura, N., Suzuki, S., 2020. Biofilms: hot spots of horizontal gene transfer (HGT) in aquatic environments, with a focus on a new HGT mechanism. *FEMS Microbiology Ecology* 96, fiaa031. https://doi.org/10.1093/femsec/fiaa031

Amaral-Zettler, L.A., Zettler, E.R., Slikas, B., Boyd, G.D., Melvin, D.W., Morrall, C.E., Proskurowski, G., Mincer, T.J., 2015. The biogeography of the Plastisphere: implications for policy. *Frontiers in Ecology and the Environment* 13, 541–546. https://doi.org/10.1890/150017

Arnold, B.J., Huang, I.-T., Hanage, W.P., 2022. Horizontal gene transfer and adaptive evolution in bacteria. *Nature Reviews Microbiology* 20, 206–218. https://doi.org/10.1038/s41579-021-00650-4

Bhagwat, G., Tran, T.K.A., Lamb, D., Senathirajah, K., Grainge, I., O'Connor, W., Juhasz, A., Palanisami, T., 2021. Biofilms enhance the adsorption of toxic contaminants on plastic microfibers under environmentally relevant conditions. *Environmental Science & Technology* 55, 8877–8887. https://doi.org/10.1021/acs.est.1c02012

Bocci, V., Galafassi, S., Levantesi, C., Crognale, S., Amalfitano, S., Congestri, R., Matturro, B., Rossetti, S., Di Pippo, F., 2024. Freshwater plastisphere: a review on biodiversity, risks, and biodegradation potential with implications for the aquatic ecosystem health. *Frontiers in Microbiology* 15. https://doi.org/10.3389/fmicb.2024.1395401

Böhning, J., Tarafder, A.K., Bharat, T.A.M., 2024. The role of filamentous matrix molecules in shaping the architecture and emergent properties of bacterial biofilms. *Biochemical Journal* 481, 245–263. https://doi.org/10.1042/BCJ20210301

Boni, W., Parrish, K., Patil, S., Fahrenfeld, N.L., 2021. Total coliform and *Escherichia coli* in microplastic biofilms grown in wastewater and inactivation by peracetic acid. *Water Environment Research* 93, 334–342. https://doi.org/10.1002/wer.1434

Bos, R.P., Kaul, D., Zettler, E.R., Hoffman, J.M., Dupont, C.L., Amaral-Zettler, L.A., Mincer, T.J., 2023. Plastics select for distinct early colonizing microbial populations with reproducible traits across environmental gradients. *Environmental Microbiology* 25, 2761–2775. https://doi.org/10.1111/1462-2920.16391

Caroppo, C., Azzaro, M., Dell'Acqua, O., Azzaro, F., Maimone, G., Rappazzo, A.C., Raffa, F., Caruso, G., 2022. Microbial biofilms colonizing plastic substrates in the Ross Sea (Antarctica). *Journal of Marine Science and Engineering* 10, 1714. https://doi.org/10.3390/jmse10111714

Caruso, G., 2020. Microbial colonization in marine environments: overview of current knowledge and emerging research topics. *Journal of Marine Science and Engineering* 8, 78. https://doi.org/10.3390/jmse8020078

Caruso, G., Maimone, G., Rappazzo, A.C., Dell'Acqua, O., Laganà, P., Azzaro, M., 2023. Microbial biofilm colonizing plastic substrates in the Ross Sea (Antarctica): first overview of community-level physiological profiles. *Journal of Marine Science and Engineering* 11, 1317. https://doi.org/10.3390/jmse11071317

Chaudhary, A., Dunn, S.T., Kelly, J., Hoellein, T.J., 2022. Plastic microbiome development in a freshwater ecosystem. *Science of the Total Environment* 848, 157697. https://doi.org/10.1016/j.scitotenv.2022.157697

Cheng, J., Wang, P., Ghiglione, J.-F., liu, L., Cai, Z., Zhou, J., Zhu, X., 2024. Bacterial pathogens associated with the plastisphere of surgical face masks and their dispersion potential in the coastal marine environment. *Journal of Hazardous Materials* 462, 132741. https://doi.org/10.1016/j.jhazmat.2023.132741

Chowdhury, A.A., Basak, N., Islam, E., 2023. Removal of uranium from water using biofilm of uranium sensitive *Methylobacterium* sp. *Journal of Hazardous Materials Advances* 10, 100296. https://doi.org/10.1016/j.hazadv.2023.100296

Cook, L.C.C., Dunny, G.M., 2014. The influence of biofilms in the biology of plasmids. *Microbiology Spectrum* 2, 2.5.29. https://doi.org/10.1128/microbiolspec.PLAS-0012-2013

Delacuvellerie, A., Benali, S., Cyriaque, V., Moins, S., Raquez, J.-M., Gobert, S., Wattiez, R., 2021. Microbial biofilm composition and polymer degradation of compostable and non-compostable plastics immersed in the marine environment. *Journal of Hazardous Materials* 419, 126526. https://doi.org/10.1016/j.jhazmat.2021.126526

Di Cesare, A., Pinnell, L.J., Brambilla, D., Elli, G., Sabatino, R., Sathicq, M.B., Corno, G., O'Donnell, C., Turner, J.W., 2021. Bioplastic accumulates antibiotic and metal resistance genes in coastal marine sediments. *Environmental Pollution* 291, 118161. https://doi.org/10.1016/j.envpol.2021.118161

Du, Y., Liu, X., Dong, X., Yin, Z., 2022. A review on marine plastisphere: biodiversity, formation, and role in degradation. *Computational and Structural Biotechnology Journal* 20, 975–988. https://doi.org/10.1016/j.csbj.2022.02.008

Elumalai, P., Gao, X., Cui, J., Kumar, A.S., Dhandapani, P., Parthipan, P., Karthikeyan, O.P., Theerthagiri, J., Kheawhom, S., Choi, M.Y., 2024. Biofilm formation, occurrence, microbial communication, impact and characterization methods in natural and anthropic systems: a review. *Environmental Chemistry Letters* 22, 1297–1326. https://doi.org/10.1007/s10311-024-01715-5

Flemming, H.-C., Van Hullebusch, E.D., Neu, T.R., Nielsen, P.H., Seviour, T., Stoodley, P., Wingender, J., Wuertz, S., 2023. The biofilm matrix: multitasking in a shared space. *Nature Reviews Microbiology* 21, 70–86. https://doi.org/10.1038/s41579-022-00791-0

Flemming, H.-C., Wingender, J., 2010. The biofilm matrix. *Nature Reviews Microbiology* 8, 623–633. https://doi.org/10.1038/nrmicro2415

Flemming, H.-C., Wingender, J., Szewzyk, U., Steinberg, P., Rice, S.A., Kjelleberg, S., 2016. Biofilms: an emergent form of bacterial life. *Nature Reviews Microbiology* 14, 563–575. https://doi.org/10.1038/nrmicro.2016.94

Galli, M., Tepsich, P., Baini, M., Panti, C., Rosso, M., Vafeiadou, A., Pantelidou, M., Moulins, A., Fossi, M.C., 2022. Microplastic abundance and biodiversity richness overlap: identification of sensitive areas in the Western Ionian Sea. *Marine Pollution Bulletin* 177, 113550. https://doi.org/10.1016/j.marpolbul.2022.113550

Gama, J.A., Fredheim, E.G.A., Cléon, F., Reis, A.M., Zilhão, R., Dionisio, F., 2020. Dominance between plasmids determines the extent of biofilm formation. *Frontiers in Microbiology* 11, 2070. https://doi.org/10.3389/fmicb.2020.02070

Ganesan, S., Ruendee, T., Kimura, S.Y., Chawengkijwanich, C., Janjaroen, D., 2022. Effect of biofilm formation on different types of plastic shopping bags: structural and physicochemical properties. *Environmental Research* 206, 112542. https://doi.org/10.1016/j.envres.2021.112542

Garcés-Ordóñez, O., Córdoba-Meza, T., Sáenz-Arias, S., Blandón, L., Espinosa-Díaz, L.F., Pérez-Duque, A., Thiel, M., Canals, M., 2024. Potentially pathogenic bacteria in the plastisphere from water, sediments, and commercial fish in a tropical coastal lagoon: an assessment and management proposal. *Journal of Hazardous Materials* 479, 135638. https://doi.org/10.1016/j.jhazmat.2024.135638

Guo, X., Sun, X., Chen, Y., Hou, L., Liu, M., Yang, Y., 2020. Antibiotic resistance genes in biofilms on plastic wastes in an estuarine environment. *Science of the Total Environment* 745, 140916. https://doi.org/10.1016/j.scitotenv.2020.140916

Howard, S.A., Carr, C.M., Sbahtu, H.I., Onwukwe, U., López, M.J., Dobson, A.D.W., McCarthy, R.R., 2023. Enrichment of native plastic-associated biofilm communities to enhance polyester degrading activity. *Environmental Microbiology* 25, 2698–2718. https://doi.org/10.1111/1462-2920.16466

Hu, X., Waigi, M.G., Yang, B., Gao, Y., 2022. Impact of plastic particles on the horizontal transfer of antibiotic resistance genes to bacterium: dependent on particle sizes and antibiotic resistance gene vector replication capacities. *Environmental Science & Technology* 56, 14948–14959. https://doi.org/10.1021/acs.est.2c00745

Jacques, M., Malouin, F., 2022. One health—one biofilm. *Veterinary Research* 53, 51. https://doi.org/10.1186/s13567-022-01067-4

Jiang, Q., Feng, M., Ye, C., Yu, X., 2022. Effects and relevant mechanisms of non-antibiotic factors on the horizontal transfer of antibiotic resistance genes in water environments: a review. *Science of the Total Environment* 806, 150568. https://doi.org/10.1016/j.scitotenv.2021.150568

Jones, J.M., Grinberg, I., Eldar, A., Grossman, A.D., 2021. A mobile genetic element increases bacterial host fitness by manipulating development. *eLife* 10, e65924. https://doi.org/10.7554/eLife.65924

Joshi, R.V., Gunawan, C., Mann, R., 2021. We are one: multispecies metabolism of a biofilm consortium and their treatment strategies. *Frontiers in Microbiology* 12. https://doi.org/10.3389/fmicb.2021.635432

Karygianni, L., Ren, Z., Koo, H., Thurnheer, T., 2020. Biofilm matrixome: extracellular components in structured microbial communities. *Trends in Microbiology* 28, 668–681. https://doi.org/10.1016/j.tim.2020.03.016

Kiran, N.S., Chandrashekar, Y., Mishra, P., 2021. Microbial remediation of organophosphates: a step forward against pesticidal ecotoxicity. *Bulletin of Environment, Pharmacology and Life Sciences* 10, 14–23.

Kiran, N.S., Yashaswini, C., Singh, S., Prajapati, B.G., 2024. Revisiting microbial exopolysaccharides: a biocompatible and sustainable polymeric material for multifaceted biomedical applications. 3 Biotech 14, 95. https://doi.org/10.1007/s13205-024-03946-3

Kirstein, I.V., Wichels, A., Gullans, E., Krohne, G., Gerdts, G., 2019. The plastisphere – uncovering tightly attached plastic "specific" microorganisms. *PLoS One* 14, e0215859. https://doi.org/10.1371/journal.pone.0215859

Koh, J., Bairoliya, S., Salta, M., Cho, Z.T., Fong, J., Neo, M.L., Cragg, S., Cao, B., 2023. Sediment-driven plastisphere community assembly on plastic debris in tropical coastal and marine environments. *Environment International* 179, 108153. https://doi.org/10.1016/j.envint.2023.108153

Laverty, A.L., Primpke, S., Lorenz, C., Gerdts, G., Dobbs, F.C., 2020. Bacterial biofilms colonizing plastics in estuarine waters, with an emphasis on Vibrio spp. and their antibacterial resistance. *PLoS One* 15, e0237704. https://doi.org/10.1371/journal.pone.0237704

Li, C., Gillings, M.R., Zhang, C., Chen, Q., Zhu, D., Wang, J., Zhao, K., Xu, Q., Leung, P.H., Li, X., Liu, J., Jin, L., 2024. Ecology and risks of the global plastisphere as a newly expanding microbial habitat. *The Innovation* 5, 100543. https://doi.org/10.1016/j.xinn.2023.100543

Li, Z., Wang, X., Wang, J., Yuan, X., Jiang, X., Wang, Y., Zhong, C., Xu, D., Gu, T., Wang, F., 2022. Bacterial biofilms as platforms engineered for diverse applications. *Biotechnology Advances* 57, 107932. https://doi.org/10.1016/j.biotechadv.2022.107932

Liang, H., de Haan, W.P., Cerdà-Domènech, M., Méndez, J., Lucena, F., García-Aljaro, C., Sanchez-Vidal, A., Ballesté, E., 2023. Detection of faecal bacteria and antibiotic resistance genes in biofilms attached to plastics from human-impacted coastal areas. *Environmental Pollution* 319, 120983. https://doi.org/10.1016/j.envpol.2022.120983

Limoli, D.H., Jones, C.J., Wozniak, D.J., 2015. Bacterial extracellular polysaccharides in biofilm formation and function. *Microbiology Spectrum* 3, 3.3.29. https://doi.org/10.1128/microbiolspec.MB-0011-2014

Lister, J.L., Horswill, A.R., 2014. *Staphylococcus aureus* biofilms: recent developments in biofilm dispersal. *Frontiers in Cellular and Infection Microbiology* 4. https://doi.org/10.3389/fcimb.2014.00178

Liu, Z., Adyel, T.M., Miao, L., You, G., Liu, S., Hou, J., 2021. Biofilm influenced metal accumulation onto plastic debris in different freshwaters. *Environmental Pollution* 285, 117646. https://doi.org/10.1016/j.envpol.2021.117646

Madsen, J.S., Burmølle, M., Hansen, L.H., Sørensen, S.J., 2012. The interconnection between biofilm formation and horizontal gene transfer. *FEMS Immunology and Medical Microbiology* 65, 183–195. https://doi.org/10.1111/j.1574-695X.2012.00960.x

Martínez-Campos, S., González-Pleiter, M., Fernández-Piñas, F., Rosal, R., Leganés, F., 2021. Early and differential bacterial colonization on microplastics deployed into the effluents of wastewater treatment plants. *Science of the Total Environment* 757, 143832. https://doi.org/10.1016/j.scitotenv.2020.143832

Messer, L.F., Wattiez, R., Matallana-Surget, S., 2024. A closer look at plastic colonisation: prokaryotic dynamics in established versus newly synthesised marine plastispheres and their planktonic state. *Environmental Pollution* 358, 124479. https://doi.org/10.1016/j.envpol.2024.124479

Miao, L., Wang, C., Adyel, T.M., Wu, J., Liu, Z., You, G., Meng, M., Qu, H., Huang, L., Yu, Y., Hou, J., 2020. Microbial carbon metabolic functions of biofilms on plastic debris influenced by the substrate types and environmental factors. *Environment International* 143, 106007. https://doi.org/10.1016/j.envint.2020.106007

Miao, L., Yu, Y., Adyel, T.M., Wang, C., Liu, Z., Liu, S., Huang, L., You, G., Meng, M., Qu, H., Hou, J., 2021. Distinct microbial metabolic activities of biofilms colonizing microplastics in three freshwater ecosystems. *Journal of Hazardous Materials* 403, 123577. https://doi.org/10.1016/j.jhazmat.2020.123577

Mirghani, R., Saba, T., Khaliq, H., Mitchell, J., Do, L., Chambi, L., Diaz, K., Kennedy, T., Alkassab, K., Huynh, T., Elmi, M., Martinez, J., Sawan, S., Rijal, G., 2022. Biofilms: formation, drug resistance and alternatives to conventional approaches. *AIMS Microbiology* 8, 239–277. https://doi.org/10.3934/microbiol.2022019

Mishra, P., Kiran, N.S., Romanholo Ferreira, L.F., Mulla, S.I., 2022. Algae bioprocess to deal with cosmetic chemical pollutants in natural ecosystems: a comprehensive review. *Journal of Basic Microbiology* 62, 1083–1097. https://doi.org/10.1002/jobm.202100467

Mishra, P., Kiran, N.S., Romanholo Ferreira, L.F., Yadav, K.K., Mulla, S.I., 2023. New insights into the bioremediation of petroleum contaminants: a systematic review. *Chemosphere* 326, 138391. https://doi.org/10.1016/j.chemosphere.2023.138391

Misic, C., Covazzi Harriague, A., 2019. Development of marine biofilm on plastic: ecological features in different seasons, temperatures, and light regimes. *Hydrobiologia* 835, 129–145. https://doi.org/10.1007/s10750-019-3934-7

Molin, S., Tolker-Nielsen, T., 2003. Gene transfer occurs with enhanced efficiency in biofilms and induces enhanced stabilisation of the biofilm structure. *Current Opinion in Biotechnology* 14, 255–261. https://doi.org/10.1016/S0958-1669(03)00036-3

Moyal, J., Dave, P.H., Wu, M., Karimpour, S., Brar, S.K., Zhong, H., Kwong, R.W.M., 2023. Impacts of biofilm formation on the physicochemical properties and toxicity of microplastics: a concise review. *Reviews of Environmental Contamination and Toxicology* 261, 8. https://doi.org/10.1007/s44169-023-00035-z

Nguyen, H.T., Lee, Y.K., Kwon, J.-H., Hur, J., 2023. Microplastic biofilms in water treatment systems: fate and risks of pathogenic bacteria, antibiotic-resistant bacteria, and antibiotic resistance genes. *Science of the Total Environment* 892, 164523. https://doi.org/10.1016/j.scitotenv.2023.164523

Oberbeckmann, S., Bartosik, D., Huang, S., Werner, J., Hirschfeld, C., Wibberg, D., Heiden, S.E., Bunk, B., Overmann, J., Becher, D., Kalinowski, J., Schweder, T., Labrenz, M., Markert, S., 2021. Genomic and proteomic profiles of biofilms on microplastics are decoupled from artificial surface properties. *Environmental Microbiology* 23, 3099–3115. https://doi.org/10.1111/1462-2920.15531

Oberbeckmann, S., Osborn, A.M., Duhaime, M.B., 2016. Microbes on a bottle: substrate, season and geography influence community composition of microbes colonizing marine plastic debris. *PLoS One* 11, e0159289. https://doi.org/10.1371/journal.pone.0159289

Okeke, E.S., Ezeorba, T.P.C., Chen, Y., Mao, G., Feng, W., Wu, X., 2022. Ecotoxicological and health implications of microplastic-associated biofilms: a recent review and prospect for turning the hazards into benefits. *Environmental Science and Pollution Research* 29, 70611–70634. https://doi.org/10.1007/s11356-022-22612-w

Otto, M., 2013. Staphylococcal infections: mechanisms of biofilm maturation and detachment as critical determinants of pathogenicity. *Annual Review of Medicine* 64, 175–188. https://doi.org/10.1146/annurev-med-042711-140023

Parrish, K., Fahrenfeld, N.L., 2019. Microplastic biofilm in fresh- and wastewater as a function of microparticle type and size class. *Environmental Science: Water Research & Technology* 5, 495–505. https://doi.org/10.1039/C8EW00712H

Patkowski, J.B., Dahlberg, T., Amin, H., Gahlot, D.K., Vijayrajratnam, S., Vogel, J.P., Francis, M.S., Baker, J.L., Andersson, M., Costa, T.R.D., 2023. The F-pilus biomechanical adaptability accelerates conjugative dissemination of antimicrobial resistance and biofilm formation. *Nature Communications* 14, 1879. https://doi.org/10.1038/s41467-023-37600-y

Pinto, M., Langer, T.M., Hüffer, T., Hofmann, T., Herndl, G.J., 2019. The composition of bacterial communities associated with plastic biofilms differs between different polymers and stages of biofilm succession. *PLoS One* 14, e0217165. https://doi.org/10.1371/journal.pone.0217165

Priya, A., Dutta, K., Daverey, A., 2022. A comprehensive biotechnological and molecular insight into plastic degradation by microbial community. *Journal of Chemical Technology & Biotechnology* 97, 381–390. https://doi.org/10.1002/jctb.6675

Purohit, J., Chattopadhyay, A., Teli, B., 2020. Metagenomic exploration of plastic degrading microbes for biotechnological application. *Current Genomics* 21, 253–270. https://doi.org/10.2174/1389202921999200525155711

Rabin, N., Zheng, Y., Opoku-Temeng, C., Du, Y., Bonsu, E., Sintim, H.O., 2015. Biofilm formation mechanisms and targets for developing antibiofilm agents. *Future Medicinal Chemistry* 7, 493–512. https://doi.org/10.4155/fmc.15.6

Rather, M.A., Gupta, K., Mandal, M., 2021. Microbial biofilm: formation, architecture, antibiotic resistance, and control strategies. *Brazilian Journal of Microbiology* 52, 1701–1718. https://doi.org/10.1007/s42770-021-00624-x

Salam, M., Zheng, H., Liu, Y., Zaib, A., Rehman, S.A.U., Riaz, N., Eliw, M., Hayat, F., Li, H., Wang, F., 2023. Effects of micro(nano)plastics on soil nutrient cycling: state of the knowledge. *Journal of Environmental Management* 344, 118437. https://doi.org/10.1016/j.jenvman.2023.118437

Sharma, D.K., Rajpurohit, Y.S., 2024. Multitasking functions of bacterial extracellular DNA in biofilms. *Journal of Bacteriology* 206, e00006-24. https://doi.org/10.1128/jb.00006-24

Sharma, S., Mohler, J., Mahajan, S.D., Schwartz, S.A., Bruggemann, L., Aalinkeel, R., 2023. Microbial biofilm: a review on formation, infection, antibiotic resistance, control measures, and innovative treatment. *Microorganisms* 11, 1614. https://doi.org/10.3390/microorganisms11061614

Shen, C., Huang, L., Xie, G., Wang, Y., Ma, Z., Yao, Y., Yang, H., 2021. Effects of plastic debris on the biofilm bacterial communities in lake water. *Water* 13, 1465. https://doi.org/10.3390/w13111465

Shi, X., Chen, Z., Wei, W., Chen, J., Ni, B.-J., 2023. Toxicity of micro/nanoplastics in the environment: roles of plastisphere and eco-corona. *Soil & Environmental Health* 1, 100002. https://doi.org/10.1016/j.seh.2023.100002

Silva, I., Rodrigues, E.T., Tacão, M., Henriques, I., 2023. Microplastics accumulate priority antibiotic-resistant pathogens: evidence from the riverine plastisphere. *Environmental Pollution* 332, 121995. https://doi.org/10.1016/j.envpol.2023.121995

Singh, S.P., Sharma, P., Bano, A., Nadda, A.K., Varjani, S., 2022. Microbial communities in plastisphere and free-living microbes for microplastic degradation: a comprehensive review. *Green Analytical Chemistry* 3, 100030. https://doi.org/10.1016/j.greeac.2022.100030

Solano, C., Echeverz, M., Lasa, I., 2014. Biofilm dispersion and quorum sensing. *Current Opinion in Microbiology* 18, 96–104. https://doi.org/10.1016/j.mib.2014.02.008

Sooriyakumar, P., Bolan, N., Kumar, M., Singh, L., Yu, Y., Li, Y., Weralupitiya, C., Vithanage, M., Ramanayaka, S., Sarkar, B., Wang, F., Gleeson, D.B., Zhang, D., Kirkham, M.B., Rinklebe, J., Siddique, K.H.M., 2022. Biofilm formation and its implications on the properties and fate of microplastics in aquatic environments: a review. *Journal of Hazardous Materials Advances* 6, 100077. https://doi.org/10.1016/j.hazadv.2022.100077

Syranidou, E., Kalogerakis, N., 2022. Interactions of microplastics, antibiotics and antibiotic resistant genes within WWTPs. *Science of the Total Environment* 804, 150141. https://doi.org/10.1016/j.scitotenv.2021.150141

Tagg, A.S., Oberbeckmann, S., Fischer, D., Kreikemeyer, B., Labrenz, M., 2019. Paint particles are a distinct and variable substrate for marine bacteria. *Marine Pollution Bulletin* 146, 117–124. https://doi.org/10.1016/j.marpolbul.2019.06.013

Teschler, J.K., Nadell, C.D., Drescher, K., Yildiz, F.H., 2022. Mechanisms underlying *Vibrio cholerae* biofilm formation and dispersion. *Annual Review of Microbiology* 76, 503–532. https://doi.org/10.1146/annurev-micro-111021-053553

Urbanek, A.K., Rymowicz, W., Mirończuk, A.M., 2018. Degradation of plastics and plastic-degrading bacteria in cold marine habitats. *Applied Microbiology and Biotechnology* 102, 7669–7678. https://doi.org/10.1007/s00253-018-9195-y

Wang, X., Li, J., Pan, X., 2024. How micro-/nano-plastics influence the horizontal transfer of antibiotic resistance genes - a review. *Science of the Total Environment* 944, 173881. https://doi.org/10.1016/j.scitotenv.2024.173881

Wang, Z., Gao, J., Zhao, Y., Dai, H., Jia, J., Zhang, D., 2021. Plastisphere enrich antibiotic resistance genes and potential pathogenic bacteria in sewage with pharmaceuticals. *Science of the Total Environment* 768, 144663. https://doi.org/10.1016/j.scitotenv.2020.144663

Wani, A.K., Naqash, N., Akhtar, N., Mir, T. ul G., Mir, B.A., Singh, R., 2024. Plastisphere: marine microbial assemblages for biodegradation of microplastics, in: Karnwal, A., Mohammad Said Al-Tawaha, A. R. (Eds.), *Microbial Applications for Environmental Sustainability*. Springer Nature, Singapore, pp. 117–136. https://doi.org/10.1007/978-981-97-0676-1_6

Weissbrodt, D.G., Shani, N., Holliger, C., 2014. Linking bacterial population dynamics and nutrient removal in the granular sludge biofilm ecosystem engineered for wastewater treatment. *FEMS Microbiology Ecology* 88, 579–595. https://doi.org/10.1111/1574-6941.12326

Wright, R.J., Erni-Cassola, G., Zadjelovic, V., Latva, M., Christie-Oleza, J.A., 2020. Marine plastic debris: a new surface for microbial colonization. *Environmental Science & Technology* 54, 11657–11672. https://doi.org/10.1021/acs.est.0c02305

Yang, Y., Liu, G., Song, W., Ye, C., Lin, H., Li, Z., Liu, W., 2019. Plastics in the marine environment are reservoirs for antibiotic and metal resistance genes. *Environment International* 123, 79–86. https://doi.org/10.1016/j.envint.2018.11.061

Yang, Y., Liu, W., Zhang, Z., Grossart, H.-P., Gadd, G.M., 2020. Microplastics provide new microbial niches in aquatic environments. *Applied Microbiology and Biotechnology* 104, 6501–6511. https://doi.org/10.1007/s00253-020-10704-x

Yoshida, S., Hiraga, K., Taniguchi, I., Oda, K., 2021. *Ideonella sakaiensis*, PETase, and MHETase: from identification of microbial PET degradation to enzyme characterization, in: Weber, G., Bornscheuer, U. T., Wei, R. (Eds.), *Methods in Enzymology, Enzymatic Plastic Degradation*. Academic Press, Cambridge, MA, pp. 187–205. https://doi.org/10.1016/bs.mie.2020.12.007

Yu, H., Zhang, Y., Tan, W., Zhang, Z., 2022. Microplastics as an emerging environmental pollutant in agricultural soils: effects on ecosystems and human health. *Frontiers in Environmental Science* 10, 855292. https://doi.org/10.3389/fenvs.2022.855292

Yu, Y., Miao, L., Adyel, T.M., Waldschläger, K., Wu, J., Hou, J., 2023. Aquatic plastisphere: interactions between plastics and biofilms. *Environmental Pollution* 322, 121196. https://doi.org/10.1016/j.envpol.2023.121196

Yuan, J., Ma, J., Sun, Y., Zhou, T., Zhao, Y., Yu, F., 2020. Microbial degradation and other environmental aspects of microplastics/plastics. *Science of the Total Environment* 715, 136968. https://doi.org/10.1016/j.scitotenv.2020.136968

Zettler, E.R., Mincer, T.J., Amaral-Zettler, L.A., 2013. Life in the "Plastisphere": microbial communities on plastic marine debris. *Environmental Science & Technology* 47, 7137–7146. https://doi.org/10.1021/es401288x

Zhai, X., Zhang, X.-H., Yu, M., 2023. Microbial colonization and degradation of marine microplastics in the plastisphere: a review. *Frontiers in Microbiology* 14, 1127308. https://doi.org/10.3389/fmicb.2023.1127308

Zhang, E., Kim, M., Rueda, L., Rochman, C., VanWormer, E., Moore, J., Shapiro, K., 2022a. Association of zoonotic protozoan parasites with microplastics in seawater and implications for human and wildlife health. *Scientific Reports* 12, 6532. https://doi.org/10.1038/s41598-022-10485-5

Zhang, G., Chen, J., Li, W., 2022b. Conjugative antibiotic-resistant plasmids promote bacterial colonization of microplastics in water environments. *Journal of Hazardous Materials* 430, 128443. https://doi.org/10.1016/j.jhazmat.2022.128443

Zheng, Z., Huang, Y., Liu, L., Wang, L., Tang, J., 2023. Interaction between microplastic biofilm formation and antibiotics: effect of microplastic biofilm and its driving mechanisms on antibiotic resistance gene. *Journal of Hazardous Materials* 459, 132099. https://doi.org/10.1016/j.jhazmat.2023.132099

11 Plastic Pollution in Macrophyte Systems

Impacts on Habitat Integrity

Hoi Shing Lo and Elena Gorokhova

11.1 INTRODUCTION

Solid waste dumped on sea shores, rivers, lakes, ponds, and wetlands causes severe environmental pollution. It is not only aesthetically unacceptable, marring the quality of these habitats, but also harmful to their inhabitants. The influx and fragmentation of anthropogenic debris result in habitat alteration, contamination with toxic chemicals, entanglement, suffocation of marine animals (Thushari and Senevirathna, 2020), alterations in bacterial communities (Seeley et al., 2020), and pathogen spread (Bowley et al., 2021), leading to cascading negative impacts on biodiversity and ecosystem stability (Ma et al., 2020). Although plastic litter is just one component of solid waste, it contributes most to the anthropogenic particles in aquatic environments (Agamuthu et al., 2019), and fragmented microplastic (MP) has emerged as a major pollution issue due to its persistence and ubiquitous occurrence in the environment.

In the last decades, surveys and analytical efforts have been undertaken to gain knowledge on the distribution of plastic debris in aquatic environments, from streams to remote oceanic islands (Napper and Thompson, 2020). As MP size overlaps with prey size in many aquatic animals (Ogonowski et al., 2018), the ecotoxicological assessment of these pollutants has largely focused on invertebrates and fish, particularly those used for human consumption (e.g., mussels, oysters, and herring; Garrido Gamarro et al., 2020) and model organisms (e.g., Daphnia and zebrafish embryos; Prokić et al., 2021). The understanding that coastal aquatic plants are often exposed to much higher levels of anthropogenic litter than nonsedentary animals has developed comparatively recently. In particular, macrophyte systems that are essential for coastal ecosystem functioning, such as seagrass and seaweed beds, have been found to retain much of the plastic debris (Esiukova et al., 2021). However, we have comparatively little understanding of the impacts of plastic litter in such systems, although these habitats are recognized for their high conservation value and multiple ecosystem services.

Plastic particles, being similar to natural particles, exert comparable effects when their sizes fall within similar ranges (Ogonowski et al., 2016, 2018, 2023). However, as plastics commonly have various chemical additives, they may cause more harm than natural particles (Maddela et al., 2023). The interactions between plastic debris and macrophytes may lead to various physical, chemical, and biological effects. For example, direct shading impedes photosynthesis, while chemical leaching from plastics, such as plasticizers, may induce various sublethal responses (J. Li et al., 2022). Beyond these direct impacts, plastic pollution can also alter the habitat properties. In particular, the accumulation of plastics fragmented to various degrees in macrophyte beds can change sediment texture, grain size distribution, and dynamics (Menicagli et al., 2021), affecting habitat stability and nutrient fluxes (Green et al., 2017). Furthermore, various microorganisms that contribute to system functioning and which occur in sediments, water and are associated with multicellular organisms are also a target for plastic-induced impacts (Lear et al., 2021). The alterations in microbial communities can have downstream effects on bacterivores, grazers, and biogeochemical processes

DOI: 10.1201/9781032717814-11

(Bakenhus et al., 2023) and change the physical environment (e.g., redox potential and porewater characteristics) of the macrophyte beds (Green et al., 2015). The complex nature of the consequences of plastic littering was only recently acknowledged in macrophyte research, underscoring the need for a unified approach to study how plastic litter in macrophyte beds affects habitat structure and ecological processes.

In this review, we outline several major types of interactions between plastic and macrophytes (Figure 11.1), involving (1) physical interference and habitat alterations due to particle retention by macrophytes; (2) chemical contamination and ecotoxicity; (3) nutrient dynamics affecting productivity; and (4) microbial communities associated with environmental compartments and plant microbiomes. These interactions can significantly impair plant physiology and growth by obstructing light and gas exchange, altering nutrient availability, and introducing toxic chemicals and novel microbial compartments. As algae and macrophytes struggle to grow and reproduce under these conditions, the entire community dependent on these primary producers and habitat engineers can be affected, ultimately threatening the biodiversity and functioning of coastal and aquatic ecosystems. Through this chapter, we advocate for a focused and unified research approach to deepen our understanding of the often-overlooked interactions between plastic pollution and macrophytes.

11.2 PHYSICAL INTERFERENCE AND HABITAT ALTERATIONS DUE TO PLASTIC RETENTION BY MACROPHYTES

Macrophytes are well-recognized as an effective habitat for particle retention in coastal ecosystems (Madsen et al., 2001). While retaining suspended particles could benefit aquatic life by providing essential nutrients, it also raises concerns about the retention of anthropogenic particles, such as plastics and other wastes. The accumulation of these potentially harmful particles on macrophytes could be detrimental to the plants, with downstream alterations in communities and their physical habitats. Therefore, understanding the particle retention by the coastal macrophytes is crucial for predicting the impacts of anthropogenic particles. This understanding includes the hydrodynamic conditions that influence particle movement and retention; the physical and chemical properties of the particles, such as size, shape, and density; the biological characteristics of the macrophytes, such as surface texture and growth patterns; and the environmental conditions that influence retention dynamics.

11.2.1 Particle Retention Dynamic in Macrophyte Beds

Hydrodynamic conditions drive particle retention on macrophytes. Water movement, including currents, waves, and tidal flows, influences the distribution and settlement of both natural and anthropogenic particles within marine ecosystems. Strong currents and turbulent waters can facilitate particle transport, increasing the likelihood of their interaction with macrophyte surfaces, particle resuspension, and deposition in macrophyte beds. High turbulence within these beds can lead to the resuspension of previously settled particles (Matsusaka, 2015), transport them through the water column (Tsai and Huang, 2019), and carry them to the surfaces of macrophytes, where they might temporarily adhere. However, this adherence is often unstable, as the same turbulent forces that facilitate initial contact can also cause detachment (Ibrahim et al., 2003), reducing overall particle accumulation on macrophyte surfaces. Conversely, low water flow conditions reduce resuspension, resulting in fewer mobilized particles.

Once settled within macrophyte beds, particles tend to be retained for longer periods and accumulate (Pluntke and Kozerski, 2003). Reduced flow diminishes the kinetic energy available to resuspend particles, allowing them to settle more easily onto both the substrate and the surfaces of macrophytes (Madsen et al., 2001). This calm environment facilitates greater sedimentation rates as particles adhere more securely to macrophyte structures and the bed itself. Therefore, stable hydrodynamic conditions promote the long-term retention of these particles, making low-flow macrophyte beds effective at trapping and holding sediments and MPs. The flow conditions due to turbulence can be empirically expressed by turbulent kinetic energy (TKE). Historically, TKE has been frequently used to model the movement of sediment particles in different habitats, including macrophyte beds (Yang et al., 2016). Recently, de Smit et al. (2021) demonstrated an exponential negative correlation between TKE and MP accumulation in macroalgae sediment. Therefore, the intensity of turbulence determines whether a plastic particle stays or leaves a macrophyte habitat, with less MP accumulation when TKE is high. Understanding these mechanisms, especially in the context of environmental changes and human activities that can alter hydrodynamic conditions and impact the retention of both sediment and MP particles in aquatic ecosystems, can help predict the behavior of plastic particles in these environments, their accumulation, and setting assessment thresholds for the overall stability of coastal habitats.

11.2.2 Physicochemical Properties Affecting Plastic Retention

The surface properties of particles, including their surface charge, size, and morphological characteristics, determine whether they exhibit attractive or repulsive behavior toward plant surfaces, sediments, and bacteria. In the environment, these parameters are dependent on the ecocorona, the layer of organic and inorganic materials that can adhere to the surface of plastics when they are exposed to ambient biotic and abiotic conditions leading to biofilm formation, chemical reactions, and physical interactions with surrounding water and sediment particles (Rummel et al., 2017). The ecocorona may increase surface hydrophobicity or hydrophilicity (assessed by wettability of biofilm, Hook et al., 2012), affecting particle interaction with water and potentially influencing how they adhere to macrophytes: a hydrophobic ecocorona could reduce the tendency of plastics to sink or adhere to macrophyte surfaces, whereas a hydrophilic ecocorona might facilitate their retention and particle aggregation (S. Liu et al., 2022). The hydrophobicity of the polymer influenced by the ecocorona depends on environmental conditions and the ability of microbiomes to form biofilms (Liu et al., 2023).

In addition, if the material incorporates additives that introduce negatively charged functional groups, such as carboxyl (–COOH) or hydroxyl (–OH) groups, it lowers the surface charge. Conversely, the surface can be modified to carry a positive charge by adding functional groups such as amines (–NH_2) (Li et al., 2020). For example, positively charged nanosized polystyrene (PS) was found more likely to adsorb onto *Chlorella* and *Scenedesmus* cells than negatively charged PS (Bhattacharya et al., 2010). This is because the cellulose in the algae exerts electrostatic repulsion on the carboxyl and sulfate groups of the negatively charged PS surface. Another study showed that positively charged particles are also more likely to attract negatively charged cell membranes, due to the presence of phospholipids and proteins (Liu et al., 2021). This attraction can result in particle uptake and internalization if the particles are small enough (typically in the size <50 nm; Giorgetti et al., 2020), potentially leading to particle toxicity.

The presence of specific functional groups can also influence ecocorona formation. For instance, hydrophilic particles with –COOH or –OH groups can interact with polar sites on biological macromolecules, facilitating sorption of organic substances, adherence, and uptake by organisms (Gaumet et al., 2010). In addition, water chemistry, e.g., pH, ionic strength, and the presence of natural organic matter, can modify the surface charges of particles and their interactions with organisms. For example, in saline waters, divalent cations of calcium and magnesium can bridge negatively charged particles and surfaces, promoting aggregation (Cai et al., 2018). Conversely, research on the interactions of MP particles with anionic clays in electrolyte solutions has found that lower ionic strength reduces the aggregation of MPs (Takács et al., 2023), whereas dissolved organics promotes it (Reichelt and Gorokhova, 2023). Furthermore, as demonstrated in estuarine environments, the increased ionic strength in saline waters significantly enhances flocculation, causing buoyant MP particles to settle more efficiently with fine-grained sediments, particularly in estuarine zones where the settling velocities are 5–21 times higher compared to freshwater, leading to the accumulation of MPs in these environments (Laursen et al., 2023).

Particle size significantly affects sedimentation within macrophyte beds. A flume study compared the retention of fine particles (<6 μm) and coarse particles (>6 μm) in a simulated seagrass canopy and concluded that fine particles are more readily retained by seagrass blades (Barcelona et al., 2021) due to greater cohesiveness. Additionally, their smaller size and lighter nature make them more susceptible to the viscosity of the surrounding water, which aids in their settlement under low-turbulence conditions (Rampall and Leighton, 1994). However, finer particles are also easier to resuspend under lower turbulence (Reeks et al., 1988). In a dense seagrass canopy, it further promotes the sedimentation of fine particles by decreasing flow velocity and turbulence, and subsequent resuspension, leading to more stable adherence to the seagrass blades and the seabed. Consequently, seagrass blades effectively trap fine particles, enhancing sediment accumulation and improving water clarity within seagrass meadows. Plastic particles in the size range of clay and silt also possess great cohesiveness (Andersen et al., 2021). Therefore, we expect that macrophyte beds tend to retain higher abundances of smaller particles compared to coarser ones, contributing to the accumulation of fine sediments and also MPs in these habitats.

The morphological characteristics of MPs are also crucial for plastic aggregation on macroalgae. Field studies have revealed that among the various shapes of MPs investigated, fibers are the most commonly retained on macrophytes, with prevalence rates reaching 100% (Table 11.1). This observation can be attributed to several factors. First, fibers are the most frequently found type of MP across water and sediment (Kye et al., 2023). This widespread distribution increases the likelihood of their presence in macrophyte beds. Second, unlike other MP shapes, such as fragments, beads, or films, fibers possess a unique morphology that promotes entanglement and adherence due to their ability to wrap around the complex structures of macroalgal thallus and seagrass leaves. This wrapping tendency is facilitated by the macrophytes' surface textures and branching patterns, which provide numerous points of attachment for the fibrous MPs (Ng et al., 2022) and prevent them from being washed away by water currents. Moreover, the flexible nature of fibers allows them to bend and twist when exposed to currents or waves (Saborowski et al., 2019), enhancing their ability to remain attached to the macrophytes. This is in contrast to more rigid plastic fragments, which may be more easily dislodged by turbulent water movements. The combination of these physical and environmental factors underscores the higher prevalence of fibrous MPs on macrophytes compared to other shapes.

The role of large plastic litter in these environments presents additional complexities and uncertainties with regard to the water flow and sediment transport in macrophyte-dominated coastal habitats. Large plastic litter, such as bottles, bags, and other debris, can potentially act as physical obstacles within the coastal zones, altering the natural hydrodynamic regime (Garrard et al., 2024). These larger items can create barriers that modify water flow patterns, leading to changes in sediment deposition and erosion processes. For instance, large plastic items can obstruct currents, causing water to slow down and drop more of its sediment load near the obstruction. This can result in localized sediment accumulation, potentially impacting the growth and distribution of macrophytes as well as other biota.

11.2.3 Role of Macrophyte Density and Morphology in Plastics Trapping

The diversity, structure, and density of macrophyte beds can modify local hydrodynamics, creating microenvironments that enhance particle retention. Observations on canopy density, specifically the number of stems in a given area, can help us understand its impact on coastal runoff and sediment resuspension. These insights are important to study the retention of plastic particles because natural and plastic particles share similarities in size, shape, and aggregation properties. Additionally, the physical processes involved in their interactions with macrophyte canopies are comparable.

TABLE 11.1
The Percentage Proportions of Fiber as MPs on Macroalgae in Field Studies

Sampling Location	Species	%Fiber	Reference
	Macroalgae		
Bodega Marine Reserve, USA	*Pelvetiopsis limitata* *Endocladia muricata*	"Most"	Saley et al. (2019)
Dunstaffnage Bay, Scotland	*Fucus vesiculosus*	92%	Ferguson (2021)
Singapore	*Padina* sp. *Sargassum ilicifolium*	97%	Seng et al. (2020)
Hong Kong	*Ulva* spp. *Gelidium pusillum* *Phycocalidia acanthophora* *Colpomenia sinuosa*	53%–67%	Ng et al. (2022)
Ria Formosa, Portugal	*Caulerpa prolifera*	100%	Cozzolino et al. (2020)
Yellow Sea, China	*Ulva prolifera*	79%–94%	Feng et al. (2020a)
Baltic Sea	*Furcellaria lumbricalis* *Coccotylus truncates* *Polysiphonia fucoides* *Ceramium tenuicorne* *Battersia arctica* *Ectocarpus siliculosus* *Cladophora glomerate* *Cladophora rupestris* *Ulva intestinalis*	93%	Esiukova et al. (2021)
Yello Sea and East China Sea, China	*Gracilaria lemaneiformis* *Chondrus ocellatus* *Ulva lactuca* *Ulva prolifera* *Saccharina japonica*	71.5%	Li et al. (2022)
	Seagrass		
Xincun and Li'an bay, China	*Enhalus acodoides*	58.6% ± 16.0%	Huang et al. (2020)
Western-Estonian archipelago in the Baltic Sea	Sediment near seagrass beds (*Zostera marina* *Zannichellia palustris* *Potamogeton perfoliatus*)	97.1%±5.1%	Kreitsberg et al. (2021)
Mallorca Island, Spain	*Posidonia oceanica* (Neptune balls)	64.9%	Sanchez-Vidal et al. (2021)
Pramuka Island, Indonesia	*Cymodocea rotundata*	30.8%	Priscilla et al. (2019)
Ria Formosa, Portugal	*Zostera noltei* *Sporobolus maritimus* *Cymodocea nodosa* *Zostera marina* *Caulerpa prolifera*	100%	Cozzolino et al. (2020)
Turneffe Atoll Marine Reserve, Belize	*Thalassia testudinum*	81%	Goss et al. (2018)
Deerness Sound, Scotland	*Zostera marina*	>69%	Jones et al. (2020)

Macroalgal species exhibit diverse morphologies, ranging from simple, flat surfaces to complex, branched, or filamentous structures carrying sticky biofilms, affecting the retention of plastic debris (Seng et al., 2020). Species with complex structures, such as *Ulva prolifera* with its distinctive tubular body, highly branched thallus (Wichard et al., 2015) and thus larger surface areas, tend to capture and retain more plastic particles than those with simpler forms (Ng et al., 2022). Also, macroalgae with fronds and blades covered in tiny projections or hairs (e.g., *Derbesia* sp. and *Acrosiphonia* sp.) further increase their capacity to capture and hold onto MPs. In contrast, species like *Sargassum* spp., which have broad, flat surfaces, are less effective at trapping plastics (Zhang et al., 2022).

The dense vegetation with a complex network of leaves and thalli acts as a physical barrier slowing down water flow (Wang et al., 2015), creating areas of lower kinetic energy facilitating particle settling (Schulz et al., 2003) and enhancing deposition of any suspended particles, including plastics (Cozzolino et al., 2020). Conversely, sparse or isolated macroalgae are less effective at trapping debris (Menicagli et al., 2021). In addition, canopy density varies not only between macrophyte habitats but also within them: at the leading edge of the meadow, both vertical velocity and TKE are high, which reduces particle deposition due to increased resuspension (Liu et al., 2022). As water flows further into the meadow, velocity and TKE decrease, leading to greater particle deposition. Therefore, the center of a macrophyte bed has a stronger ability to retain plastic particles due to its lower turbulence and higher particle retention capacity. This stabilization effect is particularly important in environments subject to frequent water movement, such as coastal and estuarine habitats, where the presence of macrophytes can significantly reduce sediment resuspension caused by tides and currents. By studying canopy density and its impact on particle retention, researchers can gain insights into the processes that influence the accumulation of both natural and anthropogenic particles. However, more studies are needed to evaluate the peculiarities of plastics in this context, including investigations into how different types of plastics interact with various macrophyte species, the effects of plastic particle shape and buoyancy on retention, and the long-term ecological impacts of plastic accumulation in these habitats.

The presence of biofilms on plant surfaces enhances their ability to trap particulates, including plastic debris. Biofilms are communities of microorganisms that adhere to surfaces and produce extracellular polymeric substances (EPSs), creating a sticky matrix that can trap plastic particles, increasing the likelihood of accumulation (Yu et al., 2023). Biofilms can also alter the buoyancy and surface properties of plastic particles (Chen et al., 2019), making them more prone to adherence. Once particles are trapped within the biofilm, they can become more difficult to dislodge, even under turbulent water conditions. The interactions between biofilms and macroalgal surfaces create microhabitats that further facilitate the retention of plastic debris, contributing to the persistence and accumulation of plastics in aquatic environments.

11.3 ECOTOXICOLOGICAL EFFECTS OF PLASTIC LITTER ON MACROPHYTES

Plastics can impact plants through various mechanisms, such as physical abrasion, leaching of toxic chemicals, and blocking sunlight. These effects can physically damage plant tissues, directly impair photosynthesis, and hinder nutrient uptake. Therefore, the biological effects of plastic pollution on macrophytes can be measured using various endpoints related to photosynthetic efficiency, biochemical reactions, growth, and reproduction. In general, the effects can be linked to physical, chemical, and (micro-)biological interactions.

11.3.1 Physical Impacts

The presence of litter in coastal areas poses significant threats to the health and stability of aquatic vegetation by obstructing sunlight, causing mechanical damage, altering water flow and sedimentation, and creating microhabitats, leading to impaired plant and microbial growth and broader ecological disruptions. For example, plastic litter can severely impact macrophytes by blocking light and smothering their surfaces. Plastic debris accumulating on the leaves and stems of macrophytes creates a physical barrier to light penetration (Wang et al., 2022), which may disrupt photosynthesis and, ultimately, plant growth. In addition to blocking light, plastic debris can interfere with gas exchange for respiration and photosynthesis, further stressing the macrophytes and leading to microhabitats with hypoxia. Such low-oxygen environments can negatively impact not only the plant's metabolic activities and overall health (Jia et al., 2023), but also facilitate the establishment of anaerobic communities of microorganisms, and alter the food webs and biogeochemical cycling (Madsen, 2011), which acts in concert with physical effects of plastics on sediment properties (Woods et al., 2021).

When plastic particles, especially MPs, in and around macrophyte beds mix with natural sediments, the physical properties of the sediment, such as porosity and permeability, change (Wang et al., 2023), which may affect the stability and nutrient availability. As macrophytes rely on stable substrates for anchoring, the presence of plastic in the sediment can hinder root growth, making it difficult for macrophytes to establish a secure hold (Hilt et al., 2006) and dislodge more easily during strong currents or storms.

Also, larger pieces of plastic debris, such as discarded fishing gear, plastic bottles, and large fragments of composite materials, can cause significant mechanical damage to macrophytes (Q. Li et al., 2022) via entangling, breaking, or shearing. The reduced structural integrity can impair nutrient uptake and photosynthesis (Ghosh et al., 2022), increase vulnerability to diseases, and decrease resilience to environmental changes (Houmani et al., 2018). Over time, this can lead to a decline in macrophyte biodiversity and productivity, not only affecting the biological ecosystem component but also creating feedback on the physical environment, i.e., increased sediment resuspension, loss of habitat, and deteriorated water quality. However, these effects are largely context-dependent; both the nature of plastics and the sediment environment interact to shape the effects and have no general drivers for the plastic–sediment interactions demonstrated in empirical studies (Waldschläger et al., 2022). Such meta-analyses, however, appear for MP-soil systems (Lozano and Rillig, 2020; Xiang et al., 2024), suggesting that the observed effects of plastics can be positive because they directly add bioavailable carbon to the soils, also reducing soil bulk density, improving aeration, and facilitating better root penetration.

Accumulated plastics can also affect the thermal properties of the habitat by absorbing and retaining heat, potentially leading to localized temperature changes that can affect macrophyte physiology and growth (Haque and Fan, 2023). While the direct evidence linking plastic accumulation to macrophyte physiology and growth is still emerging, the thermal properties of plastics and their potential to alter habitat conditions provide a plausible mechanism for such effects. Moreover, changes in temperature due to MP accumulation may influence nutrient availability and cycling in aquatic habitats (Iqbal et al., 2024), with subsequent effects on plant growth and diversity.

11.3.2 Chemical (Eco)toxicity

The chemical toxicity of plastic litter for macrophytes arises from the leaching of additives, harmful monomers and degradation products, and adsorption of persistent organic pollutants (POPs), all of which can adversely affect aquatic organisms (Okoye et al., 2022). Plastics are manufactured with various chemical additives, such as flame retardants and stabilizers, making them more flexible, durable, or resistant to degradation. These chemicals are not chemically bound to the plastic matrix and can leach into the surrounding water over time, releasing these toxic substances into the water and sediments (Sarker et al., 2020). This contamination can create an environment laden with harmful chemicals that are readily absorbed by macrophytes. Moreover, the leaching of these additives can be accelerated by various environmental factors such as temperature fluctuations, UV radiation from sunlight, and mechanical abrasion, all of which can exacerbate the release of these substances (Lan, 2022).

The substances leached from plastics upon aging, including the released monomers, can significantly interfere with the normal functioning of macrophyte cells. For instance, plasticizers and stabilizers can disrupt cellular membranes, altering their permeability and integrity. This can lead to uncontrolled ion exchange and loss of essential cellular components, which are critical for maintaining cellular homeostasis (Ghisari and Bonefeld-Jorgensen, 2009; Rodríguez de San Miguel et al., 2008). Moreover, these chemicals can inhibit enzyme activities, which are vital for numerous metabolic pathways (Zhu et al., 2019). In addition, it has been shown that leachates from high-density polyethylene (HDPE) and biodegradable plastics can inhibit the germination and growth of plants such as cotton (He et al., 2023) and Italian ryegrass (Esterhuizen et al., 2022); for instance, cotton seeds exposed to leachates from HDPE and biodegradable plastic exhibited slight oxidative stress and enzyme activity fluctuations. Similarly, leachates from new and aged polycarbonate MPs significantly reduced the germination rate and seedling growth of *Lepidium sativum* (Pflugmacher et al., 2020), showing concentration-dependent inhibition and adverse effects on root and shoot development. The chemical composition of these leachates, including additives such as bisphenol A (BPA) and their persistence in the environment, underscores the potential long-term risks to plant health (Schiavo et al., 2020). Macrophytes, being mostly submerged and consistently exposed to dissolved contaminants, have been shown to accumulate more pollutants (Fan et al., 2018). Therefore, similar or even higher phytotoxicity is expected, as leachates have been demonstrated to significantly affect macrophyte health (Mateos-Cárdenas et al., 2021).

In addition to leaching, plastics have a high affinity for hydrophobic pollutants, such as heavy metals, pesticides, and POPs (Bakir et al., 2012). As plastics accumulate these contaminants, they might pose risks to aquatic organisms, including macrophytes because when in contact with macrophytes, the adsorbed pollutants can migrate to the plant surface covered by a biofilm and then to the plant tissues. Heavy metals, for instance, can interfere with various biochemical processes within the plants, such as enzyme function, disrupting critical pathways involved in growth and metabolism (Seneviratne et al., 2019). Pesticides and POPs, even at low concentrations, can be toxic, leading to cellular damage, oxidative stress (Wang et al., 2021; Zeb et al., 2022), and interference with plant hormones (Couée et al., 2013; Liu et al., 2009). Although there are many studies describing chemical exposure effects with plastics as vectors, it is challenging to distinguish between the effects of contaminants dissolved in water and those transported by MPs when studying their impacts on algae (Aguirre-Martínez et al., 2023). Moreover, MPs can alter the toxicity of other pollutants in complex mixtures, which is often the case for coastal environments with high loads of solid and dissolved pollutants (Menéndez-Pedriza and Jaumot, 2020). For example, the combined effects of MPs and copper (Davarpanah and Guilhermino, 2015), as well as the interaction between MPs, mercury, and chlorpyrifos toxicity in algae (Pinto et al., 2023), were reported.

Regardless of the route by which plastic-derived chemicals exert their effects—whether through leachates or acting as vectors for other pollutants—photosynthesis inhibition is frequently identified as a critical mode of action. Photosynthesis, the process by which plants convert light energy into chemical energy (Sharma et al., 2020), is pivotal for the growth and vitality of all autotrophs, making it a highly ecologically relevant endpoint for both field surveys and laboratory experiments (Walker et al., 2018). Whereas physical factors affect light availability and gas exchange, heavy metals and POPs can damage the chloroplasts where photosynthesis occurs, leading to reduced chlorophyll production and impaired light absorption (Gallego et al., 1996). Reduced photosynthetic efficiency not only stunts growth, nutrient uptake (Guilizzoni, 1991), and reproductive success in macrophytes (Tomar et al., 2019) but also makes them more susceptible to other environmental stressors.

11.3.3 Microbiome-Mediated Effects on Macrophytes

Plastics in aquatic environments provide a novel surface for microbial colonization, creating what is known as the "plastisphere" (Zettler et al., 2013), a microbial community that often differs from those found on other substrates in the same habitat. These plastisphere microbes can spread and interact with surrounding microbiomes, including those of macrophytes, through natural processes like biofilm sloughing and environmental interactions.

Biofilms are complex communities of microorganisms embedded in a matrix of EPSs. Plastics, due to their durability and resistance to degradation, offer long-term substrates for biofilm development and succession (Rummel et al., 2017). Moreover, once in contact, plastic-associated microbes may interact with plants, with various effects on plant–microbiome interactions. For instance, heterotrophic communities within epiphytic biofilms can induce diseases or deformities in macrophytes. As macrophyte leaves age, an increase in the density and diversity of epiphytic bacteria has been linked to significant inward swelling, disorganization of epidermal walls, and degradation of both the epidermis and mesophyll cell walls (Rogers and Breen, 1981). While epiphytic biofilms are generally associated with pathological developments in macrophytes, certain bacterial genera within these biofilms, such as *Pseudomonas*, can have protective effects by suppressing pathogens, preventing diseases, and promoting plant growth (Zhao et al., 2017). Additionally, Xia et al. (2020) identified bacterial genera *Exiguobacterium*, *Pseudomonas*, and *Chryseobacterium* with antifungal properties in epiphytic biofilms. Notably, these genera are relatively common in plastisphere (*Exiguobacterium*: Gao and Sun (2021); *Pseudomonas*: Messer et al. (2024); *Chryseobacterium*: Joannard and Sanchez-Cid (2024)). In addition, plant-growth-promoting rhizobacteria (PGPR) form biofilms that enable effective root colonization. This enhances the plant's ability to produce beneficial hormones, absorb nutrients, and withstand stress (Ismail et al., 2020; Ren et al., 2022). However, to date, there is no evidence indicating that the aquatic plastisphere contains PGPR.

The interactions between macrophytes and their epiphytic biofilms include the uptake and recycling of essential macroelements, e.g., carbon (C), nitrogen (N), and phosphorus (P), but also trace metals and, possibly, other contaminants (Wijewardene et al., 2022). Epiphytic biofilms accumulate trace metals, pesticides, and other pollutants, which may affect contaminant uptake by the plant; however, these flows are two-sided. Jackson et al. (1994) tracked the pathway of trace metals from sediment to macrophytes and then to epiphytes, using isotopes ^{60}Co and ^{54}Mn, and found that these elements in epiphytes were primarily derived from their host plant *Myriophyllum spicatum*. As plastic particles also take up contaminants, they can be involved in these pathways in a similar fashion, although very little empirical data on these interactions are available so far.

Interactions between epiphytic biofilms and macrophytes can also influence the contaminant toxicity in the plants. Recent studies have highlighted the specific roles of epiphytic biofilms in breaking down complex compounds into simpler nutrients, mobilizing metal ions, and inducing pollutant uptake (reviewed in Srivastava et al., 2017). These interactions between macrophytes and epiphytic biofilms are crucial for the remediation and health of aquatic ecosystems. For example, epiphytic bacteria are known to accumulate mercury (Hg) and produce methylmercury (MeHg), the latter being significantly more bioavailable on macrophytes (Beauvais-Flück et al., 2018; Coelho-Souza et al., 2011). Conversely, some epiphytic bacteria can oxidize trace metal elements, rendering these oxidized compounds less toxic to macrophytes. For example, the production of biogenic manganese oxides on the leaf surfaces of *Egeria densa*, facilitated by bacterial strains *Acidovorax*, *Comamonas*, *Pseudomonas*, and *Rhizobium*, was reported by Tsuji et al., (2017), which are also found in plastisphere (Di Pippo et al., 2020; Messer et al., 2024; Nguyen et al., 2022; Wen et al., 2020).

Future research in microbiome-aware ecotoxicology (Duperron et al., 2020) of plastic pollution should focus on elucidating the complex interactions between plastisphere microbes and native macrophyte microbiomes, with particular attention to the mechanisms by which plastic-associated biofilms influence plant health and ecosystem dynamics. Investigating the dual roles of plastisphere bacteria in both promoting plant growth and inducing diseases will be crucial in understanding their overall impact. Additionally, empirical studies on the pathways of contaminant transfer between plastics, biofilms, and macrophytes are needed to assess the extent of plastic pollution's contribution to trace metal and pollutant accumulation in aquatic environments. Addressing these challenges will enhance our ability to develop ecosystem-based management of plastic pollution and mitigate its adverse effects on aquatic ecosystems.

11.4 CASCADING ECOLOGICAL EFFECTS ON COASTAL HABITATS

Habitat template theory implies that specific habitats provide characteristic conditions that facilitate the development of other species associated with them. In coastal environments, macroalgae and macrophytes provide such habitats. Therefore, plastic pollution in macrophyte habitats can trigger a series of cascading ecological effects that impact the entire ecosystem, as well as socioeconomic impacts, affecting fisheries, tourism, and coastal protection, leading to cascading ecological effects that impact the entire ecosystem (Figure 11.1).

11.4.1 Primary Productivity and Food Web Dynamics

Changes in photosynthetic efficiency induced by physical light obstruction and chemical exposure to the leachates can induce shifts in the composition and growth of primary producers, affecting herbivores and higher trophic levels (Parvez et al., 2024). Macrophytes support a diverse range of species, providing habitat structure and food resources essential for aquatic ecosystems (Thomaz and Cunha, 2010). Therefore, a reduction in macrophyte biomass leads to decreased food availability, which can result in reduced herbivore populations due to starvation or migration to other areas in search of food (Cremona et al., 2008). Consequently, this reduction in herbivores can lead to a cascading effect through the trophic levels, impacting predators that rely on these herbivores for sustenance. Moreover, contaminants from plastic pollution become bioavailable for grazers, leading to the accumulation of pathogen loads and spreading to animal hosts (Bowley et al., 2021). *Vibrio* bacteria, for example, can proliferate on plastic surfaces and transfer to marine animals, causing severe health issues and mortality (Kirstein et al., 2016). Such pathogens disrupt the health of individual organisms, cascading through populations and ecosystems, significantly disrupting food webs, biodiversity, and ecosystem stability (Takemura et al., 2014). Addressing plastic pollution is crucial to mitigating these adverse impacts on wildlife and maintaining healthy, functioning ecosystems (Ali et al., 2023). The accumulation of plastics in aquatic environments not only directly harms organisms but also indirectly affects the entire ecosystem by altering the dynamics of food webs and reducing the resilience and stability of these ecosystems.

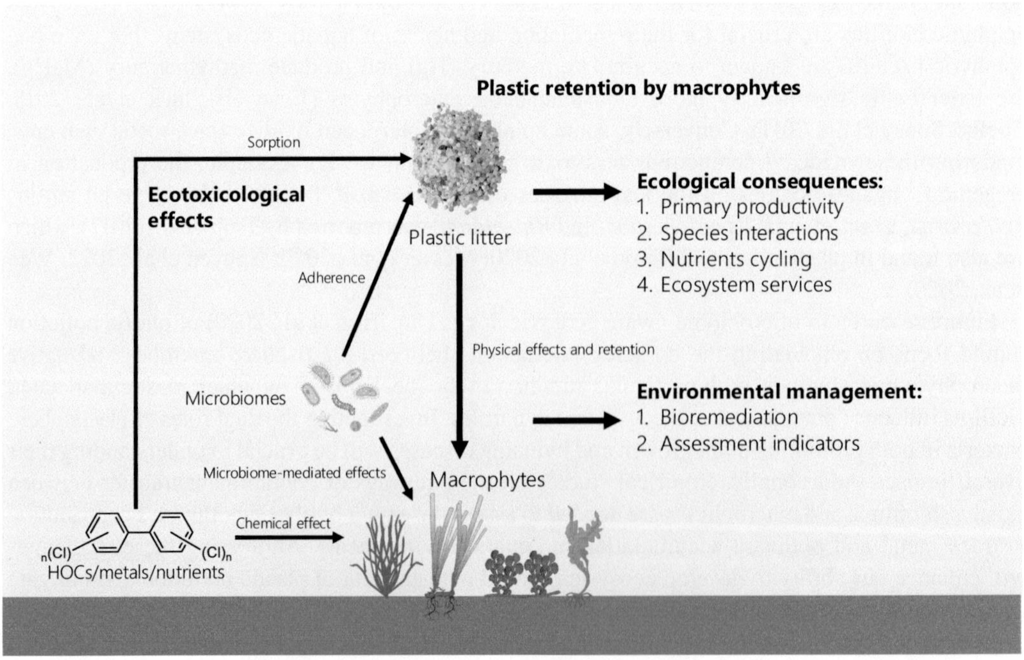

FIGURE 11.1 Overview of interactions between plastic litter and macrophytes.

11.4.2 Species Interactions

Plastic debris and MPs have profound effects on species interactions, particularly impacting macrophyte habitats across various ecosystems. In aquatic environments, plastic debris can serve as new habitats for invasive species (García-Gómez et al., 2021), which may outcompete native species for resources, altering community composition and structure. This competition can disrupt macrophyte communities, essential for maintaining healthy aquatic ecosystems. For example, PS MPs enhance the functional traits of exotic submerged macrophytes, promoting their invasion by reducing soil pH and increasing soil carbon and nitrogen content, thereby altering habitat conditions to favor invasive species over natives. PS MPs also significantly alter sediment bacterial diversity, composition, and multifunctionality, which mediates sediment properties and further enhances plant invasion. These changes create a favorable environment for invasive macrophytes and disrupt the ecological balance, impacting other organisms and overall ecosystem health (Li et al., 2024).

Macrophytes provide shelter, food, and breeding grounds for numerous aquatic organisms, including invertebrates, fish, and amphibians, supporting regional biodiversity (Thomaz, 2023). Plastic littering and macrophyte deterioration can be detrimental to these animals, particularly those that require high water quality, such as fish and invertebrates (Manca et al., 2022). If plastic pollution turns macrophyte habitats into exporters of toxic chemicals, macrophyte habitats become sources of toxic exposure for the organisms that rely on them for shelter and breeding (Tuuri and Leterme, 2023). For many fish species, macrophytes provide critical spawning grounds. The presence of plastics and their associated chemicals in these habitats can expose fish eggs and larvae to harmful substances, reducing their chances of survival and leading to decreased reproductive success (Hodkovicova et al., 2022). Similarly, invertebrates such as insects, crustaceans, and mollusks, which depend on macrophytes for shelter and food, can be affected by the toxic chemicals leached from plastics (Delaeter et al., 2022). This exposure can impair their development and reduce their populations, disrupting the entire food web. Amphibians, which use aquatic plants for breeding and larval development, are also at risk. The accumulation of toxic chemicals on macrophytes can lead to lower reproductive success and declining amphibian populations (Wu et al., 2024). This loss of biodiversity can have cascading effects on the ecosystem, as amphibians play crucial roles in controlling insect populations and serving as prey for higher trophic levels. Thus, the reduction in water purification by macrophytes has broad ecological implications.

11.4.3 Nutrient Cycling

Disruptions caused by the plastisphere can influence microbial communities, involved in the decomposition of organic matter and nutrient cycling, and lead to changes in nutrient availability for primary producers and heterotrophic microorganisms. The introduction of PS-MPs, for example, caused shifts in the abundance and activity of various bacterial taxa, creating a more dynamic, but less stable community structure (Ye et al., 2021). This alteration affected nutrient cycling, including nitrogen fixation, nitrification, and denitrification. MPs enhanced the activity of bacteria involved in these processes, promoting nitrogen-fixing bacteria that increase nitrogen availability in forms easily utilized by invasive plants (Adomako et al., 2024). This nutrient enrichment can support invasive macrophytes' rapid growth and establishment, which often have higher nutrient uptake efficiencies than native species (Peng et al., 2024). Additionally, MPs shifted the functional traits of sediment bacteria, promoting those more efficient at breaking down organic materials or thriving in nutrient-rich environments (Miao et al., 2019).

11.4.4 Ecosystem Services

Macrophyte beds provide critical ecosystem services (Thomaz, 2023), including habitat and biodiversity support, water quality improvement, carbon sequestration, and sediment stabilization.

However, plastic pollution can threaten these functions by introducing toxic chemicals and physical debris that impair plant productivity and ability to filter pollutants and excess nutrients and compromise their role in carbon sequestration and sediment stabilization. Ultimately, it can impact local economies dependent on healthy aquatic ecosystems.

Macrophytes play a crucial role in stabilizing sediments through their root systems, which bind the sediments and prevent erosion (Thomaz, 2023). This stabilization of sediments is essential for maintaining water clarity and quality, as suspended sediments can increase turbidity and decrease light penetration, which affects the photosynthesis of macrophytes and microalgae (Donohue and Garcia Molinos, 2009). A reduced density of a macrophyte bed under the combined stress of plastic pollution and its cascading effects (e.g., biological invasion and nutrient dynamics) may lead to increased erosion and deterioration of phytobenthic communities at large. Moreover, sediment resuspension can lead to the release of pollutants that are trapped in the sediments (Roberts, 2012). However, not only the pollutants but also nutrients can become bioavailable, leading to both increased exposure levels and eutrophication in the water column and posing risks to biota. Furthermore, sediment resuspension may affect water quality by increasing turbidity over larger areas, resulting in photosynthesis inhibition over large areas (Ding et al., 2017), beyond those directly polluted by plastic litter.

Macrophytes also significantly contribute to water purification by sorbing pollutants and nutrients from the water (Kochi et al., 2020). As macrophytes' health declines, their effectiveness at filtering environmental contaminants diminishes, leading to poorer water quality. Plastic leachates can disrupt the metabolic functions responsible for nutrient uptake, thus reducing the plants' efficiency in nutrient cycling (Amaneesh et al., 2023). Increased levels of nutrients can result in higher turbidity, lower oxygen levels, and the proliferation of harmful algal blooms.

11.5 USING MACROPHYTES FOR COASTAL ENVIRONMENTAL MANAGEMENT

11.5.1 Bioremediation Potential of Macrophytes

Macrophytes have attracted attention for their potential to bioremediate plastic litter, including MPs, in coastal environments and wetlands (Ali et al., 2024; Zhang et al., 2024). This involves leveraging their natural processes to absorb, accumulate, and break down environmental contaminants, including plastic pollutants, hence offering a sustainable and eco-friendly solution to marine pollution (Sfriso et al., 2021). To advance the use of macrophytes in bioremediation of plastic litter, several key research directions must be pursued.

First, it is essential to identify and characterize macrophyte species (Sfriso et al., 2021) and general morphological features (Ng et al., 2022) supporting high plastic uptake and degradation capabilities. This involves screening various species and the associated biofilms under controlled conditions to determine their efficiency in capturing different types of plastics. Second, understanding the mechanisms behind plastic degradation by microorganisms at the molecular and biochemical levels is crucial (Cai et al., 2023). This includes investigating the role of enzymes, microbial associations, and metabolic pathways involved in the breakdown of plastic polymers (McGivney et al., 2020). Third, following examples of other bioremediation workflows (Ricart et al., 2023), research should focus on optimizing the environmental conditions that enhance the bioremediation potential of macrophytes, such as hydrography, nutrient availability, salinity, and light intensity. Based on the laboratory findings, regional field studies are needed to evaluate the performance and ecological impacts of macrophytes in natural coastal settings, ensuring that their deployment does not disrupt local ecosystems. Moreover, developing biotechnological innovations, such as bioengineering techniques (Wu et al., 2016) and community manipulation (Wang et al., 2020), could further enhance the efficiency and resilience of macrophytes and the associated biofilms in plastic degradation (Ijoma et al., 2024). Finally, socioeconomic analyses are necessary to assess the feasibility and scalability of using macrophytes for coastal plastic bioremediation, including cost–benefit evaluations and potential economic incentives for stakeholders. Addressing these research areas will be pivotal in exploiting the full potential of macrophytes for mitigating plastic pollution in coastal environments.

11.5.2 Macrophytes as Indicators of Plastic Pollution

Creating and enforcing effective assessment tools and legislation to combat plastic litter in the marine environment has proven to be a significant challenge (Black et al., 2019), including the Marine Strategy Framework Directive (MSFD) (Directive 2008/56/EC), with its Descriptor 10 (Marine Litter). Macrophytes can serve as valuable indicators of plastic pollution in coastal environments, aligning with the assessment tools envisioned by the MSFD. To develop an effective indicator using macrophytes, regional data on the accumulation patterns of plastic particles within these plants are needed (Schreyers et al., 2021). By regularly monitoring and analyzing macrophytes from various coastal regions, quantitative information on the types and amounts of plastics retained by these plants can be obtained to reflect the levels of plastic contamination in specific habitats, offering a tangible measure of pollution. These data, such as the average concentration of plastics per unit mass of macrophytes, the frequency of plastic detection, and the diversity of plastic types found, can be used to develop specific metrics reflecting the pollution load (e.g., https://indicators.helcom.fi/indicator/beach-litter/). These metrics can be further standardized to create a composite indicator of plastic pollution levels and used for Descriptor 10 assessment.

Additionally, assessing the health and growth patterns of macrophytes in polluted versus nonpolluted areas (see Section 11.3) can provide insights into the ecological impacts of plastic litter. These data are currently much needed to establish threshold levels for plastic contamination in macrophytes that signify different levels of environmental health (e.g., low, moderate, and high pollution or acceptable vs unacceptable status). These thresholds can be based on baseline data from pristine environments, ecosystem modeling, and regulatory guidelines.

Macrophytes' widespread presence and sensitivity to environmental changes make them ideal candidates for monitoring programs, enabling regulatory bodies to track plastic pollution trends, evaluate the effectiveness of mitigation strategies, and enforce environmental protection measures. By integrating macrophyte-based indicators into the MSFD framework, coastal management can be more informed and responsive to the challenges posed by marine litter.

11.6 CONCLUSIONS

The complex interactions between macrophytes and plastic pollution in coastal ecosystems present significant challenges for environmental management and conservation. Here, we have explored some factors influencing plastic retention by macrophytes, the direct impacts of plastics on individual macrophyte species, and the broader ecological consequences of plastic pollution in macrophyte habitats. The impacts of plastic pollution on macrophytes cascade through the ecosystem, resulting in habitat loss, reduced primary production, and the accumulation of toxic chemicals. These effects not only diminish the ecological value of macrophyte habitats but also compromise their ability to support biodiversity, maintain food chains, and provide essential ecosystem services such as water purification. Nevertheless, macrophytes may offer a potential solution through their ability to bioremediate plastic litter. Furthermore, macrophytes can serve as valuable indicators of plastic pollution, helping to assess and monitor contamination levels in coastal environments. To fully utilize these roles, further socioeconomic analysis, research, and innovation are required, alongside the development of effective management strategies and policies to protect and enhance the ecosystem services provided by macrophytes.

REFERENCES

Adomako, M. O., Wu, J., Lu, Y., Adu, D., Seshie, V. I., & Yu, F.-H. (2024). Potential synergy of microplastics and nitrogen enrichment on plant holobionts in wetland ecosystems. *Sci. Total Environ.*, *915*, 170160. https://doi.org/10.1016/j.scitotenv.2024.170160

Agamuthu, P., Mehran, S., Norkhairah, A., & Norkhairiyah, A. (2019). Marine debris: A review of impacts and global initiatives. *Waste Manag. Res.*, *37*, 987–1002. https://doi.org/10.1177/0734242X19845041

Aguirre-Martínez, G., Carrizo, M. V., & Zenteno-Devaud, L. (2023). Microplastic particles' effects on aquatic organisms and their role as transporters of organic pollutants. *Water*, *15*, 2915. https://doi.org/10.3390/w15162915

Ali, N., Khan, M. H., Ali, M., Sidra, Ahmad, S., Khan, A., Nabi, G., Ali, F., Bououdina, M., & Kyzas, G. Z. (2023). Insight into microplastics in the aquatic ecosystem: Properties, sources, threats and mitigation strategies. *Sci. Total Environ.*, *913*, 169489. https://doi.org/10.1016/j.scitotenv.2023.169489

Ali, S. S., Elsamahy, T., Al-Tohamy, R., & Sun, J. (2024). A critical review of microplastics in aquatic ecosystems: Degradation mechanisms and removing strategies. *Environ. Sci. Ecotechnol.*, *100427*. https://doi.org/10.1016/j.ese.2024.100427

Amaneesh, C., Anna Balan, S., Silpa, P. S., Kim, J. W., Greeshma, K., Aswathi Mohan, A., Robert Antony, A., Grossart, H.-P., Kim, H.-S., & Ramanan, R. (2023). Gross negligence: Impacts of microplastics and plastic leachates on phytoplankton community and ecosystem dynamics. *Environ. Sci. Technol.*, *57*, 5–24. https://doi.org/10.1021/acs.est.2c05817

Andersen, T. J., Rominikan, S., Olsen, I. S., Skinnebach, K. H., & Fruergaard, M. (2021). Flocculation of PVC microplastic and fine-grained cohesive sediment at environmentally realistic concentrations. *Biol. Bull.*, *240*, 42–51. https://doi.org/10.1086/712929

Bakenhus, I., Jongsma, R., Michler-Kozma, D., Hölscher, L., Gabel, F., Holert, J., & Philipp, B. (2023). A domesticated photoautotrophic microbial community as a biofilm model system for analyzing the influence of plastic surfaces on invertebrate grazers in limnic environments. *Front. Microbiol.*, *14*, 1238913. https://doi.org/10.3389/fmicb.2023.1238913

Bakir, A., Rowland, S. J., & Thompson, R. C. (2012). Competitive sorption of persistent organic pollutants onto microplastics in the marine environment. *Mar. Pollut. Bull.*, *64*, 2782–2789. https://doi.org/10.1016/j.marpolbul.2012.09.010

Barcelona, A., Oldham, C., Colomer, J., Garcia-Orellana, J., & Serra, T. (2021). Particle capture by seagrass canopies under an oscillatory flow. *Coastal Eng.*, *169*, 103972. https://doi.org/10.1016/j.coastaleng.2021.103972

Beauvais-Flück, R., Slaveykova, V. I., & Cosio, C. (2018). Molecular effects of inorganic and methyl mercury in aquatic primary producers: Comparing impact to a macrophyte and a green microalga in controlled conditions. *Geosciences*, *8*, 393. https://doi.org/10.3390/geosciences8110393

Bhattacharya, P., Lin, S., Turner, J. P., & Ke, P. C. (2010). Physical adsorption of charged plastic nanoparticles affects algal photosynthesis. *J. Phys. Chem. C*, *114*, 16556–16561. https://doi.org/10.1021/jp1054759

Black, J. E., Kopke, K., & O'Mahony, C. (2019). A trip upstream to mitigate marine plastic pollution – A perspective focused on the MSFD and WFD. *Front. Mar. Sci.*, *6*, 689. https://doi.org/10.3389/fmars.2019.00689

Bowley, J., Baker-Austin, C., Porter, A., Hartnell, R., & Lewis, C. (2021). Oceanic hitchhikers – Assessing pathogen risks from marine microplastic. *Trends Microbiol.*, *29*, 107–116. https://doi.org/10.1016/j.tim.2020.06.011

Cai, L., Hu, L., Shi, H., Ye, J., Zhang, Y., & Kim, H. (2018). Effects of inorganic ions and natural organic matter on the aggregation of nanoplastics. *Chemosphere*, *197*, 142–151. https://doi.org/10.1016/j.chemosphere.2018.01.052

Cai, Z., Li, M., Zhu, Z., Wang, X., Huang, Y., Li, T., Gong, H., & Yan, M. (2023). Biological degradation of plastics and microplastics: A recent perspective on associated mechanisms and influencing factors. *Microorganisms*, *11*, 1661. https://doi.org/10.3390/microorganisms11071661

Chen, X., Xiong, X., Jiang, X., Shi, H., & Wu, C. (2019). Sinking of floating plastic debris caused by biofilm development in a freshwater lake. *Chemosphere*, *222*, 856–864. https://doi.org/10.1016/j.chemosphere.2019.02.015

Coelho-Souza, S. A., Guimarães, J. R. D., Miranda, M. R., Poirier, H., Mauro, J. B. N., Lucotte, M., & Mergler, D. (2011). Mercury and flooding cycles in the Tapajós river basin, Brazilian Amazon: The role of periphyton of a floating macrophyte (*Paspalum repens*). *Sci. Total Environ.*, *409*, 2746–2753. https://doi.org/10.1016/j.scitotenv.2011.03.028

Couée, I., Serra, A.-A., Ramel, F., Gouesbet, G., & Sulmon, C. (2013). Physiology and toxicology of hormone-disrupting chemicals in higher plants. *Plant Cell Rep., 32*, 933–941. https://doi.org/10.1007/s00299-013-1428-z

Cozzolino, L., Nicastro, K. R., Zardi, G. I., & de los Santos, C. B. (2020). Species-specific plastic accumulation in the sediment and canopy of coastal vegetated habitats. *Sci. Total Environ., 723*, 138018. https://doi.org/10.1016/j.scitotenv.2020.138018

Cremona, F., Planas, D., & Lucotte, M. (2008). Biomass and composition of macroinvertebrate communities associated with different types of macrophyte architectures and habitats in a large fluvial lake. *Fundam. Appl. Limnol., 171*(1), 119–130. https://doi.org/10.1127/1863-9135/2008/0171-0119

Davarpanah, E., & Guilhermino, L. (2015). Single and combined effects of microplastics and copper on the population growth of the marine microalgae *Tetraselmis chuii. Estuar. Coast. Shelf Sci., 167*, 269–275. https://doi.org/10.1016/j.ecss.2015.07.023

deSmit, J. C., Anton, A., Martin, C., Rossbach, S., Bouma, T. J., & Duarte, C. M. (2021). Habitat-forming species trap microplastics into coastal sediment sinks. *Sci. Total Environ., 772*, 145520. https://doi.org/10.1016/j.scitotenv.2021.145520

Delaeter, C., Spilmont, N., Bouchet, V. M. P., & Seuront, L. (2022). Plastic leachates: Bridging the gap between a conspicuous pollution and its pernicious effects on marine life. *Sci. Total Environ., 826*, 154091. https://doi.org/10.1016/j.scitotenv.2022.154091

DiPippo, F., Venezia, C., Sighicelli, M., Pietrelli, L., Di Vito, S., Nuglio, S., & Rossetti, S. (2020). Microplastic-associated biofilms in lentic Italian ecosystems. *Water Res., 187*, 116429. https://doi.org/10.1016/j.watres.2020.116429

Ding, Y., Qin, B., Deng, J., & Ma, J. (2017). Effects of episodic sediment resuspension on phytoplankton in Lake Taihu: Focusing on photosynthesis, biomass and community composition. *Aquat. Sci., 79*, 617–629. https://doi.org/10.1007/s00027-017-0523-6

Donohue, I., & Garcia Molinos, J. (2009). Impacts of increased sediment loads on the ecology of lakes. *Biol. Rev., 84*, 517–531. https://doi.org/10.1111/j.1469-185X.2009.00081.x

Duperron, S., Halary, S., Gallet, A., & Marie, B. (2020). Microbiome-aware ecotoxicology of organisms: Relevance, pitfalls, and challenges. *Front. Public Health, 8*, 407. https://doi.org/10.3389/fpubh.2020.00407

Esiukova, E. E., Lobchuk, O. I., Volodina, A. A., & Chubarenko, I. P., 2021. Marine macrophytes retain microplastics. *Mar. Pollut. Bull,. 171*, 112738. https://doi.org/10.1016/j.marpolbul.2021.112738

Esterhuizen, M., Vikfors, S., Penttinen, O.-P., Kim, Y. J., & Pflugmacher, S. (2022). *Lolium multiflorum* germination and growth affected by virgin, naturally, and artificially aged high-density polyethylene microplastic and leachates. *Front. Environ. Sci., 10*, 964230. https://doi.org/10.3389/fenvs.2022.964230

Fan, S., Liu, H., Zheng, G., Wang, Y., Wang, S., Liu, Y., Liu, X., & Wan, Y., 2018. Differences in phytoaccumulation of organic pollutants in freshwater submerged and emergent plants. *Environ. Pollut., 241*, 247–253. https://doi.org/10.1016/j.envpol.2018.05.071

Feng, Z., Zhang, T., Shi, H., Gao, K., Huang, W., Xu, J., Wang, J., Wang, R., Li, J. I., & Gao, G. (2020). Microplastics in bloom-forming macroalgae: distribution, characteristics and impacts. *Journal of Hazardous Materials, 397*, 122752.

Ferguson, N. M. (2021). Microplastic adherence to common species of macroalgae found throughout Scottish coastal zones. University of the Highlands and Islands. https://pure.uhi.ac.uk/en/studentTheses/microplastic-adherence-to-common-species-of-macroalgae-found-thro.

Gallego, S. M., Benavídes, M. P., & Tomaro, M. L. (1996). Effect of heavy metal ion excess on sunflower leaves: Evidence for involvement of oxidative stress. *Plant Sci., 121*, 151–159. https://doi.org/10.1016/S0168-9452(96)04528-1

Gao, R., & Sun, C. (2021). A marine bacterial community capable of degrading poly(ethylene terephthalate) and polyethylene. *J. Hazard. Mater., 416*, 125928. https://doi.org/10.1016/j.jhazmat.2021.125928

García-Gómez, J. C., Garrigós, M., & Garrigós, J. (2021). Plastic as a vector of dispersion for marine species with invasive potential: A review. *Front. Ecol. E, 9*, 629756. https://doi.org/10.3389/fevo.2021.629756

Garrard, S. L., Clark, J. R., Martin, N., Nelms, S. E., Botterell, Z. L. R., Cole, M., Coppock, R. L., Galloway, T. S., Green, D. S., Jones, M., Lindeque, P. K., Tillin, H. M., & Beaumont, N. J. (2024). Identifying potential high-risk zones for land-derived plastic litter to marine megafauna and key habitats within the North Atlantic. *Sci. Total Environ., 922*, 171282. https://doi.org/10.1016/j.scitotenv.2024.171282

GarridoGamarro, E., Ryder, J., Elvevoll, E. O., & Olsen, R. L. (2020). Microplastics in fish and shellfish – A threat to seafood safety? *J. Aquat. Food Prod. Technol., 29*, 417–425. https://doi.org/10.1080/10498850.2020.1739793

Gaumet, M., Gurny, R., & Delie, F. (2010). Interaction of biodegradable nanoparticles with intestinal cells: The effect of surface hydrophilicity. *Int. J. Pharm., 390*(1–2), 45–52. https://doi.org/10.1016/j.ijpharm.2009.10.008

Ghisari, M., & Bonefeld-Jorgensen, E. C. (2009). Effects of plasticizers and their mixtures on estrogen receptor and thyroid hormone functions. *Toxicol. Lett., 189*, 67–77. https://doi.org/10.1016/j.toxlet.2009.05.004

Ghosh, D., Sarkar, A., Basu, A. G., & Roy, S. (2022). Effect of plastic pollution on freshwater flora: A meta-analysis approach to elucidate the factors influencing plant growth and biochemical markers. *Water Res., 225*, 119114. https://doi.org/10.1016/j.watres.2022.119114

Giorgetti, L., Spanò, C., Muccifora, S., Bottega, S., Barbieri, F., Bellani, L., & Ruffini Castiglione, M. (2020). Exploring the interaction between polystyrene nanoplastics and *Allium cepa* during germination: Internalization in root cells, induction of toxicity and oxidative stress. *Plant Physiol. Biochem., 149*, 170–177. https://doi.org/10.1016/j.plaphy.2020.02.014

Goss, H., Jaskiel, J., & Rotjan, R. (2018). *Thalassia testudinum* as a potential vector for incorporating microplastics into benthic marine food webs. *Marine Pollution Bulletin, 135*, 1085–1089.

Green, D. S., Boots, B., Blockley, D. J., Rocha, C., & Thompson, R. (2015). Impacts of discarded plastic bags on marine assemblages and ecosystem functioning. *Environ. Sci. Technol., 49*, 5380–5389. https://doi.org/10.1021/acs.est.5b00277

Green, D. S., Boots, B., O'Connor, N. E., & Thompson, R. (2017). Microplastics affect the ecological functioning of an important biogenic habitat. *Environ. Sci. Technol., 51*, 68–77. https://doi.org/10.1021/acs.est.6b04496

Guilizzoni, P. (1991). The role of heavy metals and toxic materials in the physiological ecology of submersed macrophytes. *Aquat. Bot., 41*, 87–109. https://doi.org/10.1016/0304-3770(91)90040-C

Haque, F., & Fan, C. (2023). Fate of microplastics under the influence of climate change. *iScience, 26*, 107649. https://doi.org/10.1016/j.isci.2023.107649

He, M., Feng, Z., Xu, Y., Ding, H., Ying, C., Cai, Y., & Zhang, H. (2023). Macro- and microplastic leachates show a slightly toxic effect on seed germination of cotton. *Chemosphere, 335*, 139081. https://doi.org/10.1016/j.chemosphere.2023.139081

Hilt, S., Gross, E. M., Hupfer, M., Morscheid, H., Mählmann, J., Melzer, A., Poltz, J., Sandrock, S., Scharf, E.-M., Schneider, S., & van de Weyer, K. (2006). Restoration of submerged vegetation in shallow eutrophic lakes – A guideline and state of the art in Germany. *Limnologica, 36*, 155–171. https://doi.org/10.1016/j.limno.2006.06.001

Hodkovicova, N., Hollerova, A., Svobodova, Z., Faldyna, M., & Faggio, C. (2022). Effects of plastic particles on aquatic invertebrates and fish – A review. *Environ. Toxicol. Pharmacol., 96*, 104013. https://doi.org/10.1016/j.etap.2022.104013

Hook, A.L., Chang, C.-Y., Yang, J., Luckett, J., Cockayne, A., Atkinson, S., Mei, Y., Bayston, R., Irvine, D.J., Langer, R., Anderson, D.G., Williams, P., Davies, M.C., & Alexander, M.R., (2012). Combinatorial discovery of polymers resistant to bacterial attachment. *Nat. Biotechnol., 30*, 868–875. https://doi.org/10.1038/nbt.2316

Houmani, H., Rodríguez-Ruiz, M., Palma, J. M., & Corpas, F. J. (2018). Mechanical wounding promotes local and long distance response in the halophyte *Cakile maritima* through the involvement of the ROS and RNS metabolism. *Nitric Oxide, 74*, 93–101. https://doi.org/10.1016/j.niox.2017.06.008

Huang, Y., Xiao, X., Xu, C., Perianen, Y. D., Hu, J., & Holmer, M. (2020). Seagrass beds acting as a trap of microplastics-emerging hotspot in the coastal region? *Environmental Pollution, 257*, 113450.

Ibrahim, A. H., Dunn, P. F., & Brach, R. M. (2003). Microparticle detachment from surfaces exposed to turbulent air flow: Controlled experiments and modeling. *J. Aerosol Sci., 34*, 765–782. https://doi.org/10.1016/S0021-8502(03)00031-4

Ijoma, G. N., Lopes, T., Mannie, T., & Mhlongo, T. N. (2024). Exploring macrophytes' microbial populations dynamics to enhance bioremediation in constructed wetlands for industrial pollutants removal in sustainable wastewater treatment. *Symbiosis, 92*, 323–354. https://doi.org/10.1007/s13199-024-00981-9

Iqbal, S., Xu, J., Arif, M. S., Worthy, F. R., Jones, D. L., Khan, S., Alharbi, S. A., Filimonenko, E., Nadir, S., Bu, D., Shakoor, A., Gui, H., Schaefer, D. A., & Kuzyakov, Y. (2024). Do added microplastics, native soil properties, and prevailing climatic conditions have consequences for carbon and nitrogen contents in soil? A global data synthesis of pot and greenhouse studies. *Environ. Sci. Technol., 58*, 8464–8479. https://doi.org/10.1021/acs.est.3c10247

Ismail, N. 'Izzati, Abdullah, S. R. S., Idris, M., Kurniawan, S. B., Effendi Halmi, M. I., AL Sbani, N. H., Jehawi, O. H., & Hasan, H. A. (2020). Applying rhizobacteria consortium for the enhancement of *Scirpus grossus* growth and phytoaccumulation of Fe and Al in pilot constructed wetlands. *J. Environ. Manage., 267*, 110643. https://doi.org/10.1016/j.jenvman.2020.110643

Jackson, L. J., Rowan, D. J., Cornett, R. J., & Kalff, J. (1994). *Myriophyllum spicatum* pumps essential and nonessential trace elements from sediments to epiphytes. *Can. J. Fish. Aquat. Sci., 51*(8), 1769–1773. https://doi.org/10.1139/f94-178

Jia, L., Liu, L., Zhang, Y., Fu, W., Liu, X., Wang, Q., Tanveer, M., & Huang, L. (2023). Microplastic stress in plants: Effects on plant growth and their remediations. *Front. Plant Sci., 14*, 1226484. https://doi.org/10.3389/fpls.2023.1226484

Joannard, B., & Sanchez-Cid, C. (2024). Bacterial dynamics of the plastisphere microbiome exposed to sub-lethal antibiotic pollution. *Microbiome, 12*, 97. https://doi.org/10.1186/s40168-024-01803-2

Jones, K. L., Hartl, M. G., Bell, M. C., & Capper, A. (2020). Microplastic accumulation in a *Zostera marina* L. bed at Deerness Sound, Orkney, Scotland. *Marine Pollution Bulletin, 152*, 110883.

Kirstein, I. V., Kirmizi, S., Wichels, A., Garin-Fernandez, A., Erler, R., Löder, M., & Gerdts, G. (2016). Dangerous hitchhikers? Evidence for potentially pathogenic *Vibrio* spp. on microplastic particles. *Mar. Environ. Res., 120*, 1–8. https://doi.org/10.1016/j.marenvres.2016.07.004

Kreitsberg, R., Raudna-Kristoffersen, M., Heinlaan, M., Ward, R., Visnapuu, M., Kisand, V., Meitern, R., Kotta, J., & Tuvikene, A. (2021). Seagrass beds reveal high abundance of microplastic in sediments: A case study in the Baltic Sea. *Marine Pollution Bulletin, 168*, 112417.

Kochi, L. Y., Freitas, P. L., Maranho, L. T., Juneau, P., & Gomes, M. P. (2020). Aquatic macrophytes in constructed wetlands: A fight against water pollution. *Sustainability, 12*, 9202. https://doi.org/10.3390/su12219202

Kye, H., Kim, J., Ju, S., Lee, J., Lim, C., & Yoon, Y. (2023). Microplastics in water systems: A review of their impacts on the environment and their potential hazards. *Heliyon, 9*, e14359. https://doi.org/10.1016/j.heliyon.2023.e14359

Lan, T. (2022). Contaminant release from aged microplastic. In T. Rocha-Santos, M. F. Costa, & C. Mouneyrac (Eds.), *Handbook of Microplastics in the Environment* (pp. 543–562). Springer International Publishing. https://doi.org/10.1007/978-3-030-39041-9_11

Laursen, S. N., Fruergaard, M., Dodhia, M. S., Posth, N. R., Rasmussen, M. B., Larsen, M. N., Shilla, Dativa, Shilla, Daniel, Kilawe, J. J., Kizenga, H. J., & Andersen, T. J. (2023). Settling of buoyant microplastic in estuaries: The importance of flocculation. *Sci. Total Environ., 886*, 163976. https://doi.org/10.1016/j.scitotenv.2023.163976

Lear, G., Kingsbury, J. M., Franchini, S., Gambarini, V., Maday, S. D. M., Wallbank, J. A., Weaver, L., & Pantos, O. (2021). Plastics and the microbiome: Impacts and solutions. *Environ. Microbiome, 16*, 2. https://doi.org/10.1186/s40793-020-00371-w

Li, J., Yu, S., Yu, Y., & Xu, M. (2022). Effects of microplastics on higher plants: A review. *Bull. Environ. Contam. Toxicol., 109*, 241–265. https://doi.org/10.1007/s00128-022-03566-8

Li, M., He, L., Zhang, X., Rong, H., & Tong, M. (2020). Different surface charged plastic particles have different cotransport behaviors with kaolinite particles in porous media. *Environ. Pollut., 267*, 115534. https://doi.org/10.1016/j.envpol.2020.115534

Li, Q., Su, L., Ma, C., Feng, Z., & Shi, H. (2022). Plastic debris in coastal macroalgae. *Environ. Res., 205*, 112464. https://doi.org/10.1016/j.envres.2021.112464

Li, X., Qin, H., Tang, N., Li, X., & Xing, W. (2024). Microplastics enhance the invasion of exotic submerged macrophytes by mediating plant functional traits, sediment properties, and microbial communities. *J. Hazard. Mater., 469*, 134032. https://doi.org/10.1016/j.jhazmat.2024.134032

Liu, C., Yan, C., Sun, S., Lei, J., Nepf, H., & Shan, Y. (2022). Velocity, turbulence, and sediment deposition in a channel partially filled with a *Phragmites australis* canopy. *Water Resour. Res., 58*, e2022WR032381. https://doi.org/10.1029/2022WR032381

Liu, L., Xu, K., Zhang, B., Ye, Y., Zhang, Q., & Jiang, W. (2021). Cellular internalization and release of polystyrene microplastics and nanoplastics. *Sci. Total Environ., 779*, 146523. https://doi.org/10.1016/j.scitotenv.2021.146523

Liu, S., Junaid, M., Liao, H., Liu, X., Wu, Y., & Wang, J. (2022). Eco-corona formation and associated ecotoxicological impacts of nanoplastics in the environment. *Sci. Total Environ., 836*, 155703. https://doi.org/10.1016/j.scitotenv.2022.155703

Liu, W., Ye, J., & Jin, M. (2009). Enantioselective phytoeffects of chiral pesticides. *J. Agric. Food Chem., 57*, 2087–2095. https://doi.org/10.1021/jf900079y

Liu, X., Fang, L., Yan, X., Gardea-Torresdey, J. L., Gao, Y., Zhou, X., & Yan, B. (2023). Surface functional groups and biofilm formation on microplastics: Environmental implications. *Sci. Total Environ., 903*, 166585. https://doi.org/10.1016/j.scitotenv.2023.166585

Lozano, Y. M., & Rillig, M. C. (2020). Effects of microplastic fibers and drought on plant communities. *Environ. Sci. Technol., 54*, 6166–6173. https://doi.org/10.1021/acs.est.0c01051

Ma, H., Pu, S., Liu, S., Bai, Y., Mandal, S., & Xing, B. (2020). Microplastics in aquatic environments: Toxicity to trigger ecological consequences. *Environ. Pollut., 261*, 114089. https://doi.org/10.1016/j.envpol.2020.114089

Maddela, N. R., Kakarla, D., Venkateswarlu, K., & Megharaj, M., 2023. Additives of plastics: Entry into the environment and potential risks to human and ecological health. *J. Environ. Manage., 348*, 119364. https://doi.org/10.1016/j.jenvman.2023.119364

Madsen, E. L. (2011). Microorganisms and their roles in fundamental biogeochemical cycles. *Curr. Opin. Biotechnol., 22*, 456–464. https://doi.org/10.1016/j.copbio.2011.01.008

Madsen, J. D., Chambers, P. A., James, W. F., Koch, E. W., & Westlake, D. F. (2001). The interaction between water movement, sediment dynamics and submersed macrophytes. *Hydrobiologia, 444*, 71–84. https://doi.org/10.1023/A:1017520800568

Manca, F., Mulà, C., Gustafsson, C., Mauri, A., Roslin, T., Thomas, D. N., Benedetti-Cecchi, L., Norkko, A., & Strona, G. (2022). Unveiling the complexity and ecological function of aquatic macrophyte–animal networks in coastal ecosystems. *Biol. Rev., 97*, 1306–1324. https://doi.org/10.1111/brv.12842

Mateos-Cárdenas, A., van Pelt, F. N. A. M., O'Halloran, J., & Jansen, M. A. K. (2021). Adsorption, uptake and toxicity of micro- and nanoplastics: Effects on terrestrial plants and aquatic macrophytes. *Environ. Pollut., 284*, 117183. https://doi.org/10.1016/j.envpol.2021.117183

Matsusaka, S. (2015). High-resolution analysis of particle deposition and resuspension in turbulent channel flow. *Aerosol Sci. Technol., 49*, 739–746. https://doi.org/10.1080/02786826.2015.1066752

McGivney, E., Cederholm, L., Barth, A., Hakkarainen, M., Hamacher-Barth, E., Ogonowski, M., & Gorokhova, E. (2020). Rapid physicochemical changes in microplastic induced by biofilm formation. *Front. Bioeng. Biotechnol., 8*, 205. https://doi.org/10.3389/fbioe.2020.00205

Menéndez-Pedriza, A., & Jaumot, J. (2020). Interaction of environmental pollutants with microplastics: A critical review of sorption factors, bioaccumulation and ecotoxicological effects. *Toxics, 8*, 40. https://doi.org/10.3390/toxics8020040

Menicagli, V., Balestri, E., Vallerini, F., De Battisti, D., & Lardicci, C. (2021). Plastics and sedimentation foster the spread of a non-native macroalga in seagrass meadows. *Sci. Total Environ., 757*, 143812. https://doi.org/10.1016/j.scitotenv.2020.143812

Messer, L. F., Wattiez, R., & Matallana-Surget, S. (2024). A closer look at plastic colonisation: Prokaryotic dynamics in established versus newly synthesised marine plastispheres and their planktonic state. *Environ. Pollut., 358*, 124479. https://doi.org/10.1016/j.envpol.2024.124479

Miao, L., Wang, P., Hou, J., Yao, Y., Liu, Z., Liu, S., & Li, T. (2019). Distinct community structure and microbial functions of biofilms colonizing microplastics. *Sci. Total Environ., 650*, 2395–2402. https://doi.org/10.1016/j.scitotenv.2018.09.378

Napper, I. E., & Thompson, R. C. (2020). Plastic debris in the marine environment: History and future challenges. *Global Challenges, 4*, 1900081. https://doi.org/10.1002/gch2.201900081

Ng, K. L., Suk, K. F., Cheung, K. W., Shek, R. H. T., Chan, S. M. N., Tam, N. F. Y., Cheung, S. G., Fang, J. K.-H., & Lo, H. S. (2022). Macroalgal morphology mediates microplastic accumulation on thallus and in sediments. *Sci. Total Environ., 825*, 153987. https://doi.org/10.1016/j.scitotenv.2022.153987

Nguyen, H. T., Choi, W., Kim, E.-J., & Cho, K. (2022). Microbial community niches on microplastics and prioritized environmental factors under various urban riverine conditions. *Sci. Total Environ., 849*, 157781. https://doi.org/10.1016/j.scitotenv.2022.157781

Ogonowski, M., Gerdes, Z., & Gorokhova, E. (2018). What we know and what we think we know about microplastic effects – A critical perspective. *Curr. Opin. Environ. Sci. Health, 1*, 41–46. https://doi.org/10.1016/j.coesh.2017.09.001

Ogonowski, M., Schür, C., Jarsén, Å., & Gorokhova, E. (2016). The effects of natural and anthropogenic microparticles on individual fitness in *Daphnia magna*. *PLoS One, 11*, e0155063. https://doi.org/10.1371/journal.pone.0155063

Ogonowski, M., Wagner, M., Rogell, B., Haave, M., & Lusher, A. (2023). Microplastics could be marginally more hazardous than natural suspended solids – A meta-analysis. *Ecotoxicol. Environ. Saf., 264*, 115406. https://doi.org/10.1016/j.ecoenv.2023.115406

Okoye, C. O., Addey, C. I., Oderinde, O., Okoro, J. O., Uwamungu, J. Y., Ikechukwu, C. K., Okeke, E. S., Ejeromedoghene, O., & Odii, E. C. (2022). toxic chemicals and persistent organic pollutants associated with micro-and nanoplastics pollution. *Chem. Eng. J. Adv., 11*, 100310. https://doi.org/10.1016/j.ceja.2022.100310

Parvez, M. D. S., Ullah, H., Faruk, O., Simon, E., & Czédli, H. (2024). Role of microplastics in global warming and climate change: A review. *Water Air Soil Pollut., 235*, 201. https://doi.org/10.1007/s11270-024-07003-w

Peng, Q., Huo, B., Yang, H., Xu, Z., Mao, H., Yang, S., Dai, Y., Li, Z., & Deng, X. (2024). Increased invasion of submerged macrophytes makes native species more susceptible to eutrophication in freshwater ecosystems. *Sci. Total Environ., 909*, 168658. https://doi.org/10.1016/j.scitotenv.2023.168658

Pflugmacher, S., Sulek, A., Mader, H., Heo, J., Noh, J. H., Penttinen, O.-P., Kim, Y., Kim, S., & Esterhuizen, M. (2020). The influence of new and artificial aged microplastic and leachates on the germination of *Lepidium sativum* L. *Plants, 9*, 339. https://doi.org/10.3390/plants9030339

Pinto, E. P., Paredes, E., & Bellas, J. (2023). Influence of microplastics on the toxicity of chlorpyrifos and mercury on the marine microalgae *Rhodomonas lens. Sci. Total Environ., 857*, 159605. https://doi.org/10.1016/j.scitotenv.2022.159605

Pluntke, T., & Kozerski, H.-P. (2003). Particle trapping on leaves and on the bottom in simulated submerged plant stands. *Hydrobiologia, 506*, 575–581. https://doi.org/10.1023/B:HYDR.0000008569.29286.ec

Prokić, M. D., Gavrilović, B. R., Radovanović, T. B., Gavrić, J. P., Petrović, T. G., Despotović, S. G., & Faggio, C. (2021). Studying microplastics: Lessons from evaluated literature on animal model organisms and experimental approaches. *J. Hazard. Mater., 414*, 125476. https://doi.org/10.1016/j.jhazmat.2021.125476

Rampall, I., & Leighton, D. T. (1994). Influence of shear-induced migration on turbulent resuspension. *Int. J. Multiphase Flow, 20*, 631–650. https://doi.org/10.1016/0301-9322(94)90034-5

Reeks, M. W., Reed, J., & Hall, D. (1988). On the resuspension of small particles by a turbulent flow. *J. Phys. D: Appl. Phys., 21*, 574. https://doi.org/10.1088/0022-3727/21/4/006

Reichelt, S., & Gorokhova, E. (2023). Aggregation in experimental studies with microparticles: Experimental settings change particle size distribution during exposure. *Environ. Pollut., 336*, 122369. https://doi.org/10.1016/j.envpol.2023.122369

Ren, L., Gao, Y., Hu, Z., Jiang, X., & Yang, L. (2022). The growth of *Vallisneria natans* and its epiphytic biofilm in simulated nutrient-rich flowing water. *Water, 14*, 2236. https://doi.org/10.3390/w14142236

Ricart, A. M., Honisch, B., Fachon, E., Hunt, C. W., Salisbury, J., Arnold, S. N., & Price, N. N. (2023). Optimizing marine macrophyte capacity to locally ameliorate ocean acidification under variable light and flow regimes: Insights from an experimental approach. *PLoS One, 18*, e0288548. https://doi.org/10.1371/journal.pone.0288548

Roberts, D. A. (2012). Causes and ecological effects of resuspended contaminated sediments (RCS) in marine environments. *Environ. Int., 40*, 230–243. https://doi.org/10.1016/j.envint.2011.11.013

Rodríguezde San Miguel, E., Aguilar, J. C., & de Gyves, J. (2008). Structural effects on metal ion migration across polymer inclusion membranes: Dependence of transport profiles on nature of active plasticizer. *J. Membrane Sci., 307*, 105–116. https://doi.org/10.1016/j.memsci.2007.09.012

Rogers, K. H., & Breen, C. M. (1981). Effects of epiphyton on *Potamogeton crispus* L. leaves. *Microb. Ecol., 7*, 351–363. https://doi.org/10.1007/BF02341430

Rummel, C. D., Jahnke, A., Gorokhova, E., Kühnel, D., & Schmitt-Jansen, M. (2017). Impacts of biofilm formation on the fate and potential effects of microplastic in the aquatic environment. *Environ. Sci. Technol. Lett., 4*, 258–267. https://doi.org/10.1021/acs.estlett.7b00164

Saborowski, R., Paulischkis, E., & Gutow, L. (2019). How to get rid of ingested microplastic fibers? A straightforward approach of the Atlantic ditch shrimp *Palaemon varians. Environ. Pollut., 254*, 113068. https://doi.org/10.1016/j.envpol.2019.113068

Saley, A. M., Smart, A. C., Bezerra, M. F., Burnham, T. L. U., Capece, L. R., Lima, L. F. O., Carsh, A. C., Williams, S. L., & Morgan, S. G. (2019). Microplastic accumulation and biomagnification in a coastal marine reserve situated in a sparsely populated area. *Marine Pollution Bulletin*, 146, 54–59.

Sanchez-Vidal, A., Canals, M., De Haan, W. P., Romero, J., & Veny, M. (2021). Seagrasses provide a novel ecosystem service by trapping marine plastics. *Scientific Reports, 11*, p. 254.

Sarker, I., Moore, L. R., Paulsen, I. T., & Tetu, S. G. (2020). Assessing the toxicity of leachates from weathered plastics on photosynthetic marine bacteria *Prochlorococcus. Front. Mar. Sci., 7*. https://doi.org/10.3389/fmars.2020.571929

Schiavo, S., Oliviero, M., Chiavarini, S., & Manzo, S. (2020). Adverse effects of oxo-degradable plastic leachates in freshwater environment. *Environ. Sci. Pollut. Res., 27*, 8586–8595. https://doi.org/10.1007/s11356-019-07466-z

Schreyers, L., van Emmerik, T., Nguyen, T. L., Phung, N.-A., Kieu-Le, T.-C., Castrop, E., Bui, T.-K. L., Strady, E., Kosten, S., Biermann, L., van den Berg, S. J. P., & van der Ploeg, M. (2021). A field guide for monitoring riverine macroplastic entrapment in water hyacinths. *Front. Environ. Sci., 0*. https://doi.org/10.3389/fenvs.2021.716516

Schulz, M., Kozerski, H.-P., Pluntke, T., & Rinke, K. (2003). The influence of macrophytes on sedimentation and nutrient retention in the lower River Spree (Germany). *Water Res., 37*, 569–578. https://doi.org/10.1016/S0043-1354(02)00276-2

Seeley, M. E., Song, B., Passie, R., & Hale, R. C. (2020). Microplastics affect sedimentary microbial communities and nitrogen cycling. *Nat. Commun., 11*, 2372. https://doi.org/10.1038/s41467-020-16235-3

Seneviratne, M., Rajakaruna, N., Rizwan, M., Madawala, H. M. S. P., Ok, Y. S., & Vithanage, M. (2019). Heavy metal-induced oxidative stress on seed germination and seedling development: A critical review. *Environ. Geochem. Health, 41*, 1813–1831. https://doi.org/10.1007/s10653-017-0005-8

Seng, N., Lai, S., Fong, J., Saleh, M. F., Cheng, C., Cheok, Z. Y., & Todd, P. A. (2020). Early evidence of microplastics on seagrass and macroalgae. *Mar. Freshwater Res., 71*, 922–928. https://doi.org/10.1071/MF19177

Sfriso, A., Tomio, T., Juhmani, A.-S., Sfriso, A., Munari, C., & Mistri, M. (2021). Macrophytes: A temporary sink for microplastics in transitional water systems. *Water, 13*, 3032. https://doi.org/10.3390/w13213032

Sharma, A., Kumar, V., Shahzad, B., Ramakrishnan, M., Singh Sidhu, G. P., Bali, A. S., Handa, N., Kapoor, D., Yadav, P., Khanna, K., Bakshi, P., Rehman, A., Kohli, S. K., Khan, E. A., Parihar, R. D., Yuan, H., Thukral, A. K., Bhardwaj, R., & Zheng, B. (2020). Photosynthetic response of plants under different abiotic stresses: A review. *J. Plant Growth Regul., 39*, 509–531. https://doi.org/10.1007/s00344-019-10018-x

Srivastava, J. K., Chandra, H., Kalra, S. J. S., Mishra, P., Khan, H., & Yadav, P. (2017). Plant–microbe interaction in aquatic systems and their role in the management of water quality: A review. *Appl. Water Sci., 7*, 1079–1090. https://doi.org/10.1007/s13201-016-0415-2

Takács, D., Szabó, T., Jamnik, A., Tomšič, M., & Szilágyi, I. (2023). Colloidal interactions of microplastic particles with anionic clays in electrolyte solutions. *Langmuir, 39*, 12835–12844. https://doi.org/10.1021/acs.langmuir.3c01700

Takemura, A. F., Chien, D. M., & Polz, M. F. (2014). Associations and dynamics of *Vibrionaceae* in the environment, from the genus to the population level. *Front. Microbiol., 5*, 38. https://doi.org/10.3389/fmicb.2014.00038

Thomaz, S. M. (2023). Ecosystem services provided by freshwater macrophytes. *Hydrobiologia, 850*, 2757–2777. https://doi.org/10.1007/s10750-021-04739-y

Thomaz, S. M., & Cunha, E. R. da. (2010). The role of macrophytes in habitat structuring in aquatic ecosystems: Methods of measurement, causes and consequences on animal assemblages' composition and biodiversity. *Acta Limnol. Bras., 22*, 218–236. https://doi.org/10.4322/actalb.02202011

Thushari, G. G. N., & Senevirathna, J. D. M. (2020). Plastic pollution in the marine environment. *Heliyon, 6*, e04709. https://doi.org/10.1016/j.heliyon.2020.e04709

Tomar, R. S., Singh, B., & Jajoo, A. (2019). Effects of organic pollutants on photosynthesis. In *Photosynthesis, Productivity and Environmental Stress* (pp. 1–26). John Wiley & Sons, Ltd. https://doi.org/10.1002/9781119501800.ch1

Tsai, C. W. & Huang, S.-H. (2019). Modeling suspended sediment transport under influence of turbulence ejection and sweep events. *Water Resour. Res., 55*, 5379–5393. https://doi.org/10.1029/2018WR023493

Tsuji, K., Asayama, T., Shiraki, N., Inoue, S., Okuda, E., Hayashi, C., Nishida, K., Hasegawa, H., & Harada, E. (2017). Mn accumulation in a submerged plant *Egeria densa* (Hydrocharitaceae) is mediated by epiphytic bacteria. *Plant Cell Environ., 40*, 1163–1173. https://doi.org/10.1111/pce.12910

Tuuri, E. M. & Leterme, S. C. (2023). How plastic debris and associated chemicals impact the marine food web: A review. *Environ. Pollut., 321*, 121156. https://doi.org/10.1016/j.envpol.2023.121156

Waldschläger, K., Brückner, M. Z. M., Carney Almroth, B., Hackney, C. R., Adyel, T. M., Alimi, O. S., Belontz, S. L., Cowger, W., Doyle, D., Gray, A., Kane, I., Kooi, M., Kramer, M., Lechthaler, S., Michie, L., Nordam, T., Pohl, F., Russell, C., Thit, A., Umar, W., Valero, D., Varrani, A., Warrier, A. K., Woodall, L. C., & Wu, N. (2022). Learning from natural sediments to tackle microplastics challenges: A multidisciplinary perspective. *Earth-Sci. Rev., 228*, 104021. https://doi.org/10.1016/j.earscirev.2022.104021

Walker, B. J., Busch, F. A., Driever, S. M., Kromdijk, J., & Lawson, T. (2018). Survey of tools for measuring in vivo photosynthesis. In S. Covshoff (Ed.), *Photosynthesis: Methods and Protocols* (pp. 3–24). Springer. https://doi.org/10.1007/978-1-4939-7786-4_1

Wang, C., Jiang, L., Huang, W., Wang, C., & He, M. (2022). Light availability modulates the responses of the microalgae *Desmodesmus sp.* to micron-sized polyvinyl chloride microplastics. *Aquat. Toxicol., 249*, 106234. https://doi.org/10.1016/j.aquatox.2022.106234

Wang, C., Sun, Y., Ruan, H., & Yang, J. (2021). Toxic effects of 2,4,4′-trichlorobiphenyl (PCB-28) on growth, photosynthesis characteristics, and antioxidant defense system of *Lemna minor* L. *Plant Physiol. Biochem., 166*, 505–511. https://doi.org/10.1016/j.plaphy.2021.06.031

Wang, C., Zheng, S., Wang, P., & Hou, J. (2015). Interactions between vegetation, water flow, and sediment transport: A review. *J. Hydrodynamics, Ser. B, 27*, 24–37. https://doi.org/10.1016/S1001-6058(15)60453-X

Wang, S., Yu, J., Guo, F., Pan, G., Zhang, L., Hu, H., Lu, Y., & Dao, G. (2020). Optimization of combined submerged macrophyte planting conditions for inhibiting algae by response surface methodology. *Water, 12*, 2093. https://doi.org/10.3390/w12082093

Wang, Z., Li, W., Li, W., Yang, W., & Jing, S. (2023). Effects of microplastics on the water characteristic curve of soils with different textures. *Chemosphere, 317*, 137762. https://doi.org/10.1016/j.chemosphere.2023.137762

Wen, B., Liu, J.-H., Zhang, Y., Zhang, H.-R., Gao, J.-Z., & Chen, Z.-Z. (2020). Community structure and functional diversity of the plastisphere in aquaculture waters: Does plastic color matter? *Sci. Total Environ., 740*, 140082. https://doi.org/10.1016/j.scitotenv.2020.140082

Wichard, T., Charrier, B., Mineur, F., Bothwell, J. H., Clerck, O. D., & Coates, J. C. (2015). The green seaweed *Ulva*: A model system to study morphogenesis. *Front. Plant Sci., 6*. https://doi.org/10.3389/fpls.2015.00072

Wijewardene, L., Wu, N., Fohrer, N., & Riis, T. (2022). Epiphytic biofilms in freshwater and interactions with macrophytes: Current understanding and future directions. *Aquat. Bot., 176*, 103467. https://doi.org/10.1016/j.aquabot.2021.103467

Woods, J. S., Verones, F., Jolliet, O., Vázquez-Rowe, I., & Boulay, A.-M. (2021). A framework for the assessment of marine litter impacts in life cycle impact assessment. *Ecol. Indic., 129*, 107918. https://doi.org/10.1016/j.ecolind.2021.107918

Wu, Q., Hu, Y., Li, S., Peng, S., & Zhao, H. (2016). Microbial mechanisms of using enhanced ecological floating beds for eutrophic water improvement. *Bioresour. Technol., 211*, 451–456. https://doi.org/10.1016/j.biortech.2016.03.113

Wu, X., Yu, S., Zeng, J., Zheng, X., Ren, Z., Shu, Y., & Mai, B. (2024). Biomagnification of persistent organic pollutants (POPs) in detritivorous, phytophagous, and predatory invertebrates: How POPs enter terrestrial food web? *Sci. Total Environ., 924*, 171677. https://doi.org/10.1016/j.scitotenv.2024.171677

Xia, P., Yan, D., Sun, R., Song, X., Lin, T., & Yi, Y. (2020). Community composition and correlations between bacteria and algae within epiphytic biofilms on submerged macrophytes in a plateau lake, southwest China. *Sci. Total Environ., 727*, 138398. https://doi.org/10.1016/j.scitotenv.2020.138398

Xiang, Y., Rillig, M. C., Peñuelas, J., Sardans, J., Liu, Y., Yao, B., & Li, Y. (2024). Global responses of soil carbon dynamics to microplastic exposure: A data synthesis of laboratory studies. *Environ. Sci. Technol., 58*, 5821–5831. https://doi.org/10.1021/acs.est.3c06177

Yang, J. Q., Chung, H., & Nepf, H. M. (2016). The onset of sediment transport in vegetated channels predicted by turbulent kinetic energy. *Geophys. Res. Lett., 43*, 11261–11268. https://doi.org/10.1002/2016GL071092

Ye, G., Zhang, X., Yan, C., Lin, Y., & Huang, Q. (2021). Polystyrene microplastics induce microbial dysbiosis and dysfunction in surrounding seawater. *Environ. Int., 156*, 106724. https://doi.org/10.1016/j.envint.2021.106724

Yu, Y., Miao, L., Adyel, T. M., Waldschläger, K., Wu, J., & Hou, J. (2023). Aquatic plastisphere: Interactions between plastics and biofilms. *Environ. Pollut., 322*, 121196. https://doi.org/10.1016/j.envpol.2023.121196

Zeb, B. S., Hayat, M. T., Zeb, T., Khan, F. Y., Abbasi, H. Z., Nawaz, I., & Ebadi, A. (2022). Uptake of organic pollutants and the effects on plants. In Q. Mahmood (Ed.), *Sustainable Plant Nutrition under Contaminated Environments* (pp. 209–234). Springer International Publishing. https://doi.org/10.1007/978-3-030-91499-8_11

Zettler, E. R., Mincer, T. J., & Amaral-Zettler, L. A. (2013). Life in the "plastisphere": Microbial communities on plastic marine debris. *Environ. Sci. Technol., 47*, 7137–7146. https://doi.org/10.1021/es401288x

Zhang, S., Shen, C., Zhang, F., Wei, K., Shan, S., Zhao, Y., Man, Y. B., Wong, M. H., & Zhang, J. (2024). Microplastics removal mechanisms in constructed wetlands and their impacts on nutrient (nitrogen, phosphorus, and carbon) removal: A critical review. *Sci. Total Environ., 918*, 170654. https://doi.org/10.1016/j.scitotenv.2024.170654

Zhang, T., Wang, J., Liu, D., Sun, Z., Tang, R., Ma, X., & Feng, Z. (2022). Loading of microplastics by two related macroalgae in a sea area where gold and green tides occur simultaneously. *Sci. Total Environ., 814*, 152809. https://doi.org/10.1016/j.scitotenv.2021.152809

Zhao, B., Xing, P., & Wu, Q. L. (2017). Microbes participated in macrophyte leaf litter decomposition in freshwater habitat. *FEMS Microbiol. Ecol., 93*, fix108. https://doi.org/10.1093/femsec/fix108

Zhu, T.-K., Du, P.-P., Zeng, L.-J., Lü, H., Zhao, H.-M., Li, Y.-W., Mo, C.-H., & Cai, Q.-Y. (2019). Variation in metabolism and degradation of di-n-butyl phthalate (DBP) by high- and low-DBP accumulating cultivars of rice (*Oryza sativa* L.) and crude enzyme extracts. *Sci. Total Environ., 668*, 1117–1127. https://doi.org/10.1016/j.scitotenv.2019.03.047

12 Rhizobiome of Terrestrial Plastisphere

Kumudi Premarathna and Sumedha Madawala

12.1 INTRODUCTION

Plastics, known for their persistence and inertness, serve as substrates for microorganisms, leading to the formation of a unique ecological community on plastic debris termed the "plastisphere." This novel ecosystem, established on plastic surfaces, represents an increasingly significant habitat for microbial colonization and interaction. The plastisphere is marked by a diverse range of microorganisms that thrive on this synthetic substratum. Plastics, with their distinctive physical and chemical properties, significantly impact microbial colonization and community structure. Microbial communities that colonize the plastisphere have both beneficial and detrimental effects on their surroundings. They play a vital role in microbial evolution, nutrient cycling, and pollutant degradation and are crucial for metabolizing organic compounds, which accelerates carbon and nitrogen turnover (Li et al., 2024). This process is essential for the breakdown of plastic polymers, helping to mitigate plastic waste accumulation in the environment. However, plastisphere microbes can also pose risks to human and environmental health. They can harbor pathogenic and antibiotic-resistant microorganisms (Quilliam et al., 2023) and influence soil health by disrupting microbial communities, which affects soil fertility and plant growth. Understanding plastisphere microbial communities and their interactions is crucial for understanding ecosystem dynamics, biogeochemical cycles, and the broader implications for ecosystem sustainability amid global environmental challenges.

Recent research on the plastisphere has predominantly concentrated on marine environments, with less focus on terrestrial ecosystems (Wright et al., 2020; Joos and De Tender, 2022). However, emerging evidence suggests that the terrestrial plastisphere may be even more significant due to its environmental impacts and potential effects on human health and terrestrial ecosystems (Teuten et al., 2009; Royer et al., 2018; Rillig et al., 2019; Liu et al., 2020). With approximately 79% of the 8.3 billion metric tons of plastics produced globally ending up in landfills or terrestrial environments, soils serve as a major sink for plastic pollution. This makes the terrestrial plastisphere an ideal system for studying microbial community development and the factors shaping plastisphere microbiomes (Rohrbach et al., 2022). Given the persistence and extensive presence of microplastics in terrestrial environments and their potential to affect microbiomes, further research into their impact on terrestrial plastisphere and its implications for human health and natural ecosystems is crucial (Rohrbach et al., 2022).

The rhizosphere, the microenvironment surrounding plant roots, plays a crucial role in supporting soil microorganisms through beneficial plant interactions. Investigating the rhizobiome (the microbial community associated with plant roots) within the terrestrial plastisphere is particularly important for several reasons. The rhizobiome plays a significant role in plant growth through multiple mechanisms. Rhizobacteria and other associated microorganisms are involved in making essential nutrients available for plants, thus improving nutrient uptake and growth (Lugtenberg & Kamilova, 2009). Rhizobiome also takes part in disease suppression by outcompeting or inhibiting pathogenic organisms, thus reducing root diseases. Plant growth-promoting bacteria (PGPB) and mycoparasitic fungi, such as *Trichoderma*, produce lytic enzymes and bioactive compounds that

 DOI: 10.1201/9781032717814-12

restrict the growth of fungal, bacterial, and viral pathogens and nematodes (Orozco-Mosqueda et al., 2022). Furthermore, rhizobacteria are also known to induce systemic resistance in plants (Van Loon et al., 1998). Rhizobiome also helps plants in stress tolerance, thus increasing their resilience to abiotic stresses such as drought, salinity, and metal contamination (Moraes et al., 2021). Plants and their rhizobiome interact intimately through their respective secretions and signal-detecting compounds emitted into the rhizosphere (Orozco-Mosqueda et al., 2022). The production of organic acids by rhizobacteria influences soil chemistry and nutrient availability by altering soil pH, thus affecting nutrient uptake in plants (Khan et al., 2009). Overall, the rhizobiome optimizes plant growth and health, through many interactions and mechanisms.

Microplastics can alter microbial interactions and enzyme activities, which impacts nutrient cycling and organic matter decomposition (de Souza Machado et al., 2018a). Additionally, plastics can leach chemicals and pollutants into the soil, modifying rhizobiome conditions and potentially increasing plant susceptibility to diseases (Barnes et al., 2009). Disruptions in the rhizobiome can have broader ecological impacts, affecting soil stabilization, carbon cycling, and overall ecosystem health (Nizzetto et al., 2016). As plastics can enter the food chain through plants, understanding the rhizobiome within the plastisphere can help develop strategies to mitigate risks to human health and create effective soil remediation techniques.

12.2 PLASTIC USAGE AND POLLUTION IN TERRESTRIAL ECOSYSTEM

Plastics have become an integral part of today's society, with their usage spanning from industries to day-to-day domestic activities worldwide. Due to their versatility, durability, and cost-effectiveness, plastics are used for a wide range of industrial purposes. As a result, global production of plastics has risen sharply over the decades, driven by increasing consumer demand and industrial applications. According to global statistics, plastic production exceeded 368 million metric tons in 2019, with projections indicating further growth (Plastics Europe, 2024).

Despite this surge in production, researchers and environmentalists have expressed concerns over the environmental impacts of plastics, as they are well-known for their persistence in the environment and their contribution to pollution (Geyer et al., 2017). Despite some proportion of plastic-based waste being recycled, the majority eventually ends up in landfills, causing pollution and other environmental hazards. Countries classified as "developing" or "least developed" often lack proper mechanisms for recycling plastic waste or environmentally friendly disposal methods, leading to significant environmental pollution (Premarathna et al., 2023).

The detrimental effects of plastics on society and the environment have become increasingly significant, posing potential risks to both living organisms and human health (Yu and Flury, 2021; Weerasinghe and Madawala, 2022; Wijesooriya et al., 2023). Plastics endanger aquatic and terrestrial wildlife (Rüthi et al., 2023), contributing to an alarming accumulation of approximately 4.9 billion metric tons in terrestrial and aquatic environments as of 2015, with projections indicating a rise to 12,000 Mt by 2050 (de Souza Machado et al., 2018a). Factors such as the persistence of plastics, their single-use nature, and mismanagement of plastic waste exacerbate this environmental crisis.

The presence of plastics in soils can adversely affect soil properties and biogeochemical cycles (Zhang et al., 2024), underscoring the need for comprehensive strategies to mitigate plastic pollution and its impacts on soil health and ecosystem functions (de Souza Machado et al., 2018a). Terrestrial environments are major sinks for plastic pollution, and 80% of plastics found in aquatic environments originate from land-based sources. However, studies on plastic pollution in terrestrial ecosystems and its consequences remain inadequate (Wright et al., 2020; Joos and De Tender, 2022). Plastics, with their ability to accumulate in the environment due to low recycling rates and resistance to biodegradation, ultimately interact with plants, microbes, and animals (de Souza Machado et al., 2018b; Ya et al., 2021).

12.3 STRUCTURE AND COMPOSITION OF THE PLASTISPHERE RHIZOBIOME

The plastisphere is a novel ecological phenomenon wherein microbial communities colonize and interact with plastic debris in terrestrial and aquatic environments. Initially, the plastisphere was defined as a habitat or an environment, without considering its inhabitants. Accordingly, some used terms such as "plastisphere microbial community" or "plastisphere microbiome" when referring to its inhabitants (MacLean et al., 2021; Sun et al., 2022). According to Rillig and Lehmann (2020), the plastisphere encompasses not just the unique environment but also its inhabiting communities, similar to other ecological "spheres." Therefore, the plastisphere includes not only the microorganisms attached to the plastic surface but also those in the immediate environment that is influenced by the plastic (Rillig et al., 2023). Since its introduction, many studies, mainly on aquatic environments, have shown that microbial communities in the plastisphere differ significantly from those in the surrounding environment and other inert substrates (Kirstein et al., 2019). However, until recently, information on the terrestrial plastisphere was scarce (Rüthi et al., 2020).

The plastisphere rhizobiome, microbial communities associated with plant roots in environments contaminated with plastics, presents a unique intersection of plant–microbe interactions and plastic pollution (Figure 12.1). The composition and structure of the plastisphere rhizobiome are influenced by the presence of varying types of microplastics, which can alter the diversity and function of soil microbes. Microplastics interact intricately with the soil microbiome, which contains a complex network of microorganisms essential for nutrient cycling and maintaining soil health. Studies have shown that microplastics in soils can alter microbial community composition and activity, affecting soil fertility and other ecosystem processes (Tang, 2024; Zhang et al., 2024). Understanding these complex interactions is critical for evaluating the ecological impacts of microplastics on terrestrial ecosystems. Microplastics that enter soils through various pathways can interact with soil microorganisms and their activities in different ways: physically, chemically, and ecologically. These interactions can lead to long-term consequences.

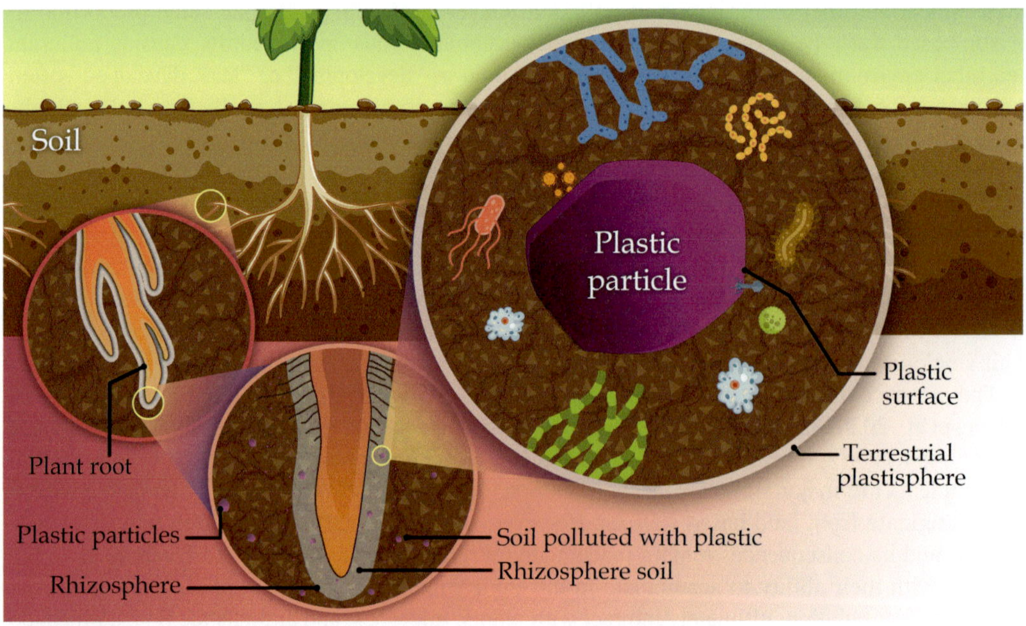

FIGURE 12.1 Illustration to depict the plastisphere rhizobiome of terrestrial ecosystems.

As plastic waste accumulates in both soil and aquatic ecosystems, it undergoes weathering and degradation, providing a substrate for diverse microorganisms. The soil plastisphere seems to differ from the aquatic plastisphere in several ways. For example, the soil plastisphere is less mobile than the aquatic plastisphere. As a result, the soil plastisphere remains static, significantly influencing the immediate environment of the plastic particle. Consequently, its colonization is strongly governed by localized processes compared to the aquatic plastisphere (Rillig et al., 2023). Plastic particles move in the soil environment through bioturbation, ingestion, and excretion by soil fauna, water movement, and human activities (Ren et al., 2021; Yu and Flury, 2021). However, the rate of movement is estimated to be 5 m in 100 years (O'Connor et al., 2019). Therefore, the role of microbes in plastic mobility is negligible, and their movement is insignificant in sampling (Rillig and Lehmann, 2020). However, Tang (2024) noted that the plastisphere can also act as a medium for transporting microorganisms over long distances, similar to the introduction of invasive species.

In recent years, research into the interaction between microplastics and microorganisms has gained prominence (Ran et al., 2024). Microplastics have been found to have detrimental effects on soil biodiversity and microbial community structure by inhibiting microbial propagation, potentially disrupting the overall stability of the soil environment. As microplastics age and decompose, they release toxic and harmful additives (e.g., plasticizers, flame retardants, and antioxidants) into the soil. These additives inhibit the abundance of sensitive microbial groups, thereby altering the composition of soil microorganisms (Ran et al., 2024). Helmberger et al. (2019) concluded that microplastics are a unique anthropogenic pollutant, and their impacts on soil biota and other ecosystem functions strongly depend on the organisms themselves, as they can influence their occurrence, form, and distribution. Despite being considered a soil pollutant, microplastics seem to incur more complex interactions with biota than other organic pollutants, as well as heavy metals and pesticides. Other known pollutants occur in molecular or ionic forms, whereas microplastics exist in varying forms, sizes, shapes, and polymer types (Helmberger et al., 2019). This variability necessitates due attention from researchers to understand their diverse effects. The diverse nature of microplastics makes them unique among pollutants in their interactions with biota. For example, the physicochemical diversity of microplastics infers more complex interactions than those of other chemical pollutants (Huerta Lwanga et al., 2017).

Studies on the soil plastisphere explore the composition and dynamics of the communities, their functional properties, and the soil environmental factors influencing them, paralleling research on aquatic environments (De Tender et al., 2017; Gong et al., 2022). Considering the basic properties of the soil plastisphere, the plastisphere community is generally known to be less diverse than in bulk soils (MacLean et al., 2021; Sun et al., 2022). Despite being less diverse compared to bulk soils, the terrestrial plastisphere hosts a range of microbial communities, with over 1,000 different species of microbes possessing specialized traits and genes (Tang, 2024). In the soil plastisphere, microbial communities influence the degradation of plastics through physical, chemical, and biological processes, determining their own persistence and fate, and subsequently influencing soil health and other ecosystem functions (Hermawan and Astuti, 2021). Studying the soil plastisphere is crucial for understanding the ecological implications of plastic pollution on terrestrial ecosystems. Furthermore, this information will assist in identifying strategies for sustainable waste management and soil conservation practices to mitigate the impacts of plastic pollution.

12.4 MICROBIAL COMMUNITIES IN THE SOIL PLASTISPHERE

Microbial communities in the plastisphere rhizobiome are typically composed of bacteria, fungi, archaea, and protozoa. The types of plastics and their associated pollutants, such as additives, determine the shifts in microbial community composition, favoring microorganisms capable of degrading plastic polymers. For example, bacterial genera such as *Pseudomonas*, *Rhodococcus*, and *Arthrobacter*, known for their biodegrading capability of plastics, could thrive in this ecosystem (Auta et al., 2017). They secrete enzymes that can degrade polymers like polyethylene (PE), polystyrene (PS), and other complex polymer materials, thus playing a crucial role in the biodegradation of plastics in the plastisphere (Shah et al., 2008).

According to other studies, bacterial phyla such as Actinobacteria, cyanobacteria, Proteobacteria, and Deinococci were more prevalent in the plastisphere in agricultural soils compared to bulk soil (Zhang et al., 2019; Luo et al., 2022), forming distinct soil microbial communities in the plastisphere compared to bulk soils (Shi et al., 2022; Li et al., 2023). Studies indicate that microplastics alter the physical properties of soil, including bulk density, water-holding capacity, and water-stable aggregates, thereby affecting microorganisms in soils (de Souza Machado et al., 2018b). Such microplastic-driven changes to soil structure directly impact soil microbial communities, which are the main drivers of nitrogen cycling, carbon processing, and other biogeochemical processes essential for human survival. High-throughput sequencing studies revealed that microplastics in soil may modify the taxonomic composition and biotic interactions within the soil microbiome (Wang et al., 2020; Hou et al., 2021; Rong et al., 2021).

The impacts of microplastics on microbial enzyme activities in soils can be either positive or negative, depending on the type, quantity, and size of microplastics. For example, microplastics such as polyvinyl chloride (PVC) and PE have been found to positively impact the abundance of Burkholderiaceae, as well as phosphatase and urease activities in acidic soils (Wijesooriya et al., 2023). However, increased levels of microplastics have been shown to significantly reduce urease and β-glucosidase activities in soils (Qian et al., 2018). Additionally, the abundance of Patescibacteria increases with the concentration of microplastics (Sun et al., 2022).

Plastisphere microbial communities differ from those colonizing other substrates in soils. Microbes in the plastisphere have shown the potential to degrade plastics and their additives and carry antimicrobial resistance genes (Wright et al., 2020). The bacteria in the terrestrial plastisphere may have different functions compared to those in the aquatic plastisphere. Cyanobacteria, for instance, can engage in symbiotic associations with other heterotrophic bacteria inhabiting the terrestrial plastisphere, thereby facilitating the establishment of other bacteria by making carbon readily available. *Pseudomonas* sp. and *Rhodococcus* sp. are also known to degrade plastics. Parasitic protists like *Spongospora* and *Anurofeca* are also known to be present in the plastisphere (Tang, 2024).

Soil fungal genera like *Alternaria*, *Penicillium*, and *Aspergillus*, which metabolize polyurethane (PU), warrant greater interest among scientists for their ability to degrade plastics and their potential metabolic pathways. Species of the genus *Mycosphaerella* are known to produce a wide variety of polymer-degrading enzymes. Burkholderiales possess a range of enzymes capable of degrading polyaromatic hydrocarbons (PAHs), considered soil pollutants (Rüthi et al., 2020).

The plastisphere is also enriched with pathogenic microbial strains that carry antibiotic-resistance genes (Keswani et al., 2016; Laganà et al., 2019; Su and Yang, 2023). In addition to being a terrestrial reservoir of potential pathogens subject to continuous change, the plastisphere microbiome has also shown the potential for emitting the greenhouse gas methane (Rohrbach et al., 2022).

The presence of plastics in soils not only influences microbial composition but also affects their functions. Microplastics alter enzyme activities involved in nutrient cycling and organic matter decomposition (Ng et al., 2018; Rillig et al., 2019). Enzymes such as cellulases, ligninases, and proteases are crucial for breaking down organic matter and releasing nutrients for plant growth. Thus, any changes to these enzymes and their activities due to plastic pollution can have downstream effects on soil health and plant growth (de Souza Machado et al., 2018b).

In addition, plastic pollution can impact microbial interactions within the rhizobiome. The presence of microplastics creates new niches for microbes, promoting the exchange of genetic material, including genes involved in the degradation of plastics (Rüthi et al., 2023). Horizontal gene transfers between microbes in the plastisphere rhizobiome can enhance their potential to degrade plastics and adapt to polluted environments.

Plastics also alter soil characteristics such as soil structure, water retention, and nutrient availability, impacting microbial communities and leading to shifts in microbial diversity, abundance, and function (Rillig, 2012; Zhang et al., 2018). Moreover, plastics introduce harmful chemicals into the soil through additives and other pollutants, which can be toxic to soil microorganisms (Teuten et al., 2009). These contaminants can influence the composition and metabolic activities of the plastisphere rhizobiome, reducing soil functionality and compromising ecosystem services.

Overall, plastic pollution in soils presents a multifaceted challenge that affects both microbial communities and their crucial functions. A more detailed understanding of how plastics interact with soil systems is essential for advancing research in this field. Similarly, understanding the composition and structure of the plastisphere rhizobiome is vital for developing strategies to mitigate the impacts of plastic pollution on plant growth and pave the way for bioremediation approaches to harness the natural potential of microbes to clean up contaminated soils (Li et al., 2024).

12.5 RHIZOBIOME ASSOCIATED WITH PLASTISPHERE

As plants extend their roots, they create a new ecosystem known as the rhizosphere, which is the part of the soil influenced and strongly attached to the roots of the plants. The rhizosphere's size, thickness, and extension depend more on the type of root system of each plant than on soil conditions (Orozco-Mosqueda et al., 2022). The collection of microorganisms interacting with plant roots is known as the rhizosphere microbiome, rhizomicrobiome, or rhizobiome (Orozco-Mosqueda et al., 2022). The rhizosphere microbiome, often referred to as the "second largest genome" of plants, plays a crucial role in detoxification and remediation of pollutants. Despite the fact that rhizobiome has many positive influences on plant growth and health, its structure and functions are less known (Ran et al., 2024).

12.6 IMPACT OF PLASTIC POLLUTION ON THE RHIZOBIOME

The increasing prevalence of plastic pollution urges scientists to understand how plastic contamination affects the rhizobiome. Plastic-contaminated soils present unique challenges and opportunities for microbes and their interactions, especially in the rhizosphere, providing a favorable environment for certain microbes to thrive, while hindering others.

1. **Impact on Soil Chemistry and Microbial Interactions:**
   Plastics influence the physical and chemical properties of the soil environment, affecting the rhizobiome (Figure 12.2). Microplastics disrupt soil structure, moisture levels, and nutrient availability (Rillig, 2012), subsequently influencing the composition and functions of microbial communities within the rhizosphere. Soil pH, a major driver of soil bacterial communities, might change the surface properties of plastics or the availability of other contaminants, such as heavy metals, present in soils (Rillig et al., 2023). These changes can be compounded by agricultural activities (e.g., manure addition and fertilization), soil contamination levels (e.g., antibiotics and metals), different source communities (e.g., plant litter and rhizosphere), and different sampling locations (e.g., ecosystem type and contamination levels), driving changes to the composition of plastisphere microbial communities (Zhu et al., 2021; Rillig et al., 2023).
2. **Shift in Microbial Populations:**
   Plastic contamination increases certain bacterial populations well-adapted to plastic-rich soils, potentially decreasing the prevalence of beneficial microbes for plant growth. For instance, fungi and bacteria capable of degrading plastics might become more prevalent, while crucial plant growth-promoting microbes might be outcompeted or diminished (Shen et al., 2022). These changes can disrupt the population balance of microbes, compromising the rhizobiome's ability to enhance plant growth and stress tolerance (Rillig et al., 2019).

3. **Effect on Symbiotic Relationships:**
   Plastic pollution can impact beneficial symbiotic relationships, such as those between plants and nitrogen-fixing bacteria, *Rhizobium* spp. Changes in soil chemistry and microbial interactions caused by plastic pollution can affect the efficiency of these beneficial symbiotic relationships, leading to reduced soil fertility and plant growth (Rillig, 2012). Also, the presence of microplastics creates new niches for microbes that are involved in degradation of plastics. The horizontal gene transfer ability between microbes housed within this microenvironment enhances their potential to degrade plastics and adapt themselves to polluted environment. Furthermore, changes in enzyme activities due to plastic pollution can also have a downstream effect on nutrient cycling and organic matter decomposition (de Souza Machado et al., 2018a; Rillig et al., 2019).

12.7 FACTORS INFLUENCING THE RHIZOBIOME IN THE TERRESTRIAL PLASTISPHERE

Soil properties such as moisture and pH are pivotal in shaping the characteristics of the soil plastisphere community (Figure 12.2). For instance, arable soils have shown that pH significantly affects bacterial community composition within the plastisphere. Soil pH not only influences the microbial community but can also alter the surface properties of microplastics and affect the availability of associated contaminants, such as heavy metals, present in the soil (Rillig et al., 2023).

Fertilizer and manure application modify soil nutrient levels, moisture, and pH, which in turn can affect the microbial communities associated with microplastics (Zhu et al., 2021). Furthermore, pesticides and herbicides may also lead to inhibit some microorganisms, leading to shifts in microbial diversity and their interactions with microplastics (Rillig et al., 2023). Different ecosystem types and contamination levels have been found to influence plastisphere microbial composition (Zhu et al., 2021; Rillig et al., 2023).

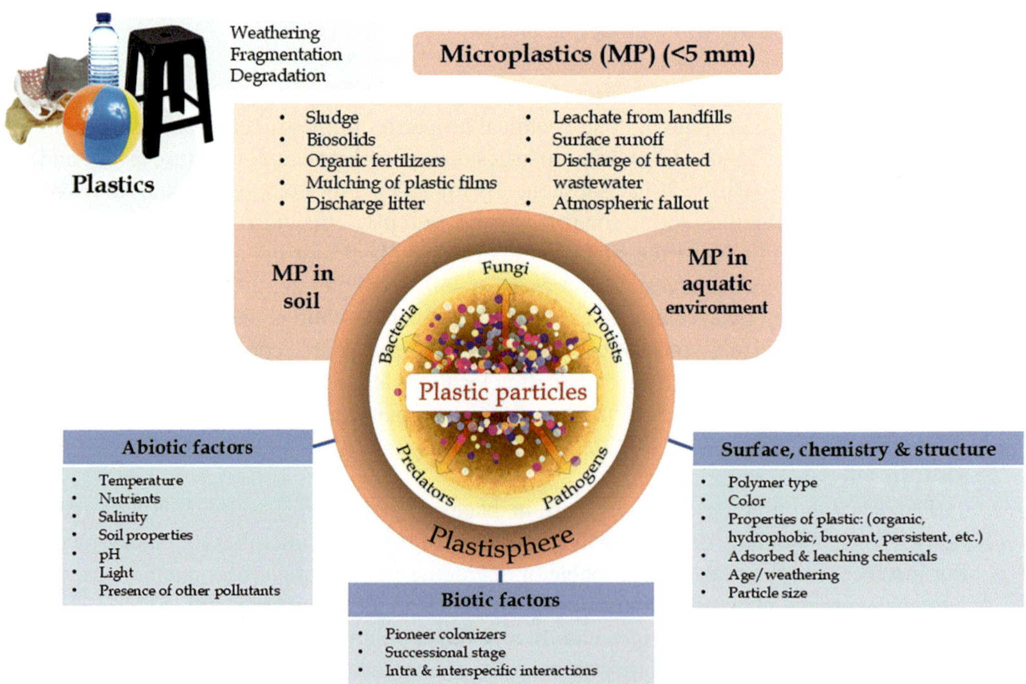

FIGURE 12.2 Factors influencing the rhizobiome of the terrestrial plastisphere.

Unique characteristics of different plastispheres, combined with localized soil environmental conditions, drive the formation and shaping of soil plastispheres and their microbial communities. Climate change and weather conditions are crucial extrinsic factors affecting the rhizobiome of the terrestrial plastisphere. Climate change, known to affect soil microbial diversity, structure, and activities, might also impact plastisphere characteristics, though the specific mechanisms remain unclear (Sun et al., 2022). The dynamic nature of the soil plastisphere, distinguishing it from its aquatic counterpart, plays a crucial role in biogeochemistry (Su and Yang, 2023). Climate change-induced extreme weather patterns can lead to long-term changes in soil properties, influencing plastisphere microbial communities over time (Su and Yang, 2023).

Plastics, composed of organic carbon compounds, tend to absorb organic matter from their surroundings, providing nutrients for microorganisms. However, as plastics weather, they can release harmful compounds that potentially affect photosynthesis and community composition. Contaminants can alter soil pH and nutrient availability, thus influencing the structure and function of microbial communities. Due to the varying polymers, additives, and aging processes, the plastisphere has become complex and diverse. Therefore, the assembly of plastisphere microbial communities is determined by a combination of these intrinsic characteristics and external environmental factors (Li et al., 2024). The microenvironmental conditions within the plastisphere are influenced by both the substrate and the physicochemical properties of the surrounding environment, fostering the recruitment of diverse microbial communities (Li et al., 2021, 2024).

Vegetation cover and its density also influence the rhizobiome of terrestrial plastispheres. Different plants exude various organic compounds into the soil, affecting microbial communities. These root-derived substances interact with microplastics, shaping their resident microbial communities (Li et al., 2024). These exudates also affect soil properties such as soil structure, pH, and soil moisture, further impacting microbial communities associated with microplastics (Rillig et al., 2023).

Urbanization and human activities are crucial extrinsic factors influencing the rhizobiome of the terrestrial plastisphere. Construction activities, pollution, and waste disposal often create altered soil properties, thus affecting the composition and abundance of microorganisms colonizing microplastics. These unique changes create distinct plastisphere microbial communities compared to natural and agricultural settings (Zhang et al., 2019; Li et al., 2024).

In conclusion, the rhizobiome of the terrestrial plastisphere is generally shaped by a complex interplay of external factors, including land use, soil characteristics, climate change, vegetation types, and human activities. All these factors directly or indirectly influence microbial community dynamics and interactions with microplastics.

12.8 EFFECTS OF MICROPLASTIC POLLUTION ON THE RHIZOBIOME

The increasing prevalence of microplastic pollution in terrestrial environments has a profound effect on the rhizobiome, as they can significantly modify microbial community composition and functions through modified soil characteristics. The introduction of microplastics into soil environments alters the ecological niches and substrates for microbial communities, such as Cyanobacteria and Basidiomycota, transitioning from rhizosphere soil to microplastic surfaces, or the "microplastisphere." This shift impacts biogeochemical processes, including biogeochemical cycles, by modifying enzymatic activities and ecological functions, including plant–microbial symbiotic associations. Additionally, interactions between microplastics and microbes can influence biogeochemical cycles through mechanisms such as leaching of dissolved organic carbon (DOC) and altered denitrification processes. Microplastics also introduce physiological stress on microorganisms due to their toxic effects (Wijesooriya et al., 2023). Research indicates that the presence of biodegradable polyhydroxyalkanoates (PHAs) not only alters soil microbial communities but also enhances microbial biomass activity within the plastisphere. This includes increased activities of enzymes such as β-glucosidase and leucine aminopeptidase, alongside reduced enzyme affinity, which contributes to higher carbon and nutrient turnover in the soil (Joos and De Tender, 2022).

Microplastic pollution tends to change microbial community structure in the rhizosphere. As plastics provide new substrates and ecological niches, microbial diversity and composition of microbial communities often tend to modify. Studies have consistently found differences in the taxonomic composition between plastisphere and bulk soil microbial communities. For instance, bacterial phyla such as Actinobacteria, Bacteroidetes, Cyanobacteria, Deinococci, Proteobacteria, and Verrucomicrobia are often enriched in the plastisphere compared to bulk soil (Shi et al., 2022; Li et al., 2023). Notably, plastic-associated microbial communities are often dominated by biofilm-forming, hydrophobic-tolerant autotrophic bacteria (Rillig et al., 2023). Field experiments in China's maize fields, characterized by a continental monsoon climate, revealed higher abundances of Acidobacteria, Alphaproteobacteria, and Firmicutes in PE plastisphere samples compared to other soil environments (Shi et al., 2022). Similarly, in agricultural fields in Hebei Province, China, using biodegradable plastics (polybutylene adipate terephthalate—PBAT and polylactic acid—PLA), Actinobacteria and Proteobacteria were notably enriched in the plastisphere (Li et al., 2023). In Germany, distinct bacterial communities were observed on PP and expanded PS microplastic particles compared to bulk soil (Kublik et al., 2022). Further research utilizing field-collected microplastics (PE, PBAT, and PLA) from a humid monsoon climate in northern China has demonstrated that biodegradable plastics can enrich bacterial taxa capable of degrading plastic polymers, such as Proteobacteria and Actinobacteria. In addition to bacterial communities, microplastics also can incur substantial changes in the community structure and diversity of soil fungal communities too (Hou et al., 2021). Fungal genera like *Aspergillus, Fusarium*, and *Penicillium* have shown tendency to utilize complex C sources that are derived from microplastics (Accinelli et al., 2020), thus showing their preference in plastic-polluted soils. Furthermore, studies also revealed that the relative abundance of mycorrhizal hyphae and other structures significantly increased in soils amended with polyester (de Souza Machado et al., 2019). Rillig et al. (2023) also observed a higher abundance of fungi belonging to Ascomycota in soils polluted with plastic.

While bacteria have been the primary focus of research, fungi also play a significant role in the soil plastisphere, potentially more so than in aquatic environments. Experiments have revealed the presence of fungal assemblages on plastic debris, with Ascomycota found to be more abundant in the plastisphere than in bulk soil (Rillig et al., 2023). Mycorrhizal fungi, which form symbiotic relationships with plants, may extend their influence from the rhizosphere into the plastisphere. These fungi can transport chemicals and organisms through their hyphae into the bulk soil, and their activity appears to be stimulated by the presence of microplastics (Rillig et al., 2023).

Additionally, it is hypothesized that biodegradable plastics, such as PLA, induce more significant changes in microbial community structure compared to nondegradable plastics like PE (Joos and De Tender, 2022). Recent findings suggest that biodegradable PHAs not only alter soil microbial communities but also increase microbial biomass activity with higher enzyme activities and lower enzyme affinity, driving greater carbon and nutrient turnover in the soil (Joos and De Tender, 2022).

The presence of microplastics can also affect key biogeochemical processes through plastic-driven changes to microbial enzymatic activities. For example, the activity of enzymes such as ß-glucosidase and leucine aminopeptidase is often increased in the plastisphere, thereby increasing the carbon and nutrient turnover in soils (Joos and De Tender, 2022). On the other hand, microplastic also hinders essential nutrient transformations like denitrification, influencing the nitrogen cycle and greenhouse emissions.

The interactions between microbes and microplastics can also influence soil properties as well as microbial activities. Microplastics absorb pollutants such as heavy metals and pesticides, which can also affect microbial physiology and dynamics. These interactions impose physiological stress on microorganisms, potentially influencing their growth and other metabolic activities. Others observed that the addition of microplastics has enhanced the relative abundance of nitrogen-fixing bacteria, particularly those belonging to the families, Burkholderiaceae and Xanthobacteriaceae (Fei et al., 2020; Yi et al., 2021; Riveros et al., 2021).

The long-term effects of microplastic pollution on the rhizobiome and soil ecosystems are imperative to understand how microplastic contamination and its persistence will influence plant and soil health, and ecosystem services over time.

12.9 RESEARCH GAPS AND FUTURE DIRECTIONS

While numerous studies have explored the composition and diversity of plastisphere microbial communities in aquatic ecosystems, limited data are available on plastisphere microbial communities in soil ecosystems and, in particular, their influence on the rhizobiome. The mechanisms shaping the community diversity and composition of the plastisphere in soils also remain unclear (Sun et al., 2022). As "plastisphere" is a relatively new term, the understanding of plastisphere microbial communities remains insufficient, particularly regarding their impact on interactions within soil–plant systems and the underlying mechanisms influencing microbial communities in microplastic-contaminated rhizosphere soils (Ran et al., 2024). Despite available studies to date, a significant knowledge gap persists concerning microbial communities in the soil plastisphere across different terrestrial ecosystems such as agricultural, urban, and forest soils. Moreover, as the rhizobiome plays a notable role in plant health and the rhizosphere, it is crucial to investigate the mechanisms and activities between the rhizobiome and the terrestrial plastisphere (Su and Yang, 2023).

Addressing challenges put forth by plastic contamination in terrestrial ecosystems requires a multifaceted approach. Understanding on interactions between plastics and the rhizobiome can help in developing strategies to mitigate the adverse effects of plastic pollution on plant and soil health. Bioremediation techniques to degrade plastics in soils and developing more resilient crop varieties to plastic-induced changes in the rhizobiome can be potential approaches to tackle this issue. Thus, future research should focus on understanding mechanisms through which plastic pollution affects the composition and functionality of the rhizobiome. The following are some of the areas that need further research;

1. Specific mechanisms that drive the abundance, diversity, and composition of microbial communities in the soil plastisphere.
2. Influence of soil types and other environmental factors on plastisphere microbial communities.
3. How plastisphere microbial communities interact with and affect soil–plant systems to influence plant health and growth, and their potential beneficial and detrimental impacts on plants.
4. Identify and characterize the functions of major microbial species in the rhizobiome of the terrestrial plastisphere and analyze their metabolites and their effects on soil and plant health.
5. Long-term effects of microplastics on soil and plant health, and potential changes to microbial community structure and function over time.
6. Bioremediation potential of specific microbial species within the rhizobiome, and genetic and enzymatic pathways involved in the biodegradation process of microplastics.
7. Interaction of microplastics with other soil pollutants, and how these interactions influence microbial community dynamics in terrestrial ecosystems.
8. Implications of microplastic contamination through agricultural practices, and suggest strategies to mitigate the adverse effects of microplastics on crop production and soil fertility status.
9. Use of more advanced technologies, such as metagenomics, metatranscriptomics, and metabolomics, to gain knowledge on microbial community structure and function when analyzing samples from soil plastisphere.

10. Finally, evidence-based policy recommendations are imperative to manage microplastic contamination in terrestrial ecosystems through promoting sustainable agricultural and land management practices, and safe waste disposal methods to minimize microplastic pollution.

ACKNOWLEDGMENT

The authors would like to acknowledge Mr. Punsara Dharaka for preparing the two illustrations for the chapter.

REFERENCES

Accinelli, C., Abbas, H.K., Bruno, V., Nissen, L., Vicari, A., Bellaloui, N., Little, N.S. and Thomas Shier, W. (2020). Persistence in soil of microplastic films from ultra-thin compostable plastic bags and implications on soil Aspergillus flavus population. *Waste Management*, 113, 312–318. https://doi.org/10.1016/j.wasman.2020.06.011

Auta, H.S., Emenike, C.U. and Fauziah, S.H. (2017). Distribution and importance of microplastics in the marine environment: A review of the sources, fate, effects, and potential solutions. *Environment International*, 102, 165–176. https://doi.org/10.1016/j.envint.2017.02.013

Barnes, D.K.A., Galgani, F., Thompson, R.C. and Barlaz, M. (2009). Accumulation and fragmentation of plastic debris in global environments. *Philosophical Transactions of the Royal Society B: Biological Sciences*, 364(1526), 1985–1998. https://doi.org/10.1098/rstb.2008.0205

de Souza Machado, A.A., Kloas, W., Zarfl, C., Hempel, S. and Rillig, M.C. (2018a). Microplastics as an emerging threat to terrestrial ecosystems. *Global Change Biology*, 24(4), 1405–1416. https://doi.org/10.1111/gcb.14020

de Souza Machado, A.A., Lau, C.W., Kloas, W., Bergmann, J., Bachelier, J.B., Faltin, E., Becker, R., Görlich, A.S. and Rillig, M.C. (2019). Microplastics can change soil properties and affect plant performance. *Environmental Science & Technology*, 53(10), 6044–6052. https://doi.org/10.1021/acs.est.9b01339

de Souza Machado, A.A., Lau, C.W., Till, J., Kloas, W., Lehmann, A., Becker, R. and Rillig, M.C. (2018b). Impacts of microplastics on the soil biophysical environment. *Environmental Science & Technology*, 52(17), 9656–9665. https://doi.org/10.1021/acs.est.8b02212

De Tender, C., Devriese, L.I., Haegeman, A., Maes, S., Vangeyte, J., Cattrijsse, A., Dawyndt, P. and Ruttink, T. (2017). Temporal dynamics of bacterial and fungal colonization on plastic debris in the North Sea. *Environmental Science & Technology*, 51(13), 7350–7360. https://doi.org/10.1021/acs.est.7b00697

Fei, Y., Huang, S., Zhang, H., Tong, Y., Wen, D., Xia, X., Wang, H., Luo, Y. and Barceló, D. (2020). Response of soil enzyme activities and bacterial communities to the accumulation of microplastics in an acid cropped soil. *Science of the Total Environment*, 707, 135634. https://doi.org/10.1016/j.scitotenv.2019.135634

Geyer, R., Jambeck, J.R. and Law, K.L. (2017). Production, use, and fate of all plastics ever made. *Science Advances*, 3(7). https://doi.org/10.1126/sciadv.1700782

Gong, W., Xing, Y., Han, L., Lu, A., Qu, H. and Xu, L. (2022). Occurrence and distribution of micro- and mesoplastics in the high-latitude nature reserve, northern China. *Frontiers of Environmental Science & Engineering*, 16(9). https://doi.org/10.1007/s11783-022-1534-7

Helmberger, M.S., Tiemann, L.K. and Grieshop, M.J. (2019). Towards an ecology of soil microplastics. *Functional Ecology*. https://doi.org/10.1111/1365-2435.13495

Hermawan, S. and Astuti, W. (2021). Analysing several ASEAN countries' policy for combating marine plastic litter. *Environmental Law Review*, 23(1), 9–22. https://doi.org/10.1177/1461452921991731

Hou, L., McMahan, C.D., McNeish, R.E., Munno, K., Rochman, C.M. and Hoellein, T.J. (2021). A fish tale: A century of museum specimens reveal increasing microplastic concentrations in freshwater fish. *Ecological Applications*. https://doi.org/10.1002/eap.2320

Huerta Lwanga, E., Mendoza Vega, J., Ku Quej, V., Chi, J. de los A., Sanchez del Cid, L., Chi, C., Escalona Segura, G., Gertsen, H., Salánki, T., van der Ploeg, M., Koelmans, A.A. and Geissen, V. (2017). Field evidence for transfer of plastic debris along a terrestrial food chain. *Scientific Reports*, 7(1). https://doi.org/10.1038/s41598-017-14588-2

Joos, L. and De Tender, C. (2022). Soil under stress: The importance of soil life and how it is influenced by (micro)plastic pollution. *Computational and Structural Biotechnology Journal*, 20, 1554–1566. https://doi.org/10.1016/j.csbj.2022.03.041

Keswani, A., Oliver, D.M., Gutierrez, T. and Quilliam, R.S. (2016). Microbial hitchhikers on marine plastic debris: Human exposure risks at bathing waters and beach environments. *Marine Environmental Research*, 118, 10–19. https://doi.org/10.1016/j.marenvres.2016.04.006

Khan, M.S., Zaidi, A. and Wani, P.A. (2009). Role of organic acids in mineral nutrition of plants. *Plant and Soil,* 321(1–2), 165–185. https://doi.org/10.1007/s11104-008-9708-7

Kirstein, I.V., Wichels, A., Gullans, E., Krohne, G. and Gerdts, G. (2019). The plastisphere – Uncovering tightly attached plastic 'specific' microorganisms. *PLoS One*, 14(4), e0215859. https://doi.org/10.1371/journal.pone.0215859

Kublik, S., Gschwendtner, S., Magritsch, T., Radl, V., Rillig, M.C. and Schloter, M. (2022). Microplastics in soil induce a new microbial habitat, with consequences for bulk soil microbiomes. *Frontiers in Environmental Science*, 10. https://doi.org/10.3389/fenvs.2022.989267

Laganà, P., Caruso, G., Corsi, I., Bergami, E., Venuti, V., Majolino, D., La Ferla, R., Azzaro, M. and Cappello, S. (2019). Do plastics serve as a possible vector for the spread of antibiotic resistance? First insights from bacteria associated to a polystyrene piece from King George Island (Antarctica). *International Journal of Hygiene and Environmental Health*, 222(1), 89–100. https://doi.org/10.1016/j.ijheh.2018.08.009

Li, C., Wang, L., Ji, S., Chang, M., Wang, L., Gan, Y. and Liu, J. (2021). The ecology of the plastisphere: Microbial composition, function, assembly, and network in the freshwater and seawater ecosystems. *Water Research*, 202, 117428. https://doi.org/10.1016/j.watres.2021.117428

Li, K., Jia, W., Xu, L., Zhang, M. and Huang, Y. (2023). The plastisphere of biodegradable and conventional microplastics from residues exhibit distinct microbial structure, network and function in plastic-mulching farmland. *Journal of Hazardous Materials*, 442, 130011. https://doi.org/10.1016/j.jhazmat.2022.130011

Li, K., Xu, L., Bai, X., Zhang, G., Zhang, M. and Huang, Y. (2024). Differential fungal assemblages and functions between the plastisphere of biodegradable and conventional microplastics in farmland. *Science of the Total Environment*, 906, 167478–167478. https://doi.org/10.1016/j.scitotenv.2023.167478

Liu, P., Zhan, X., Wu, X., Li, J., Wang, H. and Gao, S. (2020). Effect of weathering on environmental behavior of microplastics: Properties, sorption and potential risks. *Chemosphere*, 242, 125193. https://doi.org/10.1016/j.chemosphere.2019.125193

Luo, G., Jin, T., Zhang, H., Peng, J., Zuo, N., Huang, Y., Han, Y., Tian, C., Yang, Y., Peng, K. and Fei, J. (2022). Deciphering the diversity and functions of plastisphere bacterial communities in plastic-mulching croplands of subtropical China. *Journal of Hazardous Materials*, 422, 126865. https://doi.org/10.1016/j.jhazmat.2021.126865

Lugtenberg, B. and Kamilova, F. (2009). Plant-growth-promoting rhizobacteria. *Annual Review of Microbiology*, 63, 541–556. https://doi.org/10.1146/annurev.micro.62.081307.162918

MacLean, J., Mayanna, S., Benning, L.G., Horn, F., Bartholomäus, A., Wiesner, Y., Wagner, D. and Liebner, S. (2021). The terrestrial plastisphere: Diversity and polymer-colonizing potential of plastic-associated microbial communities in soil. *Microorganisms*, 9(9), 1876. https://doi.org/10.3390/microorganisms9091876

Moraes, J.D., de Souza, R.S. and Silva, C.C. (2021). Plant growth-promoting rhizobacteria and their role in improving plant resistance to abiotic stress. *Journal of Soil Science and Plant Nutrition*, 21(3), 1376–1394. https://doi.org/10.1007/s42729-021-00615-3

Ng, E.-L., Huerta Lwanga, E., Eldridge, S.M., Johnston, P., Hu, H.-W., Geissen, V. and Chen, D. (2018). An overview of microplastic and nanoplastic pollution in agroecosystems. *Science of the Total Environment*, 627, 1377–1388. https://doi.org/10.1016/j.scitotenv.2018.01.341

Nizzetto, L., Futter, M. and Langaas, S. (2016). Are agricultural soils dumps for microplastics of urban origin? *Environmental Science & Technology*, 50(20), 10777–10779. https://doi.org/10.1021/acs.est.6b04140

O'Connor, D., Pan, S., Shen, Z., Song, Y., Jin, Y., Wu, W.-M. and Hou, D. (2019). Microplastics undergo accelerated vertical migration in sand soil due to small size and wet-dry cycles. *Environmental Pollution*, 249, 527–534. https://doi.org/10.1016/j.envpol.2019.03.092

Orozco-Mosqueda, M. del C., Fadiji, A.E., Babalola, O.O., Glick, B.R. and Santoyo, G. (2022). Rhizobiome engineering: Unveiling complex rhizosphere interactions to enhance plant growth and health. *Microbiological Research*, 263, 127137. https://doi.org/10.1016/j.micres.2022.127137

Plastics Europe. (2024). *Plastics Europe Enabling a sustainable future*. Available at: https://www.plasticseurope.org/en/resources/publications/4313-plastics-facts-2020 [Accessed 3 August 2024].

Premarathna, K., Abeysundara, S.P., Gunaratne, A.M.T.A. and Madawala, H.M.S.P. (2023). Community awareness and perceptions on microplastics: A case study from Sri Lanka. *Ceylon Journal of Science*, 52(4), 425–432. https://doi.org/10.4038/cjs.v52i4.8134

Qian, H., Zhang, M., Liu, G., Lu, T., Qu, Q., Du, B. and Pan, X. (2018). Effects of soil residual plastic film on soil microbial community structure and fertility. *Water, Air, & Soil Pollution*, 229(8). https://doi.org/10.1007/s11270-018-3916-9

Quilliam, R.S., Pow, C.J., Shilla, D.A., Mwesiga, J.J., Shilla, D.A. and Woodford, L. (2023). Microplastics in agriculture – A potential novel mechanism for the delivery of human pathogens onto crops. *Frontiers in Plant Science*, 14. https://doi.org/10.3389/fpls.2023.1152419

Ran, T., Liao, H., Zhao, Y. and Li, J. (2024). Soil plastisphere interferes with soil bacterial community and their functions in the rhizosphere of pepper (Capsicum annuum L.). *Ecotoxicology and Environmental Safety*, 270, 115946–115946. https://doi.org/10.1016/j.ecoenv.2024.115946

Ren, Z., Gui, X., Xu, X., Zhao, L., Qiu, H. and Cao, X. (2021). Microplastics in the soil-groundwater environment: Aging, migration, and co-transport of contaminants – A critical review. *Journal of Hazardous Materials*, 419, 126455. https://doi.org/10.1016/j.jhazmat.2021.126455

Rillig, M.C. (2012). Microplastic in terrestrial ecosystems and the soil? *Environmental Science & Technology*, 46(12), 6453–6454. https://doi.org/10.1021/es302011r

Rillig, M.C., Kim, S.W. and Zhu, Y.-G. (2023). The soil plastisphere. *Nature Reviews Microbiology*, 22(2), 64–74. https://doi.org/10.1038/s41579-023-00967-2

Rillig, M.C. and Lehmann, A. (2020). Microplastic in terrestrial ecosystems. *Science*, 368(6498), 1430–1431. https://doi.org/10.1126/science.abb5979

Rillig, M.C., Lehmann, A., Souza Machado, A.A. and Yang, G. (2019). Microplastic effects on plants. *New Phytologist*, 223(3). https://doi.org/10.1111/nph.15794

Riveros, G., Urrutia, H., Araya, J., Zagal, E. and Schoebitz, M. (2021). Microplastic pollution on the soil and its consequences on the nitrogen cycle: A review. *Environmental Science and Pollution Research*, 29(6), 7997–8011. https://doi.org/10.1007/s11356-021-17681-2

Rohrbach, S., Gkoutselis, G., Hink, L., Weig, A.R., Obst, M., Diekmann, A., Ho, A., Rambold, G. and Horn, M.A. (2022). Microplastic polymer properties as deterministic factors driving terrestrial plastisphere microbiome assembly and succession in the field. *Environmental Microbiology*. https://doi.org/10.1111/1462-2920.16234

Rong, L., Zhao, L., Zhao, L., Cheng, Z., Yao, Y., Yuan, C., Wang, L. and Sun, H. (2021). LDPE microplastics affect soil microbial communities and nitrogen cycling. *Science of the Total Environment*, 773, 145640. https://doi.org/10.1016/j.scitotenv.2021.145640

Royer, S.-J., Ferrón, S., Wilson, S.T. and Karl, D.M. (2018). Production of methane and ethylene from plastic in the environment. *PLoS One*, 13(8), e0200574. https://doi.org/10.1371/journal.pone.0200574

Rüthi, J., Bölsterli, D., Pardi-Comensoli, L., Brunner, I. and Frey, B. (2020). The 'plastisphere' of biodegradable plastics is characterized by specific microbial taxa of alpine and arctic soils. *Frontiers in Environmental Science*, 8. https://doi.org/10.3389/fenvs.2020.562263

Rüthi, J., Cerri, M., Brunner, I., Beat Stierli, Sander, M. and Frey, B. (2023). Discovery of plastic-degrading microbial strains isolated from the alpine and Arctic terrestrial plastisphere. *Frontiers in Microbiology*, 14. https://doi.org/10.3389/fmicb.2023.1178474

Shah, A.A., Hasan, F., Hameed, A. and Ahmed, S. (2008). Biological degradation of plastics: A comprehensive review. *Biotechnology Advances*, 26(3), 246–265. https://doi.org/10.1016/j.biotechadv.2007.12.005

Shen, M., Xiong, W., Song, B., Zhou, C., Almatrafi, E., Zeng, G. and Zhang, Y. (2022). Microplastics in landfill and leachate: Occurrence, environmental behavior and removal strategies. *Chemosphere*, 305, 135325. https://doi.org/10.1016/j.chemosphere.2022.135325

Shi, Z., Xiong, L., Liu, T. and Wu, W. (2022). Alteration of bacterial communities and co-occurrence networks as a legacy effect upon exposure to polyethylene residues under field environment. *Journal of Hazardous Materials*, 426, 128126. https://doi.org/10.1016/j.jhazmat.2021.128126

Su, X. and Yang, K. (2023). Exploring the blindspot: The soil plastisphere. *Soil Ecology Letters*, 6(2). https://doi.org/10.1007/s42832-023-0209-z

Sun, Y., Shi, J., Wang, X., Ding, C. and Wang, J. (2022). Deciphering the mechanisms shaping the plastisphere microbiota in soil. *MSystems*, 7(4). https://doi.org/10.1128/msystems.00352-22

Tang, D. (2024). Terrestrial and aquatic plastisphere: Formation, characteristics, and influencing factors. *Sustainability*, 16(5), 2163–2163. https://doi.org/10.3390/su16052163

Teuten, E.L., Saquing, J.M., Knappe, D.R.U., Barlaz, M.A., Jonsson, S., Björn, A., Rowland, S.J., Thompson, R.C., Galloway, T.S., Yamashita, R., Ochi, D., Watanuki, Y., Moore, C., Viet, P.H., Tana, T.S., Prudente, M., Boonyatumanond, R., Zakaria, M.P., Akkhavong, K. and Ogata, Y. (2009). Transport and release of chemicals from plastics to the environment and to wildlife. *Philosophical Transactions of the Royal Society B: Biological Sciences*, 364(1526), 2027–2045. https://doi.org/10.1098/rstb.2008.0284

Van Loon, L.C., Bakker, P.A.H.M. and Pieterse, C.M.J. (1998). Systemic resistance induced by rhizosphere bacteria. *Annual Review of Phytopathology*, 36, 453–483. https://doi.org/10.1146/annurev.phyto.36.1.453

Wang, J., Liu, X., Dai, Y., Ren, J., Li, Y., Wang, X., Zhang, P. and Peng, C. (2020). Effects of co-loading of polyethylene microplastics and ciprofloxacin on the antibiotic degradation efficiency and microbial community structure in soil. *Science of the Total Environment*, 741, 140463. https://doi.org/10.1016/j.scitotenv.2020.140463

Weerasinghe, W.M.T.M. and Madawala, H.M.S.P. (2022). Potential impacts of two types of microplastics on *Solanum lycopersicum* L. and arbuscular mycorrhizal fungi. *Ceylon Journal of Science*, 51(2). https://doi.org/10.4038/cjs.v51i2.8008

Wijesooriya, M., Wijesekara, H., Sewwandi, M., Soysa, S., Rajapaksha, A.U., Vithanage, M. and Bolan, N. (2023). Microplastics and soil nutrient cycling. In Vithanage, M. and Prasad, M.N.V. (Eds.), *Microplastics in the Ecosphere: Air, Water, Soil, and Food*. John Wiley & Sons, Inc., Hoboken, NJ, pp. 321–338. https://doi.org/10.1002/9781119879534.ch19

Wright, R.J., Langille, M.G.I. and Walker, T.R. (2020). Food or just a free ride? A meta-analysis reveals the global diversity of the Plastisphere. *The ISME Journal*, 15(3), 789–806. https://doi.org/10.1038/s41396-020-00814-9

Ya, H., Jiang, B., Xing, Y., Zhang, T., Lv, M. and Wang, X. (2021). Recent advances on ecological effects of microplastics on soil environment. *Science of the Total Environment*, 798, 149338. https://doi.org/10.1016/j.scitotenv.2021.149338

Yi, Y.Z., Azman, S., Primus, A., Said, M.I.M. and Abideen, M.Z., 2021. Microplastic ingestion by crabs. *The 6th Proceeding of Civil Engineering. School of Civil Engineering,* Johor, Malaysia, pp. 363–375.

Yu, Y. and Flury, M. (2021). Current understanding of subsurface transport of micro- and nanoplastics in soil. *Vadose Zone Journal*, 20(2). https://doi.org/10.1002/vzj2.20108

Zhang, K., Shi, H., Peng, J., Wang, Y., Xiong, X., Wu, C. and Lam, P.K.S. (2018). Microplastic pollution in China's inland water systems: A review of findings, methods, characteristics, effects, and management. *Science of the Total Environment*, 630, 1641–1653. https://doi.org/10.1016/j.scitotenv.2018.02.300

Zhang, M., Zhao, Y., Qin, X., Jia, W., Chai, L., Huang, M. and Huang, Y. (2019). Microplastics from mulching film is a distinct habitat for bacteria in farmland soil. *Science of the Total Environment*, 688, 470–478. https://doi.org/10.1016/j.scitotenv.2019.06.108

Zhang, Y., Ma, J., Song, Y.-Q., Li, G. and O'Connor, P. (2024). Stronger deterministic processes shape the plastisphere microbiota of biodegradable microplastics compared to non-biodegradable microplastics in farmland soil. *Applied Soil Ecology*, 196, 105312–105312. https://doi.org/10.1016/j.apsoil.2024.105312

Zhu, D., Ma, J., Li, G., Rillig, M.C. and Zhu, Y.-G. (2021). Soil plastispheres as hotpots of antibiotic resistance genes and potential pathogens. *The ISME Journal*, 16(2), 521–532. https://doi.org/10.1038/s41396-021-01103-9

13 Plant Health Effects of Terrestrial Plastisphere

Manura Weerasinghe, Randima Edussuriya, Irushi Hettige, Keshani Jayathilaka, and Meththika Vithanage

13.1 MICROPLASTICS IN TERRESTRIAL ENVIRONMENT

13.1.1 Sources and Abundance of Microplastics in Terrestrial Environment

Microplastics can enter terrestrial ecosystems either directly as primary microplastics that are manufactured as micrometre-sized particles across several industries (Horton et al., 2017) or indirectly as secondary microplastics by physical, chemical, and biological fragmentation of larger plastic debris (Horton et al., 2017; Machado et al., 2018; Waldman & Rillig, 2020). Microplastic accumulation in terrestrial environments can occur through multiple sources: soil amendments such as biosolid and composts (Vithanage et al., 2021), fertilizers (Lwanga et al., 2022), plastic mulches (Crossman et al., 2020), greenhouse materials (Bläsing & Amelung, 2018), irrigation water via plastic pipes and tunnels (Bläsing & Amelung, 2018), indiscriminate disposal of solid waste; personal care products, household items, electronics in landfills (21%–42% of global plastic waste) (Galafassi et al., 2019; Nizzetto et al., 2016), sewage sludge in wastewater treatment plants (Galafassi et al., 2019), landfill leachate (Wijekoon et al., 2022), tyre wear (Luo et al., 2021), and atmospheric fallout; 0.343 million tons/year globally (Evangeliou et al., 2020), making agroecosystems the major entry route (Corradini et al., 2019) (Figure 13.1). During wastewater treatment, over 90% of microplastics are retained efficiently in sewage sludge and about 50% of sewage sludge is processed for agricultural purposes in Europe and North America. Approximately, a total input of 63,000–430,000 tons of microplastics is released annually to farmlands in Europe whereas 44,000–300,000 tons are released to farmlands in North America annually (Nizzetto et al., 2016). Agroecosystems in Chile reported severe loads of microplastics in sludge ranging from 18 to 41 particles/g (dry weight) (Corradini et al., 2019). In China, the average annual sludge-based environmental microplastic addition was estimated as 1.56×10^{14} particles/kg of dry sludge (Li et al., 2018). Sewage sludge applications in terrestrial agroecosystems alone exceed the estimated amount of 93,000–236,000 tons of microplastics exported to oceans annually making it an alarmingly high input (van Sebille et al., 2015). Fertilizer-manufacturing industries use plastic polymers to encapsulate fertilizers for slow release into the soil. Once these compounds are successfully delivered, the plastic coating of the fertilizer pellet remains in the soil. A survey by Katsumi et al. (2020) reported 6–369 mg/kg of microcapsules in all paddy fields in Japan. Due to the improper segregation of waste, most municipal soil waste is contaminated with plastics. A recent study from Gui et al. (2021) reported an abundance of 2,400 particles/kg (dry weight) of microplastics in agroecosystems in Zhejiang Province China, measuring between 50 μm and 5 mm in compost produced in rural domestic waste. Plastic mulch films, with thicknesses of 6–20 μm, have been widely used in intensive cultivation systems (Ng et al., 2018), covering approximately 180,000 km^2 of agricultural area worldwide contaminating the soil with their residues in the field (Steinmetz et al., 2016). China alone utilized over 1.4 million tons of agricultural plastic-mulching film in 2017 (Gao et al., 2019). Microplastics can also be transported over great distances by wind from surfaces such as landfills and roads, resulting in atmospheric fallout becoming a major global concern (Bläsing & Amelung, 2018).

DOI: 10.1201/9781032717814-13

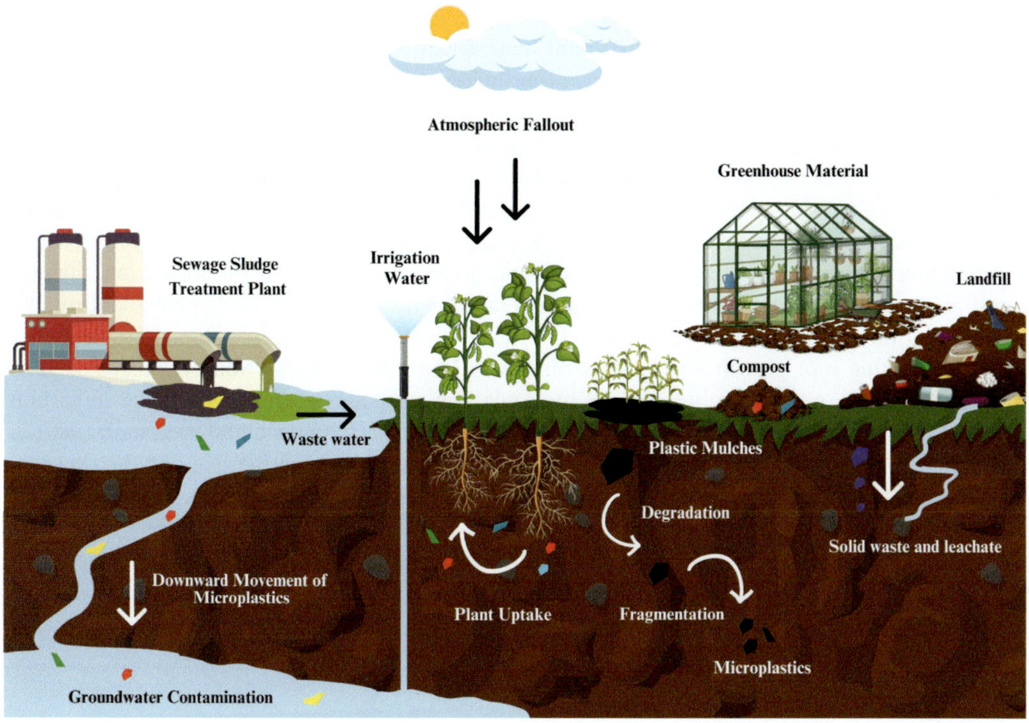

FIGURE 13.1 Sources of microplastics in the terrestrial environment (Dissanayake et al., 2022).

13.1.2 Effects of Microplastics on Terrestrial Environments

Once plastic particles enter the terrestrial environments, they can get fragmented into microplastics and then get degraded by photodegradation via UV radiation (Kumar Sen & Raut, 2015), mechanical abrasion (Zhu et al., 2019), thermal oxidation (Ziani et al., 2023), biodegradation via microbial action (Krueger et al., 2015), and interaction with soil colloids (Ren et al., 2021). The combined effects of these factors can cause ageing of microplastics and release multiple chemicals including phthalates, flame retardants, stabilizers, pigments, oligomers, and oxygenated products such as phenols, acetophenones, and carboxylated products into soil (Liu et al., 2021). Microplastics have shown interactions with a wide array of contaminants including pesticides, persistent organic pollutants, polychlorinated biphenyls (PCBs), dioxins, dichlorodiphenyltrichloroethane (DDT), and polycyclic aromatic hydrocarbons (PAHs), heavy metals, and antibiotics and may act as a vector for contaminant transfer in terrestrial environments (Vithanage & Prasad, 2023).

During the manufacturing of plastics, several hazardous substances such as plasticizers, surfactants, stabilizers, colourants, and flame retardants are added to improve their performance, flexibility, stability, and functionality (Mandal et al., 2023). Since these additives do not form any chemical bonds with the original polymer, a cocktail of toxic organic contaminants can be released over time under altered chemical conditions into the terrestrial environment (Ramanayaka et al., 2020). Phthalate-based plasticizers such as ethylhexyl phthalate, dimethyl phthalate, and diethyl phthalate are some of the most abundant toxic chemicals in microplastic-contaminated soil. In Chinese agricultural soils, an extreme high load (more than 10 kg/ha/year) of phthalate esters (PAEs) was estimated in five areas including Beijing city, Tibet, Liaoning Province, Jilin Province, and Fujian Province (Borges et al., 2019). Due to the large surface areas and hydrophobicity, microplastics can also act as a sink by adsorbing harmful organic contaminants and heavy metals (Rillig, 2012).

Polymer structure and molecular composition of microplastics also affect the adsorption of persistent organic pollutants (Wang et al., 2016).

Incorporation of microplastics into the soil matrix can trigger changes in soil physicochemical properties such as bulk density, porosity, water-holding capacity, structure, organic matter, pH, fertility, and nutrient content. These changes may negatively or positively affect plant productivity, microbial activity, bacterial composition diversity, and nutrient cycle (Rillig et al., 2024). Therefore, it's unclear yet it depends on soil physicochemical properties, soil biology, and plant type. Due to the soil porosity-altering ability, microplastics can affect the relative abundance of both aerobic and anaerobic microorganisms (Wu et al., 2020). Microplastics can also affect microbial activities by restricting soil aggregation and soil hydrophobicity (Machado et al., 2018). Moreover, changes in soil pH by the addition of microplastics have been shown to affect soil nutrient availability due to the alteration of microbial communities, consequently affecting plant growth and crop productivity by affecting root growth and seed germination (Wang et al., 2021). Soil organisms such as earthworms, mites, and collembola uptake microplastics as food, inducing a decrease in carbon biomass ingestion that can lead to energy depletion, inhibited development, and even mortality (da Costa et al., 2016). Soil animals may play a vital role in microplastic movement and distribution. Earthworms can transport microplastics in soils through adhesion and excretion, causing smaller microbeads to migrate largely downward in the soil profile (Rillig et al., 2017). The biogenic transport of microplastics in soil may lead to groundwater pollution, uptake by terrestrial plants and entry into the terrestrial food chain resulting in potential detrimental health effects for humans and other living organisms (Bank & Hansson, 2022).

13.2 TERRESTRIAL PLASTISPHERE

Plastic particles in soil interact intimately with microorganisms, likely influencing the function of ecosystems (Figure 13.2). The soil plastisphere is an ecological microenvironment conceptualized as a dynamic interface where complex interactions between microbial communities and the environment occur (Rillig et al., 2024). Soil plastisphere is not only limited to microorganisms colonizing on plastic surfaces but also includes those in soil affected by the presence of plastic particles. This unique substrate can aid microbial dispersal, and it has synergistic effects on the development, transportation, persistence, and ecology of microorganisms (Barros & Seena, 2021). The establishment and succession of soil plastisphere communities and their interactions can be influenced by biotic properties such as dominant microorganisms and interactions with surrounding microorganisms, and abiotic properties such as pH, temperature, light, soil properties, salinity, and nutrient concentrations (Tang, 2024). Interestingly, the soil plastisphere's dynamics and behaviours differ from those of its marine equivalent, where it was first identified, and it is thus likely to display unidentified characteristics concerning ecological consequences and biogeochemistry (Zhai et al., 2024). In contrast to the oceanic plastisphere, which is mobile due to currents, the soil plastisphere remains immobile (Maguire & Gardner, 2023). Therefore, microbial communities and their ecological behaviours in the soil plastisphere are highly dependent on the localized soil environment (Rillig et al., 2024). Given that soils tend to be heterogeneous, plastic particles have the ability to enter soil aggregates and be dispersed throughout distinct microenvironments within the same soil mass. Different redox reactions result in different microbial populations in soil conditions (Zhang & Liu, 2018). However, in contrast to uncolonized microplastics, the plastisphere's development modifies the physical characteristics of microplastics, hence altering their motion. The fate and movement of these pollutants in soil can be influenced by the microbial community in the plastisphere, which may have an impact on ecosystem health as well as human health if the pollutants get into the food chain (Cverenkárová et al., 2021).

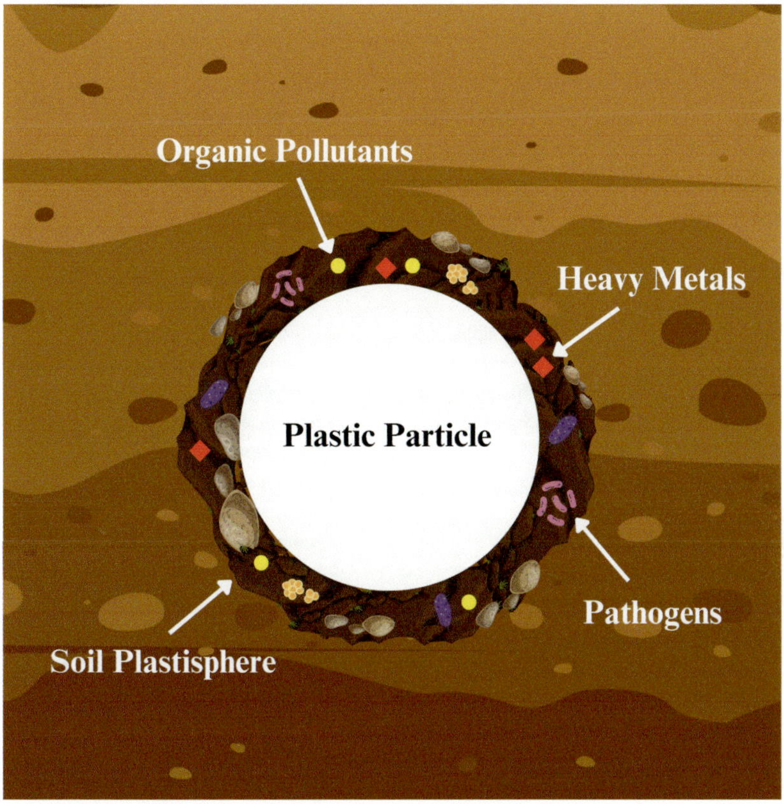

FIGURE 13.2 The composition of the soil plastisphere (Wang et al., 2022).

13.2.1 Diversity of Terrestrial Plastisphere

The terrestrial plastisphere is a habitat for numerous distinct microbial groups (Table 13.1). Different microbial species live in plastispheres than in other naturally occurring floating materials due to the unique chemical composition and a slow rate of biodegradation in plastics (Tang, 2024). It is likely that the microorganisms within the terrestrial plastisphere perform various roles. For example, bacteria that are photosynthetic are called cyanobacteria. In symbiosis with other heterotrophic bacteria populating the plastisphere through primary production and nitrogen fixation, they can be found with other phototrophs such as *Bumilleriopsis*, *Bracteacoccus*, *Tetracystis*, *Chlorochytrium*, and *Deasonia*. Their presence could facilitate the establishment of other bacteria in the plastisphere by making carbon more readily available since carbon in hard-to-degrade plastics is not easy to access (MacLean et al., 2021). Research often reveals variations in the taxonomic character of the bulk soil microbial population and the plastisphere. For instance, compared to the bulk soil, the plastisphere of farmed plastic samples contained higher concentrations of various bacterial taxa, including Actinobacteria, Bacteroidetes, Cyanobacteria, Deinococci, Proteobacteria, and Verrucomicrobia (Zhang et al., 2019). The bulk of this research, which concentrates on bacterial populations, is from China, where plastic mulching is widely employed in agriculture. The findings of Li et al. (2022) were more similar to the findings of Li et al. (2023) stating that Alpha diversities (Shannon and Simpson) were higher in firm residues from greenhouses and plastic-mulched croplands in China. However, Alpha diversities were lower in incubated soil samples from the Swiss Alps (Rüthi et al., 2020). Plastic waste in the soil can operate as a vector for plant pathogens in some situations where the plastisphere can become enriched with harmful fungi, bacteria from the genera Streptomyces

TABLE 13.1
Different Microbial Communities in Terrestrial Plastisphere

Plastic Environment	Country	Distribution	Phylum/Class/Genes	References
Polyethylene (PE), polybutylene adipate terephthalate (PBAT), and polylactic acid (PLA) in microplastics	China	PBAT/PLA plastisphere has lower alpha diversities of fungal communities	Ascomycota: Nectriaceae, Fusarium sp., Cladosporiaceae, Pleosporaceae Mortierellomycota Family: Mortierellaceae	Li et al. (2024)
Firm residues from greenhouses	China	High alpha diversities (Shannon and Simpson)	Actinobacteriota (abundant bacteria) *Nonomuraea, Blastococcus* Crenarchaeota (abundant bacteria) *Nitrososphaeraceae* Zygomycota (abundant and rare fungi) *Mortierella* Ascomycota (rare fungi) *Fusarium, Phoma, Paecilomyces*	Li et al. (2023)
PLA, PBAT, and PE incubated soil	Swiss Alps	Low Alpha diversities	Proteobacteria (PLA) *Collimonas, Rhizobiaceae* Actinobacteria (PBAT) *Nocardia, Rhodococcus, Umezawaea* Actinobacteria (PE) *Nocardia, Nocardioides, Rhodococcus*	Rüthi et al. (2020)
	Greenland		Proteobacteria (PLA) *Caulobacter, Brevundimonas* Proteobacteria (PBAT) *Aquabacterium, Acidovorax, Caulobacter* Actinobacteria (PE) *Nocardia, Nocardioides, Streptomyces*	
Poly(butylene succinate-co-butylene adipate) (PBSA) in croplands	Germany	Low bacterial and fungal richness	Ascomycota *Tetracladium, Cyphellophora, Exophiala, Plenodomus* Gammaproteobacteria *Rhizobacter, Variovorax*	Tanunchai et al. (2023)
Plastic-mulched residues from croplands	China	Alpha diversities (Simpson)	Phototroph: *Bumilleriopsis, Bracteacoccus* Parasite: *Spongospora, Anurofeca*	Li et al. (2022)
Polyethylene **(PE)**	China	N/A	Actinomycetota and Bacillota, lower proportions of Pseudomonadota, Acidobacteriota, Gemmatimonadota, and Bacteroidota	Sun et al. (2024)
Potato field and chilli field with PE and PBAT/PLA film practices	China	Soil fungal communities showed the lowest beta-diversity dispersion	Nectriaceae, Pleosporaceae, and Didymellaceae phylum Ascomycota was significantly more abundant in PBAT/PLA plastisphere	Li et al. (2024)
Abandoned landfill with PE plastic	Germany	N/A	Actinomycetospora, Arthrobacter, Rhodococcus, Rubrobacter, and Cellulosimicobium	MacLean et al. (2021)

and *Pseudomonas*, and parasitic protists like Spongospora and Anurofeca (Yoshida et al., 2016). Although research on the ecological mechanisms underlying community assembly in the plastisphere is still in its early stages, it is crucial because it holds the potential to reveal essential principles that may be widely applied.

13.3 EFFECT OF TERRESTRIAL PLASTISPHERE ON SOIL PROPERTIES

13.3.1 Influence on Soil Physicochemical Properties

Soil microbial communities and their activities are closely related to soil properties such as organic matter content, texture, and moisture content (Rillig et al., 2017). Due to the formation of soil plastispheres, they can change the physicochemical soil environment by modifying soil structure and leaching chemicals (Wang et al., 2022). When soil plastispheres are present, they can alter the soil's physical properties such as water-holding capacity, structure, and porosity and chemical properties such as pH, organic matter, and nutrients. Soil porosity controls the water distribution and aeration of soil. Therefore, high porosity due to the formation of soil plastisphere can lead to decreased water infiltration and retention affecting the plant growth (Rüthi et al., 2020). Plastispheres can also adsorb pollutants like heavy metals and organic contaminants. Leaching of these pollutants from the soil plastisphere can influence potential alteration of pH in the surrounding terrestrial environment (Chia et al., 2022). Terrestrial ecosystems such as soils are usually rich in nutrients for microorganisms, and although the plastisphere can effectively adsorb organic matter, the availability of nutrients may not be an important reason for the enrichment of microorganisms by the plastisphere (Tang, 2024). Moreover, the microbial communities within the plastisphere can influence nutrient cycling in the soil (Aralappanavar et al., 2024). However, more research on soil plastisphere is needed to fully understand the long-term effects.

13.3.2 Ecological Risks of Pathogens in Terrestrial Plastisphere

The functional composition of the plastisphere microbial community is enriched in microbial eukaryotes and bacteria, several of which can be potentially pathogenic and have been reported to carry antibiotic resistance genes (ARGs) (Kaur et al., 2022). Due to its specific concern, the enrichment of ARGs has drawn more attention from researchers. In particular, ARGs may spread by horizontal gene transfer, and then future environmental dangers are unclear given the colocation of ARG enrichment and pathogen prevalence. In recent decades, one of the main threats to global public health has been the emergence and spread of antibiotic resistance in bacteria and diseases (Zhu et al., 2022). One of the major natural habitats for bacteria that produce antibiotics is soil, which serves as a crucial reservoir for both antibiotics and environmental resistance (Séveno et al., 2002). In the past few decades, one of the biggest risks to the world's public health has emerged from resistance genes (ARGs) that can spread between bacteria through the growth of their bacterial hosts or by horizontal gene transfer to bacteria that are phylogenetically unrelated (Li et al., 2022). Given the possible serious consequences for human health, it is important to pay more attention to the enrichment of dangerous pathogens in soil plastispheres. ARGs may be taken up by soil bacteria, particularly Proteobacteria, through plasmid transfer, and subsequently relocated into the plant microbiome by spontaneous wound healing or nematode infection in the root (Shi et al., 2022). Many recognized pathogens, including *Pseudomonas aeruginosa* and staphylococci, are linked to biofilms via secreting extracellular polysaccharides and other adhesins, including proteins. Parasitic protists like Spongospora and Anurofeca are also present. In certain instances, the plastisphere could be enriched with pathogenic fungi, making plastic debris in the soil a vector for plant pathogens (Rillig et al., 2024).

13.3.3 Plastisphere–Rhizosphere Interactions

The rhizosphere, a thin layer of soil, is formed by the substances that plants release, such as mucilage and exudates, which draw microorganisms. Plants develop intricate relationships with various soil organisms, including bacteria, fungi, and protozoa (Tomazelli et al., 2024). Rhizosphere microbial populations in natural grasslands and developed pastures are shaped by the diversity of plants and root characteristics. Numerous biological processes, including the cotransport of pollutants from the environment, microplastic biodegradation, and soil carbon cycling, may be impacted by the soil plastisphere (Wang et al., 2022). In terrestrial ecosystems, soil microbial community networks can enhance the flow of carbon elements. Thus, to properly understand soil carbon cycle and microplastic–microbial interactions, the soil plastisphere must be considered (Zhurina et al., 2022). An essential component of soil organic carbon (SOC) is rhizodeposition, which generates a hotspot for rhizosphere microbial community activity, particularly for bacteria that colonize roots, such as pathogens, N-fixers, and arbuscular mycorrhizal fungi (Rillig & Lehmann, 2020). Additionally, the rhizosphere microbial community is crucial to SOC deposition, and belowground inputs of SOC are more successful than aboveground inputs for forming mineral-stabilized soil carbon (Jackson et al., 2017). Nonetheless, lack of thorough research on how the soil plastisphere reacts to soil carbon. Consequently, more investigation into the mechanisms of the soil plastisphere influencing the carbon cycle is required, with particular attention to the following factors. The function of the soil plastisphere's microbial carbon metabolism, plant development and net primary production, rhizodeposition and litter decomposition, soil carbon turnover, and greenhouse gas emissions are the first four factors (Rillig et al., 2021). The stability of rhizosphere organic carbon might be affected by the presence of microplastics in the rhizosphere. The different combination of dominant bacteria on the surfaces of various microplastics, including PE, polyamide 6 (PA), PVC, and expanded polystyrene (EPS), was revealed by a recent investigation of the various plastispheres discovered in the rhizosphere of mangroves. This information is significant for the soil carbon and sulphur cycles (Wang et al., 2022). Nevertheless, it is still unknown what precise mechanisms of the soil plastisphere are in charge of sequestering carbon and causing losses in the rhizosphere.

13.4 EFFECT OF TERRESTRIAL PLASTISPHERE ON PLANT HEALTH

Microorganisms and plastic waste contribute to both positive and negative effects of the terrestrial plastisphere on plant health. Plastisphere accumulation and aggregation close to root surfaces obstructs entrances and thereby reduces the water and nutrient uptake (Khalid et al., 2020). This also involves decreasing organic carbon content (Salam et al., 2023) and inorganic nitrogen content by approximately 30% for plants (Ingraffia et al., 2022). Plastisphere also has a detrimental effect on the mycorrhizal community by modifying the mycorrhiza's abundance and activity. This is due to the fact that the microplastic component in plastispheres modifies several soil parameters that determine the mycorrhiza's habitat space and conditions (Leifheit et al., 2021). This consequently hinders the uptake of nutrients and water, weakening plant functions and impacting plant health (Khalid et al., 2020). Plant health is also impacted by the growth of microbial communities in soil, as they use up nutrients that plants can use, which lowers plant health. In response to the impact of plastic film pollution, the plastisphere modifies the structure of soil microbial communities by dramatically reducing the abundance of *Actinomycetes* and increasing the abundance of *Proteobacteria* (Qian et al., 2018). In addition, the buildup of microplastic at root hairs inhibits the absorption of vital nutrients and water through root tips, thereby influencing plant development and fresh and dry weight (Chen et al., 2024). Furthermore, this results in reduced levels of chlorophyll, shorter roots, higher oxidative stress, and inhibited seed germination (Maity et al., 2022). This implies that the terrestrial plastisphere has a great impact on the growth and health of plants. In addition, research has demonstrated that in certain cases, the plastisphere benefits plants by facilitating the development of endophytic interactions among plants, which eliminates microorganisms that colonize the rhizosphere (Ranauda et al., 2024).

13.4.1 Plant Pathogens in Terrestrial Plastisphere

Terrestrial plastisphere may consist of over 1,000 different species of microorganisms including fungi, bacteria, algae, and other single-celled organisms. They have evolved to live in human-made plastic environments (Rillig et al., 2024). The microbial community has been supported by the plastisphere for its development, transportation, dispersal, persistence, and ecology (Tang, 2024). The terrestrial plastisphere hosts a multitude of microbial communities including autotrophs, heterotrophs, predators, and pathogens (Su & Yang, 2024). Among them, the most abundant plant pathogenic microorganisms associated with terrestrial plastisphere are fungi and bacteria. However, there are only limited studies related to the plant pathogenic functional roles of plastisphere-associated microbes.

Tang (2024) has stated that the plastisphere has a higher diversity and more balanced distribution of abundant fungi than the surroundings. In certain instances, the plastisphere could be enriched with soil-borne plant pathogenic fungi. They are making plastic debris in the soil a vector for plant pathogen transmission (Li et al., 2023). Among them, Ascomycota is the most common fungal phylum, which is associated with the terrestrial plastisphere (Tang, 2024). Under the phylum Ascomycota, there are the common genera such as *Fusarium, Paecilomyces, Tetracladium, Cladosporium*, and *Didymella* identified as plant pathogens (Table 13.2). Under the genus *Fusarium*, there are some plant pathogenic species that cause various plant diseases. *Fusarium solani* can cause root rot and lesions on leaves of potato. Also the early blight disease caused by *Fusarium solani* may result in potato yield loss. Further, vascular wilt and root rot diseases on crops such as peas and beans can

TABLE 13.2
Plant Pathogenic Microorganisms in Terrestrial Plastisphere

Domain/ Kingdom	Species	Infected Plants	Diseases and Symptoms	References
Fungi	*Fusarium solani*	Potato	Root rot and necrotic lesions. Yield losses can be occurred due to early blight disease	Bahadur (2021)
	Fusarium oxysporum	Pea plants and beans	Vascular wilt and root rot	
	Fusarium moniliforme	Corn, rice, and other cereal grains	Reduction in yield	
	Fusarium pseudograminearum	Wheat	Crown rot disease. Infected plants have stunted growth and rootlets are necrotic and rotten. Disease results yield reduction	
	Cladosporium fulvum	Tomato	Brown spots on leaves, destroying leaves and killing plants while reducing yield	Kokaeva et al. (2020) and Cheng et al. (2023)
	Cladosporium cucumerinum	Cucumber	Scab disease with water-soaked spots on leaves and runners while lesions can be seen on fruits	
	Didymella segeticola	*Camellia sinensis*	Leaf spot disease subsequently affects the quality and flavour of tea	Deng et al. (2023)
	Didymella exitialis	Wheat and barley	Leaf scorch associated with necrotic patches of leaves	Błaszczyk et al. (2021)
	Didymella glomerata	*Pistacia vera* L.	Leaf blight disease	Moral et al. (2018)
Bacteria	*Streptomyces scabies*	Potato and other taproot crops	Scab disease	Bignell et al. (2014)
	Streptomyces ipomoeae	Sweet potato	Root rot disease	Croce et al. (2021)

be seen due to the infection of *Fusarium oxysporum* (Bahadur, 2021). Some species of the genus *Cladosporium* are pathogens to various crops and they can cause economically important plant diseases. *Cladosporium fulvum* can cause brown spots on tomato leaves (Kokaeva et al., 2020). Under favourable conditions, it can cause a reduction in yield by destroying the leaves and killing plants. Another important pathogen which is known as *Cladosporium cucumerinum* can cause scab disease in cucumber globally. In this disease, water-soaked spots can be seen on leaves and runners, while lesions can be seen on fruits (Cheng et al., 2023). Further, *Didymella segeticola* can cause leaf spot disease on tea plants and subsequently affect the quality and flavour of tea by changing the composition and content of caffeine, catechins, and amino acids (Deng et al., 2023). *Pistacia vera* L. plants have been infected by leaf blight disease due to the infection of *Didymella glomerata*. Also, *Didymella exitialis* is recorded as a pathogen of wheat and barley. It can cause leaf scorch, which is associated with necrotic patches of leaves (Błaszczyk et al., 2021).

In terms of bacteria, the most commonly reported phylum is Actinobacteria and Proteobacteria on the terrestrial plastisphere (Su & Yang, 2024). Sun et al. (2022) have also identified Actinobacteria and Proteobacteria as the most dominant bacterial pathogens in polyester and polystyrene (PS)-associated plastispheres. Among them, the common plant-pathogenic genera of Actinobacteria comprise *Streptomyces*, *Nocardia*, *Mycobacterium*, and *Rodococcus*. Underground structures of diverse plant species are infected by plant pathogenic *Streptomyces* species. *Streptomyces scabies* causes common scab disease in potatoes (Bignell et al., 2014). On potatoes and other taproot crops, *Streptomyces acidiscabies* causes symptoms similar to those of *S. scabies*. Sweet potato root rot is caused by *Streptomyces ipomoeae*. It is characterized by necrosis of fibrous roots and cankers on expanded storage roots (Croce et al., 2021). The common plant pathogenic genera of proteobacteria in the terrestrial plastisphere are *Pseudomonas* and some Rhizobacter strains (Tang, 2024). Some strains of *Pseudomonas* cause diseases like bacterial blight, leaf spot, and root rot in various crops. Also, *Rhizobacter* strains have plant pathogenic potential.

13.4.2 Weakening of Plant Defences

Plants use a variety of weapons called defence mechanisms to protect themselves from infections. Biochemical processes and structural traits provide the defence mechanisms (Belete, 2018). Due to their structural features, plants have physical barriers that stop pathogens from entering and spreading, and through biochemical reactions, they produce substances that are either toxic to pathogens or that foster environments that prevent pathogen growth in plants (Belete, 2018). Plants that have experienced physical barrier damage or that are no longer able to produce substances that are toxic to pathogens or, more frequently, pathogens that are resistant to the chemicals produced by host plants are associated with a weakening of plant defence mechanisms (Pfeilmeier et al., 2016). There are several internal and external factors that contribute to the weakening process. Plant defences are the result of internal factors, primarily genetics. Weak defence mechanisms result from weak genetic conditions (Kaur et al., 2022). Plant defences can be weakened by biotic and abiotic external factors alike. Herbivory and pathogen attacks are examples of biotic factors (Bhatla et al., 2018). Nutrient shortage, water stress, high-temperature stress, and environmental pollution can all be listed as abiotic stress conditions (Lamalakshmi Devi et al., 2017). Out of all the factors mentioned above, environmental pollution is the most critical for weakening defence mechanisms. According to recent studies, plastics can impair plant defence systems through the release of harmful chemicals and the formation of plastispheres while in the soil (Rillig et al., 2024) such as the damaging of antioxidant defence system in maize plants (Sun et al., 2023) and lettuce plants (Gao et al., 2019).

The term "terrestrial plastisphere" also known as "soil plastisphere," which describes the microbial community that lives on plastic surfaces and the nearby soil surfaces of plastics, is becoming more and more recognized as a factor that has a significant impact on plant health. The soil plastisphere involves enhancing the interactions between microplastics and microbial communities in soil environments and thereby will pose ecological risks to nature (Wang et al., 2022). To affect the plants' immunity, the microbial community that lives on plastics interacts with other soil microorganisms and elements. However, as plastics weaken plants' defence systems, they have a negative impact on their health. Research has demonstrated that the buildup of plastic film residues can adversely impact crop growth and soil quality (Hegan et al., 2015). The rhizosphere is the vital point of contact between the soil matrix and plant roots where both positive and negative interactions between plants and microorganisms occur (Barra Caracciolo & Terenzi, 2021).

Plants rely on their physical barrier as their first line of defence, but the plastisphere is weakening it. Microorganisms living in the soil environment consume away at the surfaces of plastics (Rüthi, 2022). Microbes can more easily access plants as a result of the physical barrier of plastics being weakened by this degradation. By indirectly affecting defence mechanisms, this will jeopardize the health of plants (Zhou et al., 2021). The balance between good and bad bacteria is impacted by the microorganisms on plastic surfaces because they change the structure of the soil's microbial communities by varying the prevalence of microorganisms like *Actinomycetes* and *Proteobacteria* (Qian et al., 2018) (Figure 13.3). Apart from that, this soil pollution decreases soil organic matter and

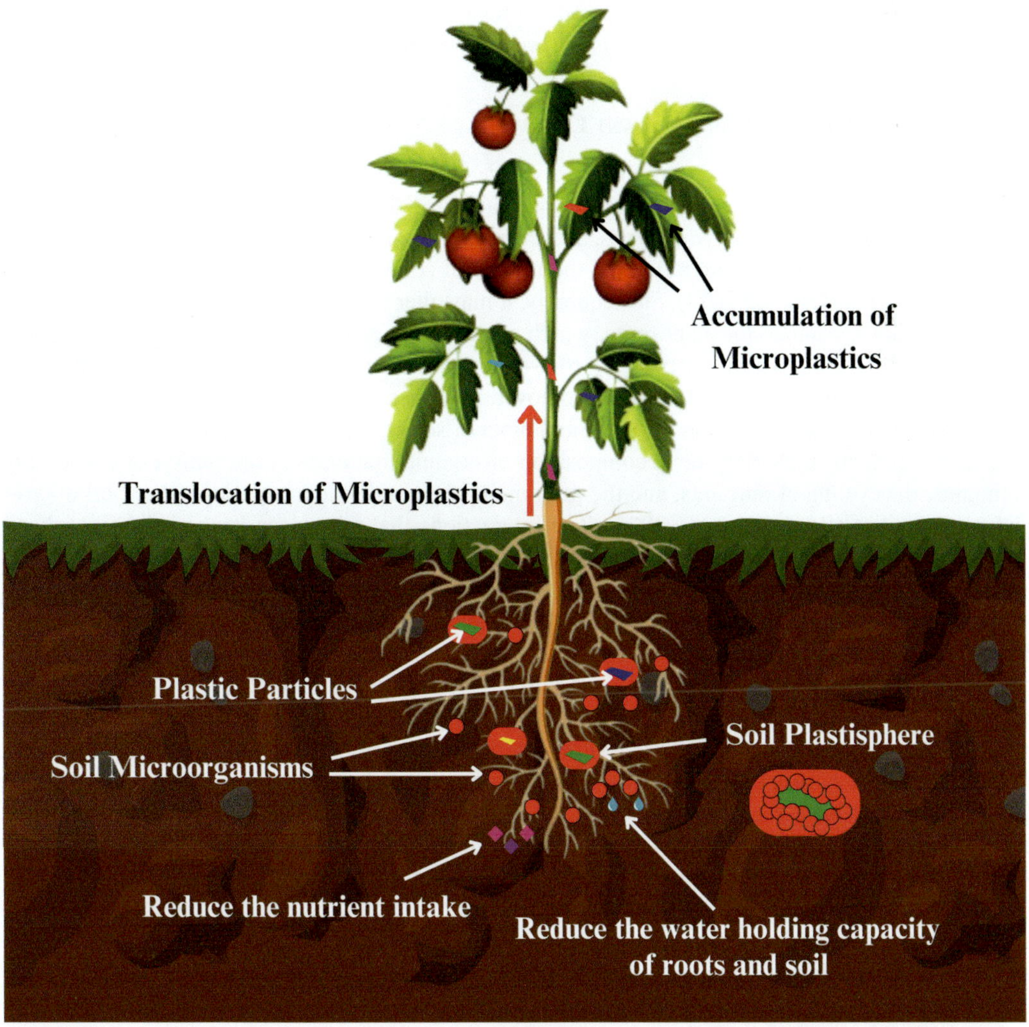

FIGURE 13.3 Weakening of plant defence mechanisms by the soil plastisphere (Yu et al., 2024).

inorganic nitrogen content by downregulating microbial genes related to soil carbon and nitrogen cycles and decreasing related enzymatic activities (Qian et al., 2018; Salam et al., 2023). Through obstructing seed pores and reducing the amount of water and nutrients that can be absorbed through accumulation and aggregation in roots, shoots, and leaves, the plastic components in soil directly affect plants (Khalid et al., 2020) (Figure 13.3). By affecting the availability of nutrients, interactions between plants and microbes, and the frequency of plant pathogens, this condition will indirectly affect the defence mechanisms of plants (Salam et al., 2023). Plant defence mechanisms may be directly or indirectly impacted by the chemicals linked to or released by plastic degradation (Gewert et al., 2015). The compounds found in plastic leachates may have an impact on plant physiology directly or on root exudates and growth. Research indicates that plastics may change the makeup and activity of soil microbial communities. The presence of plastics in soil has effects on the plants, soil organisms, and microorganisms in soil (Khalid et al., 2020); however, studies have shown that the adverse effects of plastics are mainly on the plant community and soil organisms and the effect on microbial community is less (Huo et al., 2022). Plant defence mechanisms are weakened by changes in soil microbial communities, which impact disease suppression, nutrient availability, and interactions between plants and microbes. Mycorrhizal associations, which are crucial for nutrient uptake and pathogen defence, may be hampered by plastic debris. This also influences factors mediated by mycorrhizae such as soil aggregation, water, and nutrient transport (Leifheit et al., 2021). The defence systems of plants are weakened when these associations are broken.

13.4.3 Plastisphere Effects on Plant Growth

Plastisphere as a result of weakening the plant defence mechanisms affects the plant growth. The plastisphere tends to have significant effects on plant growth, and it is highly dependent on the microplastic concentration, polymer types, and the size and shape of microplastics. The presence of plastisphere in soil results in alterations of the microbial community and the functions of soil and rhizosphere (Ranauda et al., 2024). Subsequently, it may alter water transport, nutrient availability, and penetration resistance for plant roots, which affects plant growth. Plant growth is also highly dependent on soil microbiota, particularly root-colonizing microbes such as N-fixers, pathogens, and mycorrhizal fungi (Chen et al., 2024). Moreover, this has a slightly negative impact on the plant fitness, which is influenced by the diversity and biota of soil. In fact, a multitude of ecosystem services and functions, including the decomposition of organic matter, the elimination of toxins and pollutants, the cycling of nutrients, and the promotion of plant development, depend on soil microbial communities (Ranauda et al., 2024). The interactions that take place in the soil/rhizosphere microbiome and between microbial communities and plants result in these functions and services.

Additionally, it was found that together with the effect on the microbial activity in soil, plastisphere may strongly affect root characteristics, tissue elemental composition, and plant biomass (de Souza Machado et al., 2019). Evidence suggests that terrestrial plants are capable of absorbing and accumulating plastics as small as microns (Li et al., 2020; Luo et al., 2022). Reports indicate that plastic residues hinder crop yields, with an average yield reduction of 3% for every 100 kg/ha of plastic film residue (Zhang et al., 2020). Apart from that, plants' normal growth may get affected due to the increased exposure of plants to contaminating pathogens from the plastisphere (Zhang et al., 2022). Plants that have bioaccumulated microplastic may experience a range of phytosociological consequences. Plants that have a build-up of microplastic may experience increased oxidative stress, shorter roots, lower chlorophyll levels, and inhibited seed germination (Maity et al., 2022). It was recently shown that microplastics significantly affected the photosynthetic rate, oxidative damage, and soil microbial activity of *Lactuca sativa* L. (Wang et al., 2024). The information regarding the contents of carotenoids, chlorophyll, and antioxidant enzymes also points to the detrimental effects on plant health of changed rhizosphere processes. It could adversely affect plant roots and their symbionts, thus negatively affecting plant growth. On the other hand, the adsorption of pollutants by the plastisphere may reduce the availability of pollutants to plants and soil biota, which could promote plant growth (Chen et al., 2024). The outcome of the soil-*Salvia officinalis* system suggests that the presence of plastisphere components in soil causes some advantageous interactions like the ability to form endophytic interactions between plants and the microorganisms that colonize the rhizosphere to disappear (Ranauda et al., 2024). Thus, when taking into account the effects of plastisphere, it can be stated that the terrestrial plastisphere has both positive and negative effects on the growth and health of plants.

REFERENCES

Aralappanavar, V. K., Mukhopadhyay, R., Yu, Y., Liu, J., Bhatnagar, A., Praveena, S. M., Li, Y., Paller, M., Adyel, T. M., & Rinklebe, J. (2024). Effects of microplastics on soil microorganisms and microbial functions in nutrients and carbon cycling–A review. *Science of the Total Environment, 924*, 171–435. doi:10.1016/j.scitotenv.2024.171435

Bahadur, A. (2021). Current status of fusarium and their management strategies. In Mirmajlessi, S. M. (Ed.), *Fusarium-An Overview of the Genus* (pp. 2–7). IntechOpen, London.

Bank, M. S., & Hansson, S. V. (2022). The microplastic cycle: An introduction to a complex issue. In Bank, M. S. (Ed.), *Microplastic in the environment: Pattern process* (pp. 1–16). Springer, Cham. doi:10.1021/acs.est.9b02942

Barros, J., & Seena, S. (2021). Plastisphere in freshwaters: an emerging concern. *Environmental Pollution, 290*, p. 118123.

Barra Caracciolo, A., & Terenzi, V. (2021). Rhizosphere microbial communities and heavy metals. *Microorganisms, 9*(7), 14–62. doi:10.3390/microorganisms9071462

Belete, T. (2018). Defense mechanisms of plants to insect pests: From morphological to biochemical approach. *Trends in Technical & Scientific Research, 2*(2), 30–38. doi:10.19080/TTSR.2018.02.555584

Bhatla, S. C., Lal, M. A., Kathpalia, R., Sisodia, R., & Shakya, R. (2018). Biotic stress. In Bhatla, S. C., & Lal, M. A. (Eds.), *Plant Physiology, Development and Metabolism* (pp. 1029–1095). Springer, Singapore. doi:10.1007/978–981-13–2023-1_32

Bignell, D. R. D., Fyans, J. K., & Cheng, Z. (2014). Phytotoxins produced by plant pathogenic Streptomyces species. *Journal of Applied Microbiology, 116*(2), 223–235. doi:10.1111/jam.12369

Bläsing, M., & Amelung, W. (2018). Plastics in soil: Analytical methods and possible sources. *Science of the Total Environment, 612*, 422–435. doi:10.1016/j.scitotenv.2017.08.086

Błaszczyk, L., Salamon, S., & Mikołajczak, K. (2021). Fungi inhabiting the wheat endosphere. *Pathogens, 10*(10). doi:10.3390/pathogens10101288

Chen, Y., Li, Y., Liang, X., Lu, S., Ren, J., Zhang, Y., Han, Z., Gao, B., & Sun, K. (2024). Effects of microplastics on soil carbon pool and terrestrial plant performance. *Carbon Research, 3*(1), 1–23. doi:10.1007/s44246-024-00124-1

Cheng, C.-Y., Zhang, M.-Y., Niu, Y.-C., Zhang, M., Geng, Y.-H., & Deng, H. (2023). Comparison of fungal genera isolated from cucumber plants and rhizosphere soil by using various cultural media. *Journal of Fungi, 9*(9), 934. doi:10.3390/jof9090934

Chia, R. W., Lee, J.-Y., Jang, J., Kim, H., & Kwon, K. D. (2022). Soil health and microplastics: A review of the impacts of microplastic contamination on soil properties. *Journal of Soils and Sediments, 22*(10), 2690–2705. doi:10.1007/s11368-022-03254-4

Corradini, F., Meza, P., Eguiluz, R., Casado, F., Huerta-Lwanga, E., & Geissen, V. (2019). Evidence of microplastic accumulation in agricultural soils from sewage sludge disposal. *Science of the Total Environment, 671*, 411–420. doi:10.1016/j.scitotenv.2019.03.368

Croce, V., López-Radcenco, A., Lapaz, M. I., Pianzzola, M. J., Moyna, G., & Siri, M. I. (2021). An integrative approach for the characterization of plant-pathogenic Streptomyces spp. strains based on metabolomic, bioactivity, and phylogenetic analysis. *Frontiers in Microbiology, 12*, 643792. doi:10.3389/fmicb.2021.643792

Crossman, J., Hurley, R. R., Futter, M., & Nizzetto, L. (2020). Transfer and transport of microplastics from biosolids to agricultural soils and the wider environment. *Science of the Total Environment, 724*, 138–334. doi:10.1016/j.scitotenv.2020.138334

Cverenkárová, K., Valachovičová, M., Mackuľak, T., Žemlička, L., & Bírošová, L. (2021). Microplastics in the food chain. *Life, 11*(12), 13–49. doi:10.3390/life11121349 da Costa, J. P., Santos, P. S. M., Duarte, A. C., & Rocha-Santos, T. (2016). (Nano)plastics in the environment – Sources, fates and effects. *Science of the Total Environment, 566–567*, 15–26. doi:10.1016/j.scitotenv.2016.05.041

de Souza Machado, A. A., Lau, C. W., Kloas, W., Bergmann, J., Bachelier, J. B., Faltin, E., Becker, R., Görlich, A. S., & Rillig, M. C. (2019). Microplastics can change soil properties and affect plant performance. *Environmental Science & Technology, 53*(10), 6044–6052. doi:10.1021/acs.est.9b01339

Deng, X., Yang, J., Wan, Y., Han, Y., Tong, H., & Chen, Y. (2023). Characteristics of leaf spot disease caused by Didymella species and the influence of infection on tea quality. *Phytopathology®, 113*(3), 516–527. doi:10.1094/PHYTO-06–22–0202-R

Dissanayake, P. D., Kim, S., Sarkar, B., Oleszczuk, P., Sang, M. K., Haque, M. N., Ahn, J. H., Bank, M. S., & Ok, Y. S. (2022). Effects of microplastics on the terrestrial environment: A critical review. *Environmental Research, 209*, 112734. doi:10.1016/j.envres.2022.112734

Evangeliou, N., Grythe, H., Klimont, Z., Heyes, C., Eckhardt, S., Lopez-Aparicio, S., & Stohl, A. (2020). Atmospheric transport is a major pathway of microplastics to remote regions. *Nature Communications, 11*(1), 3381. doi:10.1038/s41467-020-17201-9

Galafassi, S., Nizzetto, L., & Volta, P. (2019). Plastic sources: A survey across scientific and grey literature for their inventory and relative contribution to microplastics pollution in natural environments, with an emphasis on surface water. *Science of the Total Environment, 693*, 133–499. doi:10.1016/j.scitotenv.2019.07.305

Gao, M., Liu, Y., & Song, Z. (2019). Effects of polyethylene microplastic on the phytotoxicity of di-n-butyl phthalate in lettuce (Lactuca sativa L. var. ramosa Hort). *Chemosphere, 237*, 124–482. doi:10.1016/j.chemosphere.2019.124482

Gewert, B., Plassmann, M. M., & MacLeod, M. (2015). Pathways for degradation of plastic polymers floating in the marine environment. *Environmental Science: Processes & Impacts, 17*(9), 1513–1521. doi:10.1039/C5EM00207A

Gui, J., Sun, Y., Wang, J., Chen, X., Zhang, S., & Wu, D. (2021). Microplastics in composting of rural domestic waste: Abundance, characteristics, and release from the surface of macroplastics. *Environmental Pollution, 274*, 116–553. doi:10.1016/j.envpol.2021.116553

Hegan, D., Tong, L., Zhiquan, H., Qinming, S., & Ru, L. (2015). Determining time limits of continuous film mulching and examining residual effects on cotton yield and soil properties. *Journal of Environmental Biology, 36*(3), 677.

Horton, A. A., Walton, A., Spurgeon, D. J., Lahive, E., & Svendsen, C. (2017). Microplastics in freshwater and terrestrial environments: Evaluating the current understanding to identify the knowledge gaps and future research priorities. *Science of the Total Environment, 586*, 127–141. doi:10.1016/j.scitotenv.2017.01.190

Huo, Y., Dijkstra, F. A., Possell, M., & Singh, B. (2022). Ecotoxicological effects of plastics on plants, soil fauna and microorganisms: A meta-analysis. *Environmental Pollution, 310*, 119–892. doi:10.1016/j.envpol.2022.119892

Ingraffia, R., Amato, G., Iovino, M., Rillig, M. C., Giambalvo, D., & Frenda, A. S. (2022). Polyester microplastic fibers in soil increase nitrogen loss via leaching and decrease plant biomass production and N uptake. *Environmental Research Letters, 17*(5), 12–54. doi:10.1088/1748–9326/ac652d

Jackson, R. B., Lajtha, K., Crow, S. E., Hugelius, G., Kramer, M. G., & Piñeiro, G. (2017). The ecology of soil carbon: pools, vulnerabilities, and biotic and abiotic controls. *Annual Review of Ecology, Evolution, and Systematics, 48*, 419–445. doi:10.1146/annurev-ecolsys-112414-054234

Katsumi, N., Kusube, T., Nagao, S., & Okochi, H. (2020). The role of coated fertilizer used in paddy fields as a source of microplastics in the marine environment. *Marine Pollution Bulletin, 161*, 111–727. doi:10.1016/j.marpolbul.2020.111727

Kaur, S., Samota, M. K., Choudhary, M., Choudhary, M., Pandey, A. K., Sharma, A., & Thakur, J. (2022). How do plants defend themselves against pathogens-Biochemical mechanisms and genetic interventions. *Physiology and Molecular Biology of Plants, 28*(2), 485–504. doi:10.1007/s12298-022-01146-y

Khalid, N., Aqeel, M., & Noman, A. (2020). Microplastics could be a threat to plants in terrestrial systems directly or indirectly. *Environmental Pollution, 267*, 115653. doi:10.1016/j.envpol.2020.115653v

Kokaeva, L., Chudinova, E., Berezov, A., Yarmeeva, M., Balabko, P., Belosokhov, A., & Elansky, S. (2020). Fungal diversity in tomato (Solanum lycopersicum) leaves and fruits in Russia. *Journal of Central European Agriculture, 21*(4), 809–816. doi:10.5513/JCEA01/21.4.2869

Krueger, M. C., Harms, H., & Schlosser, D. (2015). Prospects for microbiological solutions to environmental pollution with plastics. *Applied Microbiology and Biotechnology, 99*(21), 8857–8874. doi:10.1007/s00253-015-6879-4

Kumar Sen, S., & Raut, S. (2015). Microbial degradation of low density polyethylene (LDPE): A review. *Journal of Environmental Chemical Engineering, 3*(1), 462–473. doi:10.1016/j.jece.2015.01.003

Lamalakshmi Devi, E., Kumar, S., Basanta Singh, T., Sharma, S. K., Beemrote, A., Devi, C. P., Chongtham, S. K., Singh, C. H., Yumlembam, R. A., & Haribhushan, A. (2017). Adaptation strategies and defence mechanisms of plants during environmental stress. In Ghorbanpour, M., & Varma, A. (Eds.), *Medicinal Plants and Environmental Challenges* (pp. 359–413). Springer, Cham. doi:10.1007/978-3-319-68717-9_20

Leifheit, E. F., Lehmann, A., & Rillig, M. C. (2021). Potential effects of microplastic on arbuscular mycorrhizal fungi. *Frontiers in Plant Science, 12*, 626709. doi:10.3389/fpls.2021.626709

Li, J., Liu, H., & Paul Chen, J. (2018). Microplastics in freshwater systems: A review on occurrence, environmental effects, and methods for microplastics detection. *Water Research, 137*, 362–374. doi:10.1016/j.watres.2017.12.056

Li, K., Xu, L., Bai, X., Zhang, G., Zhang, M., & Huang, Y. (2024). Differential fungal assemblages and functions between the plastisphere of biodegradable and conventional microplastics in farmland. *Science of the Total Environment, 906*, 167–478. doi:10.1016/j.scitotenv.2023.167478

Li, L., Luo, Y., Li, R., Zhou, Q., Peijnenburg, W. J. G. M., Yin, N., Yang, J., Tu, C., & Zhang, Y. (2020). Effective uptake of submicrometre plastics by crop plants via a crack-entry mode. *Nature Sustainability, 3*(11), 929–937. doi:10.1038/s41893-020-0567-9

Li, Y., Gao, W., Wang, C., & Gao, M. (2023). Distinct distribution patterns and functional potentials of rare and abundant microorganisms between plastisphere and soils. *Science of the Total Environment, 873*, 162–413. doi:10.1016/j.scitotenv.2023.162413

Li, Y., Yang, R., Guo, L., Gao, W., Su, P., Xu, Z., Xiao, H., Ma, Z., Liu, X., & Gao, P. (2022). The composition, biotic network, and assembly of plastisphere protistan taxonomic and functional communities in plastic-mulching croplands. *Journal of Hazardous Materials, 430*, 128–390. doi:10.1016/j.jhazmat.2022.128390

Liu, W., Zhang, J., Liu, H., Guo, X., Zhang, X., Yao, X., Cao, Z., & Zhang, T. (2021). A review of the removal of microplastics in global wastewater treatment plants: Characteristics and mechanisms. *Environment International, 146*, 106277. doi:10.1016/j.envint.2020.106277

Luo, Y., Li, L., Feng, Y., Li, R., Yang, J., Peijnenburg, W. J. G. M., & Tu, C. (2022). Quantitative tracing of uptake and transport of submicrometre plastics in crop plants using lanthanide chelates as a dual-functional tracer. *Nature Nanotechnology, 17*(4), 424–431. doi:10.1038/s41565-021-01063-3

Luo, Z., Zhou, X., Su, Y., Wang, H., Yu, R., Zhou, S., Xu, E. G., & Xing, B. (2021). Environmental occurrence, fate, impact, and potential solution of tire microplastics: Similarities and differences with tire wear particles. *Science of the Total Environment, 795*, 148902. doi:10.1016/j.scitotenv.2021.148902

Lwanga, E. H., Beriot, N., Corradini, F., Silva, V., Yang, X., Baartman, J., Rezaei, M., van Schaik, L., Riksen, M., & Geissen, V. (2022). Review of microplastic sources, transport pathways and correlations with other soil stressors: A journey from agricultural sites into the environment. *Chemical and Biological Technologies in Agriculture, 9*(1), 20. doi:10.1186/s40538-021-00278-9

Machado, D. S., Kloas, W., Zarfl, C., Hempel, S., & Rillig, M. C. (2018). Microplastics as an emerging threat to terrestrial ecosystems. *Global Change Biology, 24*(4), 1405–1416. doi:10.1111/gcb.14020

MacLean, J., Mayanna, S., Benning, L. G., Horn, F., Bartholomäus, A., Wiesner, Y., Wagner, D., & Liebner, S. (2021). The terrestrial plastisphere: Diversity and polymer-colonizing potential of plastic-associated microbial communities in soil. *Microorganisms, 9*(9), 1876. doi:10.3390/microorganisms9091876

Maguire, L. W., & Gardner, C. M. (2023). Fate and transport of biological microcontaminants bound to microplastics in the soil environment. *Science of the Total Environment*, 164439. doi:10.1016/j.scitotenv.2023.164439

Maity, S., Guchhait, R., Sarkar, M. B., & Pramanick, K. (2022). Occurrence and distribution of micro/nanoplastics in soils and their phytotoxic effects: A review. *Plant, Cell & Environment, 45*(4), 1011–1028. doi:10.1111/pce.14248

Mandal, A., Singh, N., Mondal, A., Talib, M., Basu, R., Biswas, M. K., & Darbha, G. K. (2023). The extent of microplastic pollution along the eastern coast of India: Focussing on marine waters, beach sand, and fish. *Marine Pollution Bulletin, 194*, 115265. doi:10.1016/j.marpolbul.2023.115265

Moral, J., Lichtemberg, P. S. F., Papagelis, A., Sherman, J., & Michailides, T. J. (2018). Didymella glomerata causing leaf blight on pistachio. *European Journal of Plant Pathology, 151*, 1095–1099. doi:10.1007/s10658-018-1422-y

Ng, E. L., Huerta Lwanga, E., Eldridge, S. M., Johnston, P., Hu, H. W., Geissen, V., & Chen, D. (2018). An overview of microplastic and nanoplastic pollution in agroecosystems. *Science of the Total Environment, 627*, 1377–1388. doi:10.1016/j.scitotenv.2018.01.341

Nizzetto, L., Futter, M., & Langaas, S. (2016). Are agricultural soils dumps for microplastics of urban origin? *Environmental Science & Technology, 50*(20), 10777–10779. doi:10.1021/acs.est.6b04140

Pfeilmeier, S., Caly, D. L., & Malone, J. G. (2016). Bacterial pathogenesis of plants: Future challenges from a microbial perspective: Challenges in bacterial molecular plant pathology. *Molecular Plant Pathology, 17*(8), 1298–1313. doi:10.1111/mpp.12427

Qian, H., Zhang, M., Liu, G., Lu, T., Qu, Q., Du, B., & Pan, X. (2018). Effects of soil residual plastic film on soil microbial community structure and fertility. *Water, Air, & Soil Pollution, 229*, 1–11. doi:10.1007/s11270-018-3916–9

Ramanayaka, S., Hettithanthri, O., Sandanayake, S., & Vithanage, M. (2020). Ecological effects of chemical contaminants adsorbed to microplastics. In Rocha-Santos, T., Costa, M. F., & Mouneyrac, C. (Eds.), *Handbook of Microplastics in the Environment* (pp. 1–31). Springer, Cham.

Ramirez, M. M. B., Ricardo, D. C., & Rendón von Osten, J. (2019). Occurrence and seasonal distribution of microplastics and phthalates in sediments from the urban channel of the Ria and coast of Campeche, Mexico. *Science of the Total Environment, 672*, 97–105. doi:10.1016/j.scitotenv.2019.03.472

Ranauda, M. A., Zuzolo, D., Maisto, M., Tartaglia, M., Scarano, P., Prigioniero, A., Sciarrillo, R., & Guarino, C. (2024). Microplastics affect soil-plant system: Implications for rhizosphere biology and fitness of sage (Salvia officinalis L.). *Environmental Pollution, 346*, 123656. doi:10.1016/j.envpol.2024.123656

Ren, Z., Gui, X., Xu, X., Zhao, L., Qiu, H., & Cao, X. (2021). Microplastics in the soil-groundwater environment: Aging, migration, and co-transport of contaminants – A critical review. *Journal of Hazardous Materials, 419*, 126455. doi:10.1016/j.jhazmat.2021.126455

Rillig, M. C. (2012). Microplastic in terrestrial ecosystems and the soil? *Environmental Science & Technology, 46*(12), 6453–6454. doi:10.1021/es302011r

Rillig, M. C., Ingraffia, R., & de Souza Machado, A. A. (2017). Microplastic incorporation into soil in agroecosystems. *Frontiers in Plant Science, 8*, 1805. doi:10.3389/fpls.2017.01805

Rillig, M. C., Kim, S. W., & Zhu, Y.-G. (2024). The soil plastisphere. *Nature Reviews Microbiology, 22*(2), 64–74. doi:10.1038/s41579-023-00967-2

Rillig, M. C., & Lehmann, A. (2020). Microplastic in terrestrial ecosystems. *Science, 368*(6498), 1430–1431. doi:10.1126/science.abb5979

Rillig, M. C., Leifheit, E., & Lehmann, J. (2021). Microplastic effects on carbon cycling processes in soils. *PLoS Biology, 19*(3), e3001130. doi:10.1371/journal.pbio.3001130

Rüthi, J. (2022). Characterization of the plastisphere microbiome and plastic-degrading microorganisms in cold terrestrial environments. Doctoral Thesis. doi:10.3929/ethz-b-000600635

Rüthi, J., Bölsterli, D., Pardi-Comensoli, L., Brunner, I., & Frey, B. (2020). The "plastisphere" of biodegradable plastics is characterized by specific microbial taxa of alpine and arctic soils. *Frontiers in Environmental Science, 8*, 562263. doi:10.3389/fenvs.2020.562263

Salam, M., Zheng, H., Liu, Y., Zaib, A., Rehman, S. A. U., Riaz, N., Eliw, M., Hayat, F., Li, H., & Wang, F. (2023). Effects of micro (nano) plastics on soil nutrient cycling: State of the knowledge. *Journal of Environmental Management, 344*, 118–437. doi:10.1016/j.jenvman.2023.118437

Séveno, N. A., Kallifidas, D., Smalla, K., van Elsas, J. D., Collard, J.-M., Karagouni, A. D., & Wellington, E. M. H. (2002). Occurrence and reservoirs of antibiotic resistance genes in the environment. *Reviews and Research in Medical Microbiology, 13*(1), 15–27. doi:10.1097/00013542-200201000-00002

Shi, Z., Xiong, L., Liu, T., & Wu, W. (2022). Alteration of bacterial communities and co-occurrence networks as a legacy effect upon exposure to polyethylene residues under field environment. *Journal of Hazardous Materials, 426*, 128126. doi:10.1016/j.jhazmat.2021.128126

Steinmetz, Z., Wollmann, C., Schaefer, M., Buchmann, C., David, J., Tröger, J., Muñoz, K., Frör, O., & Schaumann, G. E. (2016). Plastic mulching in agriculture. Trading short-term agronomic benefits for long-term soil degradation? *Science of the Total Environment, 550*, 690–705. doi:10.1016/j.scitotenv.2016.01.153

Su, X., & Yang, K. (2024). Exploring the blindspot: The soil plastisphere. *Soil Ecology Letters, 6*(2), 230209. doi:10.1007/s42832-023-0209-z

Sun, H., Shi, Y., Zhao, P., Long, G., Li, C., Wang, J., Qiu, D., Lu, C., Ding, Y., & Liu, L. (2023). Effects of polyethylene and biodegradable microplastics on photosynthesis, antioxidant defense systems, and arsenic accumulation in maize (Zea mays L.) seedlings grown in arsenic-contaminated soils. *Science of the Total Environment, 868*, 161–557. doi:10.1016/j.scitotenv.2023.161557

Sun, Y., Shi, J., Wang, X., Ding, C., & Wang, J. (2022). Deciphering the mechanisms shaping the plastisphere microbiota in soil. *Msystems, 7*(4), e00352–00322. doi:10.1128/msystems.00352-22

Sun, Y., Wu, M., Xie, S., Zang, J., Wang, X., Yang, Y., Li, C., & Wang, J. (2024). Homogenization of bacterial plastisphere community in soil: A continental-scale microcosm study. *ISME Communications, 4*(1), ycad012. doi:10.1093/ismeco/ycad012

Tang, K. H. D. (2024). Terrestrial and aquatic plastisphere: Formation, characteristics, and influencing factors. *Sustainability, 16*(5), 21–63. doi:10.3390/su16052163

Tanunchai, B., Ji, L., Schröder, O., Gawol, S. J., Geissler, A., Wahdan, S. F. M., Buscot, F., Kalkhof, S., Schulze, E.-D., & Noll, M. (2023). Fate of a biodegradable plastic in forest soil: Dominant tree species and forest types drive changes in microbial community assembly, influence the composition of plastisphere, and affect poly (butylene succinate-co-adipate) degradation. *Science of the Total Environment, 873*, 162–230. doi:10.1016/j.scitotenv.2023.162230

Tomazelli, D., dos Santos Peron, R. A., Mendes, S. D. C., Pinto, C. E., Baldissera, T. C., Baretta, D., Mendes, L. W., Goss-Souza, D., & Klauberg-Filho, O. (2024). Plant diversity and root traits shape rhizosphere microbial communities in natural grasslands and cultivated pastures. *Rhizosphere, 29*, 100–864. doi:10.1016/j.rhisph.2024.100864

van Sebille, E., Wilcox, C., Lebreton, L., Maximenko, N., Hardesty, B. D., van Franeker, J. A., Eriksen, M., Siegel, D., Galgani, F., & Law, K. L. (2015). A global inventory of small floating plastic debris. *Environmental Research Letters, 10*(12), 124006. doi:10.1088/1748–9326/10/12/124006

Vithanage, M., & Prasad, M. N. V. (2023). *Microplastics in the Ecosphere: Air, Water, Soil, and Food.* John Wiley & Sons, Hoboken, NJ.

Vithanage, M., Ramanayaka, S., Hasinthara, S., & Navaratne, A. (2021). Compost as a carrier for microplastics and plastic-bound toxic metals into agroecosystems. *Current Opinion in Environmental Science & Health, 24*, 100297. doi:10.1016/j.coesh.2021.100297

Waldman, W. R., & Rillig, M. C. (2020). Microplastic research should embrace the complexity of secondary particles. *Environmental Science & Technology, 54*(13), 7751–7753. doi:10.1021/acs.est.0c02194

Wang, C., Wang, L., Ok, Y. S., Tsang, D. C. W., & Hou, D. (2022). Soil plastisphere: Exploration methods, influencing factors, and ecological insights. *Journal of Hazardous Materials, 430*, 128503. doi:10.1016/j.jhazmat.2022.128503

Wang, C., Zhao, J., & Xing, B. (2021). Environmental source, fate, and toxicity of microplastics. *Journal of Hazardous Materials, 407*, 124357. doi:10.1016/j.jhazmat.2020.124357

Wang, J., Tan, Z., Peng, J., Qiu, Q., & Li, M. (2016). The behaviors of microplastics in the marine environment. *Marine Environmental Research, 113*, 7–17. doi:10.1016/j.marenvres.2015.10.014

Wang, W., Xie, Y., Li, H., Dong, H., Li, B., Guo, Y., Wang, Y., Guo, X., Yin, T., & Liu, X. (2024). Responses of lettuce (Lactuca sativa L.) growth and soil properties to conventional non-biodegradable and new biodegradable microplastics. *Environmental Pollution, 341*, 122897. doi:10.1016/j.envpol.2023.122897

Wijekoon, P., Koliyabandara, P. A., Cooray, A. T., Lam, S. S., Athapattu, B. C. L., & Vithanage, M. (2022). Progress and prospects in mitigation of landfill leachate pollution: Risk, pollution potential, treatment and challenges. *Journal of Hazardous Material, 421*, 126627. doi:10.1016/j.jhazmat.2021.126627

Wu, M., Yang, C., Du, C., & Liu, H. (2020). Microplastics in waters and soils: Occurrence, analytical methods and ecotoxicological effects. *Ecotoxicology and Environmental Safety, 202*, 110–910. doi:10.1016/j.ecoenv.2020.110910

Yoshida, S., Hiraga, K., Takehana, T., Taniguchi, I., Yamaji, H., Maeda, Y., Toyohara, K., Miyamoto, K., Kimura, Y., & Oda, K. (2016). A bacterium that degrades and assimilates poly (ethylene terephthalate). *Science, 351*(6278), 1196–1199. doi:10.1126/science.aad6359

Yu, Z., Xu, X., Guo, L., Jin, R., & Lu, Y. (2024). Uptake and transport of micro/nanoplastics in terrestrial plants: Detection, mechanisms, and influencing factors. *Science of the Total Environment, 907*, 168155. doi:10.1016/j.scitotenv.2023.168155

Zhai, Y., Bai, J., Chang, P., Liu, Z., Wang, Y., Liu, G., Cui, B., Peijnenburg, W., & Vijver, M. G. (2024). Microplastics in terrestrial ecosystem: Exploring the menace to the soil-plant-microbe interactions. In *TrAC Trends in Analytical Chemistry* (pp. 117–667). Elsevier Scientific Pub. Co., Amsterdam. doi:10.1016/j.trac.2024.117667

Zhang, D., Ng, E. L., Hu, W., Wang, H., Galaviz, P., Yang, H., Sun, W., Li, C., Ma, X., & Fu, B. (2020). Plastic pollution in croplands threatens long-term food security. *Global Change Biology, 26*(6), 3356–3367. doi:10.1111/gcb.15043

Zhang, G. S., & Liu, Y. F. (2018). The distribution of microplastics in soil aggregate fractions in southwestern China. *Science of the Total Environment, 642*, 12–20. doi:10.1016/j.scitotenv.2018.06.004

Zhang, J., Ren, S., Xu, W., Liang, C., Li, J., Zhang, H., Li, Y., Liu, X., Jones, D. L., & Chadwick, D. R. (2022). Effects of plastic residues and microplastics on soil ecosystems: A global meta-analysis. *Journal of Hazardous Materials, 435*, 129065. doi:10.1016/j.jhazmat.2022.129065

Zhang, M., Zhao, Y., Qin, X., Jia, W., Chai, L., Huang, M., & Huang, Y. (2019). Microplastics from mulching film is a distinct habitat for bacteria in farmland soil. *Science of the Total Environment, 688*, 470–478. doi:10.1016/j.scitotenv.2019.06.108

Zhou, J., Wen, Y., Marshall, M. R., Zhao, J., Gui, H., Yang, Y., Zeng, Z., Jones, D. L., & Zang, H. (2021). Microplastics as an emerging threat to plant and soil health in agroecosystems. *Science of the Total Environment, 787*, 147–444. doi:10.1016/j.scitotenv.2021.147444

Zhu, D., Ma, J., Li, G., Rillig, M. C., & Zhu, Y.-G. (2022). Soil plastispheres as hotspots of antibiotic resistance genes and potential pathogens. *The ISME Journal, 16*(2), 521–532. doi:10.1038/s41396-021-01103-9

Zhu, J., Zhang, Q., Li, Y., Tan, S., Kang, Z., Yu, X., Lan, W., Cai, L., Wang, J., & Shi, H. (2019). Microplastic pollution in the Maowei Sea, a typical mariculture bay of China. *Science of the Total Environment, 658*, 62–68. doi:10.1016/j.scitotenv.2018.12.192

Zhurina, M. V., Bogdanov, K. I., Gannesen, A. V., Mart'yanov, S. V., & Plakunov, V. K. (2022). Microplastics as a new ecological niche for multispecies microbial biofilms within the plastisphere. *Microbiology, 91*(2), 107–123. doi:10.1134/S0026261722020126

Ziani, K., Ioniță-Mîndrican, C. B., Mititelu, M., Neacșu, S. M., Negrei, C., Moroșan, E., Drăgănescu, D., & Preda, O. T. (2023). Microplastics: A real global threat for environment and food safety: A state of the art review. *Nutrients, 15*(3). doi:10.3390/nu15030617

14 Beyond Biofilm

The Role of Biominerals and Metals in the Plastisphere – Implications for Elemental Cycling, Biogeochemical Processes, and Human Health

Asal Peydaei, Maya Shakunt Dodhia, Debby Schmidt, Saghar Hendiani, Thomas Neu, Karina Krarup Sand, and Nicole R. Posth

14.1 INTRODUCTION

Plastic debris is widely distributed in the environment, marring our coastlines from the tropics to the arctic, entangling marine organisms, and threatening ecosystems and human health (NOAA, 2014; UN Decade, 2023; Stoett and Vince, 2019). The varied material properties and the broad size range of plastic, from the nano- to the macroscale, extend the potential for transport and interaction of this pollutant via abiotic and biologically driven processes within the complex systems of ecological and biogeochemical cycles (Mitrano et al., 2021; Rogers et al., 2020). Biofilm rapidly forms on plastic surfaces, creating a new ecological niche, termed the plastisphere, that links microbial processes to plastic degradation and transport, as well as plastic uptake by organisms along the food web (Zettler et al., 2013). The plastisphere, however, also harbors organic matter, metals, minerals, and salts reflective of the host environment (Dodhia et al., 2023, 2025; Figure 14.1). Whether in water, air, soil, or sediment, the plastisphere forms the interface between plastic and the surrounding environment (Rillig et al., 2024). Therefore, the interfacial chemistry of the plastisphere, as much as the microorganisms, drives the myriads of reactions within the biofilm, as well as between the plastic and the surrounding ecosystem. Biogeochemical processes that drive or hinder biological uptake, plastic transport, plastic stability, and even genetic transfer are all constrained by the components that build this reactive interface.

Over the past decade, interdisciplinary research integrating microbiology, mineralogy, chemistry, physics, and geology has elucidated the biogeochemically complex interface of the plastisphere. Microbial communities within the plastisphere are pivotal, facilitating the degradation, transport, and incorporation of (micro)plastics into biogeochemical cycles (Rogers et al., 2020). Emerging evidence indicates that these processes extend to mineralization and metal cycling (Dodhia et al., 2025). Furthermore, metals and minerals within the plastisphere act as nucleation sites and catalysts for various chemical reactions, functioning as both electron donors and acceptors. This dynamic interaction supports microbial element cycling and biomineralization across diverse environmental matrices.

DOI: 10.1201/9781032717814-14

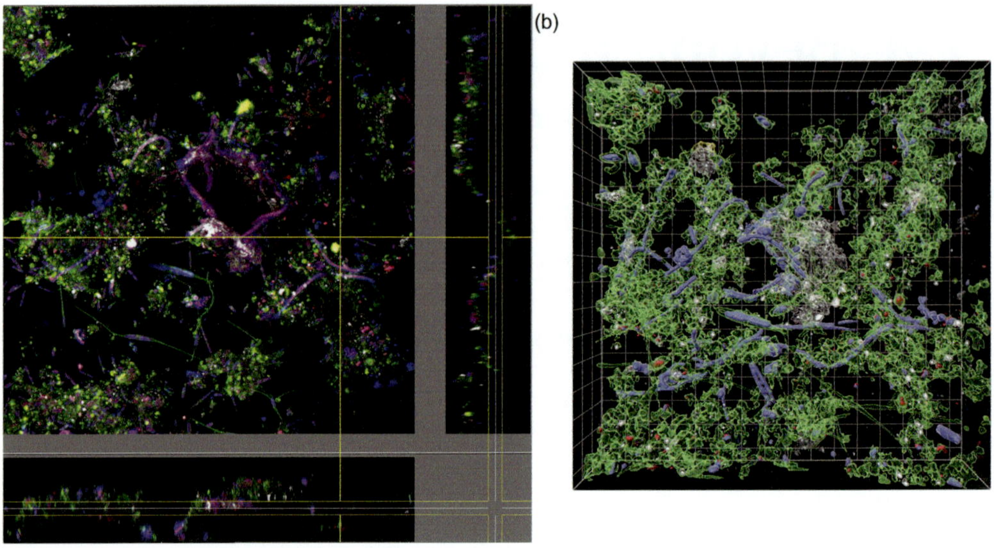

FIGURE 14.1 Confocal microscopy of an environmental biofilm on weathered polylactic acid (PLA) plastic exposed to a mesocosm tank experiment, with water from the Gulf of Bothnia exposed for 6 weeks. (a) An XYZ Projection showing a multi-channel dataset and cross-sections of the data, indicating bacteria and glycoconjugates (green), chlorophyll A pigment of phototrophs (blue), cyanobacteria (pink), matrix compounds (red) and reflective mineral constituents (grey). (b) Three-dimensional multichannel projection through the same dataset where blue represents both phototrophs and cyanobacteria. Grid size is equivalent to 20 μm.

This chapter compiles research on the occurrence of metals and minerals associated with plastics in the aquatic environment in order to consolidate our current understanding of this new substratum. The composition of the plastisphere is discussed within the context of plastic stability and biogeochemical processes across redox regimes from water to sediment. Biological and abiotic metal and mineral formation is described and set into the context of findings from recent studies across diverse locations. Finally, we link the co-occurrence of bacteria, minerals, metals, and evolutionary drivers to the propagation of antibiotic-resistance genes within this global pollutant and highlight the associated impact on human and environmental health.

14.2 PLASTISPHERE COMPOSITION AND DYNAMICS

Microbes proliferate as communities on diverse surfaces, such as natural substratum-like sediments, cobbles, wood, leaves, and macrophytes, as well as anthropogenic surfaces like plastic. Biofilms are complex communities of microorganisms ensconced within a self-produced matrix of extracellular polymeric substances (EPSs) (Figure 14.2) (Di Martino, 2018; Flemming et al., 2016; Lopez et al., 2010). Microbial consortia derive various advantages from life in a biofilm, such as shielding against ultraviolet (UV) radiation, antimicrobial agents, and environmental fluctuations (Costa et al., 2019; Erni-Cassola et al., 2020; Yin et al., 2019). Nutrients and essential elements are exchanged or made available within the biofilm community. Biofilms also play a pivotal role in modulating the fate of contaminants through processes like sequestration, biotransformation, and biodegradation (Guasch et al., 2012). In addition, they serve as reservoirs for antibiotic-resistant bacteria and pathogens (Flores-Vargas et al., 2021).

The EPS matrix of a biofilm functions as an elemental reservoir, fostering microbial growth and development (Costa et al., 2019; Flemming et al., 2023; Toyofuku et al., 2016; Vu et al., 2009). Simultaneously, the EPS matrix provides structural support, promotes attachment, retains water, and confers resistance against environmental stressors like UV radiation and biocides (Flemming and Wingender, 2010; Toyofuku et al., 2016; Vu et al., 2009). Moreover, EPS serves as an electron donor or acceptor for microorganisms, further enhancing their metabolic activities (Flemming et al., 2023). As a result, the biofilm matrix was described as an important multifunctional constituent in

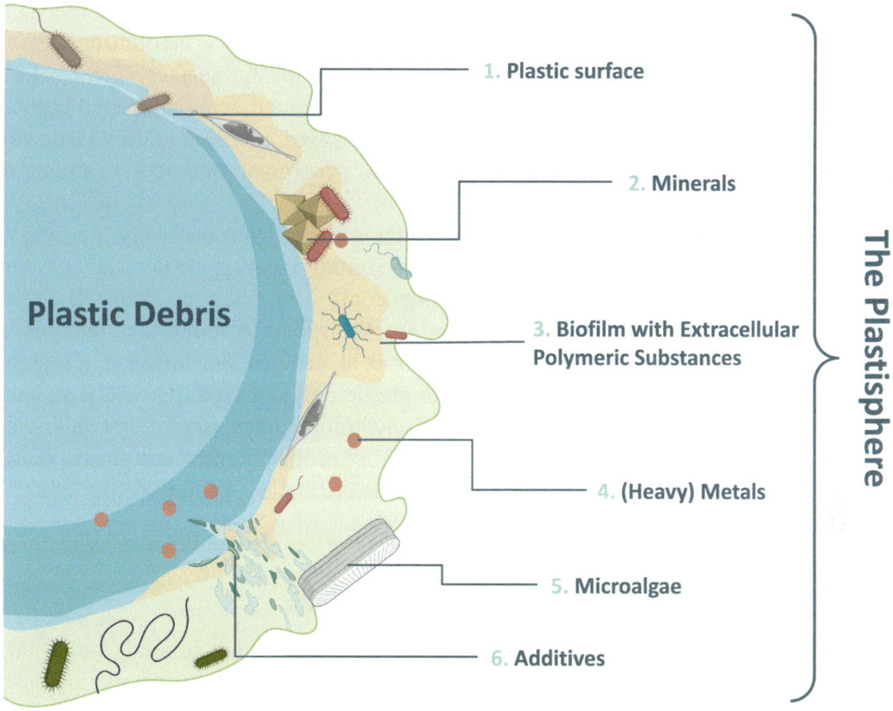

FIGURE 14.2 The plastisphere is dynamic and ever-evolving. (1) The plastic surface provides a habitat for a diverse microbial community, where microcracks and fractures offer attractive niches for microorganisms to settle. (2) Minerals attach to or form on the plastic surface, serving as nutrient sources and adding to particle mass. (3) The plastic microbiome is highly diverse and distinct from the surrounding environment, harboring unique microbial communities. EPS forms a heterogeneous, sticky matrix that enables particle adhesion and enhances biofilm resilience. (4) (Heavy) metals can accumulate in the plastic biofilm, acting as catalysts, templates, or toxins. (5) Microalgae, e.g., diatoms, attach to plastic debris, building new pathways for species distribution. (6) Additives leach from plastic particles during degradation. (Created in BioRender. Schmidt, D. (2024) https://BioRender.com/h45a629)

microbial biofilms (Neu and Lawrence, 2017). Particularly under conditions of high light and nutrient availability, the biofilm EPS matrix can become thick and adhesive, capable of trapping suspended sediments, particles, and microplastics (Figure 14.2). This enhanced mass may contribute to their deposition in sediments (Enders et al., 2019). EPS can constitute over 90% of the dry mass of a biofilm (Flemming and Wingender, 2010), forming a complex network of channels between densely packed cells that facilitate the exchange of oxygen, nutrients, and waste (Geisel et al., 2022; Quan et al., 2022).

Similar to other biofilms, biofilm formation in the plastisphere typically involves a sequence of events, including a conditioning film, microbial attachment, secretion of EPS, and microbial proliferation (Flemming et al., 2016; Zettler et al., 2013). There are three main stages that have been identified in the formation of the biofilm community on plastics; first, an initial pioneering colonization stage, followed by secondary colonization, which eventually forms into the third stage, a mature plastisphere community (Du et al., 2022). During initial colonization, pioneer bacteria from the surrounding seawater or sediments establish themselves on the plastic surface through reversible attachment and influence the subsequent microbial communities within the plastisphere matrix (De Tender et al., 2017; Shi et al., 2023). These pioneer colonies diminish the hydrophobicity of plastics over time (Lobelle and Cunliffe, 2011; Tu et al., 2020) and secrete extracellular polysaccharides that enhance the adhesion of microorganisms to the plastic surface

(Ganesh Kumar et al., 2020). Subsequently, secondary colonization utilizes various mechanisms, such as the formation of pili, adhesion proteins, and production of EPS, facilitating irreversible attachment and providing additional attachment sites on the plastic surface (Du et al., 2022; Dussud et al., 2018). Maturation of the biofilm occurs leading to a diverse microbial community with ongoing recruitment and loss of species driven by increased adhesion to the plastic surface. This process is continuously influenced by competition or synergistic interactions among different microorganisms (Lorite et al., 2011).

The physical and chemical characteristics of plastics such as hydrophobicity, particle shape, roughness, crystallinity, and surface charge influence the initial selection of bacterial communities during microbial colonization (Du et al., 2022; Rummel et al., 2017; Shi et al., 2023). However, the precise role of the plastic substratum itself in shaping plastisphere communities remains uncertain, as highlighted in a recent meta-analysis by Wright et al. (2021b). For instance, research indicates variations in community composition based on plastic polymer type, particle size, and even color (Frère et al., 2018; Harmon et al., 2024). Conversely, prior studies suggest that abiotic factors such as temperature, nutrient availability, depth, and salinity primarily influence plastisphere community structures overriding the effects of plastic-specific attributes (Amaral-Zettler et al., 2020; Oberbeckmann et al., 2018; Wright et al., 2021a,b, 2020).

14.3 METALS AND MINERALS IN THE PLASTISPHERE

Biofilm studies focused on microbe–mineral interactions demonstrate that minerals and metals provide attachment sites, act as nutrient sources, and facilitate biogeochemical degradation processes (Gorbushina, 2007; Warscheid and Braams, 2000). The mineral and metal composition of any surface thereby plays a significant role in the biogeochemical environment, the elements available for metabolic cycling, and the continuous construction of the biofilm itself. Consequently, the presence of metals can also promote metal-cycling metabolisms. Under nutrient-poor environmental conditions, this can have a further selective effect on the composition of microbial consortiums and lead to the development of distinct communities (Choe et al., 2018).

Plastic particles offer novel attachment areas and provide protection and transport for microbial life in aquatic environments. Minerals and metals on the plastic substratum and incorporated within plastic biofilms effectively alter the surface area and ionic charge of plastics and add to the physical mass, affecting particle buoyancy and transport (Leiser et al., 2021b, 2021a, 2020). Mineralization processes in aquatic environments can be abiotic, biologically induced, or biologically controlled (Figure 14.3). Abiotic minerals, such as carbonates, salts, or silicates, form through physiochemical reactions (like nucleation or supersaturation) at the plastic surface but can also adsorb or be captured by the plastic biofilm from the surrounding environment. Biologically induced mineral (BIM) formation is driven by the indirect influence of microbial activity such as metabolic activities and properties at the cell surface or within EPS, creating favorable chemical conditions for mineralization (Frankel and Bazylinski, 2003; Lowenstam, 1981).

Pyrite or goethite is often the result of such biologically induced mineralization (BIM) pathways. Metals and metal-containing minerals serve as electron acceptors and donors, enzyme cofactors, or toxic repellents, inhibiting the growth of less resilient microorganisms (Stabnikova et al., 2022). Many organisms have also developed mechanisms to deliberately control mineralization processes (BCM) for the purpose of structural support, skeletal development, or protection (Boskey, 2003). Assessing the extent to which plastic affects BCM is critical, particularly as biomineralization processes play an important role in marine element turnovers. It is also relevant to consider the significance of biomineralizing organisms for the capture, transport and storage of plastic debris in aquatic settings.

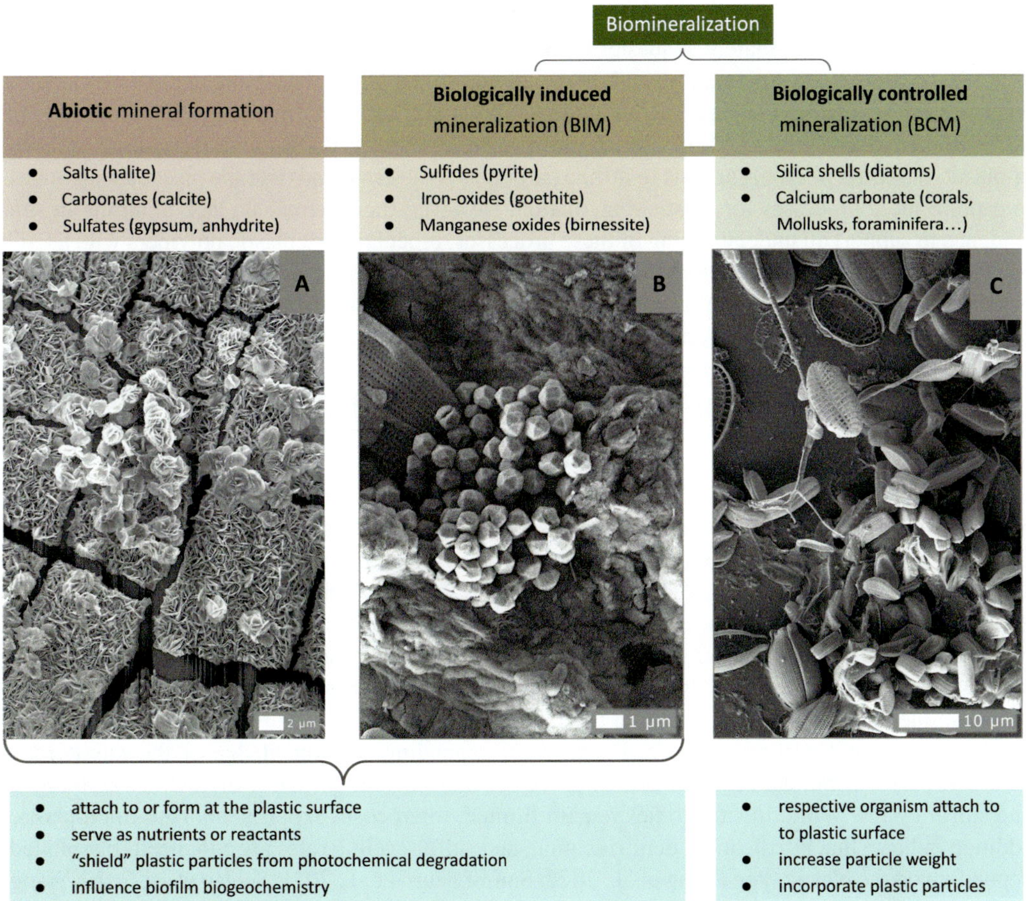

FIGURE 14.3 Different pathways of mineral formation in aquatic environments. SEM images of plastics after exposure to water (a) for 3 months and sediment (b, c) for 3 (c) and 6 (b) months at Svanemøllen Harbor, Denmark (North Harbour, Copenhagen). Showcase the attachment of different mineral types. (a) Abiotic minerals covering part of a polyethylene plastic strip. (b) Pyrite framboid embedded in the plastisphere biofilm matrix on polystyrene. (c) Attachment of microalgae (diatoms) to the plastisphere on polystyrene. (Modified from Dodhia et al., 2023).

14.3.1 Biogeochemical Cycling in the Plastisphere

Biogeochemical cycles underscore all interactions between life on Earth and the physical environment through an interplay of biotic and abiotic reactions. Through passive and active mechanisms driven by cell metabolism, microorganisms close all major element cycles (C, H, N, S, P, O, and Fe) across oxic and anoxic environments (Jørgensen, 1995). The plastisphere harbors a community of microorganisms unique from those of the surrounding environment (Amaral-Zettler et al., 2020; Li et al., 2024; Wright et al., 2021b). This distinct community is integral to plastic degradation, with numerous species now recorded to actively utilize plastic as a substratum as well as a substrate (Goudriaan et al., 2023; Roager and Sonnenschein, 2019; Rogers et al., 2020; Yoshida et al., 2016).

Of all biogeochemical cycles, plastic interactions with the carbon cycle have had the most focus and offer an illustrative example of the potential role of plastics in global element cycles. First, throughout their lifecycle, conventional petroleum-derived plastics contribute to global carbon budgets, as they are produced from fossil fuels, release greenhouse gases through their entire life cycle, and ultimately sequester carbon when they are buried in the sediment (Stubbins et al., 2021; Zhu, 2021). Second, the carbon stored within the plastic backbone and plastic-derived leachates act as a carbon source for microbial anabolic and catabolic reactions (Goudriaan et al., 2023; Romera-Castillo et al., 2018). Much plastisphere research has thus investigated the abilities of plastic-harbored microorganisms to utilize carbon as a substrate and degrade plastics with hopes to capture these processes for plastic remediation. However, there remains a lack of strong in situ evidence to support an understanding of these processes, especially for plastic polymers with a C-C backbone that have not been altered (Gambarini et al., 2021; Lear et al., 2022). To strengthen the evidence for microbial degradation studies, a combination of several factors, including composition of the plastic (including additives), description of pretreatment actions, and reliable degradation rates (>20% mass loss) are necessary (Lear et al., 2022).

In the marine environment, sediment is considered a major sink for plastic waste (Bergmann et al., 2017; Martin et al., 2022), yet our understanding of the degradative capabilities of the aquatic plastisphere in the water column still outweighs that of the sediment. Depending on the location in the ecosystem, plastic will be exposed to changing biogeochemical conditions across redox regimes (oxic, suboxic, and anoxic), as is the case as plastic transports between water and sediment. Changing redox conditions alter not only the core plastisphere microbiome but also the minerals, salts, and metals attached to the plastic and embedded within the biofilm structure (Dodhia et al., 2023). Fluctuating oxygen concentrations, decreasing light penetration, and changing microbial communities in the deeper water column and sediment all hinder the key degradative pathways for plastic as currently understood (photooxidation and thermal degradation) and indicate a need to constrain the long-term fate of plastic in these "final" sinks. Several studies have recently focused on the key element cycling of the sediment sink, expanding our knowledge of the intertwined nature of plastic-sediment biogeochemical processes and interactions associated with the nitrogen and sulfur cycles. While in situ studies remain limited, microcosm experiments with microplastic addition indicate that nitrification, denitrification, and sulfur cycling rates are enhanced or inhibited depending on the plastic type (Ding et al., 2022; Sanz-Lázaro et al., 2021; Seeley et al., 2020; Wang et al., 2023b).

14.3.2 Minerals at the Plastic Surface

The formation and occurrence of minerals in aquatic settings is dependent on the geochemical environment and redox state, as these parameters determine the ions available for both abiotic and biotic mineralization processes (Dodhia et al., 2023; Rogers et al., 2020) (Figure 14.3). In low-oxygen environments, BIM processes dominate, where microbial metabolic activities facilitate the generation of reduced species, i.e., ferrous iron ($Fe^{2=}$) and reduced sulfur compounds, enabling the formation of metal oxides and sulfides. Contrastingly, reducing conditions inhibit the precipitation of carbonates or silicates due to the increasing solubility of their respective reactants. A wide range of mineral species, such as carbonates, metal oxides, and silicates, have been reported in connection with the plastisphere (Dodhia et al., 2023; Kowalski et al., 2016; Ye and Andrady, 1991; Ye et al., 2022). Minerals such as carbonates and metal oxides can form heterogeneously on surfaces that offer a decrease in interfacial free energy and thereby decrease the energy barrier required for mineral formation (Hamm, 2005; Nielsen et al., 2020; Sand et al., 2020; Stabnikova et al., 2022).

While the surface properties of some minerals, such as their surface charge, can promote the direct attachment to the polymer surface (Corcoran, 2015; Ye et al., 2022), the presence of a biofilm on a substratum appears to be the main controlling factor for mineral adsorption (Stabnikova et al., 2022). The stickiness of EPS increases the adhesion of microplastics to natural particles, thereby enhancing their incorporation into the plastisphere (Michels et al., 2018). This is demonstrated in sinking experiments conducted by Kowalski et al. (2016) exhibiting quartz grains captured from sediments and incorporated into plastisphere biofilms. Similarly, Dodhia et al. (2023) describe the attachment of pyrite into biofilms on plastics exposed to top sediments.

14.3.3 Detection of Minerals in the Plastisphere

Minerals and metals in the plastisphere can be analyzed using visual identification of minerals, i.e., through the use of SEM coupled to elemental dispersive spectroscopy (EDS) analysis, which provides spot analyses of the elemental composition of mineral targets (Bradney et al., 2019; Tarchi et al., 2023). While useful, it does not provide a full picture of the minerals associated with micron-scale minerals found within the plastisphere and may be missed through the choice of location for analysis on the sample. Spot analysis may also result in the misidentification of minerals through the assumption of homogeneity across a visually similar inorganic layer, which may be elementally heterogeneous (Yüksel et al., 2023). Other methods to identify minerals like powder X-ray diffraction (XRD) require a bulk sample with enough volume to provide detailed information on the mineral composition and crystallinity (Ghodke et al., 2021). A large enough quantity of biofilm is often difficult to obtain from plastic samples, due to the small sample sizes typically acquired. Confocal laser scanning microscopy (CLSM) has also been used to visualize minerals in the plastisphere that informs on the spatial relationship of plastisphere components (Leiser et al., 2021b, 2021a) (Figure 14.1). Minerals are identified using the reflectance channel and may be a combination of reflective particles and minerals (Flemming et al., 2023). While CLSM allows for visualization of mineral location captured within the complex 3D architecture of the plastisphere, it does not provide further information on mineral composition. To address this, a recent cross-disciplinary study harnesses the capabilities of SEM-EDS and the use of automated software known as automated quantitative mineralogy (AQM). Automated quantitative SEM-EDS mapping locates the composition of minerals within the plastisphere and results in a 2D view identifying minerals and their orientation within the plastisphere matrix (Dodhia et al., in review). This application of a widely used mineralogical tool provides a new approach to enhance understanding of the plastisphere through expansion of the components within it. While the mineral component of the plastisphere is a new research area, it provides invaluable information for understanding the long-term interactions of plastic with the environments they settle within.

14.3.4 Mineral Impacts on Plastic and the Plastisphere

Minerals interact with plastics and the plastisphere in numerous ways: First, minerals act as a physical shield and may protect the plastic surface from photodegradation processes (Dodhia et al., 2023; Nelson et al., 2021). Second, they can increase the physical mass of the polymer, affecting particle buoyancy and downward transport (Leiser et al., 2020). Third, the incorporation of minerals into biofilms alters the plastisphere structure, enhancing biofilm heterogeneity, influencing ionic charges, and acting as a source of essential micronutrients and metal cofactors (Casar et al., 2021; Keren-Paz and Kolodkin-Gal, 2020; McLean et al., 1999). These trace nutrients and metal cofactors (required for enzymes) may be leached from the minerals through specialized microbial metabolic processes (Lepleux et al., 2012; Uroz et al., 2009). Fourth, dissolution of metal-bearing minerals can release toxic metals bound within minerals and can also produce reactive oxygen species (ROS) (Dong et al., 2022; Pal et al., 2022; Schoonen et al., 2006). Generation of ROS facilitates plastic degradation and induces oxidative stress on the microbial community (Gewert et al., 2015; Seixas et al., 2022). Contrastingly, the presence of minerals in the plastisphere modifies surface roughness and provides additional attachment sites for microbial life through increased surface area, potentially influencing the composition of the plastisphere community (Song et al., 2015). In addition, the heterogeneous nature of mineral surfaces, including the presence of cracks, fissures, and pores, can provide protective properties for microorganisms from environmental forces. Mineral surfaces also act as binding sites for both extracellular DNA and EPS aiding DNA preservation and forming a DNA reservoir (Freeman et al., 2023; Kjær et al., 2022). Further, the occurrence of solid and organic surfaces in the plastisphere facilitates horizontal gene transfer (HGT) (Sand and Jelavić, 2018).

The interactions between minerals and plastic polymers can also vary according to plastic size and shape. For instance, sediment grains have been found to become entangled in plastic fibers and nanoplastics were shown to adsorb onto goethite and hematite surfaces (Kowalski et al., 2016; Li et al., 2019; Zhang et al., 2020). This association between plastics and mineral grains may have significant implications for plastic mobilization and deposition in the water column and sediments (Laursen et al., 2023; Mueller et al., 2025). Therefore, investigating the impact of minerals within plastic biofilms and with plastics is crucial for unraveling the complexities of plastic–plastisphere ecosystem interactions.

14.3.5 Biologically Controlled Mineralization in the Plastisphere

Biomineralizing organisms interact with plastic in various ways (Figure 14.3). First, plastics have been recognized as colonization surfaces for biomineralizing organisms of various sizes. Just like bacteria or fungi, shell and skeletal-forming organisms can adhere to submerged surfaces (Tuck et al., 2022). This biofouling of plastic surfaces succeeds biofilm formation (Wright et al., 2020). Similar to mineral accumulation, the adhesion of shell-forming organisms and attachment of biomineralizing microbes can increase the mass of plastic polymers (Fazey and Ryan, 2016; Kaiser et al., 2017).

Biomineralizing organisms, such as the silica-forming diatoms, are extensively documented in plastisphere studies, although the extent to which plastics merely act as a substratum or if other benefits are derived from the polymer remains an open question. Studies investigating the influence of plastic particle size on phytoplankton indicate that photosynthetic activity may be limited or inhibited by aggregation of plastics on the organism's surface at the micro- and nanoscale (Dedman et al., 2022; Ye et al., 2023). However, at relevant environmental concentrations of micro- and nanoplastics (MNPs), no significant negative effects on diatom biomineralization have been demonstrated (Niu et al., 2021). Contrastingly, plastic particles have been found to interact with calcium carbonate-based biomineralization processes. Recent laboratory findings have demonstrated the incorporation and accumulation of MNP into biominerals, shells, and skeletons. This phenomenon is not species or organism specific but can be detected across domains in calcium carbonate-mineralizing organisms (Chen et al., 2023). In lab studies, aragonitic scleractinian corals incorporated PET plastic particles into their skeletal structure, while oysters, thick-shell mussels, and benthic foraminifera were shown to uptake MNP particles into their shells and skeletons (Han et al., 2022; Hierl et al., 2021; Joppien et al., 2022; Yu et al., 2022). The uptake of microplastic particles in oysters also increased the production of amorphous-calcium binding proteins, resulting in enhanced biomineralization (Han et al., 2022). These studies support the need for further investigation to unravel the mechanisms behind the incorporation of plastic into biominerals and explore the extent to which biominerals may act as a plastic sink (Chen et al., 2023; Matijaković et al., 2022; Soares et al., 2020, 2023).

14.3.6 Metals on Plastic

The accumulation of metals on plastic surfaces has garnered considerable attention due to its profound ecological and public health ramifications (Fischer et al., 2007; Rodrigues et al., 2022). Plastics, especially microplastics, pervade various ecosystems, acting as vectors for metal pollutants through surface adsorption processes (Liu et al., 2022). This phenomenon presents significant risks to aquatic ecosystems, as the leaching of adsorbed contaminants into water bodies can disrupt aquatic biota and bioaccumulate in trophic networks, ultimately reaching human consumers

TABLE 14.1
Recommended Maximum Permissible Metal Concentrations in Various Matrices

		Metal Concentrations (μg/g) or (μg/mL)							
	Medium of Concern	**Cd**	**Zn**	**Cu**	**Cr**	**Pb**	**Ni**	**As**	**Mn**
WHO (1996)	Permissible level in edible plants	0.02	0.06	10	1.3	2	10	—	—
WHO (2022)	Drinking water	0.003	—	2	0.05	0.01	0.07	0.01	0.08
US EPA (2015)	Drinking water	0.005	—	1.3	0.1	0.015	—	0.01	—
BIS (2012)	Drinking water	0.003	15	1.5	0.05	0.01	0.02	0.05	0.3
EU Commission (2023)	Muscle meat of fish	0.050–0.250*	—	—	—	0.3	—	—	—
	Cephalopods	1	—	—	—	0.3	—	—	—
	Crustaceans	0.5	—	—	—	0.5	—	—	—
	Bivalve molluscs	1	—	—	—	1.5	—	—	—
FAO (2023)	Natural mineral water	0.003	—	—	—	0.01	—	0.01	—
	Cephalopods	2	—	—	—	—	—	—	—
	Bivalve molluscs	2	—	—	—	—	—	—	—
US EPA (2013)	Marine sediment quality standards (dry weight)	5.1	410	390	260	450	—	57	—

(Rodrigues et al., 2022). Table 14.1 provides an overview of the recommended maximum permissible concentrations of various metals in different matrices, as proposed by several prominent regulatory bodies, including the WHO, US EPA, BIS, EU Commission, and FAO.

The mechanism of metal adsorption onto plastic surfaces involves a complex interplay of physicochemical interactions, including electrostatic attraction, surface interactions, ion exchange, hydrogen bonding, hydrophobic forces, and π–π interactions (Adeleye et al., 2024). Key factors influencing metal adsorption onto plastic surfaces include the properties of the plastics, such as polymer type, surface area, polarity, surface energy, and crystallinity (Adeleye et al., 2024; Liu et al., 2022).

Polarity and surface energy influence the interaction between plastics and metal ions, with polar plastics showing higher adsorption for metals due to stronger electrostatic interactions. For example, polyvinyl chloride (PVC) demonstrates a high adsorption capacity due to the presence of polar atomic chlorine within its molecular structure. In contrast, polyethylene (PE) and polypropylene (PP), which are nonpolar polymers, exhibit comparatively lower adsorption capacities (Liu et al., 2021, 2022). Additionally, lower crystallinity or higher amorphous content leads to rougher surfaces, enhancing adsorption by providing more active sites for metal ions (Adeleye et al., 2024).

14.3.7 Plastic Surface Characteristics and Environmental Influences

Generally, the adsorption capacity of metals onto plastics can change when the surface characteristics of plastic particles are modified, creating various active binding sites for metal ions. Aged plastics exhibit a greater surface area (Rodrigues et al., 2022), increased hydrophilicity, and more oxygen-containing functional groups compared to pristine plastics (Adeleye et al., 2024).

Functional groups such as aliphatic, aromatic, hydroxyl, carboxyl, and carbonyl prime the surface chemistry of plastics, with oxygen-containing groups like –COO and –OH significantly impacting the adsorption process through synergistic interactions and hydrogen bond formation (Adeleye et al., 2024; Stabnikova et al., 2022). Together, these changes lead to a reduction in the point of zero charge, promoting electrostatic attraction between plastics and metal cations by generating a negative charge on the plastics surface. Aged plastics also demonstrate a higher adsorption capacity for heavy metals, such as Cu^{2+} and Cd^{2+} (Liu et al., 2022). Environmental factors such as pH, salinity, and the presence of dissolved organic matter (DOM) significantly influence metal adsorption on plastics (Stabnikova et al., 2022; Wang et al., 2020).

Higher pH levels enhance the adsorption capacity by increasing the negative charge on the plastic surface, which promotes electrostatic interactions with metal cations. Salinity can impact metal adsorption by altering competition for binding sites and encouraging the formation of a denser aggregation structure (Wang et al., 2020). Furthermore, the interaction between plastics and DOM, especially humic and fulvic acids, can either inhibit or enhance metal adsorption, depending on the competitive interactions between metal ions and organic molecules (Wang et al., 2020). To provide a comprehensive overview, Table 14.2 presents a compendium of recent studies of the last 4 years on metal adsorption to plastic surfaces, highlighting the main metals discussed in the literature.

14.3.8 The Role of Biofilms in Metal Accumulation

One important function of the plastisphere is its ability to accumulate and retain metals from the surrounding environment. This process is influenced by the biofilm composition and surrounding water chemistry (Stabnikova et al., 2022). Studies have demonstrated the efficiency of biofilms in general for biosorbing heavy metals in different aquatic environments. For example, biofilms in the Metro River in Indonesia showed significant biosorption capabilities for copper (II), with a maximum absorption capacity of 2.14 mg per wet gram of biofilm (Kurniawan et al., 2022). Similarly, biofilms in streams affected by mining activities exhibit enhanced metal accumulation, influenced by the local water chemistry, which alters the bioavailability and interaction of metals with the biofilm matrix (Laderriere et al., 2021).

Similarly, the plastisphere has been found to influence metal accumulation in estuarine waters. This process not only impacts the local metal dynamics but also influences the microbial community structure within the biofilms, which can further affect metal cycling and ecological interactions in these environments (Richard et al., 2019). Seasonal variability in metal adsorption by biofilms, however, has been demonstrated by the differing trends in metal accumulation, i.e., in the Eziama River. During the rainy season, iron was the most accumulated metal, whereas zinc dominated in the dry season, highlighting the dynamic nature of biofilm interactions with metals based on environmental conditions (Kanu et al., 2006).

14.3.9 Impact of Metal Accumulation on Plastic Degradation

Polymer degradation is a complex process where the physical or chemical properties of a polymer change due to exposure to environmental factors such as light, heat, moisture, chemical conditions, or biological activity. Degradation leads to the breaking of polymer bonds and subsequent chemical transformations. The impact of polymer degradation is evident in various material properties. Mechanically, degraded polymers may exhibit reduced strength and flexibility, leading to crazing (the formation of fine cracks on the surface), cracking, and erosion. Optically, degradation can cause discoloration, affecting material transparency or color consistency (Shashoua, 2016). Electrically, changes in the polymer structure can alter its conductive properties, making it less effective in electronic applications. Additionally, polymer degradation can lead to phase separation or delamination, where different components of a composite material begin to separate, weakening the overall structure. The chemical changes involved in degradation often include bond scission, where the long chains of polymers break into smaller segments, chemical transformations that alter the material composition, and the formation of new functional groups that can change the polymer behavior and properties (Shah, 2021). The degradation pathways for plastics in the environment are primarily driven by photodegradation, thermo-oxidative degradation, and biodegradation. Photodegradation, initiated by exposure to UV radiation from sunlight, breaks

TABLE 14.2

A Compendium of Recent Studies on Metal Adsorption to Plastic Surfaces, Highlighting the Main Metals Discussed

Study (First Author, year)	Sample Source	Microplastic (MP) Analyzed	Metal Concentration (µg/g)									Metal Extraction Method	Metal Detection Method
			Fe	Mn	Cu	Zn	Pb	Cd	Cr	Ni	As		
Abbasi et al. (2024)	Sediment along the shores, Lublin, Poland	PE	61.0–696.2	0.49–11.32	0.76–211.90	3.1–46.8	0.31–1.48	0	0.40–12.70	0.17–5.16		BCR sequential extraction	ICP-MS
		PET	70.0–1,341.2	4.59–53.57	0.93–3.47	5.3–164.7	0.29–1.72	0.03	1.22–3.76	0.21–0.93			
		PP	473.9–543.6	10.33–11.02	1.98–27.58	26.3–30.1	2.02–6.76	0	2.16–2.50	1.24			
	Water, Lublin, Poland	PE	11.1–5,205.3	0.72–221.70	0.01–116.05	0.9–188.1	0.10–9.34	0	0.30–19.68	0.41–7.98			
		PET	32.1–1,431.5	2.51–2,286.98	0.09–6.79	3.3–30.8	0.21–3.54	0	0.29–5.77	1.24–1.59			
		PP	292.6–5,081	7.11–482.88	0.77–213.83	28.6–930.3	0.56–41.01	1.44–9.86	1.46–30.55	0.79–28.07			
Lenoble et al., 2024	Port of Galway (Ireland)	LDPE		0.0019	0.0094		0.009	0.0018	0.0083	0.0005	0.083	NaOH digestio n	ICPMS
		PET		0.0006	0.0027		0.0012	0.0001	0.0015	0.0002	0.007		
		PP		0.0011	0.004		0.0027	0.0001	0.0032	0.0003	0.036		
	Port of Ares (Spain)	LDPE		0.0095	0.0139		0.00077	0.0013	0.0156	0.0025	1,322		
		PET		0.1066	0.0145		0.00165	0.0011	0.0296	0.0038	1,286		
		PP		0.0329	0.0163		0.00175	0.0039	0.0227	0.0029	1,858		
	Villefranche Bay (France	LDPE		0.000008	0.0003		0.00179	0.00003	0.0007	0.0003	0.075		
		PET		0.000001	0.0008		0.00001	0.00004	0.0008	0.0003	0.055		
		PP		<LOD	<LOD		<LOD	<LOD	<LOD	<LOD			
	Toulon Bay (France)	LDPE		0.01341	0.00685		0.0179	0.00011	0.00828	0.00054	0.219		
		PET		0.01127	0.01452		0.0377	0.00024	0.01007	0.00043	0.408		
		PP		0.01884	0.00859		0.0298	0.00014	0.00879	0.00037	0.40		

(Continued)

TABLE 14.2 (*Continued*)
A Compendium of Recent Studies on Metal Adsorption to Plastic Surfaces, Highlighting the Main Metals Discussed

Study (First Author, year)	Sample Source	Microplastic (MP) Analyzed	Metal Concentration (µg/g)									Metal Extraction Method	Metal Detection Method
			Fe	Mn	Cu	Zn	Pb	Cd	Cr	Ni	As		
Dodhia et al., 2023	Oxic Sediment, Svanemøllen Harbor, Denmark	PE	84.2	342	3.4	23.9			0.4			HNO3 extraction	ICP-MS
		PS	435.9	<LOD	21.6	113.2			1.7				
	Anoxic Sediment, Svanemøllen Harbor, Denmark	PE	1.9	0.3	<LOD	0.7			0.1				
		PS	3.4	0.2	3.1	5.6			0.2				
Carnati et al. (2023)	Lake Como, Italy	PE	128.9–79.255	1.69–9.488	30.418–0.783	27.252–95.497	0.193–0.834	0.041–0.072	0.69–2	1424–0.762	0.018–0.093	Acid digestion	ICP-MS
		PET	62.419–74.752	0.485–0.998	0.281–0.4	2.566–4.197	0.16–0.302	0.003–0.117	0.479–1.311	0.308–1.004	0.000–0.052		
		PP	49.605–195.178	0.596–4.431	0.446–0.93	609.941–2.289	0.607–0.125	0.023–0.01	0.272–0.934	0.207–0.561	0.208–<LOD		
Campanale et al. (2022)	Freshwater, Ofanto river, Italy	Unidentified MPs	822.0–3,711.0	42.5–437.3	16.6–111.1	106.6–540.3	5.4–418.4	0.1–814.4	1.1–94.1	3.1–14.9		Microwave-assisted acid digestion	ICP-MS
Fajković et al. (2022)	Croatian Adriatic coastal area	PP			0.204–16.593	0.950–10.884	0.347–63.904	0.015–1.115				Acid digestion (HNO3. HClO4)	DPASV
Squadrone et al. (2022)	Mediterranean Sea (zooplankton ingested plastic samples)	Unidentified MPs	14.1–17.9	1.58–1.62	0.91–1.09	4.34–4.66	1.49–1.51	0.032–0.034	6.992–7.008	3.11–3.29	0.392–0.408	HNO3:HCl (1:3) extraction	ICP-MS
Souza et al. (2022)	Sandy beaches, coast of Bahia, Brazil	Plastic pellets	7.7–192		0.07–6.95	0.7–38.0	0.06–1.13	0.02–0.59		0.02–2.22			

(Continued)

TABLE 14.2 (*Continued*)

A Compendium of Recent Studies on Metal Adsorption to Plastic Surfaces, Highlighting the Main Metals Discussed

Study (First Author, year)	Sample Source	Microplastic (MP) Analyzed	Metal Concentration (μg/g)									Metal Extraction Method	Metal Detection Method
			Fe	Mn	Cu	Zn	Pb	Cd	Cr	Ni	As		
Pasumpon and Vasudevan (2022)	Thondi coast, Palk Bay, India	PA	0.0444–0.0446	0.0053–0.0055	0.0011–0.0015	0.0129–0.0133	0.0020–0.0024	0.0029–0.0033	0.0127–0.0129	0.0725–0.0727		Acid digestion (HNO3. HCl)	ICP-MS. SEM-EDX
Fred-Ahmadu et al. (2022)	Lagoon sediments, Gulf of Guinea, Nigeria	Unidentified MPs	2.85–30.35	0.11–1.77	0.07–0.11	0.07–0.10	0.15–0.85	0.04–0.16	0.07–0.10	0.04–0.09	0.08–0.11	Nitric acid digestion	ICP-OES
	beach sediments, Gulf of Guinea (Nigeria)	Unidentified MPs	1.04–13.49	0.05–5.28	0.06–0.24	12.05–31.47	0.08–0.56	0.04–0.10	0.06–0.18	0.03–0.07	0.09–0.12		
Sarkar et al. (2021)	Wastewater canals (water and sediment), Natural wastewater treatment wetland, East Kolkata, India	Mixed PE and PET			23.94–44.50	85.86–335.96	0.03–0.04	2.11–5.64	16.45–69.16	7.94–12.35	1.16–2.90	Triacids mixture (HNO3: HClO4: H2SO4. 10:4:1)	ICP-MS
	Treatment ponds (water and sediment)	Mixed PE and PET			0.29–119.59	1.88–1,191.52	0.04–104.63	0.65–5.78	26.26–342.28	12.43–75.77	1.56–4.51		

The table predominantly focuses on PE, PP, and polyethylene terephthalate (PET), with two accounts for polystyrene (PS) and polyamide (PA).

down plastic polymers by causing the formation of free radicals, which subsequently lead to chain scission and the weakening of the material (Shashoua, 2016).

Plastics exhibit significant resistance to degradation due to their stable chemical structure and hydrophobic nature. The primary reason for their durability lies in the carbon–carbon bonds that form the backbone of many synthetic polymers such as PE, PP, and polystyrene (PS). These strong covalent bonds are difficult for microorganisms to break down. Additionally, the high molecular weight and crystallinity of these materials hinder microbial access and enzymatic attack. The hydrophobic surfaces of plastics further repel water, which is essential for the hydrolytic processes that many microorganisms use to degrade organic materials (Mohanan et al., 2020).

Unlike natural organic materials, which decompose through microbial and fungal action, plastics are not easily metabolized by microorganisms due to their synthetic origin and complex structure. This biological inertness means that plastics undergo slow, but continuous degradation processes over time (Zhang et al., 2020). In environments where plastic particles are no longer exposed to photochemical degradation, such as in deep water or sediments, transition metals bound to the plastic surface have been discussed as potential catalysts facilitating plastic degradation (Shashoua., 2016). These metals accelerate the breakdown of polymer chains through various oxidative and reductive processes (Lenstra et al., 2022). The presence of transition metals in sediments can influence the degradation pathways of plastics by facilitating redox reactions that break down polymer chains. This process is particularly relevant in anoxic conditions, where microbial activity can release transition metals from sediments, further contributing to the catalytic degradation of plastics (Saup et al., 2017; Yang et al., 2022).

14.4 IMPLICATIONS OF THE PLASTISPHERE FOR ECOSYSTEM FUNCTIONING AND GLOBAL HEALTH

The interaction of plastics with metals, minerals, and microbes has created a dynamic and novel interface in the environment. Beyond the perturbations of elemental cycles, concentrations of toxic metals, and heightened risk to animal life and biodiversity, the plastisphere poses a multifaceted threat to human health through the enrichment and transport of pathogens (Lear et al., 2022; Ormsby et al., 2023), facilitation of antibiotic resistance (AR), and potential entry into the food chain. The evolutionary forces shaping the microbial communities in the plastisphere involve selection pressures from the biological and chemical environment as well as deterministic processes driven by the unique physical and chemical properties of the plastisphere substrates (Chen et al., 2020; Feng et al., 2023; Wang et al., 2020).

The presence of metals helps select for metal-cycling bacteria that utilize and thrive with the resource. Further, metals and minerals act as additional reaction and attachment sites for further biofilm and niche expansion and subsequent attachment sites for minerals and metals. In contrast to the benefits of metal-cycling metabolisms, however, the concentration of metals and minerals in the plastisphere also presents concerns for ecosystems and human health across scales. First, high concentrations of metals can exceed threshold values and pose a health risk for ecosystems and humans (Table 14.1). Metals such as lead, cadmium, and mercury are toxic to marine organisms even at low concentrations (Rahman and Singh, 2019). Their presence in the plastisphere can lead to a range of adverse effects, including impaired growth, reproduction, and development, as well as increased mortality rates (Banaee et al., 2019; Jinhui et al., 2019; Wen et al., 2018; Yan et al., 2020).

The extended transport time and transport distance of plastics increases the risk of exposure beyond the plastisphere itself. Metals are toxic to many life forms and if ingested by marine life, the organism itself can suffer as well as higher organisms when the organism enters the human food chain, a risk enhanced by biomagnification (Liu et al., 2018; Mansour and Sidky, 2002). Further, when the plastisphere reaches coastal areas and beaches the exposure to the food chain is enhanced as well as the risk of propagating the pollution to terrestrial or coastal ecosystems via deposition of the polluted plume. Besides the toxicity caused by the metals, the high occurrence of pathogens and antibiotic-resistant bacteria in the thriving niches within the plastisphere poses a threat to human health (Imran et al., 2019).

Several studies highlight the plastisphere as a reservoir for antimicrobial-resistant bacteria (ARB) and antibiotic-resistant genes (ARGs) (Gillings and Stokes, 2012; Stevenson et al., 2024; Tuvo et al., 2023; Zadjelovic et al., 2023). One of the primary drivers behind the evolution of antimicrobial resistance within and between bacterial communities is HGT. In HGT, ARGs are transferred between similar or phylogenetically distinct species. Compared to bacteria in suspension, wastewater communities, and natural aggregates, however, bacteria in the plastisphere seem to have a higher transfer frequency of antibiotic genes (Arias-Andres et al., 2018; Feng et al., 2023; Wang et al., 2023a). The transfer of ARGs mainly occurs via three pathways: conjugation, transformation, or transduction. In conjugation, DNA is transferred from one bacterium to another; in transduction, ARG is transferred between a virus and a bacterium. In transformation, bacteria acquire ARG from the environment including suspended DNA and DNA adsorbed to surfaces.

The reasons behind the enhanced evolution of AR in the plastisphere are most likely caused by a multitude of processes acting in concert. First, within the plastisphere there are elevated levels of both AR and heavy metals (Joppien et al., 2022; Liu et al., 2021). The presence of AB and heavy metals creates an evolutionary driving force, selecting for ARB and ARGs within microbial communities. Second, the persistence and dispersal capabilities of plastics, particularly MNPs, enable long-term and widespread distribution of ARB and ARGs across ecosystems. This continuous transport promotes the spread and evolution of AR by exposing diverse microbial communities to selective pressures over time and space (Li et al., 2024; Zhu et al., 2022).

Further, the interaction of bacteria with microplastics and minerals can induce stress for the organism (Stevenson et al., 2024), which acts in concert with the pressures induced by the presence of antibiotics and heavy metals. Consequently, the bacteria heighten their stress response, increase mutation rates, and enhance ARG uptake (Zhang et al., 2024). In addition, the complexity of microbial networks within the plastisphere in comparison to natural environments is found to vary depending on environment; i.e., the plastisphere in freshwater is less complex than those in seawater environments (Li et al., 2021). Further, simulations of species extinction in the plastisphere testing the stability of the networks therein suggest a less robust system compared to those in natural environments (Li et al., 2024). A reduced complexity can lead to the dominance of specific ARB, as there are fewer competitive interactions to inhibit their proliferation (Li et al., 2024).

Evidently, a wide range of ecological niches are developing in the plastisphere that create a continuous cycle of adsorption of solids, EPSs, bacteria, and higher organisms facilitating biofilm formation and niche expansion. The nutrient-rich and ever-expanding surface area in the plastisphere enhances the adsorption capacity of extracellular DNA, bacteria, and viruses (as well as other adsorbents) and increases the probability for mechanisms of HGT. As a consequence of all of these factors, the plastisphere serves as a hotspot for ARG propagation and ARG amplification.

Compared to geological sedimentary particles, such as rock debris and minerals, the transport of microplastic in sedimentary systems is distinctly spatiotemporally. Due to their apparent durability and large surface area to mass ratio, microplastics can linger much longer in surface waters and sedimentary systems than geologic sediment normally does. The longer time in the water column allows dispersal of the plastic to locations far from source and will conceptually have wide-ranging implications for propagation of AR and ARG to distant environments.

14.5 CONCLUSION AND OUTLOOK

The plastisphere is a dynamic biofilm that harbors organic matter, metals, minerals, and salts reflective of the host environment. It represents a unique microbial habitat shaped by a combination of selection pressures from contaminants, deterministic assembly processes, dispersal of plastics in the environment, and the diverse physical and chemical properties of plastic and associated substrates. The presence of certain minerals and metals is irrevocably linked to the prevailing redox conditions, while their accumulation in the plastisphere is predominantly impacted by the thickness of the biofilm (Dodhia et al., 2023; Kielemoes et al., 2002). Metals and minerals act as catalytic templates, evolutionary drivers, and structural fragments within specialized microbial communities, and we have yet to define these processes in the plastisphere. In addition, our understanding of uptake mechanisms, the influence of minerals and metals at different stages of plastisphere formation, and to what degree they determine microbial community composition and gene transfer is still limited. Therefore, a number of recommendations and open questions remain to enhance our understanding of the role of minerals and metals in the plastisphere.

1. The properties and interactions in the plastisphere are strongly impacted by the conditions of the surrounding environment. It is therefore advisable to assess the influence of metals and minerals in different settings, i.e., water column, deep water, and sediment as facilitators for processes like transport, degradation, gene transfer, and storage in future research. Particularly in sediment, there remains a limited understanding of interactions related to plastic fate, even though it is considered the largest plastic sink in aquatic environments.
2. Furthermore, the incorporation of plastic in carbonaceous biominerals presents a recently identified, potentially underestimated transport path for plastic into the sediment sink. Potential impacts of plastic on the calcium carbonate-based biomineralization processes and the respective organism must be assessed.
3. Plastics pose a series of health-related concerns for humans and ecosystems that are still largely unconstrained. Areas of study should address: (1) how the transport of plastics introduces pollutants and microorganisms to new environments, potentially disrupting local ecosystems and spreading contaminants and pathogens; (2) how the concentration of metals in the plastisphere induces bioaccumulation in tissue of marine organisms at all trophic levels and magnifies up the food chain; (3) how the presence of heavy metals in the plastisphere exposes a wide range of organisms and ecosystems with implications for neurological organ function; (4) the role of the plastisphere as a vector for both invasive species pathogens and AR; and (5) how enhanced levels of antibiotics and high selection pressures within the plastisphere select for resistant microbes, which in combination with the spatiotemporal distribution of the plastisphere represents an emergent risk of high global concern (Figure 14.4).

Future research addressing these gaps in knowledge will help tackle the mechanisms behind these ecosystem functions and health issues to better constrain the environmental fate of this pollutant as well as identify novel approaches to remediation and materials development.

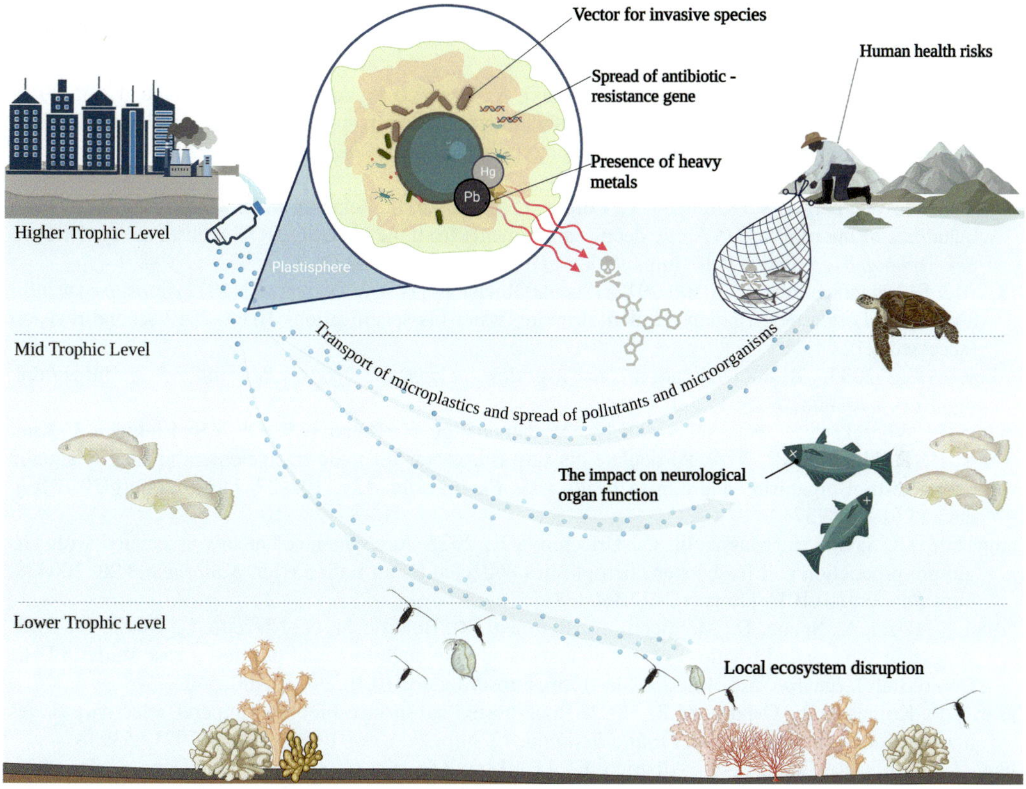

FIGURE 14.4 Overview of the health-related concerns associated with plastics in the environment. (Created in BioRender. Schmidt, D. (2024) https://BioRender.com/ l40q780).

ACKNOWLEDGMENTS

NRP, AP, and DS thank the Carlsberg Foundation (Semper Ardens grant no. CF22-0963) for their support in the study of minerals and metals in the plastisphere. KKS and SH thank the VILLUM Foundation (grant no. 25352) for their research support on propagation of AR. NRP and MSD further thank the VILLUM Foundation (grant no. 15397) for their support of research into the biogeochemical processes in the plastisphere. KKS, NRP, SH, and MSD thank Geocenter Danmark for support for research into ARG propagation into the plastisphere (4-2023). KKS and NP thank the Novo Nordisk Foundation (Grant NNF23OC0086712) for supporting transdisciplinary scientific explorations. The authors further acknowledge the support of the H2020-INFRIA-AQUACOSM-plus Transnational Access program (grant no. 871081) and the Umeå Marine Sciences Centre (UMF) for our research at their mesocosm facility highlighted herein.

REFERENCES

Abbasi, S., Turner, A., Jędruchniewicz, K., Różyło, K., Oleszczuk, P., 2024. Exchangeable versus residual metals in naturally aged plastic litter. *Environ. Sci. Pollut. Res.* 31, 24197–24206. https://doi.org/10.1007/s11356-024-32683-6

Adeleye, A.T., Bahar, M.M., Megharaj, M., Fang, C., Rahman, M.M., 2024. The unseen threat of the synergistic effects of microplastics and heavy metals in aquatic environments: a critical review. *Curr. Pollut. Rep.* https://doi.org/10.1007/s40726-024-00298-7

Amaral-Zettler, L.A., Zettler, E.R., Mincer, T.J., 2020. Ecology of the plastisphere. *Nat. Rev. Microbiol.* 18, 139–151. https://doi.org/10.1038/s41579-019-0308-0

Arias -Andres, M., Klümper, U., Rojas-Jimenez, K., Grossart, H.-P., 2018. Microplastic pollution increases gene exchange in aquatic ecosystems. *Environ. Pollut.* 237, 253–261. https://doi.org/10.1016/j.envpol.2018.02.058

Banaee, M., Soltanian, S., Sureda, A., Gholamhosseini, A., Haghi, B.N., Akhlaghi, M., Derikvandy, A., 2019. Evaluation of single and combined effects of cadmium and micro-plastic particles on biochemical and immunological parameters of common carp (Cyprinus carpio). *Chemosphere* 236, 124335. https://doi.org/10.1016/j.chemosphere.2019.07.066

Bergmann, M., Wirzberger, V., Krumpen, T., Lorenz, C., Primpke, S., Tekman, M.B., Gerdts, G., 2017. High quantities of microplastic in Arctic deep-sea sediments from the HAUSGARTEN observatory. *Environ. Sci. Technol.* 51, 11000–11010. https://doi.org/10.1021/acs.est.7b03331

BIS, 2012. BIS specifications (IS 10500-2012) (Second Revision) [WWW Document] (2012) https://www.indiawaterportal.org/articles/indian-standard-drinking-water-bis-specifications-10500-2012-second-revision (accessed 6.12.24).

Boskey, A.L., 2003. Biomineralization: an overview. *Connect. Tissue Res.* 44, 5–9. https://doi.org/10.1080/03008200390152007

Bradney, L., Wijesekara, H., Palansooriya, K.N., Obadamudalige, N., Bolan, N.S., Ok, Y.S., Rinklebe, J., Kim, K.-H., Kirkham, M.B., 2019. Particulate plastics as a vector for toxic trace-element uptake by aquatic and terrestrial organisms and human health risk. *Environ. Int.* 131, 104937. https://doi.org/10.1016/j.envint.2019.104937

Campanale, C., Savino, I., Massarelli, C., Uricchio, V.F., 2022. Assessment of metals associated with virgin pre-production and freshwater microplastics collected by an Italian river. *NanoImpact* 28, 100438. https://doi.org/10.1016/j.impact.2022.100438

Carnati, S., Pozzi, A., Spanu, D., Monticelli, D., Bettinetti, R., Boldrocchi, G., Nizzetto, L., Binda, G., 2023. Assessing sources and fractions of metals associated with environmental plastics: a case study in Lake Como (Italy). *Environ. Sci. Adv.* 2, 1746–1756. https://doi.org/10.1039/D3VA00254C

Casar, C.P., Kruger, B.R., Osburn, M.R., 2021. Rock-hosted subsurface biofilms: mineral selectivity drives hotspots for intraterrestrial life. *Front. Microbiol.* 12. https://doi.org/10.3389/fmicb.2021.658988

Chen, C.-F., Ju, Y.-R., Chen, C.-W., Albarico, F.P.J.B., Lim, Y.C., Ke, C., Cheng, Y.-R., Dong, C.-D., 2023. Microplastics in coral reef sediments underestimated? They may hide in biominerals. *Sci. Total Environ.* 899, 165708. https://doi.org/10.1016/j.scitotenv.2023.165708

Chen, G., Feng, Q., Wang, J., 2020. Mini-review of microplastics in the atmosphere and their risks to humans. *Sci. Total Environ.* 703, 135504. https://doi.org/10.1016/j.scitotenv.2019.135504

Choe, Y.-H., Kim, M., Woo, J., Lee, M.J., Lee, J.I., Lee, E.J., Lee, Y.K., 2018. Comparing rock-inhabiting microbial communities in different rock types from a high arctic polar desert. *FEMS Microbiol. Ecol.* 94, fiy070. https://doi.org/10.1093/femsec/fiy070

Corcoran, P.L., 2015. Benthic plastic debris in marine and fresh water environments. *Environ. Sci. Process. Impacts* 17, 1363–1369. https://doi.org/10.1039/C5EM00188A

Costa, R.A., De Lira, J.V., Aragão, M.F., 2019. Biofilm-formation by drug-resistant Staphylococcus aureus from cow milk. *J. Consum. Prot. Food Saf.* 14, 63–69. https://doi.org/10.1007/s00003-018-1186-5

De Tender, C., Devriese, L.I., Haegeman, A., Maes, S., Vangeyte, J., Cattrijsse, A., Dawyndt, P., Ruttink, T., 2017. Temporal dynamics of bacterial and fungal colonization on plastic debris in the North Sea. *Environ. Sci. Technol.* 51, 7350–7360. https://doi.org/10.1021/acs.est.7b00697

Dedman, C.J., Christie-Oleza, J.A., Fernández-Juárez, V., Echeveste, P., 2022. Cell size matters: nano- and micro-plastics preferentially drive declines of large marine phytoplankton due to co-aggregation. *J. Hazard. Mater.* 424, 127488. https://doi.org/10.1016/j.jhazmat.2021.127488

Di Martino, P., 2018. Extracellular polymeric substances, a key element in understanding biofilm phenotype. *AIMS Microbiol.* 4, 274–288. https://doi.org/10.3934/microbiol.2018.2.274

Ding, J., Meng, F., Chen, H., Chen, Q., Hu, A., Yu, C.-P., Chen, L., Lv, M., 2022. Leachable additives of tire particles explain the shift in microbial community composition and function in coastal sediments. *Environ. Sci. Technol.* 56, 12257–12266. https://doi.org/10.1021/acs.est.2c02757

Dodhia, M.S., Rogers, K.L., Fernández-Juárez, V., Carreres-Calabuig, J.A., Löscher, C.R., Tisserand, A.A., Keulen, N., Riemann, L., Shashoua, Y., Posth, N.R., 2023. Microbe-mineral interactions in the Plastisphere: coastal biogeochemistry and consequences for degradation of plastics. *Front. Mar. Sci.* 10. https://doi.org/10.3389/fmars.2023.1134815

Dodhia, M.S., Fruergaard, M., Shilla, D., Shilla, D.J., Posth, N.R., Keulen, N., 2025. Automated quantitative mineralogy and plastics–A new view of the plastic surface in aquatic environments. Mar. Pollut. Bull. 211, p. 117441.

Dong, H., Huang, L., Zhao, L., Zeng, Q., Liu, X., Sheng, Y., Shi, L., Wu, G., Jiang, H., Li, F., Zhang, L., Guo, D., Li, G., Hou, W., Chen, H., 2022. A critical review of mineral–microbe interaction and co-evolution: mechanisms and applications. *Natl. Sci. Rev.* 9, nwac128. https://doi.org/10.1093/nsr/nwac128

Du, Y., Liu, X., Dong, X., Yin, Z., 2022. A review on marine plastisphere: biodiversity, formation, and role in degradation. *Comput. Struct. Biotechnol. J.* 20, 975–988. https://doi.org/10.1016/j.csbj.2022.02.008

Dussud, C., Meistertzheim, A.L., Conan, P., Pujo-Pay, M., George, M., Fabre, P., Coudane, J., Higgs, P., Elineau, A., Pedrotti, M.L., Gorsky, G., Ghiglione, J.F., 2018. Evidence of niche partitioning among bacteria living on plastics, organic particles and surrounding seawaters. *Environ. Pollut.* 236, 807–816. https://doi.org/10.1016/j.envpol.2017.12.027

Enders, K., Käppler, A., Biniasch, O., Feldens, P., Stollberg, N., Lange, X., Fischer, D., Eichhorn, K.-J., Pollehne, F., Oberbeckmann, S., Labrenz, M., 2019. Tracing microplastics in aquatic environments based on sediment analogies. *Sci. Rep.* 9, 15207. https://doi.org/10.1038/s41598-019-50508-2

Erni -Cassola, G., Wright, R.J., Gibson, M.I., Christie-Oleza, J.A., 2020. Early colonization of weathered polyethylene by distinct bacteria in marine coastal seawater. *Microb. Ecol.* 79, 517–526. https://doi.org/10.1007/s00248-019-01424-5

EU Commission 2023. Regulation - 2023/915 - EN - EUR-Lex [WWW Document], n.d. URL https://eur-lex.europa.eu/eli/reg/2023/915/oj (accessed 6.12.24).

FAO Codex 2023. General Standard for Contaminants and Toxins in Food and Feed - CODEX STAN 193–1995 | Knowledge for policy [WWW Document] (2023) https://knowledge4policy.ec.europa.eu/publication/general-standard-contaminants-toxins-food-feed-codex-stan-193-1995_en (accessed 6.15.24).

Fajković, H., Cukrov, N., Frančišković-Bilinski, S., Huljek, L., Kostanjšek, I., Kwokal, Ž., Matekalo, R., Pikelj, K., Cuculić, V., 2022. Correlation of metals and degraded marine (micro)plastic litter in geologically similar coastal areas with different anthropogenic characteristics. *Mar. Pollut. Bull.* 183, 114041. https://doi.org/10.1016/j.marpolbul.2022.114041

Fazey, F.M.C., Ryan, P.G., 2016. Biofouling on buoyant marine plastics: an experimental study into the effect of size on surface longevity. *Environ. Pollut.* 210, 354–360. https://doi.org/10.1016/j.envpol.2016.01.026

Feng, Y., Sun, J.-W., Shi, W.-W., Duan, J.-L., Sun, X.-D., Feng, L.-J., Wang, Q., Gan, Y.-D., Yuan, X.-Z., 2023. Microplastics exhibit accumulation and horizontal transfer of antibiotic resistance genes. *J. Environ. Manage.* 336, 117632. https://doi.org/10.1016/j.jenvman.2023.117632

Fischer, A.C., Kroon, J.J., Verburg, T.G., Teunissen, T., Wolterbeek, H.T., 2007. On the relevance of iron adsorption to container materials in small-volume experiments on iron marine chemistry: 55Fe-aided assessment of capacity, affinity and kinetics. *Mar. Chem.* 107, 533–546. https://doi.org/10.1016/j.marchem.2007.08.004

Flemming, H.-C., Van Hullebusch, E.D., Neu, T.R., Nielsen, P.H., Seviour, T., Stoodley, P., Wingender, J., Wuertz, S., 2023. The biofilm matrix: multitasking in a shared space. *Nat. Rev. Microbiol.* 21, 70–86. https://doi.org/10.1038/s41579-022-00791-0

Flemming, H.-C., Wingender, J., 2010. The biofilm matrix. *Nat. Rev. Microbiol.* 8, 623–633. https://doi.org/10.1038/nrmicro2415

Flemming, H.-C., Wingender, J., Szewzyk, U., Steinberg, P., Rice, S.A., Kjelleberg, S., 2016. Biofilms: an emergent form of bacterial life. *Nat. Rev. Microbiol.* 14, 563–575. https://doi.org/10.1038/nrmicro.2016.94

Flores -Vargas, G., Bergsveinson, J., Lawrence, J.R., Korber, D.R., 2021. Environmental biofilms as reservoirs for antimicrobial resistance. *Front. Microbiol.* 12, 766242. https://doi.org/10.3389/fmicb.2021.766242

Frankel, R.B., Bazylinski, D.A., 2003. Biologically induced mineralization by bacteria. *Rev. Mineral. Geochem.* 54, 95–114. https://doi.org/10.2113/0540095

Fred -Ahmadu, O.H., Ayejuyo, O.O., Tenebe, I.T., Benson, N.U., 2022. Occurrence and distribution of micro(meso)plastic-sorbed heavy metals and metalloids in sediments, Gulf of Guinea coast (SE Atlantic). *Sci. Total Environ.* 813, 152650. https://doi.org/10.1016/j.scitotenv.2021.152650

Freeman, C.L., Dieudonné, L., Agbaje, O.B.A., Žure, M., Sanz, J.Q., Collins, M., Sand, K.K., 2023. Survival of environmental DNA in sediments: mineralogic control on DNA taphonomy. *Environ. DNA* 5, 1691–1705. https://doi.org/10.1002/edn3.482

Frère, L., Maignien, L., Chalopin, M., Huvet, A., Rinnert, E., Morrison, H., Kerninon, S., Cassone, A.-L., Lambert, C., Reveillaud, J., Paul-Pont, I., 2018. Microplastic bacterial communities in the Bay of Brest: influence of polymer type and size. *Environ. Pollut.* 242, 614–625. https://doi.org/10.1016/j.envpol.2018.07.023

Gambarini, V., Pantos, O., Kingsbury, J.M., Weaver, L., Handley, K.M., Lear, G., 2021. Phylogenetic distribution of plastic-degrading microorganisms. *mSystems* 6. https://doi.org/10.1128/msystems.01112-20

Ganesh Kumar, A., Anjana, K., Hinduja, M., Sujitha, K., Dharani, G., 2020. Review on plastic wastes in marine environment – biodegradation and biotechnological solutions. *Mar. Pollut. Bull.* 150, 110733. https://doi.org/10.1016/j.marpolbul.2019.110733

Geisel, S., Secchi, E., Vermant, J., 2022. The role of surface adhesion on the macroscopic wrinkling of biofilms. *eLife* 11, e76027. https://doi.org/10.7554/eLife.76027

Gewert, B., Plassmann, M.M., MacLeod, M., 2015. Pathways for degradation of plastic polymers floating in the marine environment. *Environ. Sci. Process. Impacts* 17, 1513–1521. https://doi.org/10.1039/C5EM00207A

Ghodke, S.A., Maheshwari, U., Gupta, S., Sonawane, S.H., Bhanvase, B.A., 2021. Nanomaterials for adsorption of pollutants and heavy metals: introduction, mechanism, and challenges, in: O.V. Kharissova, L.M. Torres-Martínez, B.I. Kharisov (eds.), Handbook of Nanomaterials for Wastewater Treatment. Elsevier, pp. 343–366. https://doi.org/10.1016/B978-0-12-821496-1.00032-5

Gillings, M.R., Stokes, H.W., 2012. Are humans increasing bacterial evolvability? Trends *Ecol. E* 27, 346–352. https://doi.org/10.1016/j.tree.2012.02.006

Gorbushina, A.A., 2007. Life on the rocks: life on the rocks. *Environ. Microbiol.* 9, 1613–1631. https://doi.org/10.1111/j.1462-2920.2007.01301.x

Goudriaan, M., Morales, V.H., van der Meer, M.T.J., Mets, A., Ndhlovu, R.T., van Heerwaarden, J., Simon, S., Heuer, V.B., Hinrichs, K.-U., Niemann, H., 2023. A stable isotope assay with 13C-labeled polyethylene to investigate plastic mineralization mediated by Rhodococcus ruber. *Mar. Pollut. Bull.* 186, 114369. https://doi.org/10.1016/j.marpolbul.2022.114369

Guasch, H., Garcia, X., Urrea-Clos, G., Bañeras, L., 2012. Changes in the microbial communities along the environmental gradient created by a small Fe spring. *Freshw. Sci.* 31, 599–609. https://doi.org/10.1899/10-127.1

Hamm, C.E., 2005. The evolution of advanced mechanical defenses and potential technological applications of diatom shells. *J. Nanosci. Nanotechnol.* 5, 108–119. https://doi.org/10.1166/jnn.2005.023

Han, Z., Jiang, T., Xie, L., Zhang, R., 2022. Microplastics impact shell and pearl biomineralization of the pearl oyster Pinctada fucata. *Environ. Pollut.* 293, 118522. https://doi.org/10.1016/j.envpol.2021.118522

Harmon, S.M., Chen, Q., Ma, C., Ji, M., Yan, X., Ji, R., Shi, H., 2024. The effects of microplastic pollution on aquatic organisms, in: E.Y. Zeng (ed.), *Microplastic Contamination in Aquatic Environments*. Elsevier, pp. 355–379. https://doi.org/10.1016/B978-0-443-15332-7.00001-6

Hierl, F., Wu, H.C., Westphal, H., 2021. Scleractinian corals incorporate microplastic particles: identification from a laboratory study. *Environ. Sci. Pollut. Res.* 28, 37882–37893. https://doi.org/10.1007/s11356-021-13240-x

Imran, M., Das, K.R., Naik, M.M., 2019. Co-selection of multi-antibiotic resistance in bacterial pathogens in metal and microplastic contaminated environments: an emerging health threat. *Chemosphere* 215, 846–857. https://doi.org/10.1016/j.chemosphere.2018.10.114

Jinhui, S., Sudong, X., Yan, N., Xia, P., Jiahao, Q., Yongjian, X., 2019. Effects of microplastics and attached heavy metals on growth, immunity, and heavy metal accumulation in the yellow seahorse, Hippocampus kuda Bleeker. *Mar. Pollut. Bull.* 149, 110510. https://doi.org/10.1016/j.marpolbul.2019.110510

Joppien, M., Westphal, H., Chandra, V., Stuhr, M., Doo, S.S., 2022. Nanoplastic incorporation into an organismal skeleton. *Sci. Rep.* 12, 14771. https://doi.org/10.1038/s41598-022-18547-4

Jørgensen, B.B., 1995. Die Mikrowelt der Meeresbakterien. *Naturwissenschaften* 82, 269–278. https://doi.org/10.1007/BF01134524

Kaiser, D., Kowalski, N., Waniek, J.J., 2017. Effects of biofouling on the sinking behavior of microplastics. *Environ. Res. Lett.* 12, 124003. https://doi.org/10.1088/1748-9326/aa8e8b

Kanu, I., Achi, O.K., Ezeronye, O.U., Anyanwu, E.C., 2006. Seasonal variation in bacterial heavy metal biosorption in water samples from Eziama river near soap and brewery industries and the environmental health implications. *Int. J. Environ. Sci. Technol.* 3, 95–102. https://doi.org/10.1007/BF03325912

Keren-Paz, A., Kolodkin-Gal, I., 2020. A brick in the wall: discovering a novel mineral component of the biofilm extracellular matrix. *New Biotechnol.* 56, 9–15. https://doi.org/10.1016/j.nbt.2019.11.002

Kielemoes, J., Bultinck, I., Storms, H., Boon, N., Verstraete, W., 2002. Occurrence of manganese-oxidizing microorganisms and manganese deposition during biofilm formation on stainless steel in a brackish surface water. *FEMS Microbiol. Ecol.* 39, 41–55. https://doi.org/10.1111/j.1574-6941.2002.tb00905.x

Kjær, K.H., Winther Pedersen, M., De Sanctis, B., De Cahsan, B., Korneliussen, T.S., Michelsen, C.S., Sand, K.K., Jelavić, S., Ruter, A.H., Schmidt, A.M.A., Kjeldsen, K.K., Tesakov, A.S., Snowball, I., Gosse, J.C., Alsos, Inger G., Wang, Y., Dockter, C., Rasmussen, M., Jørgensen, M.E., Skadhauge, B., Prohaska, A., Kristensen, J.Å., Bjerager, M., Allentoft, M.E., Coissac, E., PhyloNorway Consortium, Rouillard, A., Simakova, A., Fernandez-Guerra, A., Bowler, C., Macias-Fauria, M., Vinner, L., Welch, J.J., Hidy,

A.J., Sikora, M., Collins, M.J., Durbin, R., Larsen, N.K., Willerslev, E., 2022. A 2-million-year-old ecosystem in Greenland uncovered by environmental DNA. *Nature* 612, 283–291. https://doi.org/10.1038/s41586-022-05453-y

Kowalski, N., Reichardt, A.M., Waniek, J.J., 2016. Sinking rates of microplastics and potential implications of their alteration by physical, biological, and chemical factors. *Mar. Pollut. Bull.* 109, 310–319. https://doi.org/10.1016/j.marpolbul.2016.05.064

Kurniawan, A., Ulfa, S.M., Chamidah, C., 2022. The biosorption of copper(II) using a natural biofilm formed on the stones from the Metro River, Malang City, Indonesia. *Int. J. Microbiol.* 2022, 1–6. https://doi.org/10.1155/2022/9975333

Laderriere, V., Le Faucheur, S., Fortin, C., 2021. Exploring the role of water chemistry on metal accumulation in biofilms from streams in mining areas. *Sci. Total Environ.* 784, 146986. https://doi.org/10.1016/j.scitotenv.2021.146986

Laursen, S.N., Fruergaard, M., Dodhia, M., Posth, N., Rasmussen, M., Larsen, M., Shilla, D., Shilla, D., Kilawe, J., Kizenga, H., Andersen, T., 2023. Settling of buoyant microplastic in estuaries: the importance of flocculation. *Sci. Total Environ.* 886, 163976. https://doi.org/10.1016/j.scitotenv.2023.163976

Lear, G., Maday, S.D.M., Gambarini, V., Northcott, G., Abbel, R., Kingsbury, J.M., Weaver, L., Wallbank, J.A., Pantos, O., 2022. Microbial abilities to degrade global environmental plastic polymer waste are overstated. *Environ. Res. Lett.* 17, 043002. https://doi.org/10.1088/1748-9326/ac59a7

Leiser, R., Jongsma, R., Bakenhus, I., Möckel, R., Philipp, B., Neu, T.R., Wendt-Potthoff, K., 2021a. Interaction of cyanobacteria with calcium facilitates the sedimentation of microplastics in a eutrophic reservoir. *Water Res.* 189, 116582. https://doi.org/10.1016/j.watres.2020.116582

Leiser, R., Schumann, M., Dadi, T., Wendt-Potthoff, K., 2021b. Burial of microplastics in freshwater sediments facilitated by iron-organo flocs. *Sci. Rep.* 11, 24072. https://doi.org/10.1038/s41598-021-02748-4

Leiser, R., Wu, G.-M., Neu, T.R., Wendt-Potthoff, K., 2020. Biofouling, metal sorption and aggregation are related to sinking of microplastics in a stratified reservoir. *Water Res.* 176, 115748. https://doi.org/10.1016/j.watres.2020.115748

Lenoble, V., Cindrić, A.-M., Briand, J.-F., Pedrotti, M.L., Lacerda, A.L., Muniategui-Lorenzo, S., Fernández-González, V., Moscoso-Pérez, C.M., Andrade-Garda, J.M., Casotti, R., Murano, C., Donnarumma, V., Frizzi, S., Hannon, C., Joyce, H., Nash, R., Frias, J., 2024. Bioaccumulation of trace metals in the plastisphere: awareness of environmental risk from a European perspective. *Environ. Pollut.* 348, 123808. https://doi.org/10.1016/j.envpol.2024.123808

Lenstra, W.K., Van Helmond, N.A.G.M., Żygadłowska, O.M., Van Zummeren, R., Witbaard, R., Slomp, C.P., 2022. Sediments as a source of iron, manganese, cobalt and nickel to continental shelf waters (Louisiana, Gulf of Mexico). *Front. Mar. Sci.* 9, 811953. https://doi.org/10.3389/fmars.2022.811953

Lepleux, C., Turpault, M.P., Oger, P., Frey-Klett, P., Uroz, S., 2012. Correlation of the abundance of betaproteobacteria on mineral surfaces with mineral weathering in forest soils. *Appl. Environ. Microbiol.* 78, 7114–7119. https://doi.org/10.1128/AEM.00996-12

Li, C., Gillings, M.R., Zhang, C., Chen, Q., Zhu, D., Wang, J., Zhao, K., Xu, Q., Leung, P.H., Li, X., Liu, J., Jin, L., 2024. Ecology and risks of the global plastisphere as a newly expanding microbial habitat. *The Innovation* 5, 100543. https://doi.org/10.1016/j.xinn.2023.100543

Li, C., Wang, Lifei, Ji, S., Chang, M., Wang, Longfei, Gan, Y., Liu, J., 2021. The ecology of the plastisphere: microbial composition, function, assembly, and network in the freshwater and seawater ecosystems. *Water Res.* 202, 117428. https://doi.org/10.1016/j.watres.2021.117428

Li, M., He, L., Zhang, M., Liu, X., Tong, M., Kim, H., 2019. Cotransport and deposition of iron oxides with different-sized plastic particles in saturated quartz sand. *Environ. Sci. Technol.* 53, 3547–3557. https://doi.org/10.1021/acs.est.8b06904

Liu, S., Shi, J., Wang, J., Dai, Y., Li, H., Li, J., Liu, X., Chen, X., Wang, Z., Zhang, P., 2021. Interactions between microplastics and heavy metals in aquatic environments: a review. *Front. Microbiol.* 12, 652520. https://doi.org/10.3389/fmicb.2021.652520

Liu, Y., Liu, G., Yuan, Z., Liu, H., Lam, P.K.S., 2018. Heavy metals (As, Hg and V) and stable isotope ratios (δ13C and δ15N) in fish from Yellow River Estuary, China. *Sci. Total Environ.* 613–614, 462–471. https://doi.org/10.1016/j.scitotenv.2017.09.088

Liu, Z., Adyel, T.M., Wang, Z., Wu, J., Liu, J., Miao, L., Hou, J., 2022. Effects of biofilms on trace metal adsorption on plastics in freshwater systems. *Int. J. Environ. Res. Public. Health* 19, 13752. https://doi.org/10.3390/ijerph192113752

Lobelle, D., Cunliffe, M., 2011. Early microbial biofilm formation on marine plastic debris. *Mar. Pollut. Bull.* 62, 197–200. https://doi.org/10.1016/j.marpolbul.2010.10.013

Lopez, D., Vlamakis, H., Kolter, R., 2010. Biofilms. *Cold Spring Harb. Perspect. Biol.* 2, a000398–a000398. https://doi.org/10.1101/cshperspect.a000398

Lorite, G.S., Rodrigues, C.M., De Souza, A.A., Kranz, C., Mizaikoff, B., Cotta, M.A., 2011. The role of conditioning film formation and surface chemical changes on Xylella fastidiosa adhesion and biofilm evolution. *J. Colloid Interface Sci.* 359, 289–295. https://doi.org/10.1016/j.jcis.2011.03.066

Lowenstam, H.A., 1981. Minerals formed by organisms. *Science* 211, 1126–1131. https://doi.org/10.1126/science.7008198

Mansour, S.A., Sidky, M.M., 2002. Ecotoxicological studies. 3. Heavy metals contaminating water and fish from Fayoum Governorate. *Egypt. Food Chem.* 78, 15–22. https://doi.org/10.1016/S0308-8146(01)00197-2

Martin, C., Young, C.A., Valluzzi, L., Duarte, C.M., 2022. Ocean sediments as the global sink for marine micro- and mesoplastics. *Limnol. Oceanogr. Lett.* 7, 235–243. https://doi.org/10.1002/lol2.10257

Matijaković Mlinarić, N., Selmani, A., Brkić, A.L., Njegić Džakula, B., Kralj, D., Kontrec, J., 2022. Exposure of microplastics to organic matter in waters enhances microplastic encapsulation into calcium carbonate. *Environ. Chem. Lett.* 20, 2235–2242. https://doi.org/10.1007/s10311-022-01433-w

McLean, R.J.C., Fuqua, C., Siegele, D.A., Kirkland, B.L., Adams, J.L., Whiteley, M., 1999. Biofilm growth and illustrations of its role in mineral formation. Microb. Biofilms. *New Frontiers Proceedings of the 8th International Symposium on Microbial Ecology Atlantic Canada Society for Microbial Ecology, Halifax, Canada.*

Michels, J., Stippkugel, A., Lenz, M., Wirtz, K., Engel, A., 2018. Rapid aggregation of biofilm-covered microplastics with marine biogenic particles. *Proc. R. Soc. B Biol. Sci.* 285, 20181203. https://doi.org/10.1098/rspb.2018.1203

Mitrano, D.M., Wick, P., Nowack, B., 2021. Placing nanoplastics in the context of global plastic pollution. *Nat. Nanotechnol.* 16, 491–500. https://doi.org/10.1038/s41565-021-00888-2

Mohanan, N., Montazer, Z., Sharma, P.K., Levin, D.B., 2020. Microbial and enzymatic degradation of synthetic plastics. *Front. Microbiol.* 11, 580709. https://doi.org/10.3389/fmicb.2020.580709

Müller, S., Fiutowski, J., Rasmussen, M.B., Zunic, T.B., Rubahn, H.-G., Posth, N.R. 2025. Nanoplastic in aqueous environments: The role of chemo-electric properties for nanoplastic-mineral interaction. *Sci. Total Environ.* 964, 178529. https://doi.org/10.1016/j.scitotenv.2025.178529.

Nelson, T.F., Reddy, C.M., Ward, C.P., 2021. Product formulation controls the impact of biofouling on consumer plastic photochemical fate in the ocean. *Environ. Sci. Technol.* 55, 8898–8907. https://doi.org/10.1021/acs.est.1c02079

Neu, T.R., Lawrence, J.R., 2017. The extracellular matrix - an intractable part of biofilm systems, in: H.-C., Flemming, T.R., Neu, J. Wingender (eds.), *The Perfect Slime*. IWA Publishing, pp. 25–60.

Nielsen, A.R., Jelavić, S., Murray, D., Rad, B., Andersson, M.P., Ceccato, M., Mitchell, A.C., Stipp, S.L.S., Zuckermann, R.N., Sand, K.K., 2020. Thermodynamic and kinetic parameters for calcite nucleation on peptoid and model scaffolds: a step toward nacre mimicry. *Cryst. Growth Des.* 20, 3762–3771. https://doi.org/10.1021/acs.cgd.0c00029

Niu, Z., Vandegehuchte, M.B., Catarino, A.I., Everaert, G., 2021. Environmentally relevant concentrations and sizes of microplastic do not impede marine diatom growth. *J. Hazard. Mater.* 409, 124460. https://doi.org/10.1016/j.jhazmat.2020.124460

NOAA [WWW Document], 2014. NASA GISS NASA News Feature Releases NASA NOAA Find 2014 Warmest Year Mod. Rec. URL https://www.giss.nasa.gov/research/news/20150116/ (accessed 1.23.23).

Oberbeckmann, S., Kreikemeyer, B., Labrenz, M., 2018. Environmental factors support the formation of specific bacterial assemblages on microplastics. *Front. Microbiol.* 8, 2709. https://doi.org/10.3389/fmicb.2017.02709

Ormsby, M.J., White, H.L., Metcalf, R., Oliver, D.M., Quilliam, R.S., 2023. Clinically important E. coli strains can persist, and retain their pathogenicity, on environmental plastic and fabric waste. *Environ. Pollut.* 326, 121466. https://doi.org/10.1016/j.envpol.2023.121466

Pal, A., Bhattacharjee, S., Saha, J., Sarkar, M., Mandal, P., 2022. Bacterial survival strategies and responses under heavy metal stress: a comprehensive overview. *Crit. Rev. Microbiol.* 48, 327–355. https://doi.org/10.1080/1040841X.2021.1970512

Pasumpon, N., Vasudevan, S., 2022. Baseline study of trace metal concentrations in abandoned, lost or otherwise discarded fishing gear along Thondi coast, Palk Bay, India. *J. Sea Res.* 182, 102189. https://doi.org/10.1016/j.seares.2022.102189

Quan, K., Hou, J., Zhang, Z., Ren, Y., Peterson, B.W., Flemming, H.-C., Mayer, C., Busscher, H.J., Van Der Mei, H.C., 2022. Water in bacterial biofilms: pores and channels, storage and transport functions. *Crit. Rev. Microbiol.* 48, 283–302. https://doi.org/10.1080/1040841X.2021.1962802

Rahman, Z., Singh, V.P., 2019. The relative impact of toxic heavy metals (THMs) (arsenic (As), cadmium (Cd), chromium (Cr)(VI), mercury (Hg), and lead (Pb)) on the total environment: an overview. *Environ. Monit. Assess.* 191, 419. https://doi.org/10.1007/s10661-019-7528-7

Richard, H., Carpenter, E.J., Komada, T., Palmer, P.T., Rochman, C.M., 2019. Biofilm facilitates metal accumulation onto microplastics in estuarine waters. *Sci. Total Environ.* 683, 600–608. https://doi.org/10.1016/j.scitotenv.2019.04.331

Rillig, M.C., Kim, S.W., Zhu, Y.-G., 2024. The soil plastisphere. *Nat. Rev. Microbiol.* 22, 64–74. https://doi.org/10.1038/s41579-023-00967-2 McLean, R. J., et al. "Biofilm growth and illustrations of its role in mineral formation." *Microbial Biosystems: New Frontiers Proceedings of the 8th International Symposium on Microbial Ecology Atlantic Canada Society for Microbial Ecology, Halifax, Canada.* 1999.

Roager, L., Sonnenschein, E.C., 2019. Bacterial candidates for colonization and degradation of marine plastic debris. *Environ. Sci. Technol.* 53, 11636–11643. https://doi.org/10.1021/acs.est.9b02212

Rodrigues, J.P., Duarte, A.C., Santos-Echeandía, J., 2022. Interaction of microplastics with metal(oid)s in aquatic environments: what is done so far? *J. Hazard. Mater. Adv.* 6, 100072. https://doi.org/10.1016/j.hazadv.2022.100072

Rogers, K.L., Carreres-Calabuig, J.A., Gorokhova, E., Posth, N.R., 2020. Micro-by-micro interactions: how microorganisms influence the fate of marine microplastics. *Limnol. Oceanogr. Lett.* 5, 18–36. https://doi.org/10.1002/lol2.10136

Romera-Castillo, C., Pinto, M., Langer, T.M., Álvarez-Salgado, X.A., Herndl, G.J., 2018. Dissolved organic carbon leaching from plastics stimulates microbial activity in the ocean. *Nat. Commun.* 9, 1430. https://doi.org/10.1038/s41467-018-03798-5

Rummel, C.D., Jahnke, A., Gorokhova, E., Kühnel, D., Schmitt-Jansen, M., 2017. Impacts of biofilm formation on the fate and potential effects of microplastic in the aquatic environment. *Environ. Sci. Technol. Lett.* 4, 258–267. https://doi.org/10.1021/acs.estlett.7b00164

Sand, K.K., Jelavić, S., 2018. Mineral facilitated horizontal gene transfer: a new principle for evolution of life? *Front. Microbiol.* 9. https://doi.org/10.3389/fmicb.2018.02217

Sand, K.K., Jelavić, S., Dobberschütz, S., Ashby, P.D., Marshall, M.J., Dideriksen, K., Stipp, S.L.S., Kerisit, S.N., Friddle, R.W., DeYoreo, J.J., 2020. Mechanistic insight into biopolymer induced iron oxide mineralization through quantification of molecular bonding. *Nanoscale Adv.* 2, 3323–3333. https://doi.org/10.1039/D0NA00138D

Sanz-Lázaro, C., Casado-Coy, N., Beltrán-Sanahuja, A., 2021. Biodegradable plastics can alter carbon and nitrogen cycles to a greater extent than conventional plastics in marine sediment. *Sci. Total Environ.* 756, 143978. https://doi.org/10.1016/j.scitotenv.2020.143978

Sarkar, D.J., Das Sarkar, S., Das, B.K., Sahoo, B.K., Das, A., Nag, S.K., Manna, R.K., Behera, B.K., Samanta, S., 2021. Occurrence, fate and removal of microplastics as heavy metal vector in natural wastewater treatment wetland system. *Water Res.* 192, 116853. https://doi.org/10.1016/j.watres.2021.116853

Saup, C.M., Williams, K.H., Rodríguez-Freire, L., Cerrato, J.M., Johnston, M.D., Wilkins, M.J., 2017. Anoxia stimulates microbially catalyzed metal release from Animas River sediments. *Environ. Sci. Process. Impacts* 19, 578–585. https://doi.org/10.1039/C7EM00036G

Schmidt, D. (2024). Plastisphere. Created in BioRender. https://BioRender.com/h45a629

Schoonen, M.A.A., Cohn, C.A., Roemer, E., Laffers, R., Simon, S.R., O'Riordan, T., 2006. Mineral-induced formation of reactive oxygen species. *Rev. Mineral. Geochem.* 64, 179–221. https://doi.org/10.2138/rmg.2006.64.7

Seeley, M.E., Song, B., Passie, R., Hale, R.C., 2020. Microplastics affect sedimentary microbial communities and nitrogen cycling. *Nat. Commun.* 11, 2372. https://doi.org/10.1038/s41467-020-16235-3

Seixas, A.F., Quendera, A.P., Sousa, J.P., Silva, A.F.Q., Arraiano, C.M., Andrade, J.M., 2022. Bacterial response to oxidative stress and RNA oxidation. *Front. Genet.* 12, 821535. https://doi.org/10.3389/fgene.2021.821535

Shah, S.B., 2021. Heavy metals in the marine environment—an overview, in: S.B. Shah (ed.), *Heavy Metals in Scleractinian Corals*. Springer International Publishing, Cham, pp. 1–26. https://doi.org/10.1007/978-3-030-73613-2_1

Shashoua, Y., 2016. *Conservation of Plastics: Materials Science, Degradation and Preservation*. Routledge, London, New York.

Shi, X., Chen, Z., Wei, W., Chen, J., Ni, B.-J., 2023. Toxicity of micro/nanoplastics in the environment: roles of plastisphere and eco-corona. *Soil Environ. Health* 1, 100002. https://doi.org/10.1016/j.seh.2023.100002

Soares, M.O., Matos, E., Lucas, C., Rizzo, L., Allcock, L., Rossi, S., 2020. Microplastics in corals: an emergent threat. *Mar. Pollut. Bull.* 161, 111810. https://doi.org/10.1016/j.marpolbul.2020.111810

Soares, M.O., Rizzo, L., Ximenes Neto, A.R., Barros, Y., Martinelli Filho, J.E., Giarrizzo, T., Rabelo, E.F., 2023. Do coral reefs act as sinks for microplastics? *Environ. Pollut.* 337, 122509. https://doi.org/10.1016/j.envpol.2023.122509

Song, F., Koo, H., Ren, D., 2015. Effects of material properties on bacterial adhesion and biofilm formation. *J. Dent. Res.* 94, 1027–1034. https://doi.org/10.1177/0022034515587690

Souza, L.A., Santos, A.C.S.S., Leão, J.M., Schaeppi, O.C., Hatje, V., 2022. Occurrence and contents of trace metals and rare earth elements on plastic pellets. *Mar. Pollut. Bull.* 185, 114261. https://doi.org/10.1016/j.marpolbul.2022.114261

Squadrone, S., Pederiva, S., Bezzo, T., Sartor, R.M., Battuello, M., Nurra, N., Griglione, A., Brizio, P., Abete, M.C., 2022. Microplastics as vectors of metals contamination in Mediterranean Sea. *Environ. Sci. Pollut. Res.* 29, 29529–29534. https://doi.org/10.1007/s11356-021-13662-7

Stabnikova, O., Stabnikov, V., Marinin, A., Klavins, M., Vaseashta, A., 2022. The role of microplastics biofilm in accumulation of trace metals in aquatic environments. *World J. Microbiol. Biotechnol.* 38, 117. https://doi.org/10.1007/s11274-022-03293-6

Stoett, P., Vince, J., 2019. The plastic–climate nexus: Linking science, policy, and justice, in: Harris, P.G. (Ed.), *Climate change and ocean governance: Politics and policy for threatened seas*. Cambridge: Cambridge University Press, pp. 345–361. https://doi.org/10.1017/9781108502238.021

Stevenson, E.M., Buckling, A., Cole, M., Lindeque, P.K., Murray, A.K., 2024. Selection for antimicrobial resistance in the plastisphere. *Sci. Total Environ.* 908, 168234. https://doi.org/10.1016/j.scitotenv.2023.168234

Stubbins, A., Law, K.L., Muñoz, S.E., Bianchi, T.S., Zhu, L., 2021. Plastics in the Earth system. *Science* 373, 51–55. https://doi.org/10.1126/science.abb0354

archi, M., Zaaboub, N., Alsubih, M., Brik, B., Martins, M.V.A., Aleya, L., Trabelsi, L., 2023. Microalgae colonization and trace element accumulation on the plastisphere of marine plastic debris in Monastir Bay (Eastern Tunisia). *Environ. Sci. Pollut. Res.* 30, 32427–32451. https://doi.org/10.1007/s11356-022-23930-9

Toyofuku, M., Inaba, T., Kiyokawa, T., Obana, N., Yawata, Y., Nomura, N., 2016. Environmental factors that shape biofilm formation. *Biosci. Biotechnol. Biochem.* 80, 7–12. https://doi.org/10.1080/09168451.2015.1058701

Tu, C., Chen, T., Zhou, Q., Liu, Y., Wei, J., Waniek, J.J., Luo, Y., 2020. Biofilm formation and its influences on the properties of microplastics as affected by exposure time and depth in the seawater. *Sci. Total Environ.* 734, 139237. https://doi.org/10.1016/j.scitotenv.2020.139237

Tuck, B., Watkin, E., Somers, A., Machuca, L.L., 2022. A critical review of marine biofilms on metallic materials. *NPJ Mater. Degrad.* 6, 1–12. https://doi.org/10.1038/s41529-022-00234-4

Tuvo, B., Scarpaci, M., Bracaloni, S., Esposito, E., Costa, A.L., Ioppolo, M., Casini, B., 2023. Microplastics and antibiotic resistance: the magnitude of the problem and the emerging role of hospital wastewater. *Int. J. Environ. Res. Public. Health* 20, 5868. https://doi.org/10.3390/ijerph20105868

UN Decade, 2023. Ocean Decade. URL https://oceandecade.org/challenges/ (accessed 6.12.24).

Uroz, S., Calvaruso, C., Turpault, M.-P., Frey-Klett, P., 2009. Mineral weathering by bacteria: ecology, actors and mechanisms. *Trends Microbiol.* 17, 378–387. https://doi.org/10.1016/j.tim.2009.05.004

USEPA, O., 2015. National Primary Drinking Water Regulations [WWW Document]. URL https://www.epa.gov/ground-water-and-drinking-water/national-primary-drinking-water-regulations (accessed 6.12.24).

Vu, B., Chen, M., Crawford, R.J., Ivanova, E.P., 2009. Bacterial extracellular polysaccharides involved in biofilm formation. *Molecules* 14, 2535–2554. https://doi.org/10.3390/molecules14072535

Wang, H., Xu, K., Wang, J., Feng, C., Chen, Y., Shi, J., Ding, Y., Deng, C., Liu, X., 2023a. Microplastic biofilm: an important microniche that may accelerate the spread of antibiotic resistance genes via natural transformation. *J. Hazard. Mater.* 459, 132085. https://doi.org/10.1016/j.jhazmat.2023.132085

Wang, H., Yang, Q., Li, D., Wu, J., Yang, S., Deng, Y., Luo, C., Jia, W., Zhong, Y., Peng, P., 2023b. Stable isotopic and metagenomic analyses reveal microbial-mediated effects of microplastics on sulfur cycling in coastal sediments. *Environ. Sci. Technol.* 57, 1167–1176. https://doi.org/10.1021/acs.est.2c06546

Wang, Z., Hua, P., Dai, H., Li, R., Xi, B., Gui, D., Zhang, J., Krebs, P., 2020. Influence of surface properties and antecedent environmental conditions on particulate-associated metals in surface runoff. *Environ. Sci. Ecotechnol.* 2, 100017. https://doi.org/10.1016/j.ese.2020.100017

Warscheid, T., Braams, J., 2000. Biodeterioration of stone: a review. *Int. Biodeterior. Biodegrad.* 46, 343–368. https://doi.org/10.1016/S0964-8305(00)00109-8

Wen, B., Jin, S.-R., Chen, Z.-Z., Gao, J.-Z., Liu, Y.-N., Liu, J.-H., Feng, X.-S., 2018. Single and combined effects of microplastics and cadmium on the cadmium accumulation, antioxidant defence and innate immunity of the discus fish (Symphysodon aequifasciatus). *Environ. Pollut.* 243, 462–471. https://doi.org/10.1016/j.envpol.2018.09.029

WHO2022. Guidelines for drinking-water quality: fourth edition incorporating the first and second addenda [WWW Document] (2022) https://www.who.int/publications/i/item/9789240045064 (accessed 6.15.24).

Wright, R.J., Bosch, R., Langille, M.G.I., Gibson, M.I., Christie-Oleza, J.A., 2021a. A multi-OMIC characterisation of biodegradation and microbial community succession within the PET plastisphere. *Microbiome* 9, 141. https://doi.org/10.1186/s40168-021-01054-5

Wright, R.J., Erni-Cassola, G., Zadjelovic, V., Latva, M., Christie-Oleza, J.A., 2020. Marine plastic debris: a new surface for microbial colonization. *Environ. Sci. Technol.* 54, 11657–11672. https://doi.org/10.1021/acs.est.0c02305

Wright, R.J., Langille, M.G.I., Walker, T.R., 2021b. Food or just a free ride? A meta-analysis reveals the global diversity of the Plastisphere. *ISME J.* 15, 789–806. https://doi.org/10.1038/s41396-020-00814-9

Yan, W., Hamid, N., Deng, S., Jia, P.-P., Pei, D.-S., 2020. Individual and combined toxicogenetic effects of microplastics and heavy metals (Cd, Pb, and Zn) perturb gut microbiota homeostasis and gonadal development in marine medaka (Oryzias melastigma). *J. Hazard. Mater.* 397, 122795. https://doi.org/10.1016/j.jhazmat.2020.122795

Yang, Y., Chen, J., Chen, Z., Yu, Z., Xue, J., Luan, T., Chen, S., Zhou, S., 2022. Mechanisms of polystyrene microplastic degradation by the microbially driven Fenton reaction. *Water Res.* 223, 118979. https://doi.org/10.1016/j.watres.2022.118979

Ye, S., Andrady, A.L., 1991. Fouling of floating plastic debris under Biscayne Bay exposure conditions. *Mar. Pollut. Bull.* 22, 608–613. https://doi.org/10.1016/0025-326X(91)90249-R

Ye, S., Rao, M., Xiao, W., Zhou, J., Li, M., 2023. The relative size of microalgal cells and microplastics determines the toxicity of microplastics to microalgae. *Process Saf. Environ. Prot.* 169, 860–868. https://doi.org/10.1016/j.psep.2022.11.077

Ye, X., Cheng, Z., Wu, M., Hao, Y., Lu, G., Hu, B.X., Mo, C., Li, Q., Wu, J., Wu, J., 2022. Effects of clay minerals on the transport of polystyrene nanoplastic in groundwater. *Water Res.* 223, 118978. https://doi.org/10.1016/j.watres.2022.118978

Yin, W., Wang, Y., Liu, L., He, J., 2019. Biofilms: the microbial "protective clothing" in extreme environments. *Int. J. Mol. Sci.* 20, 3423. https://doi.org/10.3390/ijms20143423

Yoshida, S., Hiraga, K., Takehana, T., Taniguchi, I., Yamaji, H., Maeda, Y., Toyohara, K., Miyamoto, K., Kimura, Y., Oda, K., 2016. A bacterium that degrades and assimilates poly(ethylene terephthalate). *Science* 351, 1196–1199. https://doi.org/10.1126/science.aad6359

Yu, Y., Tian, D., Han, Y., Huang, L., Tang, Y., Zhang, W., Zhou, W., Shi, W., Yu, Y., Liu, G., 2022. Impacts of microplastics and carbamazepine on the shell formation of thick-shell mussels and the underlying mechanisms of action. *Sci. Total Environ.* 838, 156442. https://doi.org/10.1016/j.scitotenv.2022.156442

Yüksel, B., Şen, N., Ögünç, G.I., Erdoğan, A., 2023. Elemental profiling of toxic and modern primers using ICP-MS, SEM-EDS, and XPS: an application in firearm discharge residue investigation. *Aust. J. Forensic Sci.* 55, 529–546. https://doi.org/10.1080/00450618.2022.2043436

Zadjelovic, V., Wright, R.J., Borsetto, C., Quartey, J., Cairns, T.N., Langille, M.G.I., Wellington, E.M.H., Christie-Oleza, J.A., 2023. Microbial hitchhikers harbouring antimicrobial-resistance genes in the riverine plastisphere. *Microbiome* 11, 225. https://doi.org/10.1186/s40168-023-01662-3

Zettler, E.R., Mincer, T.J., Amaral-Zettler, L.A., 2013. Life in the "Plastisphere": microbial communities on plastic marine debris. *Environ. Sci. Technol.* 47, 7137–7146. https://doi.org/10.1021/es401288x

Zhang, J., Ma, W., Li, Y., Zhong, D., Zhou, Z., Ma, J., 2024. The resistance change and stress response mechanisms of chlorine-resistant bacteria under microplastic stress in drinking water distribution system. *Environ. Pollut.* 124331. https://doi.org/10.1016/j.envpol.2024.124331

Zhang, Y., Luo, Y., Guo, X., Xia, T., Wang, T., Jia, H., Zhu, L., 2020. Charge mediated interaction of polystyrene nanoplastic (PSNP) with minerals in aqueous phase. *Water Res.* 178, 115861. https://doi.org/10.1016/j.watres.2020.115861

Zhu, D., Ma, J., Li, G., Rillig, M.C., Zhu, Y.-G., 2022. Soil plastispheres as hotspots of antibiotic resistance genes and potential pathogens. *ISME J.* 16, 521–532. https://doi.org/10.1038/s41396-021-01103-9

Zhu, X., 2021. The plastic cycle – an unknown branch of the carbon cycle. *Front. Mar. Sci.* 7. https://doi.org/10.3389/fmars.2020.609243

15 Engineering Microbiosphere for Degradation of Plastics

Uvin Eksith Senadheera, Madushika Sewwandi, Jasintha Jayasanka, and Meththika Vithanage

15.1 INTRODUCTION

Unorganized plastic-based waste management has led to plastic waste accumulation in oceans, inland waters, lakes, and land (Rhodes, 2019). The unprecedented plastic accumulation in the natural environment threatens the planet earth's sustainability (Jambeck et al., 2015). Most plastic waste discharged into the ecosystem is chemically stable, corrosion-resistant, and recalcitrant to biodegradation (Yang et al., 2020). Plastics do not entirely represent an inert surface (Shen et al., 2019). Plastics often absorb nutrients and residual organic matter from the surrounding environment. These nutrients and residual organic matter furnish substrates for microbial biofilm formation on these synthetic particulates. The microbiosphere formed on a relatively new ecological niche of plastics comprises multifarious microbial communities such as autotrophs, heterotrophs, symbionts, and predators attached to the plastic surfaces (Arias-Andres et al., 2018). Increasing abundance and ecological pressure on these plastic-dwelling microbes have forced them to consume plastic substrate as a carbon source for metabolism. Therefore, recent studies have focused on harnessing the plastic degradability of plastic-dwelling microbes for ecologically safe plastic bioremediation (Rummel et al., 2017). Biological degradation methods by plastic-dwelling algae, bacteria, and fungi are attractive tools for cost-effective and eco-friendly approaches. Figure 15.1 gives an overview of the colonization, proliferation, and degradation of plastic by a plastic-degrading microbiosphere (PDM).

The most established approach to utilize the plastic-degradation capacity of plastic-dwelling microbes is to use artificial and natural plastic-degrading consortia with limited microbes isolated from naturally occurring PDM. Top-down and bottom-up strategies for the creation of such plastic-degrading consortia with limited microbes have constraints, such as restricted efficiency, inhibitory effects of plastic additives, toxins, antagonistic microbial effects, lack of biofilm formation, control, Division of Labor (DOL), spatial organization, and complexity of natural ecosystems, making it challenging to deploy a self-sustaining plastic-degrading consortium for plastic biodegradation (Rüthi et al., 2023). Nonetheless, engineering of an entire self-sustaining PDM is feasible. This PDM should comprise a multitude of plastic-degrading strains representing heterotrophs, saprotrophs, and autotrophs alongside biofilm formation, community-defense enhancers, and surfactant producers. These microbes can work synergistically with reduced metabolic burden and improved DOL to ultimately achieve an ecologically viable plastic biodegradation.

By researching the plastisphere at cellular, community, and molecular levels, it is feasible to engineer a PDM for plastic biodegradation (Bhatt et al., 2021). In recent studies and reviews on microbiosphere-mediated plastic biodegradation, only a few studies have addressed plastic biodegradation by focusing on engineering an entire self-sustaining PDM (Patrício Silva et al., 2021). In the book chapter, the authors delve into crucial design features and engineering strategies to be considered while engineering a durable and environmentally feasible PDM (Leong et al., 2014).

DOI: 10.1201/9781032717814-15

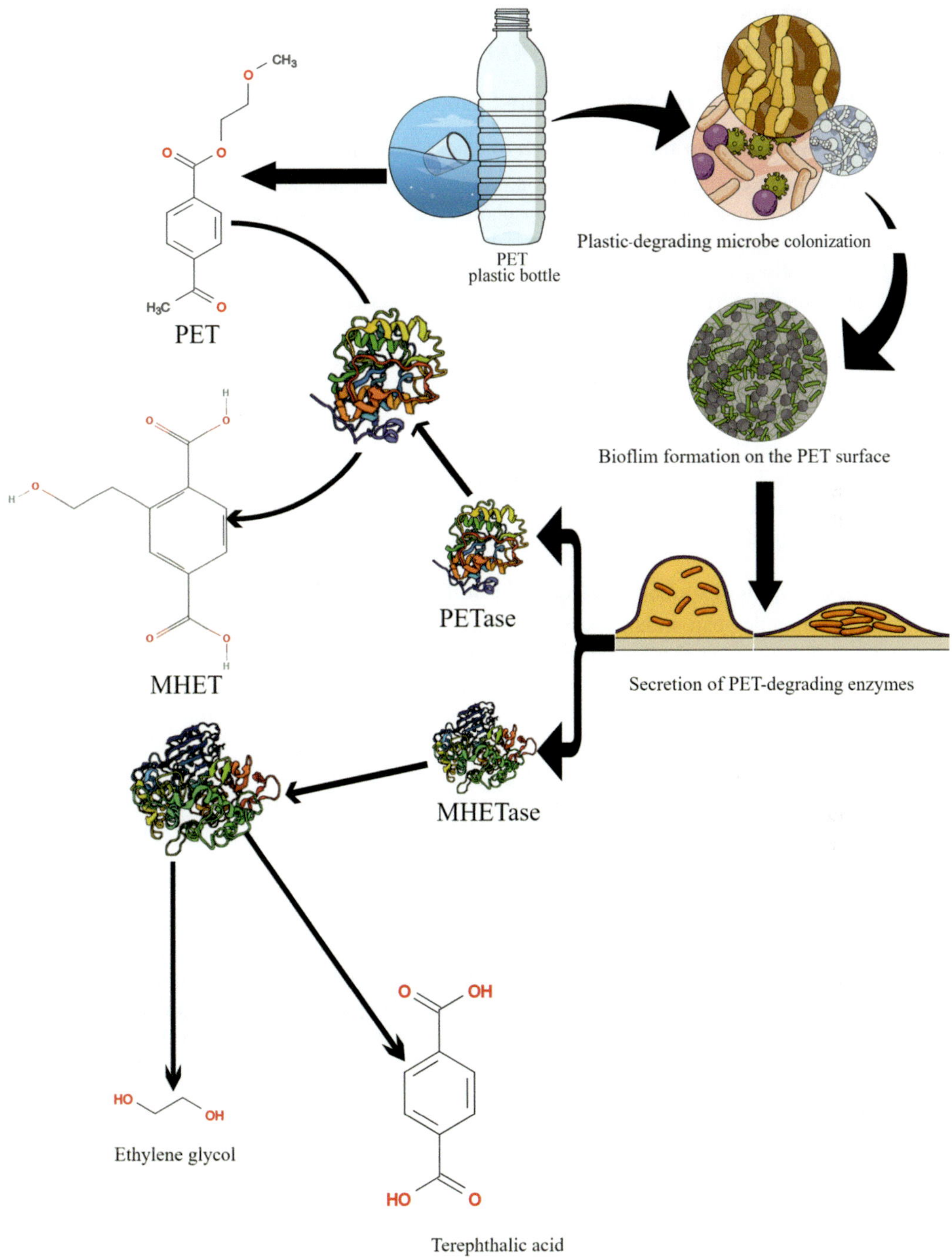

FIGURE 15.1 The colonization, proliferation, and establishment of a natural plastic-degrading microbiosphere on PET (polyethylene terephthalate) as an example. PET is subsequently degraded into MHET (mono(2-hydroxyethyl) terephthalic), ethylene glycol and terephthalic acid by PETase (polyethylene Terephthalate hydrolase) and MHETase (mono(2-hydroxyethyl) terephthalic hydrolase), respectively.

15.2 ENGINEERING MICROBIOSPHERE ECOSYSTEMS FOR ENHANCED PLASTIC BIODEGRADATION

15.2.1 Primary Design Features of a Self-Sustaining PDM

15.2.1.1 A Solid Platform for the Growth of Microbiosphere Microbes

The core component of the microbiosphere community is the plastic waste itself. Therefore, the initiation point for engineering the microbiosphere would be a solid platform (SP) to harbor the plastic-degrading microbes. The SP provides a durable, colonizable, and resilient surface to facilitate biofilm formation. The SP should be easily degradable. It should not persist in the natural environment for extended periods and lead to pernicious interactions with other macro and microbial life nature. The SP must be engineered with an environmentally friendly material similar to conventional plastic. Polyhydroxy butyrate (PHB) can be successfully used as an SP for the microbiosphere microbes to colonize and proliferate (Leong et al., 2014). PHB polymer can serve as the carbon and energy source for microbiosphere microorganisms. Polycaprolactone and polylactic acid (PLA) can be used to synthesize the SP for the microbiosphere microorganisms due to their biodegradability (Wang et al., 2020). The SP material can be customized by mixing different additives, such as compost and sand, to make it suitable for the ecosystem that PDM is being used. In addition, the SP provides an uninterrupted stream of nutrients to the microbiosphere to sustain growth and cell viability.

15.2.1.2 Development of a Self-Sustainable Biofilm-Based PDM on the SP

Microbial communities can endure extreme climatic conditions ranging from several months to several years. The engineered microbiosphere community should provide many beneficial traits for self-sustainability and enhanced plastic degradation. It should enable efficient distribution of labor and reduce the metabolic burden imposed on a single microbial strain of the microbiosphere. One of the methods to be used to grow and sustain a self-sustainable microbial community for plastic degradation is modular co-culture engineering (Zhang and Wang, 2016). One of the benefits of modular co-culture engineering is that it enables reduced metabolic burden and improves the overall bioconversion of plastic into nontoxic metabolites (Johns et al., 2016). While engineering the plastic-degrading, self-sustaining microbial population, a few crucial factors should be considered. They are engineering various interspecies and intraspecies interactions of plastic degraders, construction of spatiotemporal dynamics, maintenance and modeling of community-wide functional robustness, and development of population control mechanisms. Moreover, three options must be considered while engineering the self-sustainable plastic-degrading microbial community. They are two- or three-way auxotrophy, syntrophy, and cross-feeding approaches. Community-wide robustness can be maintained by discouraging freeloaders or cheaters via segregation. It is essential to have biocontainment measures such as dead man or passcode kill switches for the microbial community to be used in natural ecosystems (Pande et al., 2015).

The self-sustaining microbial community should be engineered by centralizing around the activity of plastic-degrading heterotrophs. The microbiosphere should be engineered as the plastic hydrolyzing extracellular enzyme secretion as the driving force in plastic degradation. Plastic-degrading heterotrophs can utilize plastics as their sole carbon source. The plastic-degrading extracellular enzymes secreted by the heterotrophs break down long carbon chains in plastics into oligomers. These oligomers can be taken inside the cells and degraded into monomers for carbon and energy. Due to the vast nature of plastic-degrading heterotrophic microorganisms within the microbiosphere, it is practically impossible to experimentally test all combinations of heterotrophic plastic degraders while engineering a self-sustaining consortium. Therefore, possible heterotrophic plastic degraders in the microbiosphere can be filtered by simulation to test their compatibility in silico. The Flexible Synthetic Consortium Optimization (FLYCOP) tool can be used for designing and optimizing heterotrophic plastic-degrading communities in the microbiosphere, using a stochastic local search approach instead of a trial-and-error approach. It uses multiple consortium configurations and can be effectively used for detailed evaluation before engineering the actual in vivo experiments on plastic degradation. A microbial community modeler (MCM) can also be used in engineering a synthetic self-sustainable heterotrophic consortium in plastic degradation from the microbiosphere (García-Jiménez et al., 2018). Using in silico tools in engineering synthetic consortiums circumvents the trial-and-error efforts and the demand for chemical optimization.

Furthermore, a hybrid microbiosphere comprising indigenous and natural heterotrophic microorganisms can be engineered to have enhanced plastic degradation. However, synthetic heterotrophic microbial consortiums can undergo mutations over time, giving rise to nonproductive phenotypes, and decrease the population-level performances. Therefore, care must be taken to reduce the mutations during the engineering of synthetic microbial consortiums. Heterotrophic plastic-degrading strains in the microbiosphere were demonstrated to be exceptionally amendable hosts for genetic engineering for enhanced plastic degradation. Genetic engineering involves the manipulation of the genetic material of the microbiosphere to increase its plastic-degradation efficiency. The chief objective of genetic engineering is to isolate the genes involved in plastic-degradation pathways, modify the isolated genes, and clone them in appropriate microbiosphere hosts using techniques such as site-directed mutagenesis—polymerase chain reaction (PCR), and antisense RNA technology (Hussain et al., 2018).

Several regulatory obstacles hamper using genetically engineered heterotrophic microorganisms in on-site plastic biodegradation. Even though genetically engineered heterotrophic microorganisms have shown high efficacy in laboratory conditions, a limited number of on-site validations are done to evaluate their efficiency (Honjo et al., 2019). To reduce the negative impact of genetically engineered plastic-degrading heterotrophic microorganisms, tools such as antibiotic genes-free genetic engineering tools and suicide gene systems can be introduced. After complete plastic biodegradation, programmed cell death circuits can eradicate the genetically engineered microbes from ecosystems (Sedlmayer et al., 2018).

Incorporating plastic-degrading heterotrophic fungal isolates synergistically with the plastic-degrading heterotrophic bacteria is a successful way to engineer the microbiosphere for enhanced plastic biodegradation (Sánchez, 2020). Engineering the self-sustaining plastic-degrading microbial community with heterotrophic fungal-bacterial conjugate creates two types of plastic biodegradation within the microbiosphere. First is the physical deterioration of plastics by fungi. Producing acidic and alkaline products by plastic-degrading heterotrophic fungi can further enhance plastic degradation. Incorporating heterotrophic plastic-degrading fungi can form crevices and cracks in the plastic structure through mycelial growth into the plastic to enhance plastic biodegradation (Sánchez, 2020).

In instances where an adequate source of carbon is not readily available or in scarcity of plastic polymers, it is essential to include autotrophs such as *Flavobacteriaceae* and *Rhodobacteriaceae*, which can convert inorganic CO_2 into organic molecules in the presence of light that the heterotrophic plastic degraders can use to thrive (Malik et al., 2023). Plastic-degrading autotrophs can be used to supply oxygen and nitrogen circulation and rapid proliferation of plastic-degrading heterotrophs. Plastic-degrading cyanobacterial species such as *Oscillatoria subbrevis* and *Phormidium lucidum* can be used within the microbiosphere to continuously supply oxygen and fix nitrogen within the plastic-degrading microbial community. The photoautotrophic production of oxygen cause affects the rapid proliferation of aerobic plastic degraders within the community (Sarmah and Rout, 2020). Plastic-degrading cyanobacteria and microalgae can also be used within microbiosphere engineering to remove toxic metabolites and pollutants from the vicinity of the microbial community. Plastic-degrading autotrophs such as purple sulfur bacteria such as *Chromatium* sp. and *Ectothiorhodospira* sp. can be effectively used to recycle vital nutrients such as nitrogen and sulfur within the plastic-degrading microbial community without leaching them out of the microbiosphere during harsh climatic conditions (Lear et al., 2021). The plastic-degrading heterotrophs in the self-sustaining microbial community can provide the additional CO_2 from plastic metabolism. The symbiotic relationship engineered within the microbiosphere can confer a survival advantage in natural ecosystems.

15.2.1.3 Improving the Cellular Attachment of the Self-Sustaining PDM into the SP

The affinity of the cells of plastic-degrading microbes can be attributed to the chemical additives and contaminants within plastic waste. However, the cellular attachment of the engineered consortium of heterotrophs autotrophs and supplementary microbes to the plastic surfaces can be facilitated in two ways. First is the attachment of the self-sustaining PDM via cellular appendages such as pili, fimbriae, curli fibers, and flagella, and the other is the extracellular matrix formation called biofilm, which is often initiated by physical attachment of cellular appendages (Figure 15.2a). Biofilm formation is the most stable and long-lasting one of the two ways to attain the cellular attachment of plastics to the self-sustaining plastic-degrading microbial community (Flemming et al., 2016). The core structure of the biofilm formation of the self-sustaining plastic-degrading microbial community is formed by extracellular polysaccharide substances (EPS). EPS comprises proteins, lipids, deoxyribonucleic acid (DNA), and polysaccharides generating adhesive forces such as the London dispersion force and hydrogen bonds with the plastics. The self-sustaining plastic-degrading microbial community should be engineered to achieve a higher rate of biofilm formation for the rapid

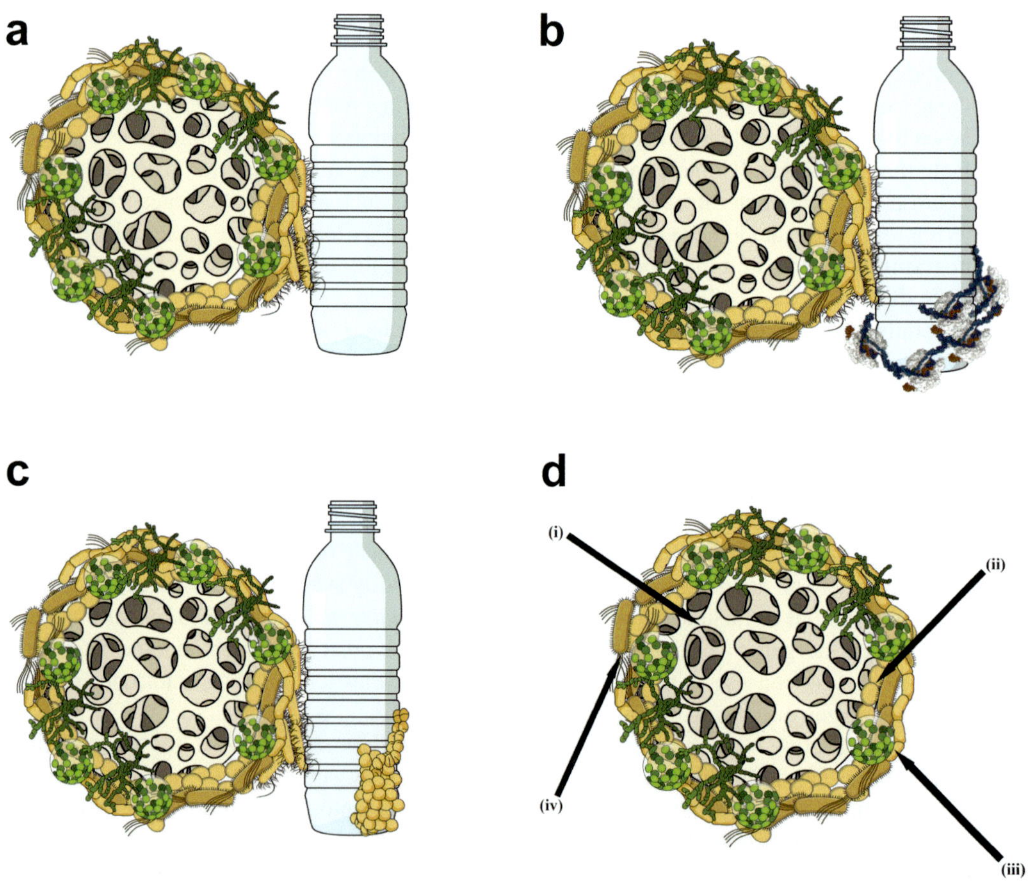

FIGURE 15.2 (a) Attachment of the engineered plastic-degrading microbiosphere (PDM) onto the plastic surface via physical appendages or biofilm. (b) Secretion of improvised plastic-degrading enzymes (PDEs) into the plastic substrate by the PDM enhances plastic biodegradation. (c) Elimination of the plastic-dwelling pathogens by the engineered PDM. (d) An overview of the engineered PDM with proposed primary design features: (i) biodegradable SP (solid platform) to accommodate the PDM, (ii) plastic-degrading microbes with metabolic pathways to degrade plastic, metabolites, and toxins, (iii) attachment with specialized appendages to the plastic surface, and (iv) self-sustaining microbial community with heterotrophs, phototrophs, and saprotrophs.

attachment of the community to plastic surfaces. Introduction of EPS formation enhancement additives will accelerate the biofilm formation on the plastic surface within the microbiosphere. Mineral oil can be used for EPS secretion enhancement during low-density polyethylene (LDPE) biodegradation assays by *Pseudomonas* sp. AKS2 (Tribedi and Sil, 2012). Engineering the self-sustaining microbial community by incorporating actinomycetal species can effectively produce extracellular polymeric compounds such as mucopolysaccharides enriched in N-acetylglucosamine and improve the adhesion of the self-sustaining plastic-degrading microbial community to plastics. In addition, plastic-degrading fungal isolates can increase the adhesion of the self-sustaining PDM to plastic surfaces. Incorporating fungal isolates in the microbiosphere allows the secretion of hydrophobic proteins capable of adhesion to the hydrophobic polymers (Piscitelli et al., 2017). Due to these proteins' strong adhesion and surface activity, the fungal mycelia can allow the self-sustaining PDM to attach firmly to the plastic surface. Supplementary, plastic-degrading bacteria can be modified to display plastic-degrading enzymes (PDEs) that can enhance the attachment of the cells to the plastic surface and immobilize the substrate. The membrane-anchored polyethylene terephthalate (PET) esterase displayed on the cell surface of *Rhodococcus pyridinivorans* P23 can be effectively used to immobilize the plastic substrate for the microbiosphere to degrade it (Guo et al., 2023).

15.2.1.4 Enhancing Enzyme-Mediated Depolymerization of Plastic by the Self-Sustaining PDM

The slow turnover and the poor thermostability of the enzyme systems of the PDEs of the members of the microbiosphere are some of the limiting factors that reduce the efficacy of microbiosphere-mediated plastic biodegradation. Therefore, the performance of the enzymes of the PDM should be improved by protein engineering. Synthetic biology approaches can be used to redesign the existing PDEs in PDMs. Using computational biology, enzyme-mediated depolymerization can be streamlined by facilitating the theoretical design steps utilizing environmental sequence data. Using strategies such as Affinity Analysis-based Dynamic Docking (ADD), researchers can engineer enzymes such as TurboPETase, which can outperform natural PET hydrolase achieving a complete PET degradation (Su et al., 2023; Wu et al., 2023). Machine learning approaches such as the plastic enzyme degradation (PED) model can be effectively used to predict enzyme activity in various plastic biodegradation processes with high accuracy. This can be used to screen out the most suitable PDEs from the natural pool of PDEs in PDM engineering (Zurier and Goddard, 2023).

15.2.1.4.1 Improving the PDE Thermostability in the PDM

Improving the PDE thermostability during approaching or increasing the high glass transition temperature (T_g) increases the mobility and flexibility of the polymer chains (Wei and Zimmermann, 2017). Improved T_g assists their attachment to the enzyme surface and access to the active sites, eventually increasing plastic biodegradation. The thermostability of the PDEs in the engineered PDM can be increased by adding disulfide links and salt bridges (Then et al., 2016; Numoto et al., 2023). Since proteins rely on disulfide and salt bridges to fold into a global shape that provides thermostability, introducing disulfide bonds by switching out residues at a metal bonding site increases the PDE thermostability. For example, introducing a disulfide bond into TfCut2 esterase from *Thermobifida fusca* by D204C and E253C mutations into the calcium-binding region can increase protein denaturation temperature and the plastic-hydrolysis rate (Then et al., 2016). Introduction of hydrogen bonds in the region responsible for maintaining a stable conformation of PDEs improves thermostability. Introduction of hydrogen bonds preserves a higher-order protein conformation by enhancing the structural stability and extreme temperature tolerance of PETase (Son et al., 2019). The thermostability of PETase was greatly enhanced by hydrogen bonds formed through water between S121E and N172 residues in the 6e7 connecting loop region.

15.2.1.4.2 Improving the Strength of Plastic Substrate Attachment to PDE Active Sites

Plastic depolymerization efficiency can be enhanced by improving the contact between the plastic substrate and the PDE active site. One of the approaches for strengthening the affinity of plastic-substrate attachment to PDE active sites is the widening of the active site aperture (Fang et al., 2023; Wu et al., 2024). A higher PET and polyamide (PA) depolymerization rate was displayed in the L182A mutant of *Fusarium solani* cutinase due to a large active site opening due to amino acid mutation (Araújo et al., 2007). A similar technique was used to enhance the PET and Poly Butylene Succinate-co-Adipate (PBSA) degradation using PETase, Cut190, *Pseudomonas aestusnigri* hydrolase, and (mono(2-hydroxyethyl) terephthalic) MHETase (Bollinger et al., 2020). Increasing the active site size does not always equate to elevated plastic degradation. Another potential strategy is to increase the hydrophobicity of the substrate-binding moieties in the active sites. Improving the hydrophobicity might be advantageous for plastic binding and degradation efficiency, as displayed with PETase (Ma et al., 2018).

15.2.1.4.3 Improving the Substrate–Enzyme Interactions

PDE efficiency can be enhanced by modifying the surface characteristics via enzyme–substrate interactions as hydrophobic and electrostatic interactions between the substrate and amino acid moieties of the enzyme surface govern the substrate binding process. Therefore, the strategy can be executed by adding heterologous binding molecules to PDEs. For example, carbohydrate-binding molecules (CBMs) can bind to various plastic polymers (Ribitsch et al., 2017). The CBMs derived from *Trichoderma reesei* fused to cutinase Thc_Cut1 were used to improve PET hydrolysis via the improved binding capacity of the enzyme to PET via CBMs.

15.2.1.4.4 Optimization of Auxiliary PDE Functionalities

The activity of PDEs can be inhibited via products or intermediates of plastic biodegradation, and it can be circumvented by changing the architecture of the active site. For example, modification of the substrate binding groove of TfCut2 through G62A mutation resulted in a 5.5-fold decrease in the binding constant for the inhibition hydrolysis product of MHET, thus improving overall PET degradation (Wei et al., 2016). In addition, the fusion of PDE with an additional enzyme capable of hydrolyzing product intermediates can also be used for enhanced plastic biodegradation in the microbiosphere. For example, the introduction of an additional Thc_Cut1 enzyme to hydrolyze PAs) enhances promiscuous amidase activity (Biundo et al., 2019). Residues that prevent water from interacting with transition states were chosen for mutations, for a 6–15-fold increase in hydrolytic activity. Table 15.1 shows different PDE engineering strategies for enhanced plastic degradation within an engineered PDM. Figures 15.3 and 15.2b give an overview of protein engineering strategies for enhancement of PDE-mediated plastic biodegradation.

15.2.1.5 Introduction of Auxiliary Microbes for the Supportive Roles in the PDM

Plastic-degrading fungal cultures can engineer a supportive matrix of mycelium for the adhesion, attachment, and protection of the plastic-degrading heterotrophic and autotrophic microbiosphere. Utilization of fungi such as *Aspergillus*, *Fusarium*, and *Penicillium* to secrete hydrophobic proteins can be used to penetrate the polymeric surfaces of the plastics and facilitate the deeper burial of plastic-degrading heterotrophs to continue the biodegradation process. Co-culturing of the plastic-degrading fungi creates a filament growth of the mycelia on the SP that could protect the heterotrophic and autotrophic population of the plastic degraders from UV radiation, and mycelia, provide an intercalated network for the growth of the microbiosphere. Dedicated microorganisms to remove and recycle dead organic matter within the microbiosphere should be incorporated into the self-sustaining plastic-degrading microbial community. Fungi such as *Chytridiomycota*, *Basidiomycota*, and *Ascomycota* can be used for the role of saprotrophs in the self-sustaining microbial community (Zeghal et al., 2021).

TABLE 15.1
Potential Protein Engineering Strategies with Applications on Microbiosphere Engineering for Enhanced Plastic Biodegradation

Type of PDE	Isolated Microbe	Plastic Substrate	Engineering Strategy and Modifications Induced	Results	References
PETase	*Indonella sakaiensis*	PET	Construction of a chimeric bifunctional enzyme fused with MHETase.	Increased catalytic activity. Greater than thrice that of the native enzyme.	Knott et al. (2020)
Cutinase	*T. cellulosilytica*	PA	Increasing the number of polar motifs at I179A, I179N, and I179Q positions in the active site.	Elevated the amidase activity by 15-fold on the model polyamide substrate.	Barth et al. (2016)
Esterase	*Clostridium botulinum*	PET	Modulation of the surface hydrophobicity via truncation of 17 residues at the N-terminal.	The modified enzyme showed increased hydrolysis compared to the wild-type enzyme.	Biundo et al. (2019)
PETase	*I. sakaiensis*	PET and PEF	Limiting the opening size active site cleft at S238F/W159H.	Enhancement of the PET and PEF hydrolysis	Austin et al. (2018)
PET hydrolase	*Pseudomonas aestusnigri*	PET	Increasing the size of the active site cleft Y250S. Maximizing the size of the cleft in the active site. Y250S	Acquired the ability to hydrolyze crystalline PET from commercial plastic containers with increased decomposition rates.	Bollinger et al. (2020)
PHB depolymerase	*Ralstonia pickettii T1*	PHB	Changing the hydrophobicity of the surface Y443F	Elevated enzymatic activity compared to the wild type of enzymes	Hiraishi et al. (2010)
PETase	*I. sakaiensis*	PET	Increasing the hydrophobicity of the enzyme's active site at L88F and I179F, respectively.	Elevated hydrolytic efficiency is 2.1 and 22.5 times, respectively, compared to the wild type of enzyme.	Ma et al. (2018)
Lipase	*Thermomyces lanuginosus*	PCL	Development of a bifunctional chimeric enzyme tethered to a cutinase.	Enhanced PCL hydrolysis by a factor of 13.3 compared to the endogenous enzyme.	Tan et al. (2012)

Note: PDEs, plastic-degrading enzymes; PA, polyamide; PET, polyethylene terephthalate, PEF, polyethylene furanoate; PCL, polycaprolactone; PHB, polyhydroxy butyrate.

15.2.1.6 Microbial Strain Uses in PDM for Plastic Additive Bioremediation

Different chemicals added to plastics to enhance the properties of plastics, such as fillers, flame retardants, antioxidants, impact modifiers, heat stabilizers, and UV-light stabilizers, also need to be biodegraded to reduce their ecological toxicity. It is essential to incorporate microbes capable of degrading plastic additives such as diethylhexyl phthalate, bisphenol, or organic pollutants, e.g., textile dyes and polycyclic aromatic hydrocarbons. In engineering a PDM, plastic and toxin-degrading microbial species would be essential for completely bioremediating plastic in natural ecosystems (Karkanorachaki et al., 2022). For example, microbes such as *Pseudomonas alloputida* REBP7

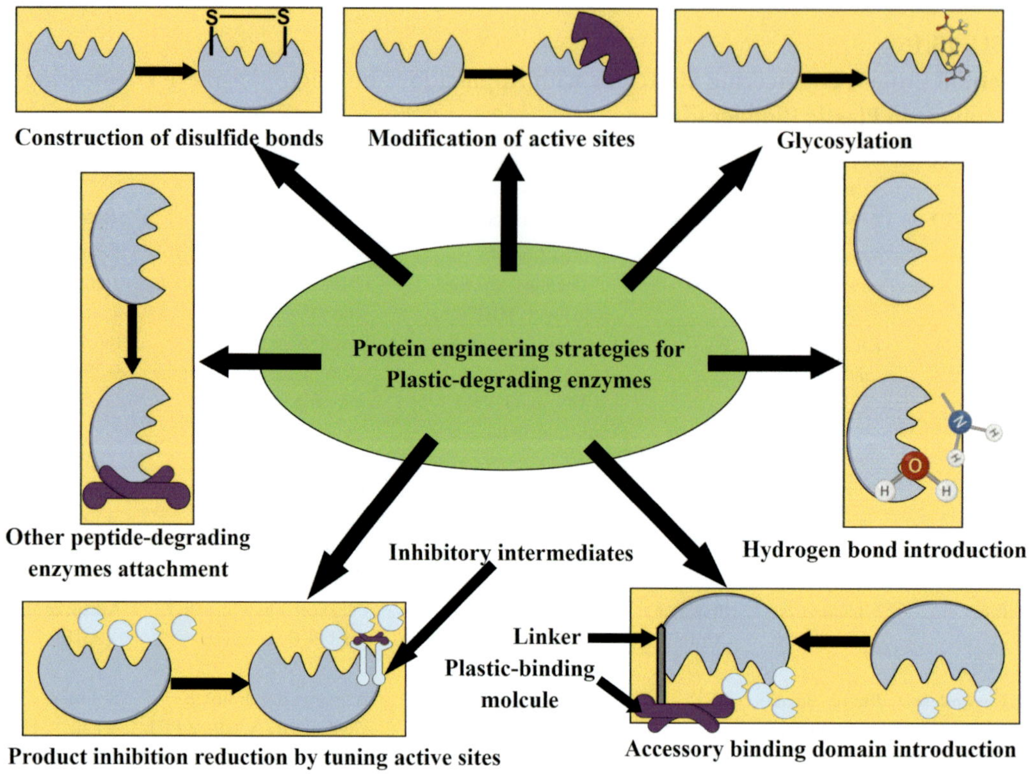

FIGURE 15.3 Protein engineering strategies to improve the functionality of PDEs (plastic-degrading enzymes) in the engineered plastic-degrading microbiosphere.

and *Pseudomonas aeruginosa* REBP5 can be used as additive degraders on LDPE (Rüthi et al., 2023). In addition, *Neodevriesia* and *Lachnellula* strains can be successfully used in cold climate conditions during the deployment of the PDM making the PDM climate resilient (Chofifawati et al., 2023).

15.2.1.7 Developing Pathogen Defense in Self-Sustaining PDM Ecosystems

The engineered self-sustaining PDM risks harboring pathogens such as *Vibrio cholera*. These pathogens pose a risk as they can act as opportunistic colonizers of the microbiosphere rather than the core stable members of the engineered microbiosphere. Another concern is that the engineered self-sustaining PDM might enrich the antimicrobial resistance genes through gene transfer and increase the chances of certain members of the microbiosphere acquiring wide-spectrum drug resistance. Therefore, it is crucial to incorporate microbial strains that might help reduce the engineered microbiosphere's pathogenicity, such as *Radiolaria*, Choanoflagellates, and *Micromonas* sp. (Dey et al., 2022). Incorporating "predatory" microbes in the self-sustaining PDM can reduce pathogens' unwanted colonization and infiltration. Integrating antipathogenic features in the self-sustaining PDM can reduce the spread of infectious diseases. For example, predatory microbes such as *Bdellovibrio bacteriovorus* can be effectively incorporated into the microbiosphere, reducing the risk of pathogenic colonization of *Vibrio cholera* (Richards et al., 2012) (Figure 15.2c). Figure 15.2d gives an overview of the engineered PDM for enhanced plastic biodegradation with primary design features.

15.3 ENGINEERING STRATEGIES FOR SUCCESSFUL FORMULATION OF PDM

15.3.1 DOL in the Engineered PDM

DOL enables microbial members to accomplish tasks more efficiently than monocultures. From an ecological perspective, DOL can be defined by specializing in different cells and cooperating with them to provide an inclusive effect on the microbiosphere. During the engineering of a PDM, DOL is the separation of tasks, irrespective of the ecological and evolutionary consequences (Figure 15.4). However, when properly engineering the PDM, the DOL can facilitate the microbiosphere succession in natural ecosystems in multiple ways.

15.3.1.1 Improving Functionality via the Specialization of Plastic Degraders of the PDM

DOL enables the organization of the pathways of plastic degradation by compartmentalizing different steps of plastic degradation into different members of the microbiosphere. The separation caused by DOL is vital when multiple plastic biodegradation processes cannot exist in the same cell. The specialization of a single plastic-degrading microbial member of the engineered microbiosphere in a subset of an entire plastic biodegradation pathway can be adopted to engineer a robust PDM. For example, utilization of two *Pseudomonas putida* strains for the biodegradation of PET hydrolysate via catabolic specialization, where each strain specializes in biodegradation of terephthalic acid (TPA) and ethylene glycol (EG), respectively, to achieve complete assimilation of substrate, the reduced levels of catabolic cross-talk and accelerated biodegradation of PET hydrolysates can be stated (Bao et al., 2023). The above-mentioned approach reduces the PDM complexity, maintains coexistence, reduces the competition between different microbes of the engineered PDM, and ensures long-term survivability of the PDM (Figure 15.3). The DOL, via specialization, allows the biodegradation of a diverse number of plastic polymers. For example, DOL among *Rhodococcus* spp. and *Hydrogenophaga* spp., via specialization of plastic catabolism allows the biodegradation of polyesters, PAs, and polyurethanes (Schaerer et al., 2023a).

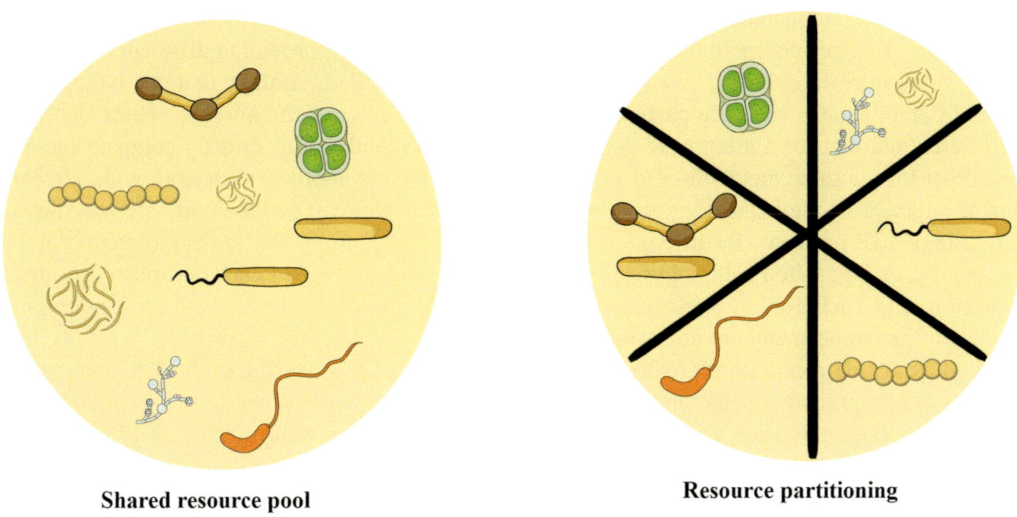

FIGURE 15.4 Adequate resource partitioning in the plastic-degrading microbiosphere ensures coexistence and reduction of competition among different plastic degraders for resources.

15.3.1.2 Reduction of the Metabolic Burden in the Engineered PDM

The maintenance of heterogeneous components such as plasmids and enzymes can impose a metabolic burden on plastic-degrading microbes of PDM. Gene expression can funnel away host resources such as ATP, nucleotides, and amino acids, which are subsequently unavailable for the biofilm formation and proliferation of colonies in the microbiosphere. Metabolic burden can cause a reduction of the fitness of the engineered microbiosphere while losing plastic biodegradation efficiency and functional mutations. Since the engineered PDM is a complex system, a single plastic-degrading member containing all the heterogeneous components, such as plastic-degrading gene-holding plasmids, might not grow due to the severe burden. DOL should be used to enable more complex plastic biodegradation pathways in the microbiosphere. Using microbes to utilize the intermediate products from another member of the microbiosphere is suitable to reduce the metabolic burden. Using aromatic compound-degrading microbiota mainly comprising *Rhodococcus* spp. and *Hydrogenophaga* to use intermediate products produced from the members of the microbiosphere for metabolic needs reduce the metabolic burden on a single member (Schaerer et al., 2023a).

15.3.1.3 Reduction of Engineering Complexity of the Engineered PDM

The compartmentalization of cellular processes into different plastic-degrading populations can elevate the modularity of the engineered PDM. Individual models in the microbiosphere can be modified, tuned, or replaced independently in a plug-and-play manner without implications for other functions. Depending on the complexity of the PDM, DOL's utilization might become more straightforward to control than that of a single cell as each individual of the PDM would contain only a subset of the total complexity of the microbiosphere. The strategy proposed by Salzano et al. (2021) can be used in DOL to reduce the complexity of the microbiosphere. The strategy is based on utilizing external control inputs, assuming that each cell in the microbiosphere is endowed with a reversible, bistable memory mechanism. The authors provide an analytical framework to engineer external feedback control measures to balance the ratio between microbial members whose memory is stabilized at either equilibrium linked to different phenotypes. A genetic toggle-switch can control the stability and robustness of the control laws in silico implementation of memory elements.

15.3.2 Optimized Resources Allocation

Optimized resource allocation can engineer a robust PDM by selecting plastic-degrading microbial strains with complementary degradation capabilities and enhancing the overall plastic biodegradation. To avoid competition exclusion of species in the PDM, providing different ecological niches in the microbiosphere is essential (Harcombe et al., 2014). A strategy that can be used for niche differentiation is resource partitioning, where each member of the microbiosphere depends on different nutrients or a mixture of nutrients. No two microbial strains directly compete against each other for the same metabolites (e.g., plastic-degradation metabolites) (Schaerer et al., 2023b). During engineering the PDM, the separate reservoirs of nutrients can be inherently presented on a self-sustained SP, presented by different members of the self-sustaining plastic-degrading microbial community, or segregated using physical barriers. In a well-engineered PDM, optimized resource allocation can be achieved by engineering the metabolic profiles of the plastic-degrading microbial strains so that some members are orthogonal for at least one primary metabolite, typically the plastic-degradation metabolites (Arias-Andres et al., 2018, 2019). Ensuring each PDM has exclusive access to a critical nutrient and optimized resource allocation is crucial to reducing competition among the engineered PDM members. The strategy can be implemented in PDMs about carbon heterogeneity. A nature-inspired-plastic-degrading bacterial consortium comprising strains of *Actinomycetales*, *Burkholderiales*, *Sphingobacteriales*, *Rhizobiales*, and *Rhodobacterales* can degrade the mixture of polyether–polyurethane–acrylate (PE-PU-A) co-polymer and xenobiotic additives of N-methyl 2-pyrrolidone, isopropanol, and glycol ethers (Gaytán et al., 2020). The strains in the plastic-degrading consortia showed a diauxic growth and eventually oxidized the additive first and then the co-polymer. The degradation was not dependent on the type of carbon source (co-polymer or the additives), and two primary carbon sources were used independently, leading to the biodegradation of the co-polymer–additive conjugate.

15.3.3 Enhancement of Chemical–Symbiotic Relationships

Chemical interactions between the members of the PDM can incentivize and stabilize coexistence by conjugating the survival of different plastic-degrading species. The chemicals that can be exchanged are mostly metabolites directly influencing the growth of the microbiosphere via cellular metabolism or inducing gene expression to promote the survival of the PDM through cell-to-cell signaling (Figure 15.5e). One of the chemical–symbiotic interactions that can be used to engineer a robust PDM is commensalism, where one plastic-degrading microbial population benefits from another population while receiving no harm or benefit in return. For example, biodegradation using four-species plastic-degrading microbial consortia on PET. Two engineered *Bacillus subtilis* secrete PET hydrolase and monohydroxyethyl terephthalate hydrolase for the initial degradation of PET. The accumulation of PET biodegradation metabolites such as TPA and EG inhibited the proliferation of the PET-degrading *B. subtilis* species. *Rhodococcus jostii* and *Pseudomonas putida* were introduced to the microbial consortium to mitigate the inhibition caused by TPA and EG on PET-degrading *B. subtilis* species. The two engineered *B. subtilis* species could use PET as the sole carbon source, and the *Rhodococcus jostii* and *Pseudomonas putida* could reduce the inhibition caused by PET-degrading by-products (Qi et al., 2021). The latter species in the consortium directly benefit from the two strains of *B. subtilis* but contribute to the overall PET biodegradation efficiency.

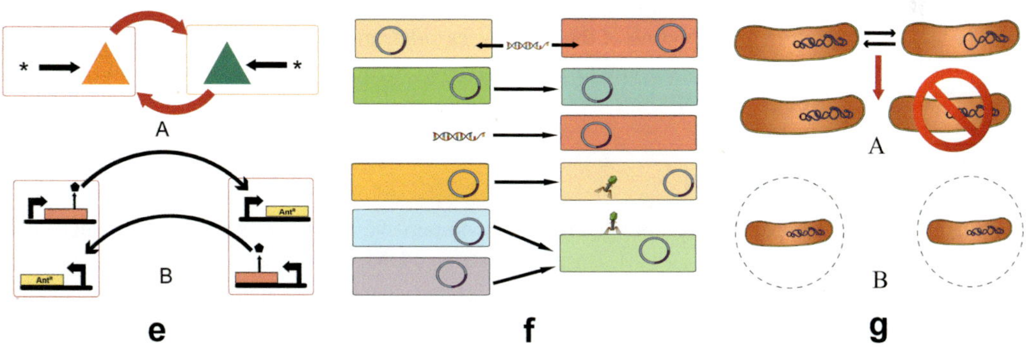

FIGURE 15.5 (e) Spatial organization separates different plastic-degrading microbiosphere (PDM) populations into distinct niches and maintains coexistence. (A) In a well-mixed PDM, without any stabilizing forces, the competition for resources results in competitive exclusion of least-fit populations. (B) Segregation of antagonistic populations into their own microenvironments reduces the competition in the PDM and maintains stable coexistence of the two species. (f) Horizontal gene transfer (HGT) maintains diversity and modulates coexistence among different plastic degraders. HGT maintains genetic diversity in the PDM, promoting diversification and specialization of different species into different niches of the microbiosphere and thus coexistence. (g) Chemically induced interactions among plastic degraders in the engineered plastic-degrading microbiosphere (PDM) can stabilize the coexistence of different populations. (A) Cross-feeding where different populations in the engineered PDM synthesize and share nutrients with each other outweigh the species competition and ensure the survival of all PDM populations. (B) Cell–cell communication through small molecules also ensures coexistence of PDM populations. For example, cells can mediate the survival of other PDM populations via gene circuit.

Another relationship that can be used to maintain the coexistence of the members of the engineered PDM is mutualism. Cross-feeding is a common strategy to achieve mutualism, where microbial members rely on metabolites produced by other members of the microbiosphere. Mutualism can be used to rapidly establish the secondary plastic-degrading colonizers during biofilm formation on the plastic polymers and increase the biomass required for the plastic biodegradation on the biofilm. For example, secondary colonizers such as *Flavobacteriaceae* (Bacteroidetes) proliferate using the organic molecules synthesized by the primary colonizers of the biofilm such as *Alphaproteobacteria* and *Gammaproteobacteria* on plastic surface biofilms (Dussud et al., 2018; Zhang et al., 2022). The mutualistic relationship between plastic-degrading strains can be used to accelerate plastic biodegradation. For example, *Arthrobacter* sp. and *Streptomyces* sp. suggest that PE biodegradation is enhanced due to the mutualistic activity and co-metabolism of plastic-degradation products. Furthermore, mutualism also enhanced biofilm formation (Han et al., 2020).

15.3.4 Horizontal Gene Transfer

Horizontal gene transfer (HGT) is considered the driving force of microbial evolution. The mobile genes typically encode genetic information for novel degradation capabilities, antibiotic resistance, etc. Shuffling these genes enhances diversification and specialization in a PDM by forming different phenotypes that can coexist within the microbiosphere (Figure 15.5f). HGT can maintain robust coexistence of the plastic-degrading populations in the engineered microbiosphere. For example, by maintaining cooperative traits, HGT can promote coexistence in plastic-degrading communities. Incorporating biofilm-inducing plastic-degrading microbial strains into the engineered PDM enhances the HGT to form valuable phenotypes. Best-suited strategies to identify plastic degraders to be incorporated into the PDM with the potential ability to inherit genes from other plastic-degrading microorganisms through HGT is using bioinformatics prediction analysis of gene products of potential plastic degraders retrieved from omics approaches, databases, and functional analysis (Zampolli et al., 2022). Identification of genes responsible for plastic biodegradation, release of the identified plastic-catabolic genes into the extracellular environment/biofilm in the form of DNA short strands, and plasmid and flooding of the plastic-degradation biofilms of the engineered microbiosphere with plastic catabolic genes can promote the HGT to enhance plastic degradation (French et al., 2020; Edwards et al., 2022).

15.3.5 Spatial Organization

The absence of forces promoting coexistence and competitive exclusion prevents the microbial populations from forming a stable PDM. Based on natural PDM, the microbial communities in the engineered PDM can be distributed with spatially defined structures (Figure 15.5g). The spatial localization of the engineered PDM is vital in biofilm formation. Spatial organization can reduce the stress of the PDM in harsh conditions. The construction of an ambient medium or an SP (as discussed above) can be used better to organize the spatial organization of the PDMs. For example, hydrogel can be used to immobilize the plastic degraders of the microbiosphere and assist in the material exchange of the plastic-degrading communities and also has to preserve the members of the PDM (Johnston et al., 2020). Devising culture devices can also help in spatial–temporal distribution and control of the plastic-degrading members of the engineered microbiosphere. For example, microfluidic technology can be used for regulation of different plastic-degrading microbial strains and improves the overall control of the PDM (Wang et al., 2019).

15.4 RESEARCH GAPS AND NOVEL INSIGHTS

It is unclear whether engineered PDM uses additives as the preferred carbon source over plastic and creates false positives and whether various additives negatively affect microbe-mediated plastic biodegradation due to their toxicity. While engineering the microbiosphere, it is essential to circumvent the negative impacts of additives on plastic biodegradation. The integration of heterotrophs, autotrophs, saprotrophs, and supplementary microorganisms in the engineered microbiosphere with both additives and plastic biodegradability has yet to be thoroughly investigated, especially in terrestrial ecosystems. Even though the "bottom-up" engineering of the PDM is primarily focused on incorporating previously identified microbial strains with enhanced plastic biodegradation to improve their efficiency. However, the lack of understanding of essential functional genes, naturally occurring microbial symbiotic processes, and the metabolic mechanisms of these artificially incorporated microbes restrict the construction of a fully functional, self-sustaining, PDM. The approach's limitations include the metabolic burden due to the arbitrary engineering attempts into engineered combinations and molecular mechanisms and antagonism between the selected microbial strains.

The inability of traditional microbial culture techniques to directly isolate a vast majority of efficient plastic degraders is hindering the development of a self-sustaining PDM in a natural ecosystem. The current literature lacks crucial information and protocols related to reported plastic-degrading microbes. For example, the location and the conditions of the isolation of a potential plastic degrader suitable for a self-sustaining PDM community, strength of evidence for plastic biodegradation, lack of deposited strains of plastic-degrading strains in culture banks, and false or inaccurate taxonomic classification hinder the use of an identified plastic-degrading microbe to be used in microbiosphere engineering.

REFERENCES

Araújo, R. et al. (2007) Tailoring cutinase activity towards polyethylene terephthalate and polyamide 6,6 fibers, *Journal of Biotechnology,* 128(4), 849–857. https://doi.org/10.1016/j.jbiotec.2006.12.028

Arias-Andres, M. et al. (2018) Microplastics: New substrates for heterotrophic activity contribute to altering organic matter cycles in aquatic ecosystems, *Science of the Total Environment,* 635, 1152–1159. https://doi.org/10.1016/j.scitotenv.2018.04.199

Arias-Andres, M., Rojas-Jimenez, K. and Grossart, H.-P. (2019) Collateral effects of microplastic pollution on aquatic microorganisms: An ecological perspective, *TrAC Trends in Analytical Chemistry,* 112, 234–240. https://doi.org/10.1016/j.trac.2018.11.041

Austin, H.P. et al. (2018) Characterization and engineering of a plastic-degrading aromatic polyesterase, *Proceedings of the National Academy of Sciences of the United States of America,* 115(19). https://doi.org/10.1073/pnas.1718804115

Bao, T. et al. (2023) Engineering microbial division of labor for plastic upcycling, *Nature Communications,* 14(1). https://doi.org/10.1038/s41467-023-40777-x

Barth, M. et al. (2016) A dual enzyme system composed of a polyester hydrolase and a carboxylesterase enhances the biocatalytic degradation of polyethylene terephthalate films, *Biotechnology Journal,* 11(8), 1082–1087. https://doi.org/10.1002/biot.201600008

Bhatt, P. et al. (2021) Microplastic contaminants in the aqueous environment, fate, toxicity consequences, and remediation strategies, *Environmental Research,* 200, 111762. https://doi.org/10.1016/j.envres.2021.111762

Biundo, A. et al. (2019) Switched reaction specificity in polyesterases towards amide bond hydrolysis by enzyme engineering, *RSC Advances,* 9(62), 36217–36226. https://doi.org/10.1039/c9ra07519d

Bollinger, A. et al. (2020) A novel polyester hydrolase from the marine bacterium Pseudomonas aestusnigri – Structural and functional insights, *Frontiers in Microbiology,* 11. https://doi.org/10.3389/fmicb.2020.00114

Chofifawati, A. et al. (2023) Potential of biological agents (Pseudomonas sp.) in plastic waste biodegradation process, *Jurnal Biolokus*, 5(2), 114. https://doi.org/10.30821/biolokus.v5i2.1192

Dey, S. et al. (2022) Plastisphere community assemblage of aquatic environment: Plastic-microbe interaction, role in degradation and characterization technologies, *Environmental Microbiome,* 17(1). https://doi.org/10.1186/s40793-022-00430-4

Dussud, C. et al. (2018) Colonization of non-biodegradable and biodegradable plastics by marine microorganisms, *Frontiers in Microbiology,* 9. https://doi.org/10.3389/fmicb.2018.01571

Edwards, S. et al. (2022) Microbial consortia and mixed plastic waste: Pangenomic analysis reveals potential for degradation of multiple plastic types via previously identified PET degrading bacteria, *International Journal of Molecular Sciences,* 23(10), 5612. https://doi.org/10.3390/ijms23105612

Fang, Y. et al. (2023) High-efficiency depolymerization/degradation of polyethylene terephthalate plastic by a whole-cell biocatalyst, *3 Biotech,* 13(5). https://doi.org/10.1007/s13205-023-03557-4

Flemming, H.-C. et al. (2016) Biofilms: An emergent form of bacterial life, *Nature Reviews. Microbiology,* 14(9), 563–575. https://doi.org/10.1038/nrmicro.2016.94

French, K.E., Zhou, Z. and Terry, N. (2020) Horizontal 'gene drives' harness indigenous bacteria for bioremediation, *Scientific Reports,* 10(1). https://doi.org/10.1038/s41598-020-72138-9

García-Jiménez, B., García, J.L. and Nogales, J. (2018) FLYCOP: Metabolic modeling-based analysis and engineering microbial communities, *Bioinformatics,* 34(17), i954–i963. https://doi.org/10.1093/bioinformatics/bty561

Gaytán, I. et al. (2020) Degradation of recalcitrant polyurethane and xenobiotic additives by a selected landfill microbial community and its biodegradative potential revealed by Proximity Ligation-Based metagenomic analysis, *Frontiers in Microbiology,* 10. https://doi.org/10.3389/fmicb.2019.02986

Guo, W. et al. (2023) Biodegradation of PET by the membrane-anchored PET esterase from the marine bacterium Rhodococcus pyridinivorans P23, *Communications Biology,* 6(1). https://doi.org/10.1038/s42003-023-05470-1

Han, Y.-N. et al. (2020) Greater biofilm formation and increased biodegradation of polyethylene film by a microbial consortium of Arthrobacter sp. and Streptomyces sp., *Microorganisms,* 8(12), 1979. https://doi.org/10.3390/microorganisms8121979

Harcombe, W.R. et al. (2014) Metabolic resource allocation in individual microbes determines ecosystem interactions and spatial dynamics, *Cell Reports,* 7(4), 1104–1115. https://doi.org/10.1016/j.celrep.2014.03.070

Hiraishi, T., Komiya, N. and Maeda, M. (2010) Y443F mutation in the substrate-binding domain of extracellular PHB depolymerase enhances its PHB adsorption and disruption abilities, *Polymer Degradation and Stability,* 95(8), 1370–1374. https://doi.org/10.1016/j.polymdegradstab.2010.01.022

Honjo, H. et al. (2019) Synthetic microbial consortium with specific roles designated by genetic circuits for cooperative chemical production, *Metabolic Engineering,* 55, 268–275. https://doi.org/10.1016/j.ymben.2019.08.007

Hussain, I. et al. (2018) Microbe and plant assisted-remediation of organic xenobiotics and its enhancement by genetically modified organisms and recombinant technology: A review, *Science of the Total Environment,* 628–629, 1582–1599. https://doi.org/10.1016/j.scitotenv.2018.02.037

Jambeck, J.R. et al. (2015) Plastic waste inputs from land into the ocean, *Science,* 347(6223), 768–771. https://doi.org/10.1126/science.1260352

Johns, N.I. et al. (2016) Principles for designing synthetic microbial communities, *Current Opinion in Microbiology,* 31, 146–153. https://doi.org/10.1016/j.mib.2016.03.010

Johnston, T.G. et al. (2020) Compartmentalized microbes and co-cultures in hydrogels for on-demand bioproduction and preservation, *Nature Communications,* 11(1). https://doi.org/10.1038/s41467-020-14371-4

Karkanorachaki, K. et al. (2022) Nanoplastic generation from secondary PE microplastics: Microorganism-induced fragmentation, *Microplastics,* 1(1), 85–101. https://doi.org/10.3390/microplastics1010006

Knott, B.C. et al. (2020) Characterization and engineering of a two-enzyme system for plastics depolymerization, *Proceedings of the National Academy of Sciences of the United States of America,* 117(41), 25476–25485. https://doi.org/10.1073/pnas.2006753117

Lear, G. et al. (2021) Plastics and the microbiome: Impacts and solutions, *Environmental Microbiome,* 16(1). https://doi.org/10.1186/s40793-020-00371-w

Leong, Y.K. et al. (2014) Current trends in polyhydroxyalkanoates (PHAs) biosynthesis: Insights from the recombinant Escherichia coli, *Journal of Biotechnology,* 180, 52–65. https://doi.org/10.1016/j.jbiotec.2014.03.020

Ma, Y. et al. (2018) Enhanced poly(ethylene terephthalate) hydrolase activity by protein engineering, *Engineering,* 4(6), 888–893. https://doi.org/10.1016/j.eng.2018.09.007

Malik, N. et al. (2023) Recent advances in the omics-based assessment of microbial consortia in the plastisphere environment: Deciphering the dynamic role of hidden players, *Process Safety and Environmental Protection/Transactions of the Institution of Chemical Engineers. Part B, Process Safety and Environmental Protection/Chemical Engineering Research and Design/Chemical Engineering Research & Design,* 176, 207–225. https://doi.org/10.1016/j.psep.2023.06.013

Numoto, N., Kamiya, N. and Oda, M. (2023) Improvement of the thermostability and activity of the PET-degrading enzyme Cut190 toward a detailed understanding and application of the enzymatic reaction mechanism, in ACS *Symposium Series*, pp. 89–100. https://doi.org/10.1021/bk-2023-1450.ch006

Pande, S. et al. (2015) Metabolic cross-feeding via intercellular nanotubes among bacteria, *Nature Communications,* 6(1). https://doi.org/10.1038/ncomms7238

Piscitelli, A. et al. (2017) Applications of functional amyloids from fungi: Surface modification by class I hydrophobins, *Biomolecules,* 7(4), 45. https://doi.org/10.3390/biom7030045

Qi, X. et al. (2021) Evaluation of PET degradation using artificial microbial consortia, *Frontiers in Microbiology,* 12. https://doi.org/10.3389/fmicb.2021.778828

Rhodes, C.J. (2019) Solving the plastic problem: From cradle to grave, to reincarnation, *Science Progress,* 102(3), 218–248. https://doi.org/10.1177/0036850419867204

Ribitsch, D. et al. (2017) Small cause, large effect: Structural characterization of cutinases from Thermobifida cellulosilytica, *Biotechnology and Bioengineering,* 114(11), 2481–2488. https://doi.org/10.1002/bit.26372

Richards, G.P. et al. (2012) Predatory bacteria as natural modulators of vibrio parahaemolyticus and vibrio vulnificus in seawater and oysters, *Applied and Environmental Microbiology*, 78(20), 7455–7466. https://doi.org/10.1128/aem.01594-12

Rummel, C.D. et al. (2017) Impacts of biofilm formation on the fate and potential effects of microplastic in the aquatic environment, *Environmental Science & Technology Letters,* 4(7), 258–267. https://doi.org/10.1021/acs.estlett.7b00164

Rüthi, J. et al. (2023) Discovery of plastic-degrading microbial strains isolated from the alpine and Arctic terrestrial plastisphere, *Frontiers in Microbiology,* 14. https://doi.org/10.3389/fmicb.2023.1178474

Salzano, D., Fiore, D. and Di Bernardo, M. (2021) Controlling reversible cell differentiation for labor division in microbial consortia, *bioRxiv (*Cold Spring Harbor Laboratory) [Preprint]. https://doi.org/10.1101/2021.08.03.454926

Sánchez, C. (2020) Fungal potential for the degradation of petroleum-based polymers: An overview of macro- and microplastics biodegradation, *Biotechnology Advances,* 40, 107501. https://doi.org/10.1016/j.biotechadv.2019.107501

Sarmah, P. and Rout, J. (2020) Role of algae and cyanobacteria in bioremediation: Prospects in polyethylene biodegradation, in Elsevier *eBooks,* pp. 333–349. https://doi.org/10.1016/b978-0-12-819311-2.00022-x.

Schaerer, L. et al. (2023a) Coexistence of specialist and generalist species within mixed plastic derivative-utilizing microbial communities, *Microbiome,* 11(1). https://doi.org/10.1186/s40168-023-01645-4

Schaerer, L.G. et al. (2023b) Versatile microbial communities rapidly assimilate ammonium hydroxide-treated plastic waste, *Journal of Industrial Microbiology & Biotechnology*, 50(1). https://doi.org/10.1093/jimb/kuad008

Sedlmayer, F., Aubel, D. and Fussenegger, M. (2018) Synthetic gene circuits for the detection, elimination and prevention of disease, *Nature Biomedical Engineering,* 2(6), 399–415. https://doi.org/10.1038/s41551-018-0215-0

Shen, M. et al. (2019) Micro(nano)plastics: Unignorable vectors for organisms, *Marine Pollution Bulletin,* 139, 328–331. https://doi.org/10.1016/j.marpolbul.2019.01.004

Silva, A.L.P. et al. (2021) An urgent call to think globally and act locally on landfill disposable plastics under and after covid-19 pandemic: Pollution prevention and technological (Bio) remediation solutions, *Chemical Engineering Journal,* 426, 131201. https://doi.org/10.1016/j.cej.2021.131201

Son, H.F. et al. (2019) Rational protein engineering of thermo-stable PETase from Ideonella sakaiensis for highly efficient PET degradation, *ACS Catalysis,* 9(4), 3519–3526. https://doi.org/10.1021/acscatal.9b00568

Su, T. et al. (2023) Dynamic docking assisted engineering of hydrolase for efficient PET depolymerization, *Research Square (Research Square)* [Preprint]. https://doi.org/10.21203/rs.3.rs-2349308/v1

Tan, L.-T. et al. (2012) Directed evolution of poly[(R)-3-hydroxybutyrate] depolymerase using cell surface display system: Functional importance of asparagine at position 285, *Applied Microbiology and Biotechnology,* 97(11), 4859–4871. https://doi.org/10.1007/s00253-012-4366-8

Then, J. et al. (2016) A disulfide bridge in the calcium binding site of a polyester hydrolase increases its thermal stability and activity against polyethylene terephthalate, *FEBS Open Bio,* 6(5), 425–432. https://doi.org/10.1002/2211-5463.12053

Tribedi, P. and Sil, A.K. (2012) Low-density polyethylene degradation by Pseudomonas sp. AKS2 biofilm, *Environmental Science and Pollution Research International,* 20(6), 4146–4153. https://doi.org/10.1007/s11356-012-1378-y

Wang, C. et al. (2019) A novel microbe consortium, nano-visible light photocatalyst and microcapsule system to degrade PAHs, *Chemical Engineering Journal,* 359, 1065–1074. https://doi.org/10.1016/j.cej.2018.11.077

Wang, G.X. et al. (2020) Seawater-degradable polymers—Fighting the marine plastic pollution, *Advanced Science,* 8(1). https://doi.org/10.1002/advs.202001121

Wei, R. and Zimmermann, W. (2017) Microbial enzymes for the recycling of recalcitrant petroleum-based plastics: How far are we?, *Microbial Biotechnology,* 10(6), 1308–1322. https://doi.org/10.1111/1751-7915.12710

Wei, R. et al. (2016) Engineered bacterial polyester hydrolases efficiently degrade polyethylene terephthalate due to relieved product inhibition, *Biotechnology and Bioengineering,* 113(8), 1658–1665. https://doi.org/10.1002/bit.25941

Wu, B. et al. (2023) Deep learning-aided redesign of a hydrolase for near 100% PET depolymerization under industrially relevant conditions, *Research Square (Research Square)* [Preprint]. https://doi.org/10.21203/rs.3.rs-2465520/v1

Wu, Y. et al. (2024) Opportunities and challenges for plastic depolymerization by biomimetic catalysis, *Chemical Science,* 15(17), 6200–6217. https://doi.org/10.1039/d4sc00070f

Yang, Y. et al. (2020) Microplastics provide new microbial niches in aquatic environments, *Applied Microbiology and Biotechnology,* 104(15), 6501–6511. https://doi.org/10.1007/s00253-020-10704-x

Zampolli, J. et al. (2022) Genome-based exploration of Rhodococcus species for plastic-degrading genetic determinants using bioinformatic analysis, *Microorganisms,* 10(9), 1846. https://doi.org/10.3390/microorganisms10091846

Zeghal, E. et al. (2021) The potential role of marine fungi in plastic degradation – A review, *Frontiers in* Mar*ine Science,* 8. https://doi.org/10.3389/fmars.2021.738877

Zhang, H. and Wang, X. (2016) Modular co-culture engineering, a new approach for metabolic engineering, *Metabolic Engineering,* 37, 114–121. https://doi.org/10.1016/j.ymben.2016.05.007

Zhang, S.-J. et al. (2022) The structure and assembly mechanisms of plastisphere microbial community in natural marine environment, *Journal of Hazardous Materials,* 421, 126780. https://doi.org/10.1016/j.jhazmat.2021.126780

Zurier, H.S. and Goddard, J.M. (2023) A high-throughput expression and screening platform for applications-driven PETase engineering, *Biotechnology and Bioengineering*, 120(4), 1000–1014. https://doi.org/10.1002/bit.28319

16 Environmental Health Impact of Plastisphere

Tatenda Dalu, Collins Oduro, Naicheng Wu, Franco Teixeira de Mello, Edmond Sanganyado, Rana Al-Jaibachi, Matthew Elmo Gruwell, Conrad Sparks, and Pule Mpopetsi

16.1 INTRODUCTION

Plastics, the primary constituents of the plastisphere, have been found in every corner of the globe, from the poles to the equator, across oceans, rivers, lakes and even in remote mountainous regions, thus underscoring their pervasive nature (Dalu et al., 2019; Li et al., 2021a; Wang et al., 2023; Nava et al., 2023). Studies (e.g., Barros and Seena, 2021; Nava et al., 2023; Li et al., 2024a) have demonstrated that microplastics, plastic fragments that are smaller than 5 mm, can be found in high concentrations in rivers, lakes, reservoirs and drinking water sources, indicating their widespread distribution and persistence in freshwater environments. Their presence in freshwater systems is particularly concerning because of their direct implications for water quality and aquatic life (Bucci et al., 2020; Ghosh et al., 2022). The plastisphere's emergence in freshwater ecosystems marks a significant and troubling development in the Anthropocene, the era dominated by human impact on climate and environmental conditions. As a branch of research, the freshwater plastisphere is a fairly new field of study that is at present mostly focused on broad, metagenomic/genetic diversity studies. Understanding the freshwater plastisphere involves defining its nature, assessing its widespread distribution and recognising its implications for human and environmental health (Di Pippo et al., 2022; Bocci et al., 2024). Initially identified in marine environments, the plastisphere comprises diverse microorganisms colonising plastics (Zettler et al., 2013; Barros and Seena, 2021; Li et al., 2021a). This concept has recently extended to freshwater environments, with the revelation that plastics provide a foundation for unique microbial communities, distinct from those in ambient water and sediments (Zettler et al., 2013; McCormick et al., 2014; Amaral-Zettler et al., 2020). These communities, consisting of bacteria, fungi, algae and other microbes, engage in intricate interactions with plastic surfaces, significantly affecting plastic degradation and environmental health (Oberbeckmann et al., 2015; Barros and Seena, 2021).

The colonisation of microplastics by microbial communities, formation of plastisphere, raises significant environmental and human health concerns (Figure 16.1; Barros and Seena, 2021). Microplastics in freshwater systems harbour pathogenic bacteria and viruses, including antibiotic-resistant strains, posing risks to aquatic organisms and humans (Wright and Kelly, 2017). On the environmental front, the plastisphere has been shown to alter the natural microbial communities, potentially disrupting nutrient cycles and facilitating the spread of invasive species and pathogens (Zettler et al., 2013; Wu et al., 2019; Kirstein et al., 2019). These disruptions can affect ecosystem functioning, biodiversity and freshwater ecosystems' services. The ingestion of microplastic plastisphere by aquatic organisms can lead to the bioaccumulation of pathogens and chemicals up the food chain, ultimately affecting human health through contaminated water and seafood consumption (Witsø et al., 2023; Li et al., 2024b). Thus, the plastisphere represents a potential vector for transmitting pathogens (Metcalf et al., 2022; Bocci et al., 2024).

DOI: 10.1201/9781032717814-16

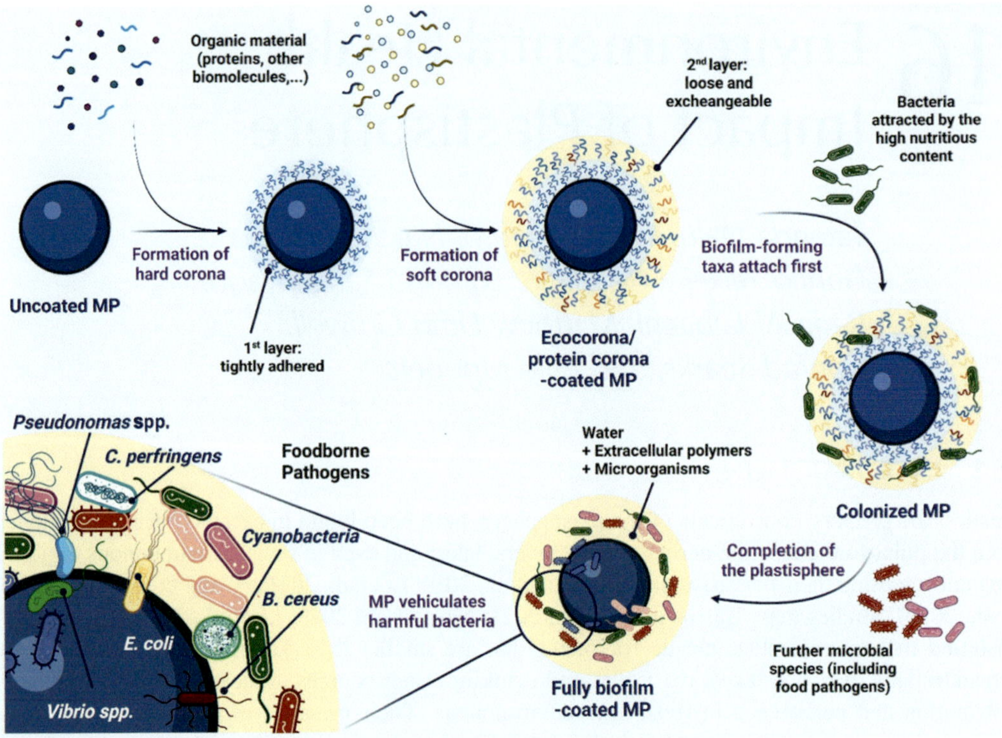

FIGURE 16.1 Step-by-step formation of plastisphere on microplastics in aquatic environments. (Adapted from Tavelli et al., 2022.)

Plastisphere microbes can be influenced by local and geographical aspects, so it may also functionally influence differences in the carbon metabolic activities of freshwater plastid assemblages from different locations (Miao et al., 2020). Within bacteria, *Acidobacteria*, *Actinobacteria*, *Bacteroidetes*, *Firmicutes*, *Planctomycetes*, *Proteobacteria* and *Verrucomicrobia* were the bacterial phyla most frequently detected on plastic particles in freshwaters (Bocci et al., 2024). In arctic system plastisphere records (i.e., microbial communities colonising polyhydroxybutyrate, high-density polyethylene and low-density polyethylene), the bacterial community was characterised by genera typical of polar environments such as *Mycoplana*, *Erythromicrobium* and *Rhodoferax*, with species capable of metabolising recalcitrant chemicals (the eukaryotic communities were characterised by the presence of ciliates of the genera *Stentor*, *Vorticella* and *Uroleptus* and the algae *Cryptomonas*, *Chlamydomonas*, *Tetraselmis* and *Epipyxis* and the fungal communities, the most abundant genera were *Betamyces*, *Cryptococcus*, *Arrhenia* and *Paranamyces* (González-Pleiter et al., 2021).

The degradation processes within the plastisphere, which leads to the release of toxic chemical additives from plastics, represent a significant toxicological threat that contributes to the contamination of freshwater environments and magnifies the potential for adverse health outcomes (Li et al., 2021b). This degradation, intertwined with the capacity of microplastics to act as reservoirs for hazardous substances, highlights the need for an integrated research approach to decipher the plastisphere's complex interactions and cumulative effects (Nabi et al., 2023). Additionally, plastisphere microbiomes linked to the breakdown of organic compounds can alter the nitrogen cycle in freshwater environments, increasing the emission of nitrite—a toxic substance—and releasing nitrous oxide (greenhouse gases) (Li et al., 2024b). For example, a high abundance of denitrifiers in freshwater plastisphere can lead to elevated concentrations of both nitrite and nitrous oxide, which can cause significant environmental impacts (Li et al., 2024b). Therefore, the freshwater plastisphere environment presents a complex interplay of environmental, ecological and health-related challenges.

As the infiltration of microplastics into freshwater systems continues unabated, a thorough understanding of the plastisphere's dynamics, facilitated by interdisciplinary research, is imperative. Such efforts are crucial, not only for unravelling the ecological and health risks associated with the freshwater plastisphere but also for informing effective strategies to mitigate the pervasive influence of microplastics on natural and human-made ecosystems.

16.2 FUNCTIONAL ROLES OF THE PLASTISPHERE COMMUNITY

Research focusing on the environmental impacts of microbial microplastic communities (i.e., plastisphere) is still in its early stages. However, evidence is mounting regarding the differences in microbial community compositions found on microplastics (i.e., freshwater plastisphere) compared to the surrounding environment (McCormick et al., 2016; Hoellein et al., 2017; Barros and Seena, 2021; Di Pippo et al., 2022; Bocci et al., 2024). With increasing information on this research topic, attention should also be directed towards determining the effects of the plastisphere on freshwater ecosystems, as well as its potential impacts on humans (Bocci and Seena, 2021). There is also potential evidence for such plastisphere communities to provide a niche for plastic-degrading microbes and the potential use of such microbes to combat pollution (Li et al., 2022). In this section, we will highlight recent research on some of the most pressing concerns associated with freshwater plastisphere communities, such as the increase in microbial pathogen species, the spread of antibiotic-resistance genes (ARGs) and mobile genetic elements (MGEs) and the potential for microplastic communities to harbour viral pathogens. Of the topics mentioned above, great attention is paid to the potential spread of pathogenic microbes through microplastic microbial communities (Figure 16.2). The field

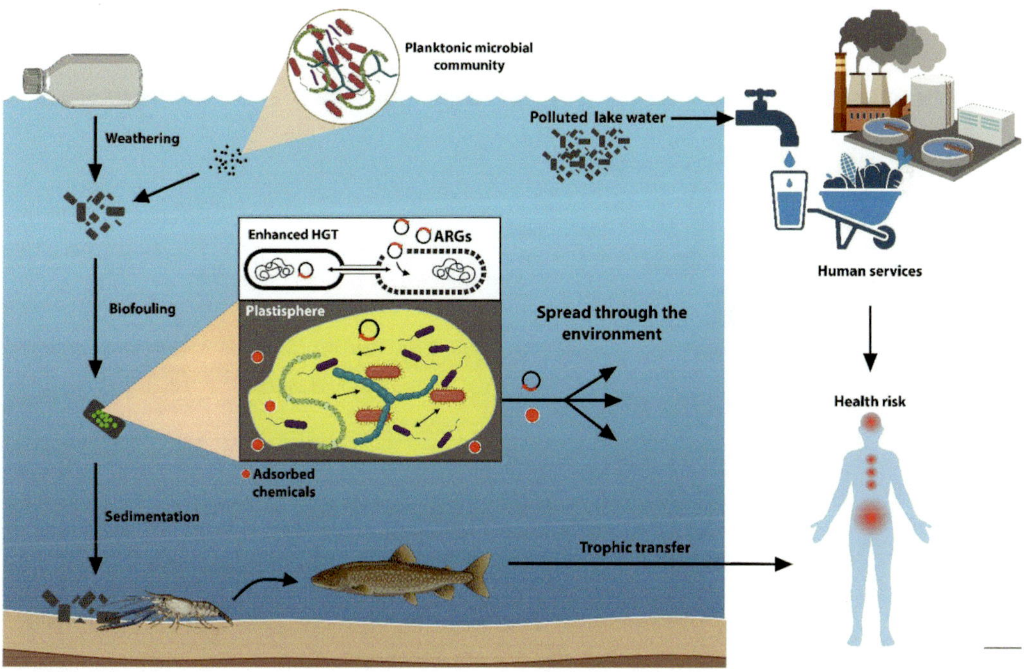

FIGURE 16.2 Plastisphere and plastics pathways in which they may affect the natural environment and human health. Plastics released in freshwater generally act as vectors for microbial organisms and chemicals adsorbed on their surfaces. The close relationship between the biofilm microbial organisms leads to enhanced horizontal gene transfer (HGT) with the potential spreading of antibiotic-resistance genes through the plastisphere members. As ubiquitous pollutants, plastic debris travels throughout the environment, leading to the spread of antibiotic resistance and adsorbed chemicals into freshwater, before finally reaching services provided to human health and society. The persistent plastic debris can also enter the trophic chain. (Adapted from Bocci et al., 2024.)

of metagenomics makes broad surveys of bacterial and eukaryote species in an environment accessible and biodiversity surveys are a good place to start exploring an environmental niche, especially one that has recently been created through human manufacturing (Wang et al., 2020; Barros and Seena, 2021; Di Pippo et al., 2022; Bocci et al., 2024). For example, using a high-throughput sequencing approach on samples from three rivers, Li et al. (2022) compared the relative abundance of microbial functions in the freshwater plastisphere to the surrounding environment and found significant differences between the two, including multiple bacterial and fungal functions that are pathogenic.

16.3 POTENTIAL HEALTH IMPACTS

Numerous studies (e.g., Wright et al., 2021; Tavelli et al., 2022; Witsø et al., 2023) have reported the presence of microbial food infectants on a range of microplastic particles (see Figure 16.3). Vibrio bacteria (i.e., *Vibrio alginolyticus*, *Vibrio cholerae*, *Vibrio parahaemolyticus* and *Vibrio vulnificus*) is an important microbial component in early microplastic colonisation and has frequently been found to be a major plastisphere community component (Kirstein et al., 2016; Frère et al., 2018; Silva et al., 2019; Laverty et al., 2020). For example, *Vibrio* bacteria made up 24% of the biofilm communities, and the Vibrionaceae family up to 20% of the bacterial population on floating plastics (Delacuvellerie et al., 2022). However, most of these studies that have detected Vibrio bacteria have been conducted within marine or estuarine environments. Similarly, fungi (i.e., Chytridiomycota and Cryptomycota) and fungal-like organisms (e.g., members affiliated to Peronosperales (Oomycetes)), which can have saprotrophic or parasitic lifestyle, have also been recorded in freshwater microplastic-associated biofilms (Di Pippo et al., 2022). Despite these documented records of bacteria and fungi in freshwater plastisphere, multiple hazards associated with the presence of particular pathogens such as *Vibrio*, *Aeromonas*, *Clostridium* and *Pseudomonas*

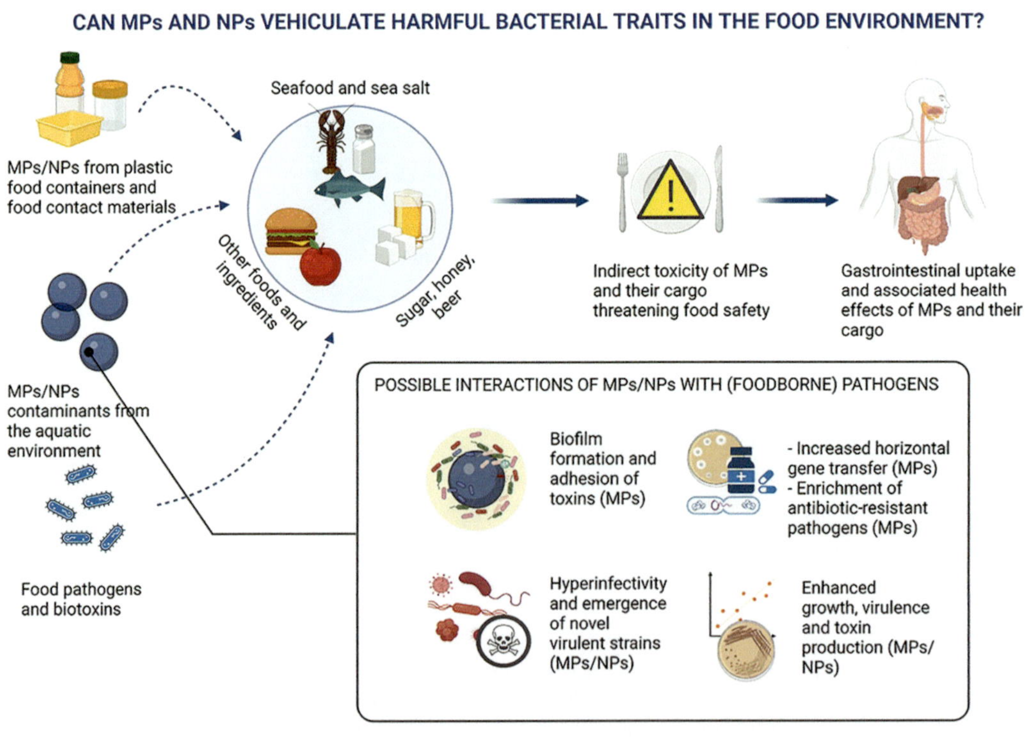

FIGURE 16.3 Pathways in which harmful microbial pathogens can be introduced from the natural environment into the food chain and ultimately to humans. (Adapted from Tavelli et al., 2022.)

in the plastisphere have mostly been described from the marine environment (Tavelli et al., 2022). Thus, attention also needs to be extended towards pathogens in freshwater plastisphere.

The most common food pathogens on larger plastic debris may also be of concern, with *Bacillus cereus* being a good example which has been isolated from plastic litter derived from the food packaging sector (Schmid et al., 2021). Biofilm formation on plastic litter suggests the possibility of also attaching to microplastics, which offers an additional contamination route. The pathogen *Listeria monocytogenes*, which has been identified in the plastisphere, occasionally causes listeriosis outbreaks mostly due to biofilm formation in food-processing plants (Silva et al., 2019; Nowak et al., 2021; Tavelli et al., 2022). For most of these microbial pathogenic species, the possibility of being dispersed into food products via freshwater microplastic colonisation deserves further research. Therefore, we highlight major knowledge gaps with regard to both the mechanisms and the possible health outcomes when it comes to the freshwater plastisphere. Foodstuff biofilm-coated microplastics may pose several risks to food safety (Tavelli et al., 2022), hence further research is essential to determine the freshwater plastisphere extent on human health.

16.4 EXPOSURE ROUTES—INGESTION, INHALATION AND DERMAL EXPOSURE

Ingestion is a major route of aquatic animal exposure to microplastics, as the plastisphere can resemble plankton, a food source for many fish and invertebrates (Mateos-Cárdenas et al., 2022; Mariani et al., 2023). Filter feeders such as mussels can passively consume microplastics suspended in water, with predators further concentrating microplastics as they prey on contaminated organisms (Al-Jaibachi et al., 2018; Wazne et al., 2023). In contrast, inhalation is another pathway where microscopic plastic fragments can be airborne, especially near breaking waves, and inhaled by surrounding organisms including humans (Pérez-Guevara et al., 2022). Additionally, dermal exposure, in which microplastics are in direct contact with an animal's skin, can allow for plastic particles (and thus plastisphere) to enter the bodies, ultimately causing harm to amphibians and humans with permeable skin (Kutralam-Muniasamy et al., 2024). This exposure can in turn cause gut blockages, hinder nutrient uptake and even induce inflammation and mortality (Figure 16.4). It is important to note that although both in vitro and animal studies have confirmed the microplastic uptake by gut

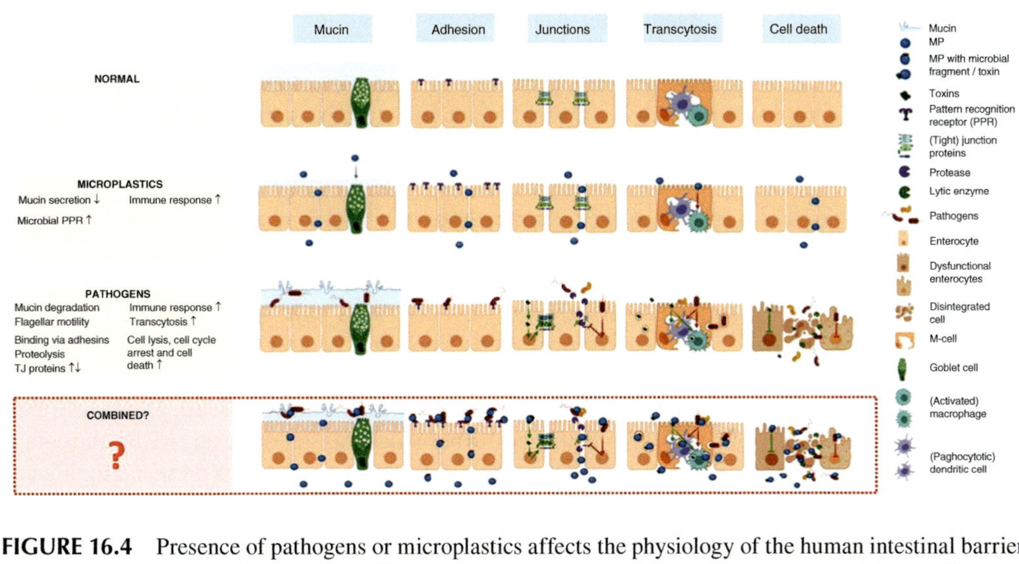

FIGURE 16.4 Presence of pathogens or microplastics affects the physiology of the human intestinal barrier on many levels (e.g., mucin secretion, adhesion, paracellular transport, transcytosis and cell death). (Adapted from Tavelli et al., 2022.)

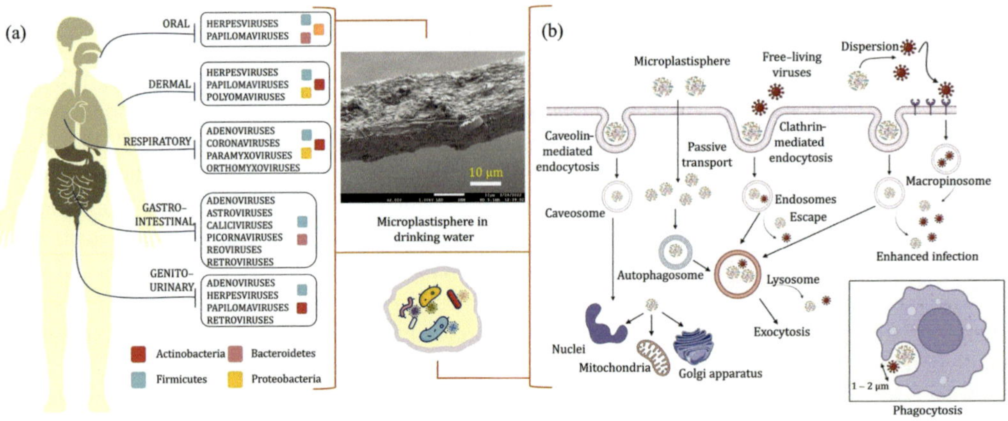

FIGURE 16.5 Plastisphere microbes and their viral human hosts including several processes involved in interactions: (a) potential virus–microbiota interactions with examples of natural hosts and (b) proposed primary routes for intracellular processing and the fate of internalised plastisphere-carrying viruses by mammalian cells. (Adapted from Kutralam-Muniasamy et al., 2024.)

epithelium, evidence in humans is still scarce (Tavelli et al., 2022). Thus, this research needs to be explored to better understand the implications of plastisphere on human health.

Humans are infected by various viruses, and human enteric and respiratory viruses have been found on plastisphere (reviewed in Kutralam-Muniasamy et al., 2024). This poses a potential risk to human health. For example, a recent study found that drinking water contained biofilm-coated microplastic particles in Mexico City, highlighting real-life exposure to microplastics and freshwater plastisphere (Figure 16.5; Pérez-Guevara et al., 2022). Mammalian cells can interact with the plastisphere, where microplastics can be internalised through clathrin-coated and caveolae-mediated micropinocytosis, endocytosis and/or passive penetration (Liu et al., 2021). Therefore, membrane-pierced plastisphere allows for direct viral entrance at the cytoplasmic level. Once inside the mammalian cells, viruses associated with the plastisphere can escape endosomes and/or lysosomes and spread to the cytoplasm, reaching organelles such as the Golgi bodies, mitochondria and nucleus (Figure 16.4; Kutralam-Muniasamy et al., 2024). Furthermore, it is important to highlight that plastisphere extracellular assemblages can break down, causing the release of viruses and/or virus-infected cells into the surrounding environment. This mostly happens when microplastic biofilms are loosely attached and are easily disrupted by fluid shear forces. Once released, these viral particles can enter various cell types through macropinocytosis and phagocytosis, potentially spreading infection (Uribe-Querol and Rosales, 2020).

16.5 POTENTIAL OF MICROPLASTICS TO SERVE AS VECTORS FOR ANTIBIOTIC-RESISTANCE GENES

The high levels of released antibiotics in freshwater environments have led to increases in ARGs in organisms living in these ecosystems and the prevalence of ARGs in freshwater microbes has increased selection for resistance in bacterial communities (Nnadozie and Odume, 2019). The ARGs, existing as either intracellular or extracellular DNA (Syranidou and Kalogerakis, 2022), can facilitate antibiotic resistance in bacteria and other pathogens. Antibiotic-resistant bacteria (ARB), those that can survive and reproduce in the presence of antibiotics, and pathogenic bacteria present a significant threat to human and animal health by compromising the efficacy of clinical treatment of pathogens and bacterial diseases (Allen et al., 2010; Zhang et al., 2022). Although the prevalence of ARGs and ARB have been documented in freshwater ecosystems (Winkworth, 2013; Nnadozie and Odume, 2019; Sabri et al., 2020), much more research needs to be done pertaining to their occurrence, types and spread in these environments (Allen et al., 2010, Nnadozie and Odume, 2019). However, one freshwater environment in which ARGs and ARB are commonly prevalent are locations with high pollution of antibiotics, locations such as wastewater treatment plants (WWTPs) (Nnadozie and Odume, 2019; Zhao et al., 2024) or other areas with high levels of anthropogenic impacts and urbanisation (Li et al., 2015; Li et al., 2022). Research focused on the interactions of ARGs/ARB and microplastic surfaces in freshwater environments is crucial to understanding the impacts of plastic pollution in facilitating antibiotic resistance in bacteria and the spread of ARB pathogens (Zhuang and Wang, 2023). This means that research into the freshwater plastisphere resistome, ARGs found on plastic surfaces in freshwater environments, is a critical area of research that is just beginning to be explored (Yang et al., 2020; Syranidou and Kalogerakis, 2022; Zhu et al., 2023).

16.6 INCREASED PREVALENCE OF ARGS IN THE PLASTISPHERE VS. SURROUNDING ENVIRONMENT

In aquatic environments, plastic debris or particles can potentially provide a substrate for forming biofilm (i.e., plastisphere), by attracting different and diverse microbial species, including ARB, that colonise or attach to the plastic substrate (Zettler et al., 2013; Song et al., 2020). One area of concern is that the substrate offered by microplastics allows for a higher concentration of ARGs than the surrounding water though some of the data has been contradictory. Bocci et al. (2024) reviewed work by multiple groups (Wu et al., 2019; Wang et al., 2020; Xu et al., 2022) and concluded that there were no significant differences between ARGs and MGEs in the plastisphere and surrounding biofilms as it pertained to abundance and diversity (Bocci et al., 2024). Xu et al. (2022) found a decrease in ARG abundance associated with microplastics in comparison to surrounding water though they did find a selective increase in ARGs that provide resistance to Rifamycin and Vancomycin (Xu et al., 2022). However, other studies have shown that microplastic debris can selectively enhance ARGs within biofilms in aquatic environments (Syranidou and Kalogerakis, 2022; Li et al., 2022) and that there is potential for plastic debris, including microplastics, to act as vectors for pathogens and antibacterial resistance genes in freshwater environments (Song et al., 2020; Li et al., 2022; Zadjelovic et al., 2023). Additionally, correlations have been observed between the abundance of ARGs and plastisphere substrates, between certain bacterial assemblages, microplastics and ARG prevalence (Su et al., 2021), and ARG abundance in polyethylene biofilms over other plastic types (Guo et al., 2020). Furthermore, other studies have uncovered higher abundances of ARGs and ARB on plastic substrates (i.e., polyethylene and polypropylene) compared to water or sediment/sand, thus indicating the potential for the plastisphere to act as ARG reservoirs and the dissemination of antibiotic-resistant priority pathogens (Guo et al., 2020; Silva et al., 2023). Concerning size, data suggests that plastic substrate size does have a significant influence on ARG abundance or diversity in aquatic environments (Yang et al., 2019; Syranidou and Kalogerakis, 2022).

One important answer to the question of ARGs in the plastisphere vs. surrounding water came from a recent study by Guruge et al. (2024), which demonstrated that there is significant potential for antibiotic resistance being facilitated by microplastics. They used metagenomics to test for ARGs, biocide/metal resistance genes (BMRGs) and MGEs on microplastics and the surrounding water in two rivers near Tokyo Japan. They quantified the difference of relative ARG, BMRG and MGE abundance between the plastisphere and freshwater using contigs per kilobase of gene per million (RPKM). Analyses identified 1110 ARG-encoding genes, of which 387 were exclusive to the plastisphere, 395 were only found in water and 327 were found in both. However, the abundances between the two test groups found that the plastisphere samples were 3.46-fold higher for ARGs (Guruge et al., 2024). Of the ARGs found, in total they conferred resistance to 25 drug types. Most of the ARGs were MDRGs (multidrug resistance genes), and the second most common were beta-lactams followed by glycopeptide antibiotics (Guruge et al., 2024). Data on the BMRGs found 260 genes only in microplastic samples, 17 genes only in water samples and 69 genes found in both types and data on MGE did not find significant differences between the groups (Guruge et al., 2024). This study clearly shows that there is potential for a much higher concentration of ARGs associated with the plastisphere and additional research into such trends is warranted.

16.7 MECHANISMS FOR INCREASED PREVALENCE OF ARG IN THE PLASTISPHERE

Aggregation of these diverse microbial species on plastic substrates, usually in close proximity to each other, can create an important sink of ARGs (Bank et al., 2020) and facilitate and enhance the exchange of genetic material between these microbes. This occurs mostly through horizontal gene transfer (HGT) and can include the exchange of ARGs between the plastisphere microbial species, thus promoting ARB that may be pathogenic in aquatic environments, including in freshwater (Imran et al., 2019; Syranidou and Kalogerakis, 2022). Shi et al. (2023) outlined the following four points that enhance the spread of ARGs in plastisphere microbial structure ecosystem as compared to surrounding water: (1) facilitated growth of ARB, (2) changes to microbial ecosystem structure, (3) enhanced HGT of ARG through conjugation and transformation due to the high density of microorganisms and (4) absorbed pollutants to provide selective pressure on microbial ecosystems in the plastisphere. Other factors that contribute to the increase in HGT of ARG include a rough surface, which can enhance conjugation (Gu et al., 2017), enhanced colonisation (Shi et al., 2022), or increased transfer frequency in resistomes from microplastic-associated bacteria (Li et al., 2021). Also of note, Xu et al. (2022) found an increase in co-occurrence between ARGs and MGEs showing that the plastisphere can propagate HGT of ARGs located on plasmids (Xu et al., 2022).

16.8 DIRECT ASSESSMENT OF HUMAN-ASSOCIATED HEALTH RISKS ARG IN THE PLASTISPHERE

Recent work of Zhu et al. (2023) made efforts to directly assess the human-associated health risks of ARGs (HRA) that can be intracellular or extracellular in the bacterial flora of the plastisphere, specifically microplastics. They employed a metagenomic approach to identify and quantify the ARGs in five different microplastic sources. The work identified 660 ARGs known to enable resistance to 21 types of antibiotics of which 77 were MDRG and constituted 58% of the total ARGs in the plastisphere. Of the original 660 ARGs found, 394 individual ARGs were linked to beta-lactam antibiotic resistance (Zhu et al., 2023). Also of concern, many of the ARGs co-occurred with MGEs and were located on plasmids meaning that they can readily undergo HGT, and they picked up many human pathogenic bacteria such as *Acinetobacter baumannli*, *Pseudomonas koreensis* and *Acinetobacter johnsonii* (Zhu et al., 2023). Of the five microplastics that Zhu et al. (2023) compared for HRA, polyethylene had the highest risk. This chapter represents the first of its kind, as it aims to pinpoint the direct health risks associated with microbes and ARGs in the freshwater plastisphere showing a great need to continue in similar based research to truly understand the danger associated with microplastic pollution and the potential of such debris to transmit harmful microbes and antibiotic resistance.

16.9 POTENTIAL FOR MICROPLASTICS TO ADSORB AND TRANSPORT POLLUTANTS

A major concern regarding the plastisphere is its potential to disperse anthropogenic pollutants and invasive or pathogenic species in freshwater systems. The colonisation of microplastics depends on their physical and chemical characteristics and those of the environment (Barros and Seena, 2021). Characteristics of microplastics that affect colonisation include roughness, hydrophobicity, charge and buoyancy (Oberbeckmann et al., 2015; Hossain et al., 2019). Different additives incorporated in plastics (e.g., dyes, plasticisers, stabilisers and flame retardants) also influence colonisation by favouring microorganisms that can use these additives as a source of nutrients (Wen et al., 2015). Even colour can affect the community structure and functional diversity of the plastisphere. For instance, biofilms colonising blue microplastics were observed to have higher microbial diversity than those on transparent or yellow microplastics (Wen et al., 2020).

Due to their higher surface-to-volume ratio and sorption potential, microplastics can be effective in transportation and accumulating persistent organic pollutants (POPs) and metals from surrounding waters (Lee et al., 2014), such as polychlorinated biphenyls (PCBs), polycyclic aromatic hydrocarbon (PAHs) and organochlorine pesticides (p,p'–DDT, p,p'–DDD, p,p'–DDE, gamma–HCH) (Campanale et al., 2020). Some studies (e.g., Li et al., 2018) have shown the adsorption of antibiotics on polyamide (PA) in freshwater environments, thus demonstrating the potential of microplastics as vectors of toxic compounds and their role in determining the fate of these compounds in freshwater environments.

16.10 CONCLUSIONS

Microplastics can be considered a new microbial niche that serves as hotspots for microbial interactions, potentially increasing risks to freshwater ecosystems and human health (Arias-Andres et al., 2018; Eckert et al., 2018; Wang et al., 2020; Junaid et al., 2022a,b). They are also considered hotspots for resistance genes (Lu et al., 2019), attributed to biofilm formation favouring vertical and HGT (Arias-Andres et al., 2018). Their capabilities to act as sources and vectors for pathogenic bacteria, mainly associated with the genera *Vibrio*, *Pseudomonas* and *Acinetobacter* (Junaid et al., 2022a), and viruses (e.g., noro- and adenovirus), further pose risk to human health (Moresco et al., 2021; Tumwesigye et al., 2023; Witsø et al., 2023). The spread of plastisphere in freshwater environments is an emerging concern, as this can allow for microplastics to be transported to novel environments.

REFERENCES

Al-Jaibachi, R., Cuthbert, R.N., and Callaghan, A., 2018. Up and away: ontogenic transference as a pathway for aerial dispersal of microplastics. *Biology Letters*, 14(9), 20180479.

Allen, H.K., Donato, J., Wang, H.H., Cloud-Hansen, K.A., Davies, J., and Handelsman, J., 2010. Call of the wild: antibiotic resistance genes in natural environments. *Nature Reviews Microbiology*, 8(4), 251–259.

Amaral-Zettler, L.A., Zettler, E.R., and Mincer, T.J., 2020. Ecology of the plastisphere. *Nature Reviews Microbiology*, 18(3), 139–151.

Arias-Andres, M., Klümper, U., Rojas-Jimenez, K., and Grossart, H.P., 2018. Microplastic pollution increases gene exchange in aquatic ecosystems. *Environmental Pollution*, 237, 253–261.

Barros, J., and Seena, S., 2021. Plastisphere in freshwaters: an emerging concern. *Environmental Pollution*, 290, 118123.

Bank, M.S., Ok, Y.S., and Swarzenski, P.W., 2020. Microplastic's role in antibiotic resistance. *Science*, 369(6509), 1315–1315.

Bocci, V., Galafassi, S., Levantesi, C., Crognale, S., Amalfitano, S., Congestri, R., Matturro, B., Rossetti, S., and Di Pippo, F., 2024. Freshwater plastisphere: a review on biodiversity, risks, and biodegradation potential with implications for the aquatic ecosystem health. *Frontiers in Microbiology*, 15, 1395401.

Bucci, K., Tulio, M., and Rochman, C.M., 2020. What is known and unknown about the effects of plastic pollution: a meta-analysis and systematic review. *Ecological Applications*, 30(2), e02044.

Campanale, C., Dierkes, G., Massarelli, C., Bagnuolo, G., and Uricchio, V.F., 2020. A relevant screening of organic contaminants present on freshwater and pre-production microplastics. *Toxics*, 8(4), 100.

Dalu, T., Malesa, B., and Cuthbert, R.N., 2019. Assessing factors driving the distribution and characteristics of shoreline macroplastics in a subtropical reservoir. *Science of the Total Environment*, 696, 133992.

Delacuvellerie, A., Ballerini, T., Frère, L., Matallana-Surget, S., Dumontet, B., and Wattiez, R., 2022. From rivers to marine environments: a constantly evolving microbial community within the plastisphere. *Marine Pollution Bulletin*, 179, 113660.

Di Pippo, F., Crognale, S., Levantesi, C., Vitanza, L., Sighicelli, M., Pietrelli, L., Di Vito, S., Amalfitano, S., and Rossetti, S., 2022. Plastisphere in lake waters: microbial diversity, biofilm structure, and potential implications for freshwater ecosystems. *Environmental Pollution*, 310, 119876.

Eckert, E.M., Di Cesare, A., Kettner, M.T., Arias-Andres, M., Fontaneto, D., Grossart, H.P. and Corno, G., 2018. Microplastics increase impact of treated wastewater on freshwater microbial community. *Environmental Pollution*, 234, 495–502.

Frère, L., Maignien, L., Chalopin, M., Huvet, A., Rinnert, E., Morrison, H., Kerninon, S., Cassone, A.L., Lambert, C., Reveillaud, J., and Paul-Pont, I., 2018. Microplastic bacterial communities in the Bay of Brest: influence of polymer type and size. *Environmental Pollution*, 242, 614–625.

Ghosh, D., Sarkar, A., Basu, A.G., and Roy, S., 2022. Effect of plastic pollution on freshwater flora: a meta–analysis approach to elucidate the factors influencing plant growth and biochemical markers. *Water Research*, 225, 119114.

González-Pleiter, M., Velázquez, D., Casero, M.C., Tytgat, B., Verleyen, E., Leganés, F., Rosal, R., Quesada, A., and Fernández-Piñas, F., 2021. Microbial colonizers of microplastics in an Arctic freshwater lake. *Science of the Total Environment*, 795, 148640.

Gu, H., Kolewe, K. W., and Ren, D., 2017. Conjugation in *Escherichia coli* biofilms on poly (dimethylsiloxane) surfaces with microtopographic patterns. *Langmuir*, 33(12), 3142–3150.

Guo, X.P., Sun, X.L., Chen, Y.R., Hou, L., Liu, M., and Yang, Y., 2020. Antibiotic resistance genes in biofilms on plastic wastes in an estuarine environment. *Science of the Total Environment*, 745, 140916.

Guruge, K.S., Goswami, P., Kanda, K., Abeynayaka, A., Kumagai, M., Watanabe, M., and Tamamura-Andoh, Y., 2024. Plastiome: plastisphere-enriched mobile resistome in aquatic environments. *Journal of Hazardous Materials*, 471, 134353.

Hoellein, T.J., McCormick, A.R., Hittie, J., London, M.G., Scott, J.W. and Kelly, J.J., 2017. Longitudinal patterns of microplastic concentration and bacterial assemblages in surface and benthic habitats of an urban river. *Freshwater Science*, 36(3), 491–507.

Hossain, M.R., Jiang, M., Wei, Q., and Leff, L.G., 2019. Microplastic surface properties affect bacterial colonization in freshwater. *Journal of Basic Microbiology*, 59(1), 54–61.

Imran, M., Das, K.R., and Naik, M.M., 2019. Co–selection of multi–antibiotic resistance in bacterial pathogens in metal and microplastic contaminated environments: an emerging health threat. *Chemosphere*, 215, 846–857.

Junaid, M., Liu, S., Liao, H., Liu, X., Wu, Y., and Wang, J., 2022a. Wastewater plastisphere enhances antibiotic resistant elements, bacterial pathogens, and toxicological impacts in the environment. *Science of the Total Environment*, 841, 156805.

Junaid, M., Siddiqui, J.A., Sadaf, M., Liu, S., and Wang, J., 2022b. Enrichment and dissemination of bacterial pathogens by microplastics in the aquatic environment. *Science of the Total Environment*, 830, 154720.

Kirstein, I.V., Wichels, A., Gullans, E., Krohne, G., and Gerdts, G., 2019. The plastisphere – Uncovering tightly attached plastic "specific" microorganisms. *PLOS ONE*, 14(4), e0215859.

Kirstein, I.V., Kirmizi, S., Wichels, A., Garin-Fernandez, A., Erler, R., Löder, M., and Gerdts, G., 2016. Dangerous hitchhikers? Evidence for potentially pathogenic *Vibrio* spp. on microplastic particles. *Marine Environmental Research*, 120, 1–8.

Kutralam-Muniasamy, G., Shruti, V.C., and Pérez-Guevara, F., 2024. Plastisphere–hosted viruses: a review of interactions, behavior, and effects. *Journal of Hazardous Materials*, 472, 134533.

Laverty, A.L., Primpke, S., Lorenz, C., Gerdts, G., and Dobbs, F.C., 2020. Bacterial biofilms colonizing plastics in estuarine waters, with an emphasis on *Vibrio* spp. and their antibacterial resistance. *PLOS ONE*, 15(8), e0237704.

Lee, H., Shim, W.J., and Kwon, J.H. 2014. Sorption capacity of plastic debris for hydrophobic organic chemicals. *Science of the Total Environment*, 470, 1545–1552.

Li, B., Yang, Y., Ma, L., Ju, F., Guo, F., Tiedje, J. M., and Zhang, T. 2015. Metagenomic and network analysis reveal wide distribution and co-occurrence of environmental antibiotic resistance genes. *The ISME Journal*, 9(11), 2490–2502.

Li, C., Wang, L., Ji, S., Chang, M., Wang, L., Gan, Y., and Liu, J., 2021a. The ecology of the plastisphere: microbial composition, function, assembly, and network in the freshwater and seawater ecosystems. *Water Research*, 202, 117428.

Li, C., Busquets, R., and Campos, L. C. 2021b. What do we know about the effectiveness of microplastic removal technologies? *Environmental Pollution*, 268, 115806.

Li, C., Li, X., Bank, M.S., Dong, T., Fang, J.K.H., Leusch, F.D., Rillig, M.C., Wang, J., Wang, L., Xia, Y., and Xu, E.G., 2024a. The "microplastome"–a holistic perspective to capture the real–world ecology of microplastics. *Environmental Science and Technology*, 58(9), 4060–4069.

Li, C., Gillings, M.R., Zhang, C., Chen, Q., Zhu, D., Wang, J., Zhao, K., Xu, Q., Leung, P.H., Li, X., and Liu, J., 2024b. Ecology and risks of the global plastisphere as a newly expanding microbial habitat. *The Innovation*, 5(1),100543.

Li, H., Luo, Q., Zhao, S., Zhao, P., Yang, X., Huang, Q., and Su, J. 2022. Watershed urbanization enhances the enrichment of pathogenic bacteria and antibiotic resistance genes on microplastics in the water environment. *Environmental Pollution*, 313, 120185.

Li, J., Zhang, K., and Zhang, H., 2018. Adsorption of antibiotics on microplastics. *Environmental Pollution*, 237, 460–467.

Li, R., Zhu, L., Yang, K., Li, H., Zhu, Y.G., and Cui, L., 2021. Impact of urbanization on antibiotic resistome in different microplastics: evidence from a large-scale whole river analysis. *Environmental Science and Technology*, 55(13), 8760–8770.

Liu, L., Xu, K., Zhang, B., Ye, Y., Zhang, Q., and Jiang, W., 2021. Cellular internalization and release of polystyrene microplastics and nanoplastics. *Science of the Total Environment*, 779, 146523.

Lu, J., Zhang, Y., Wu, J., and Luo, Y., 2019. Effects of microplastics on distribution of antibiotic resistance genes in recirculating aquaculture system. *Ecotoxicology and Environmental Safety*, 184, 109631.

Mariani, F., Di Lernia, D., Venditti, I., Pelella, E., Muzzi, M., Di Giulio, A., and Ceschin, S. (2023). Trophic transfer of microplastics from producer (*Lemna minuta*) to primary consumer (*Cataclysta lemnata*) in a freshwater food chain. *Science of the Total Environment*, 891, 164459.

Mateos-Cárdenas, A., Moroney, A. von der G., van Pelt, F.N.A.M., O'Halloran, J., and Jansen, M.A.K., 2022. Trophic transfer of microplastics in a model freshwater microcosm; lack of a consumer avoidance response. *Food Webs*, 31, e00228.

McCormick, A., Hoellein, T.J., Mason, S.A., Schluep, J., and Kelly, J.J., 2014. Microplastic is an abundant and distinct microbial habitat in an urban river. *Environmental Science and Technology*, 48(20), 11863–11871.

McCormick, A.R., Hoellein, T.J., London, M.G., Hittie, J., Scott, J.W. and Kelly, J.J., 2016. Microplastic in surface waters of urban rivers: concentration, sources, and associated bacterial assemblages. *Ecosphere*, 7(11), p. e01556.

Metcalf, R., Oliver, D.M., Moresco, V., and Quilliam, R.S., 2022. Quantifying the importance of plastic pollution for the dissemination of human pathogens: the challenges of choosing an appropriate 'control' material. *Science of the Total Environment*, 810, 152292.

Moresco, V., Oliver, D.M., Weidmann, M., Matallana-Surget, S., and Quilliam, R.S., 2021. Survival of human enteric and respiratory viruses on plastics in soil, freshwater, and marine environments. *Environmental Research*, 199, 111367.

Miao, L., Wang, C., Adyel, T.M., Wu, J., Liu, Z., You, G., Meng, M., Qu, H., Huang, L., Yu, Y., and Hou, J., 2020. Microbial carbon metabolic functions of biofilms on plastic debris influenced by the substrate types and environmental factors. *Environment International*, 143, 106007.

Nabi, I., Zaheer, M., Jin, W., and Yang, L., 2023. Biodegradation of macro- and micro-plastics in environment: a review on mechanism, toxicity, and future perspectives. *Science of the Total Environment*, 858, 160108.

Nava, V., Chandra, S., Aherne, J., Alfonso, M.B., Antão-Geraldes, A.M., Attermeyer, K., Bao, R., Bartrons, M., Berger, S.A., Biernaczyk, M., and Bissen, R., 2023. Plastic debris in lakes and reservoirs. *Nature*, 619(7969), 317–322.

Nnadozie, C.F., and Odume, O.N., 2019. Freshwater environments as reservoirs of antibiotic resistant bacteria and their role in the dissemination of antibiotic resistance genes. *Environmental Pollution*, 254, 113067.

Nowak, J., Visnovsky, S.B., Pitman, A.R., Cruz, C.D., Palmer, J., Fletcher, G.C., and Flint, S., 2021. Biofilm formation by Listeria monocytogenes 15G01, a persistent isolate from a seafood–processing plant, is influenced by inactivation of multiple genes belonging to different functional groups. *Applied and Environmental Microbiology*, 87(10), e02349–20.

Oberbeckmann, S., Löder, M.G.J., and Labrenz, M. (2015). Marine microplastic–associated biofilms – A review. *Environmental Chemistry*, 12(5), 551.

Pérez-Guevara, F., Roy, P.D., Elizalde-Martínez, I., Kutralam-Muniasamy, G., and Shruti, V.C., 2022. Human exposure to microplastics from urban decentralized pay-to-fetch drinking-water refill kiosks. *Science of the Total Environment*, 848, 157722.

Sabri, N., Schmitt, H., Van der Zaan, B., Gerritsen, H.W., Zuidema, T., Rijnaarts, H.H.M, and Langenhoff, A.A.M., 2020. Prevalence of antibiotics and antibiotic resistance genes in a wastewater effluent–receiving river in the Netherlands. *Journal of Environmental Chemical Engineering*, 8(1), 102245.

Schmid, P.J., Maitz, S., and Kittinger, C., 2021. *Bacillus cereus* in packaging material: molecular and phenotypical diversity revealed. *Frontiers in Microbiology*, 12, 698974.

Shi, X., Xia, Y., Wei, W., and Ni, B.J. 2022. Accelerated spread of antibiotic resistance genes (ARGs) induced by non-antibiotic conditions: roles and mechanisms. *Water Research*, 224, 119060.

Shi, X., Chen, Z., Wei, W., Chen, J., and Ni, B.J. 2023. Toxicity of micro/nanoplastics in the environment: roles of plastisphere and eco-corona. *Soil and Environmental Health*, 1(1), 100002.

Silva, I., Rodrigues, E.T., Tacão, M., and Henriques, I., 2023. Microplastics accumulate priority antibiotic–resistant pathogens: evidence from the riverine plastisphere. *Environmental Pollution*, 332, 121995.

Silva, M.M., Maldonado, G.C., Castro, R.O., de Sá Felizardo, J., Cardoso, R.P., Dos Anjos, R.M., and de Araújo, F.V., 2019. Dispersal of potentially pathogenic bacteria by plastic debris in Guanabara Bay, RJ, Brazil. *Marine Pollution Bulletin*, 141, 561–568.

Song, J., Jongmans-Hochschulz, E., Mauder, N., Imirzalioglu, C., Wichels, A, and Gerdts, G., 2020. The travelling particles: investigating microplastics as possible transport vectors for multidrug resistant *E. coli* in the Weser estuary (Germany). *Science of the Total Environment*, 720, 137603.

Su, Y., Zhang, Z., Zhu, J., Shi, J., Wei, H., Xie, B., and Shi, H., 2021. Microplastics act as vectors for antibiotic resistance genes in landfill leachate: the enhanced roles of the long–term aging process. *Environmental Pollution*, 270, 116278.

Syranidou, E., and Kalogerakis, N., 2022. Interactions of microplastics, antibiotics and antibiotic resistant genes within WWTPs. *Science of the Total Environment*, 804, 150141.

Tavelli, R., Callens, M., Grootaert, C., Abdallah, M.F., and Rajkovic, A., 2022. Foodborne pathogens in the plastisphere: can microplastics in the food chain threaten microbial food safety? *Trends in Food Science and Technology*, 129, 1–10.

Tumwesigye, E., Nnadozie, C.F., Akamagwuna, F.C., Noundou, X.S., Nyakairu, G.W., and Odume, O.N., 2023. Microplastics as vectors of chemical contaminants and biological agents in freshwater ecosystems: current knowledge status and future perspectives. *Environmental Pollution*, 330, 121829.

Uribe-Querol, E., and Rosales, C., 2020. Phagocytosis: our current understanding of a universal biological process. *Frontiers in Immunology*, 11, 1066.

Wang, J., Qin, X., Guo, J., Jia, W., Wang, Q., Zhang, M., and Huang, Y., 2020. Evidence of selective enrichment of bacterial assemblages and antibiotic resistant genes by microplastics in urban rivers. *Water Research*, 183, 116113.

Wang, Y., Liu, G., Wang, Y., Mu, H., Shi, X., Wang, C., and Wu, N., 2023. The global trend of microplastic research in freshwater ecosystems. *Toxics*, 11(6), 539.

Wazne, M., Mermillod-Blondin, F., Vallier, M., Hervant, F., Dumet, A., Nel, H.A., Kukkola, A., Krause, S., and Simon, L., 2023. Microplastics in freshwater sediments impact the role of a main bioturbator in ecosystem functioning. *Environmental Science and Technology*, 57(8), 3042–3052.

Wen, B., Liu, J.H., Zhang, Y., Zhang, H.R., Gao, J.Z., and Chen, Z.Z., 2020. Community structure and functional diversity of the plastisphere in aquaculture waters: does plastic color matter? *Science of the Total Environment*, 740, 140082.

Wen, G., Kötzsch, S., Vital, M., Egli, T., and Ma, J., 2015. BioMig a method to evaluate the potential release of compounds from and the formation of biofilms on polymeric materials in contact with drinking water. *Environmental Science and Technology*, 49(19), 11659–11669.

Winkworth, C.L., 2013. Antibiotic resistance genes in freshwater biofilms along a whole river. *Journal of Water and Health*, 11(2), 186–198.

Witsø, I.L., Basson, A., Vinje, H., Llarena, A.K., Bringas, C.S., Aspholm, M., Wasteson, Y., and Myrmel, M., 2023. Freshwater plastispheres as a vector for foodborne bacteria and viruses. *Environmental Microbiology*, 25(12), 2864–2881.

Wright, R.J., Langille, M.G., and Walker, T.R., 2021. Food or just a free ride? A meta-analysis reveals the global diversity of the Plastisphere. *The ISME Journal*, 15(3), 789–806.

Wright, S.L., and Kelly, F.J., 2017. Plastic and human health: a micro issue? *Environmental Science and Technology*, 51(12), 6634–6647.

Wu, X., Pan, J., Li, M., Li, Y., Bartlam, M., and Wang, Y., 2019. Selective enrichment of bacterial pathogens by microplastic biofilm. *Water Research*, 165, 114979.

Xu, C., Lu, J., Shen, C., Wang, J., and Li, F. 2022. Deciphering the mechanisms shaping the plastisphere antibiotic resistome on riverine microplastics. *Water Research*, 225, 119192.

Yang, K., Chen, Q.L., Chen, M.L., Li, H.Z., Liao, H., Pu, Q., Zhu, Y.G., and Cui, L., 2020. Temporal dynamics of antibiotic resistome in the plastisphere during microbial colonization. *Environmental Science and Technology*, 54(18), 11322–11332.

Yang, Y., Liu, G., Song, W., Ye, C., Lin, H., Li, Z., and Liu, W., 2019. Plastics in the marine environment are reservoirs for antibiotic and metal resistance genes. *Environment International*, 123, 79–86.

Zadjelovic, V., Wright, R.J., Borsetto, C., Quartey, J., Cairns, T.N., Langille, M.G., Wellington, E.M., and Christie-Oleza, J.A., 2023. Microbial hitchhikers harbouring antimicrobial-resistance genes in the riverine plastisphere. *Microbiome*, 11(1), 225.

Zettler, E.R., Mincer, T.J., and Amaral–Zettler, L.A., 2013. Life in the "plastisphere": microbial communities on plastic marine debris. *Environmental Science and Technology*, 47(13), 7137–7146.

Zhang, Z., Zhang, Q., Wang, T., Xu, N., Lu, T., Hong, W., Penuelas, J., Gillings, M., Wang, M., Gao, W., and Qian, H., 2022. Assessment of global health risk of antibiotic resistance genes. *Nature Communications*, 13(1), 1553.

Zhao, K., Li, C., and Li, F., 2024. Research progress on the origin, fate, impacts and harm of microplastics and antibiotic resistance genes in wastewater treatment plants. *Scientific Reports*, 14(1), 9719.

Zhu, L., Li, R., Yang, K., Xu, F., Lin, C., Chen, Q., Zhu, D., Sun, Q., Zhu, Y.G., and Cui, L., 2023. Quantifying health risks of plastisphere antibiotic resistome and deciphering driving mechanisms in an urbanizing watershed. *Water Research*, 245, 120574.

Zhuang, S., and Wang, J., 2023. Interaction between antibiotics and microplastics: recent advances and perspective. *Science of the Total Environment*, 897, 165414.

17 Insights into Microbial Enzyme-Mediated Hydrolysis of Fossil-Based Polymers in Plastisphere

Prospects for Environmental Sustainability

Sujata Dey, Ajaya Kumar Rout, Koushik Ghosh, and Bijay Kumar Behera

17.1 INTRODUCTION

The term "plastic" includes a broad spectrum of synthetically manufactured long-chain polymeric molecules, progressively ousting naturally occurring materials across various sectors and everyday applications. Fossil-based plastics have become indispensable to our survival, with production rates rapidly escalating to meet our daily requirements. The surge in plastic production is primarily driven by the synthetic manufacturing of approximately 50% of single-use plastics. Commonly utilized fossil-based plastics across diverse applications (Figure 17.1) include polyethylene (PE) (30.3%), polyethylene terephthalate (PET) (8.4%), polypropylene (PP) (19.7%), polystyrene (PS) (6.1%), polyurethane (PU) (7.8%), and polyvinyl chloride (PVC) (9.6%). Reports indicate a staggering production of approximately 8.3 billion tons of plastic, with a significant portion (79%) ending up in landfills (Kamboj, 2016). The fossil-based polymers are synthesized from petroleum, natural gas, or coal.

Over time, synthetic plastics break down into microplastics, which accumulate in terrestrial and aquatic ecosystems, potentially acting as carriers for organic pollutants and pathogens. Current disposal methods for synthetic plastic waste, such as incineration, ocean dumping, or landfilling, are not environmentally friendly and may result in secondary pollution. Therefore, researchers are actively seeking effective plastic management strategies. Pyrolysis has garnered attention as a recycling method, although it may produce undesirable by-products. As research into degradation technologies continues, biodegradation of fossil-based plastics has emerged as a sustainable approach, with discoveries of microorganisms capable of metabolizing these plastics in recent years. Several microorganisms, encompassing bacteria and fungi, are identified as capable of breaking down fossil-based plastics, predominantly observed under controlled laboratory conditions (Table 17.1) (Jumaah, 2017). Plasticizers and chemical additives incorporated during synthetic plastic manufacturing serve as potential carbon and energy sources for microbes (Gu, 2017). Due to the robust chemical bonds in polymers, they aren't readily metabolized by bacteria residing on plastic surfaces. Hence, allowing sufficient time for microbial growth to utilize the carbon within the plastic is a crucial step in fossil-based plastic degradation (Matjašič et al., 2021). In natural environments, microorganisms attach to the surface of plastics and develop a biofilm. Plastics serve as a robust matrix that supports the colonization of various microorganisms, which can gradually affect the integrity of the polymer. This complex interaction between plastics and microbes plays a critical

DOI: 10.1201/9781032717814-17

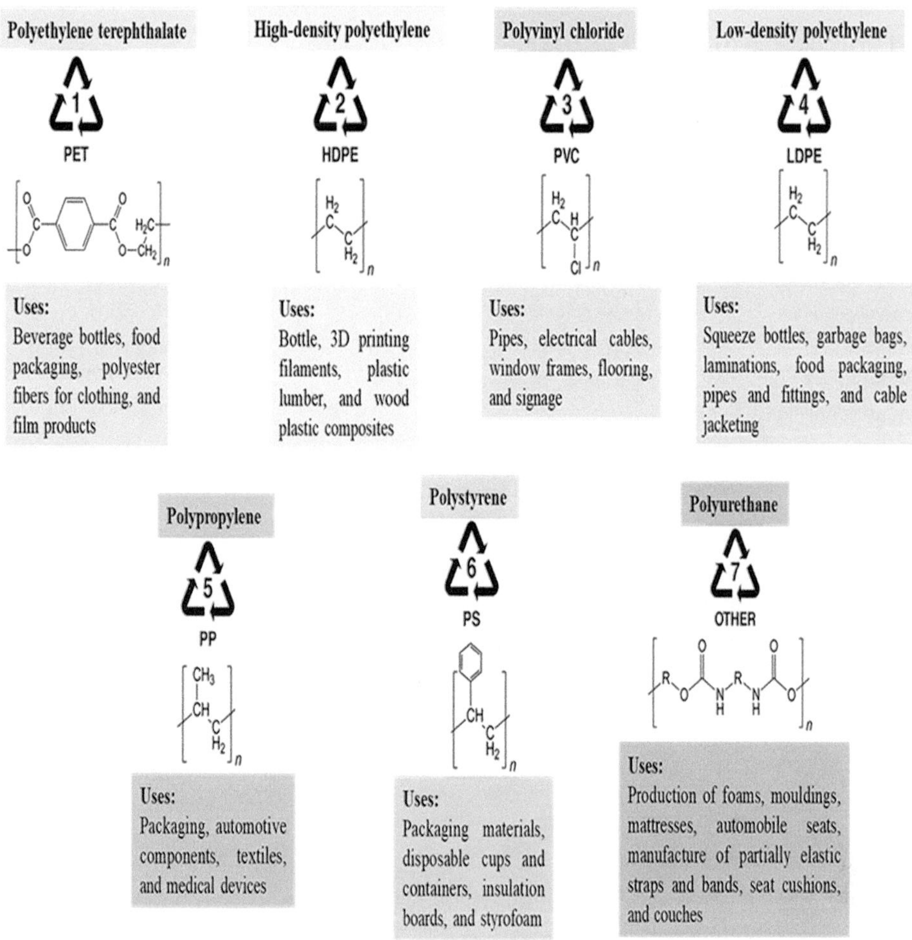

FIGURE 17.1 Commonly used fossil-based plastics and their diverse applications.

TABLE 17.1
Microorganisms and Enzymes Capable of Metabolizing Several Fossil-Based Plastics

Microbes	Enzymes	Plastics	References
Candida antarctica	Lipases	PET	Carniel et al. (2017)
Pichia pastoris		PET	Gao et al. (2017)
Aspergillus tubingenesis		PU	Khan et al. (2017)
Candida antarctica		PU	Gautam et al. (2007)
Candida rugosa		PU	Yoon et al. (2002)
Aspergillus fumigatus S45	Esterases	PU	Osman et al. (2017)
Aspergillus tubingenesis		PU	Khan et al. (2017)
Humicola insolens (HiC)	Cutinases	PET	Skamnioti et al. (2008)
Fusarium solani pisi (FsC)		PET	Vertommen et al. (2005)
Thermobifida fusca (TfH)		PET	Müller et al. (2005)
Thielavia terrestris CAU709		PU	Dimarogona et al. (2015)
Trichoderma DIA-T	Ureases	PU	Loredo-Treviño et al. (2011)

(*Continued*)

TABLE 17.1 (*Continued*)
Microorganisms and Enzymes Capable of Metabolizing Several Fossil-Based Plastics

Microbes	Enzymes	Plastics	References
Rhodococcus ruber	Laccases	PE	Sivan et al. (2006)
Cochliobolus sp.		PS	Chaudhary and Vijayakumar (2020)
Aspergillus flavus		HDPE	Zhang et al. (2020)
Pleurotus ostreatus		LDPE	Gómez-Méndez et al. (2018)
Trichoderma harzianum		PE	Sowmya et al. (2014)
Trichoderma harzianum	Manganese peroxidases	PE	Sowmya et al. (2014)
Pleurotus ostereatus	Lignin peroxidases	LDPE	Gómez-Méndez et al. (2018)
Pseudomonas sp. E4	Alkane monooxygenase	LMWPE	Gyung Yoon et al. (2012)
Pseudomonas aeruginosa E7		LMWPE	Jeon and Kim (2015)

Polyethylene terephthalate (PET); polyurethane (PU); polyethylene (PE); polystyrene (PS); high-density polyethylene (HDPE); low-density polyethylene (LDPE); low molecular weight polyethylene (LMWPE)

role in promoting the degradation of fossil-based plastics (Figure 17.2). While microbial colonization does not necessarily lead to the sinking or immediate breakdown of plastic particles, it does help facilitate their enzymatic degradation (Parida et al., 2024; Thakur et al., 2023; Sharma, 2018).

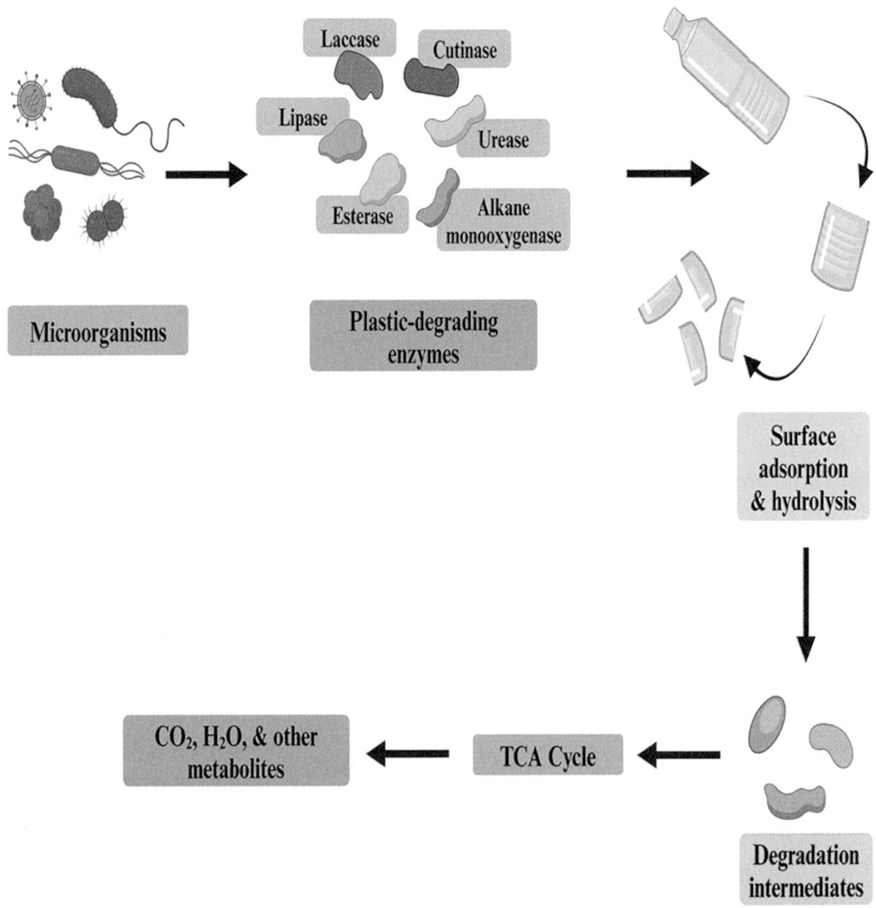

FIGURE 17.2 Microbial enzyme-mediated degradation pathway of fossil-based plastics.

Research on microbial enzyme-mediated biodegradation remains limited, with several unresolved challenges. This review briefly highlights structural insights, potential biotechnological applications, challenges, and prospects of plastic-degrading enzymes (PDEs).

17.2 ENZYMES INVOLVED IN FOSSIL-BASED PLASTICS DEGRADATION

17.2.1 Hydrolases

Plastic, characterized by its high stability as a polymer and insolubility in water (Mueller, 2006), possesses carbon chains that are exceptionally resilient and resistant to degradation (Goldman, 2010). The complex macromolecules of plastic present challenges for microbial utilization (Mueller, 2006). Consequently, extracellular hydrolytic enzymes effectively break ester or amide bonds in plastics (Krueger et al., 2015). Hydrolases encompass various enzymes such as lipases, proteases, carboxylesterases/esterases, cutinases, and ureases, each exhibiting distinct substrate preferences and interfacial activities.

17.2.1.1 Lipases (EC 3.1.1.3)

They are external enzymes known as triacylglycerol acyl hydrolases, which produce free fatty acids and glycerol from long-chain triglycerides (Singh and Mukhopadhyay, 2012). They exhibit water insolubility due to a domain covering their hydrophobic active center (Eberl et al., 2009). Typical examples include *Pseudomonas* sp., *Candida* sp., *Humicola* sp., and *Thermomyces lanuginosus*, which possess lipase activity capable of degrading PET (Andersen et al., 1999).

17.2.1.2 Proteases (EC 3.4.21.112)

Another category of hydrolases, proteases, are recognized for their role in catalyzing hydrolytic reactions. Several species have been documented to produce proteases capable of degrading various types of plastics. *Aspergillus* sp., *Humicola* sp., *Phaenarochete* sp., *Rhizopus* sp., *Trichoderma* sp., *Penicillium* sp., *Alternaria* sp., *Mucor* sp., and *Fusarium* sp. among others, have been identified as producers of proteases involved in plastic degradation (Barth et al., 2015). *P. chlororaphis* and *P. fluorescens* have been observed to target PU for protease degradation. Furthermore, *Arthrobotrys oligospora* has been found to produce serine protease, facilitating the degradation of PLA.

17.2.1.3 Esterases/Carboxylesterases (EC 3.1.1.1)

Hydrolases possess the ability to break down and synthesize ester bonds. Organic solvents demonstrate notable specificity in terms of both the location and orientation of their actions and exhibit resilience (Satti and Shah, 2020). Both fungal and bacterial hydrolases play key roles in plastic breakdown. Specifically, they contribute to the degradation of plastics like PU, which feature hydrolyzable ester bonds (Roy et al., 2021). Furthermore, it has been observed that the combined application of amidase and esterase enzymes enhances the breakdown of carbamate bonds, surpassing the degradation efficiency of either enzyme used individually for targeting PU (Magnin et al., 2019).

17.2.1.4 Cutinases (EC 3.1.1.74)

The enzymes belonging to the α/β hydrolase family are extracellular serine esterases (Skamnioti et al., 2008). These enzymes feature an active site characterized by a catalytic triad consisting of histidine, serine, and aspartic acid (Martinez et al., 1992). Unlike some other enzymes, cutinases lack a hydrophobic lid, which renders their active sites sufficiently spacious for larger substrates. This structural feature also helps them to cleave the ester linkages within synthetic polyesters. Distinct cutinases with PET-degrading ability have been isolated and identified from different sources, including *Fusarium solani pisi* (FsC) (Vertommen et al., 2005), *Humicola insdins* (HiC) (Ronkvist et al., 2009), and *T. fusca* (TfH) (Müller et al., 2005). Studies have shown that FsC and HiC are particularly effective in degrading low crystallinity PET films, achieving a weight loss of up to 97% within 96 hours (Ronkvist et al., 2009). Moreover, a PU-degrading cutinase was isolated and identified from the thermophilic fungus *Thielavia terrestris* CAU709 (Dimarogona et al., 2015).

17.2.1.5 Ureases (EC 3.5.1.5)

Loredo-Treviño et al. (2011) isolated certain fungi from plastic, which demonstrated the capability to thrive on PU. Urease activity was assessed by introducing phenolic compounds into the medium and quantifying their utilization to determine biodegradation levels. The study revealed that the majority of the fungi exhibited urease activity (95.45%), surpassing several other enzymes in this aspect. It was concluded that urease might be the primary fungal enzyme responsible for PU degradation.

17.2.2 Oxidoreductases

Hydrolytic enzymes are capable of degrading plastic heteroatoms within their primary chain structure (Krueger et al., 2015). Plastics such as PE, PS, PVC, and PP are polyolefins distinguished by inert C–C backbone structures lacking functional groups which make up over 60% of the market share. Consequently, these plastics cannot undergo hydrolysis via conventional biological enzymes. Instead, they can solely be disintegrated through oxidation into smaller molecules, which may subsequently be assimilated microbes (Plastics Europe, 2015). As of now, only a limited number of enzymes capable of substantially oxidizing these high-energy structures have been identified (Inderthal et al., 2021).

17.2.2.1 Laccases (EC 1.10.3.2)

They are present in various organisms including plants, fungi, and bacteria containing multinuclear copper, categorized within the polyphenol oxidase family, exhibiting broad activity across a range of temperatures and pH levels (Desai and Nityanand, 2011). Laccases are notably recognized as key enzymes involved in the HDPE breakdown (Othman et al., 2021). Their mechanism involves depolymerization and oxidation of the amorphous regions of the polymer chain (Kang et al., 2019). Notably, a laccase reported from *Rhodococcus rubber* C208 demonstrates effective degradation of UV-treated PE films when copper is present (Sivan et al., 2006). Additionally, a laccase sourced from *Cochliobolus* sp. is found effective in PVC biodegradation (Chaudhary and Vijayakumar, 2020).

17.2.2.2 Peroxidases (EC 1.11.1)

They represent a category of oxidoreductases (heme-containing) characterized by their wide substrate range and redox potentials. The elevated redox potential facilitates the reduction of peroxidases and oxidation of diverse organic and inorganic substrates (Ikehata et al., 2004; Erden et al., 2009). Three distinct classes of peroxidases have been identified: class I encompasses intracellular peroxidases, class II includes extracellular fungal peroxidases, and class III consists of extracellular plant peroxidases. Fungal peroxidases from class II, such as lignin peroxidase (LiP) and manganese peroxidase (MnP), are capable of breaking down plastics. Nwogu (2012) utilized various mushroom species secreting extracellular enzymes to degrade PE. Analysis revealed oxidative degradation of PE by *Pleurotus tuber-regium*, with manganese peroxidase (Mnp) identified as the primary enzyme involved, potentially complemented by laccases.

17.2.2.3 Alkane Hydroxylases (AHs)/Monooxygenases (EC 1.14.15.3)

They are critical enzymes utilized by anaerobic bacteria for alkane degradation (anaerobic digestion), and they play a pivotal role in the breakdown of straight-chain alkanes involving the enzymes of the β-oxidation pathway within the AH system pathway (Jeon and Kim, 2016). Monooxygenases constitute the primary enzymes within AH systems. Many microbial species, including bacteria such as *Pseudomonas putida*, *Anoxybacillus* sp., and *Geobacillus* sp. and fungi like *Trichospora* sp. and *Candida* sp. renowned for their ability to degrade PE, also exhibit proficiency in hydrolyzing linear n-alkanes (Prenafeta-Boldú et al., 2018). AHs vary significantly in type and quantity across diverse bacteria, reflecting the diversity in the carbon chain lengths of alkane molecules (Jeon and Kim, 2016). Various bacterial genomes encode distinct AHs, ranging from alkB1 to alkB5 (Takei et al., 2008). The primary step involves the hydroxylation of carbon–carbon (-C-C-) bonds to yield 10 or 20 alcohols, followed by oxidation of alcohols to ketones or aldehydes, eventually leading to the formation of hydrophilic carboxylic acids (Álvarez-Barragán et al., 2016). Microbial enzymatic action leads to a substantial reduction in carbonyl groups, resulting in the formation of carboxylic acids.

17.3 RECENT TRENDS IN ENZYME ENGINEERING

Enzymes, which are essentially proteins, can have their catalytic efficiency for plastic degradation enhanced through protein engineering techniques like directed evolution or rational design (Zhu et al., 2022).

Directed evolution simulates natural evolution by repeatedly diversifying genes and screening or selecting for proteins with improved or novel functions (Wang et al., 2021). However, the use of directed evolution for engineering PDEs is constrained by the lack of effective high-throughput methods for gene diversification and screening (Zhu et al., 2022; Wang et al., 2021). Consequently, only a few studies have employed directed evolution in this context. Recently, Apitius et al. (2019) developed an ultrahigh-throughput esterase A-based *Escherichia coli* cell surface display system, which allows for the screening and evolution of polymer-binding peptides. This innovative approach significantly enhanced the binding properties (by 12-fold) of adhesion promoters targeting whole-cell attachment to polymer surfaces like PET. Despite this advancement, there is still a limited amount of research focused on using directed evolution for PDE mutations, highlighting the need for further studies to advance the field.

Rational protein engineering, on the other hand, relies on the detailed knowledge of the protein's structure and function, obtained through computational modeling and simulations. This approach involves modifying the structural and mechanistic features of the enzyme to enhance its thermal stability, improve substrate binding at the active site, strengthen enzyme–substrate interactions, and fine-tune the active sites to optimize enzymatic performance (Zhu et al., 2022).

17.4 BIOTECHNOLOGICAL APPLICATIONS

The degradation of plastics through microbial action is a multifaceted process that entails the involvement of numerous enzymes to achieve the complete transformation of organic carbon into CO_2. While various microbial enzymes have been utilized in plastic biodegradation, the degrading pathway remains unclear. To elucidate the enzymatic pathways acting on plastic polymers, it is essential to screen them from diverse microbial strains. This necessitates the development of various strategies that shift the focus from microbes to the role of enzymes in the mechanism of plastic degradation. Several techniques can be employed to screen and identify enzymes involved in the plastic-degradation pathway.

Genome mining has emerged as a key approach for screening novel enzymes with desired properties. This method involves genomic DNA extraction and generating metagenomic data necessary for designing enzymes with specific characteristics (Lorenz and Eck, 2005). There are two primary methods for screening and producing novel enzymes from the metagenome: sequence-based screening and function-based screening (DeCastro et al., 2016). Sequence-based screening involves predicting genes and their corresponding enzymes by sequencing metagenomic DNA and comparing them with conserved sequences of target enzyme homologs (Ferrandi et al., 2015). Function-based screening identifies novel genes and enzymes with specific properties (Zhang et al., 2014). In this technique, positive colonies comprising target enzymes are directly isolated from metagenomic libraries by activity detection, followed by the cloning and characterization of the target enzymes.

Mass spectrometry (MS)-based proteomics offers another important method for discovering novel enzymes, employing bottom-up and top-down proteomics (Zhang et al., 2014). In bottom-up proteomics, a protein mixture is isolated and then undergoes enzymatic (trypsin) or chemical digestion to produce peptides. Then, these peptides are separated using either one-dimensional or multidimensional liquid chromatography. Finally, the obtained mixture is then identified by bioinformatics tools such as MASCOT (Perkins et al., 1999). On the contrary, top-down proteomics involves prefractionation techniques such as isoelectric focusing for protein mixture samples, followed by their analysis using MS or other advanced methodologies to identify individual components (Zhou et al., 2007). Nano-liquid chromatography tandem MS is frequently employed to discover specific enzymes, bacterial enzymatic cascades, and lignin-degrading enzymes (Sorigué et al., 2017).

Plastic degradation is a complex process facilitated by combined enzymatic actions. Metagenomics and proteomics stand out as prominent and widely utilized techniques for discovering novel enzymes in vitro. However, relying solely on one approach may not suffice. Therefore, an integrated strategy that incorporates multi-omics is essential for identifying enzymatic degradation pathways and comprehending their mechanisms. This integrated approach offers improved efficiency in identifying both cultivable and uncultivable biological samples. By adopting this integrated "metaomic" approach, researchers have successfully identified ten novel enzymes (including six putative PETase-like enzymes and four MHETase-like enzymes) within a marine microbial consortium. These enzymes play a crucial role in the synergistic biodegradation of aromatic-aliphatic plastic polymers (Pinnell and Turner, 2019; Meyer-Cifuentes et al., 2020).

17.5 FACTORS AFFECTING BIODEGRADATION OF PLASTICS

Plastic biodegradation refers to the catabolic process that enables the decomposition of plastics (Tschan et al., 2012). Several environmental factors contribute significantly to the degradation of plastics (Wackett and Hershberger, 2001). Tschan et al. (2012) elucidated plastic biodegradation as the catabolic process vital for breaking down plastics. Several environmental factors depicted in Figure 17.3 significantly contribute to the deterioration, degradation, and eventual depletion of plastics (Wackett and Hershberger; 2001).

17.5.1 UV Radiation

Since a significant amount of plastic waste ends up in water bodies, floating plastic particles in the sunlit (photic) zone are exposed to various factors like wave-induced shear stress, solar UV radiation, heating, and microbial colonization (Vaksmaa et al., 2021). In contrast, microbial enzymatic activity becomes the dominant factor influencing plastic degradation in the deeper, lightless (aphotic) zones (Parida et al., 2024; Zhang et al., 2021). Research on the impact of UV light on microplastic degradation has demonstrated that the wavelength of UV light, the presence of additives, and the material

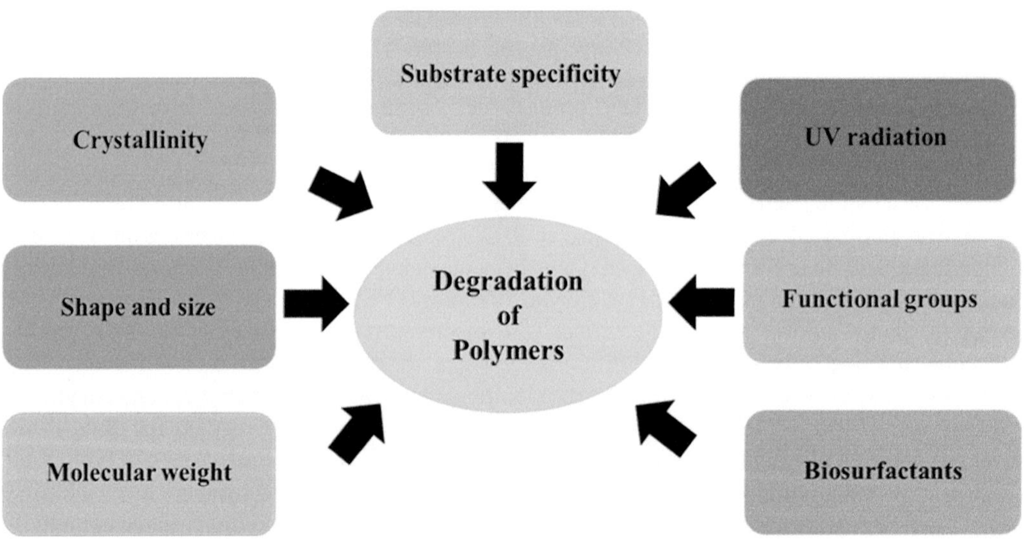

FIGURE 17.3 Environmental factors responsible for the deterioration, degradation, and depletion of fossil-based plastics.

composition of the plastic particles all affect the levels of surface oxidation caused by UV exposure (Burrows et al., 2024). UV irradiation reduces the size of floating plastics over time as prolonged photoaging produces reactive oxygen species (Wu et al., 2023). The rate of plastic photooxidation differs greatly depending on environmental location, resulting in varying fragmentation rates. This variability is driven by the unique UV exposure conditions in different geographic regions.

17.5.2 Crystallinity

Plastic polymers consist of both crystalline and amorphous regions, with the amorphous parts providing flexibility. As crystallinity increases, the mobility of the polymer backbone decreases, reducing the accessibility of polymer chains to enzymatic degradation (Mohanan et al., 2020). The crystallinity degree (Xc) of a polymer represents the ratio of crystalline regions to amorphous regions within the material. Crystalline regions have a structured and orderly arrangement where polymer chains are aligned in a regular pattern, whereas amorphous regions are characterized by a lack of order and a more random structure. The Xc significantly affects the physical properties of a material, such as its density, thermal stability, and resistance to both chemical and enzymatic degradation. Techniques like X-ray diffraction and differential scanning calorimetry are commonly used to measure crystallinity (Kong and Hay, 2003; Chukhchin et al., 2016; Uzun, 2023). There is a strong relationship between the polymer's crystallinity and its enzymatic breakdown rate; higher levels of crystallinity are typically linked to slower enzymatic degradation (Thomsen et al., 2023).

17.5.3 Molecular Weight

The resistance of plastics to degradation is largely due to their long-chain molecular structure. As the polymer chain lengthens, the number of bonds that enzymes must cleave increases. Additionally, high molecular weight polymers are less permeable across cell membranes, making it difficult for microorganisms to absorb them. To use plastics as a carbon source, they must first be broken down by enzymes into smaller monomers and oligomers from the longer polymer chains. These smaller fragments can then be absorbed by the cell and utilized as a carbon source for metabolism (Fotopoulou and Karapanagioti, 2017; Urbanek et al., 2018; Ciuffi et al., 2024). Molecular weight is a key factor in assessing enzymatic plastic degradation.

17.5.4 Substrate Specificity

The substrate specificity of microbes is a pivotal characteristic influencing plastic biodegradation (Wei and Zimmermann; 2017). Molecular weight is also a critical determinant of biodegradability as they are inversely related to each other.

17.5.5 Shape and Size of Polymer

The shape and size of polymer molecules is a crucial factor in the biodegradation process (Perotti et al., 2016). Polymer molecules with larger surface areas degrade more rapidly compared to those with smaller surface areas, underscoring the importance of defined shape and size in the biodegradation process.

17.5.6 Biosurfactants

Biosurfactants often serve as toxin-free agents that enhance the biodegradability of plastics. The addition of biosurfactants promotes faster biodegradation in both fossil-based and bio-polymers.

17.5.7 Functional Groups

Functional groups present in biosurfactants facilitate accelerated polymer biodegradation under optimal temperature, pH, and salt conditions when incubated with biosurfactants.

17.6 ADVANTAGES AND LIMITATIONS

Microbes and eukaryotic colonizers secrete extracellular hydrolytic enzymes, which play a key role in facilitating the biodegradation process (Kiran et al., 2022). The production of biofilms further supports the formation of hetero-aggregates, where microplastic particles become adhesive due to the presence of extracellular hydrolytic enzymes within the biofilm matrix (Wu et al., 2019). The high efficiency of these microorganisms in degrading plastics is believed to result from the evolutionary adaptation of multifunctional enzyme systems and distinctive extracellular digestion processes (Mohanan et al., 2020). Bacteria utilize enzymes for metabolic activities, enabling them to naturally transform various complex chemical structures, including pollutants, into simpler ones, even though most plastics degrade slowly and inefficiently at lower temperatures (Nguyen et al., 2023). The development of bacterial biofilms can be influenced by factors such as material surface roughness, surface free energy, hydrophobicity, and electrostatic interactions (Nguyen et al., 2023). Bacteria have the potential to evolve new enzymes capable of degrading plastics by adapting to new carbon sources (Oberbeckmann et al., 2014). This adaptability has led to the suggestion that marine bacteria could be a promising tool for the biodegradation and eventual elimination of plastic waste (Nguyen et al., 2023).

However, there are several often-overlooked aspects of microbial enzyme-mediated polymer degradation. For instance, physical, mechanical, and chemical weathering of plastics is a critical preliminary step before microbial degradation can occur. Additionally, the colonization of plastics by microbes does not necessarily indicate enzymatic breakdown. Even after colonization, microorganisms tend to favor additives over polymers as their carbon source. The enzymatic degradation of fossil-based plastics is limited to only a few types of polymers, and large amounts of enzymes would be needed for effective environmental degradation. While industrial bioprocesses for recycling and upcycling plastics are already in place, the search for innovative methods to identify plastic-active enzymes is still ongoing.

17.7 CONCLUSION

The widespread use of fossil-based polymers in numerous products worldwide has resulted in significant environmental pollution. While significant progress has been made in demonstrating biodegradation as a potential solution, enzyme-efficient degradation is still under research. Several challenges such as low degradation efficiency and limited enzyme resources stipulate for further exploration. The stability and efficiency of studied enzymes are also insufficient for practical production applications. This has spurred global research efforts to create innovative eco-friendly degradation approaches. Employing biodegradable plastics is a crucial waste management strategy aimed at fostering a sustainable environment. Bacteria and fungi along with their enzymes are key players in enhancing the enzymatic degradation of fossil-based polymers. Research efforts should primarily focus on isolating, screening, and producing PDEs at both pilot and industrial scales, as well as exploring related genetic engineering approaches to enhance plastic biodegradation, thereby benefiting the environment on a large scale. Additionally, investigating the ecological impact of plastic-degradation products can provide theoretical support for selecting scientifically sound environmentally friendly plastic-degradation methods. Hence, the development of various new environmentally friendly materials is crucial to reduce carbon dioxide emissions into the atmosphere.

REFERENCES

Álvarez-Barragán J, Domínguez-Malfavón L, Vargas-Suárez M, González-Hernández R, Aguilar-Osorio G, Loza-Tavera H. Biodegradative activities of selected environmental fungi on a polyester polyurethane varnish and polyether polyurethane foams. Kivisaar M, editor. *Applied and Environmental Microbiology* [Internet]. 2016 Sep;82(17):5225–35. Available from: https://dx.doi.org/10.1128/aem.01344-16

Andersen, BK, Borch, K, Abo, M, Damgaard, B. 1999. U.S. Patent No. 5,997,584. U.S. Patent and Trademark Office, Washington, DC.

Apitius L, Rübsam K, Jakesch C, Jakob F, Schwaneberg U. Ultrahigh-throughput screening system for directed polymer binding peptide evolution. *Biotechnology and Bioengineering* [Internet]. 2019 May 3;116(8):1856–67. Available from: https://dx.doi.org/10.1002/bit.26990

Barth M, Wei R, Oeser T, Then J, Schmidt J, Wohlgemuth F, et al. Enzymatic hydrolysis of polyethylene terephthalate films in an ultrafiltration membrane reactor. *Journal of Membrane Science* [Internet]. 2015 Nov;494:182–7. Available from: https://dx.doi.org/10.1016/j.memsci.2015.07.030

Burrows S, Colwell J, Costanzo S, Kaserzon S, Okoffo E, Ribeiro F, et al. UV sources and plastic composition influence microplastic surface degradation: Implications for plastic weathering studies. *Journal of Hazardous Materials Advances* [Internet]. 2024 May;14:100428. Available from: https://dx.doi.org/10.1016/j.hazadv.2024.100428

Carniel A, Valoni É, Nicomedes J, Gomes A da C, Castro AM de. Lipase from *Candida antarctica* (CALB) and cutinase from *Humicola insolens* act synergistically for PET hydrolysis to terephthalic acid. *Process Biochemistry* [Internet]. 2017 Aug;59:84–90. Available from: https://dx.doi.org/10.1016/j.procbio.2016.07.023

Chaudhary AK, Vijayakumar RP. Studies on biological degradation of polystyrene by pure fungal cultures. *Environment, Development and Sustainability* [Internet]. 2020;22(5):4495–508. Available from: https://dx.doi.org/10.1007/s10668-019-00394-5

Chukhchin DG, Malkov AV, Tyshkunova IV, Mayer LV, Novozhilov EV. Diffractometric method for determining the degree of crystallinity of materials. *Crystallography Reports* [Internet]. 2016 May;61(3):371–5. Available from: https://dx.doi.org/10.1134/s1063774516030081

Ciuffi B, Fratini E, Rosi L. Plastic pretreatment: The key for efficient enzymatic and biodegradation processes. *Polymer Degradation and Stability* [Internet]. 2024 Apr;222:110698. Available from: https://dx.doi.org/10.1016/j.polymdegradstab.2024.110698

Desai SS, Nityanand C. Microbial laccases and their applications: A review. *Asian Journal of Biotechnology* [Internet]. 2011 Feb 15;3(2):98–124. Available from: https://dx.doi.org/10.3923/ajbkr.2011.98.124

DeCastro ME, Rodríguez-Belmonte E, González-Siso MI. Metagenomics of thermophiles with a focus on discovery of novel thermozymes. *Frontiers in Microbiology* [Internet]. 2016 Sep 27;7. Available from: https://dx.doi.org/10.3389/fmicb.2016.01521

Dimarogona M, Nikolaivits E, Kanelli M, Christakopoulos P, Sandgren M, Topakas E. Structural and functional studies of a *Fusarium oxysporum* cutinase with polyethylene terephthalate modification potential. *Biochimica et Biophysica Acta (BBA) - General Subjects* [Internet]. 2015 Nov;1850(11):2308–17. Available from: https://dx.doi.org/10.1016/j.bbagen.2015.08.009

Eberl A, Heumann S, Brückner T, Araujo R, Cavaco-Paulo A, Kaufmann F, et al. Enzymatic surface hydrolysis of poly(ethylene terephthalate) and bis(benzoyloxyethyl) terephthalate by lipase and cutinase in the presence of surface active molecules. *Journal of Biotechnology* [Internet]. 2009 Sep;143(3):207–12. Available from: https://dx.doi.org/10.1016/j.jbiotec.2009.07.008

Erden E, Ucar MC, Gezer T, Pazarlioglu NK. Screening for ligninolytic enzymes from autochthonous fungi and applications for decolorization of Remazole Marine Blue. *Brazilian Journal of Microbiology* [Internet]. 2009 Jun;40(2):346–53. Available from: https://dx.doi.org/10.1590/s1517-83822009000200026

Ferrandi EE, Sayer C, Isupov MN, Annovazzi C, Marchesi C, Iacobone G, et al. Discovery and characterization of thermophilic limonene-1,2-epoxide hydrolases from hot spring metagenomic libraries. *The FEBS Journal* [Internet]. 2015 Jun 16;282(15):2879–94. Available from: https://dx.doi.org/10.1111/febs.13328

Fotopoulou KN, Karapanagioti HK. Degradation of various plastics in the environment. *Hazardous Chemicals Associated with Plastics in the Marine Environment* [Internet]. 2017;71–92. Available from: https://dx.doi.org/10.1007/698_2017_11

Gao A, Shen H, Zhang H, Feng G, Xie K. Hydrophilic modification of polyester fabric by synergetic effect of biological enzymolysis and non-ionic surfactant, and applications in cleaner production. *Journal of Cleaner Production* [Internet]. 2017 Oct;164:277–87. Available from: https://dx.doi.org/10.1016/j.jclepro.2017.06.214

Gautam R, Bassi AS, Yanful EK. *Candida rugosa* lipase-catalyzed polyurethane degradation in aqueous medium. *Biotechnology Letters* [Internet]. 2007 Apr 21;29(7):1081–6. Available from: https://dx.doi.org/10.1007/s10529-007-9354-1

Goldman AS. Carbon–carbon bonds get a break. *Nature* [Internet]. 2010 Jan 27;463(7280):435–6. Available from: https://dx.doi.org/10.1038/463435a

Gómez-Méndez LD, Moreno-Bayona DA, Poutou-Piñales RA, Salcedo-Reyes JC, Pedroza-Rodríguez AM, Vargas A, et al. Biodeterioration of plasma pretreated LDPE sheets by *Pleurotus ostreatus*. Mukherjee A, editor. *PLOS ONE* [Internet]. 2018 Sep 13;13(9):e0203786. Available from: https://dx.doi.org/10.1371/journal.pone.0203786

Gu JD. Biodegradability of plastics: The pitfalls. *Applied Environmental Biotechnology* [Internet]. 2017;2(2):59–61. Available from: https://dx.doi.org/10.26789/aeb.2017.01.008

Gyung Yoon M, Jeong Jeon H, Nam Kim M. Biodegradation of polyethylene by a soil bacterium and AlkB cloned recombinant cell. *Journal of Bioremediation & Biodegradation* [Internet]. 2012;03(04). Available from: https://dx.doi.org/10.4172/2155-6199.1000145

Ikehata K, Buchanan ID, Smith DW. Recent developments in the production of extracellular fungal peroxidases and laccases for waste treatment. *Journal of Environmental Engineering and Science* [Internet]. 2004 Jan 1;3(1):1–19. Available from: https://dx.doi.org/10.1139/s03-077

Inderthal H, Tai SL, Harrison STL. Non-hydrolyzable plastics – An interdisciplinary look at plastic bio-oxidation. *Trends in Biotechnology* [Internet]. 2021 Jan;39(1):12–23. Available from: https://dx.doi.org/10.1016/j.tibtech.2020.05.004

Jeon HJ, Kim MN. Comparison of the functional characterization between alkane monooxygenases for low-molecular-weight polyethylene biodegradation. *International Biodeterioration & Biodegradation* [Internet]. 2016 Oct;114:202–8. Available from: https://dx.doi.org/10.1016/j.ibiod.2016.06.012

Jeon HJ, Kim MN. Functional analysis of alkane hydroxylase system derived from *Pseudomonas aeruginosa* E7 for low molecular weight polyethylene biodegradation. *International Biodeterioration & Biodegradation* [Internet]. 2015 Sep;103:141–6. Available from: https://dx.doi.org/10.1016/j.ibiod.2015.04.024

Jumaah OS. Screening of plastic degrading bacteria from dumped soil area. *IOSR Journal of Environmental Science, Toxicology and Food Technology* [Internet]. 2017 May;11(05):93–8. Available from: https://dx.doi.org/10.9790/2402-1105029398

Kamboj, M. Degradation of plastics for clean environment. *International Journal of Advanced Research in Engineering and Applied Sciences*. 2016;5(3):1–10.

Kang BR, Kim SB, Song HA, Lee TK. Accelerating the biodegradation of high-density polyethylene (HDPE) using *Bjerkandera adusta* TBB-03 and lignocellulose substrates. *Microorganisms* [Internet]. 2019 Aug 31;7(9):304. Available from: https://dx.doi.org/10.3390/microorganisms7090304

Khan S, Nadir S, Shah ZU, Shah AA, Karunarathna SC, Xu J, et al. Biodegradation of polyester polyurethane by *Aspergillus tubingensis*. *Environmental Pollution* [Internet]. 2017 Jun;225:469–80. Available from: https://dx.doi.org/10.1016/j.envpol.2017.03.012

Kiran BR, Kopperi H, Venkata Mohan S. Micro/nano-plastics occurrence, identification, risk analysis and mitigation: Challenges and perspectives. *Reviews in Environmental Science and Bio/Technology* [Internet]. 2022 Jan 27;21(1):169–203. Available from: https://dx.doi.org/10.1007/s11157-021-09609-6

Kong Y, Hay JN. The enthalpy of fusion and degree of crystallinity of polymers as measured by DSC. *European Polymer Journal* [Internet]. 2003 Aug;39(8):1721–7. Available from: https://dx.doi.org/10.1016/s0014-3057(03)00054-5

Krueger MC, Harms H, Schlosser D. Prospects for microbiological solutions to environmental pollution with plastics. *Applied Microbiology and Biotechnology* [Internet]. 2015 Aug 30;99(21):8857–74. Available from: https://dx.doi.org/10.1007/s00253-015-6879-4

Loredo-Treviño A, García G, Velasco-Téllez A, Rodríguez-Herrera R, Aguilar CN. Polyurethane as substrate for fungal strains. *Advances in Bioscience and Biotechnology* [Internet]. 2011;02(02):52–8. Available from: https://dx.doi.org/10.4236/abb.2011.22009

Lorenz P, Eck J. Metagenomics and industrial applications. *Nature Reviews Microbiology* [Internet]. 2005 Jun;3(6):510–6. Available from: https://dx.doi.org/10.1038/nrmicro1161

Magnin A, Pollet E, Perrin R, Ullmann C, Persillon C, Phalip V, et al. Enzymatic recycling of thermoplastic polyurethanes: Synergistic effect of an esterase and an amidase and recovery of building blocks. *Waste Management* [Internet]. 2019 Feb;85:141–50. Available from: https://dx.doi.org/10.1016/j.wasman.2018.12.024

Martinez C, De Geus P, Lauwereys M, Matthyssens G, Cambillau C. *Fusarium solani* cutinase is a lipolytic enzyme with a catalytic serine accessible to solvent. *Nature* [Internet]. 1992 Apr;356(6370):615–8. Available from: https://dx.doi.org/10.1038/356615a0

Matjašič T, Simčič T, Medveššek N, Bajt O, Dreo T, Mori N. Critical evaluation of biodegradation studies on synthetic plastics through a systematic literature review. *Science of the Total Environment* [Internet]. 2021 Jan;752:141959. Available from: https://dx.doi.org/10.1016/j.scitotenv.2020.141959

Meyer-Cifuentes IE, Werner J, Jehmlich N, Will SE, Neumann-Schaal M, Öztürk B. Synergistic biodegradation of aromatic-aliphatic copolyester plastic by a marine microbial consortium. *Nature Communications* [Internet]. 2020 Nov 13;11(1). Available from: https://dx.doi.org/10.1038/s41467-020-19583-2

Mohanan N, Montazer Z, Sharma PK, Levin DB. Microbial and enzymatic degradation of synthetic plastics. *Frontiers in Microbiology* [Internet]. 2020 Nov 26;11. Available from: https://dx.doi.org/10.3389/fmicb.2020.580709

Mueller RJ. Biological degradation of synthetic polyesters—Enzymes as potential catalysts for polyester recycling. *Process Biochemistry* [Internet]. 2006 Oct;41(10):2124–8. Available from: https://dx.doi.org/10.1016/j.procbio.2006.05.018

Müller R, Schrader H, Profe J, Dresler K, Deckwer W. Enzymatic degradation of poly(ethylene terephthalate): Rapid hydrolyse using a hydrolase from *T. fusca*. *Macromolecular Rapid Communications* [Internet]. 2005 Aug 22;26(17):1400–5. Available from: https://dx.doi.org/10.1002/marc.200500410

Nguyen LH, Nguyen BS, Le DT, Alomar TS, AlMasoud N, Ghotekar S, et al. A concept for the biotechnological minimizing of emerging plastics, micro- and nano-plastics pollutants from the environment: A review. *Environmental Research* [Internet]. 2023 Jan;216:114342. Available from: https://dx.doi.org/10.1016/j.envres.2022.114342

Nwogu N. Capability of selected mushrooms to biodegrade polyethylene. *Mycosphere* [Internet]. 2012;3(4):455–62. Available from: https://dx.doi.org/10.5943/mycosphere/3/4/9

Oberbeckmann S, Loeder MGJ, Gerdts G, Osborn AM. Spatial and seasonal variation in diversity and structure of microbial biofilms on marine plastics in Northern European waters. *FEMS Microbiology Ecology* [Internet]. 2014 Nov 1;90(2):478–92. Available from: https://dx.doi.org/10.1111/1574-6941.12409

Osman M, Satti SM, Luqman A, Hasan F, Shah Z, Shah AA. Degradation of polyester polyurethane by *Aspergillus* sp. strain S45 isolated from soil. *Journal of Polymers and the Environment* [Internet]. 2017 Feb 6;26(1):301–10. Available from: https://dx.doi.org/10.1007/s10924-017-0954-0

Othman AR, Hasan HA, Muhamad MH, Ismail N 'Izzati, Abdullah SRS. Microbial degradation of microplastics by enzymatic processes: A review. *Environmental Chemistry Letters* [Internet]. 2021 Mar 3;19(4):3057–73. Available from: https://dx.doi.org/10.1007/s10311-021-01197-9

Parida D, Katare K, Ganguly A, Chakraborty D, Konar O, Nogueira R, et al. Molecular docking and metagenomics assisted mitigation of microplastic pollution. *Chemosphere* [Internet]. 2024 Mar;351:141271. Available from: https://dx.doi.org/10.1016/j.chemosphere.2024.141271

Perkins DN, Pappin DJC, Creasy DM, Cottrell JS. Probability-based protein identification by searching sequence databases using mass spectrometry data. *Electrophoresis* [Internet]. 1999 Dec 1;20(18):3551–67. Available from: https://dx.doi.org/10.1002/(sici)1522-2683(19991201)20:18<3551::aid-elps3551>3.0.co;2-2

Perotti G, Kijchavengkul T, Auras R, Constantino V. Nanocomposites based on cassava starch and chitosan-modified clay: Physico-mechanical properties and biodegradability in simulated compost soil. *Journal of the Brazilian Chemical Society* [Internet]. 2016; Available from: https://dx.doi.org/10.21577/0103-5053.20160213

Pinnell LJ, Turner JW. Shotgun metagenomics reveals the benthic microbial community response to plastic and bioplastic in a coastal marine environment. *Frontiers in Microbiology* [Internet]. 2019 Jun 7;10. Available from: https://dx.doi.org/10.3389/fmicb.2019.01252

Plastics Europe. 2015. Plastics-The Facts 2015. An Analysis of European Plastics Production. Demand and Waste Data, 2015. https://plasticseurope.org/knowledge-hub/plastics-the-facts-2015/.

Prenafeta-Boldú FX, de Hoog GS, Summerbell RC. Fungal communities in hydrocarbon degradation. Microbial communities utilizing hydrocarbons and lipids: Members. *Metagenomics and Ecophysiology* [Internet]. 2018;1–36. Available from: https://dx.doi.org/10.1007/978-3-319-60063-5_8-1

Ronkvist ÅM, Xie W, Lu W, Gross RA. Cutinase-catalyzed hydrolysis of poly(ethylene terephthalate). *Macromolecules* [Internet]. 2009 Jul 2;42(14):5128–38. Available from: https://dx.doi.org/10.1021/ma9005318

Roy R, Mukherjee G, Das Gupta A, Tribedi P, Sil AK. Isolation of a soil bacterium for remediation of polyurethane and low-density polyethylene: A promising tool towards sustainable cleanup of the environment. *3 Biotech* [Internet]. 2021 Jan;11(1). Available from: https://dx.doi.org/10.1007/s13205-020-02592-9

Satti SM, Shah AA. Polyester-based biodegradable plastics: An approach towards sustainable development. *Letters in Applied Microbiology* [Internet]. 2020 Mar 18;70(6):413–30. Available from: https://dx.doi.org/10.1111/lam.13287

Sharma SR. Bioremediation of polythenes and plastics: A microbial approach. *Nanotechnology in the Life Sciences* [Internet]. 2018;97–114. Available from: https://dx.doi.org/10.1007/978-3-030-02369-0_6

Singh AK, Mukhopadhyay M. Overview of fungal lipase: A review. *Applied Biochemistry and Biotechnology* [Internet]. 2012;166(2):486–520. Available from: https://dx.doi.org/10.1007/s12010-011-9444-3

Sivan A, Szanto M, Pavlov V. Biofilm development of the polyethylene-degrading bacterium *Rhodococcus ruber*. *Applied Microbiology and Biotechnology* [Internet]. 2006 Mar 14;72(2):346–52. Available from: https://dx.doi.org/10.1007/s00253-005-0259-4

Skamnioti P, Furlong RF, Gurr SJ. Evolutionary history of the ancient cutinase family in five filamentous *Ascomycetes* reveals differential gene duplications and losses and in *Magnaporthe grisea* shows evidence of sub-and neo-functionalization. *New Phytologist* [Internet]. 2008 Oct 15;180(3):711–21. Available from: https://dx.doi.org/10.1111/j.1469-8137.2008.02598.x

Sorigué D, Légeret B, Cuiné S, Blangy S, Moulin S, Billon E, et al. An algal photoenzyme converts fatty acids to hydrocarbons. *Science* [Internet]. 2017 Sep;357(6354):903–7. Available from: https://dx.doi.org/10.1126/science.aan6349

Sowmya HV, Ramalingappa, Krishnappa M, Thippeswamy B. Degradation of polyethylene by *Trichoderma harzianum*—SEM, FTIR, and NMR analyses. *Environmental Monitoring and Assessment* [Internet]. 2014 Jul 23;186(10):6577–86. Available from: https://dx.doi.org/10.1007/s10661-014-3875-6

Takei D, Washio K, Morikawa M. Identification of alkane hydroxylase genes in *Rhodococcus* sp. strain TMP2 that degrades a branched alkane. *Biotechnology Letters* [Internet]. 2008 Apr 15;30(8):1447–52. Available from: https://dx.doi.org/10.1007/s10529-008-9710-9

Thakur B, Singh J, Singh J, Angmo D, Vig AP. Biodegradation of different types of microplastics: Molecular mechanism and degradation efficiency. *Science of the Total Environment* [Internet]. 2023 Jun;877:162912. Available from: https://dx.doi.org/10.1016/j.scitotenv.2023.162912

Thomsen TB, Almdal K, Meyer AS. Significance of poly(ethylene terephthalate) (PET) substrate crystallinity on enzymatic degradation. *New Biotechnology* [Internet]. 2023 Dec;78:162–72. Available from: https://dx.doi.org/10.1016/j.nbt.2023.11.001

Tschan MJL, Brulé E, Haquette P, Thomas CM. Synthesis of biodegradable polymers from renewable resources. *Polymer Chemistry* [Internet]. 2012;3(4):836–51. Available from: https://dx.doi.org/10.1039/c2py00452f

Urbanek AK, Rymowicz W, Mirończuk AM. Degradation of plastics and plastic-degrading bacteria in cold marine habitats. *Applied Microbiology and Biotechnology* [Internet]. 2018 Jul 11;102(18):7669–78. Available from: https://dx.doi.org/10.1007/s00253-018-9195-y

Uzun İ. Methods of determining the degree of crystallinity of polymers with X-ray diffraction: A review. *Journal of Polymer Research* [Internet]. 2023 Sep 28;30(10). Available from: https://dx.doi.org/10.1007/s10965-023-03744-0

Vaksmaa A, Hernando-Morales V, Zeghal E, Niemann H. Microbial degradation of marine plastics: Current state and future prospects. *Biotechnology for Sustainable Environment* [Internet]. 2021;111–54. Available from: https://dx.doi.org/10.1007/978-981-16-1955-7_5

Vertommen MAME, Nierstrasz VA, Veer M van der, Warmoeskerken MMCG. Enzymatic surface modification of poly(ethylene terephthalate). *Journal of Biotechnology* [Internet]. 2005 Dec;120(4):376–86. Available from: https://dx.doi.org/10.1016/j.jbiotec.2005.06.015

Wackett LP, Douglas Hershberger C. *Biocatalysis and Biodegradation* [Internet]. ASM Press; 2001. Available from: https://dx.doi.org/10.1128/9781555818036

Wang Y, Xue P, Cao M, Yu T, Lane ST, Zhao H. Directed evolution: Methodologies and applications. *Chemical Reviews* [Internet]. 2021 Jul 23;121(20):12384–444. Available from: https://dx.doi.org/10.1021/acs.chemrev.1c00260

Wei R, Zimmermann W. Microbial enzymes for the recycling of recalcitrant petroleum-based plastics: how far are we? *Microbial Biotechnology* [Internet]. 2017 Mar 28;10(6):1308–22. Available from: https://dx.doi.org/10.1111/1751-7915.12710

Wu P, Huang J, Zheng Y, Yang Y, Zhang Y, He F, et al. Environmental occurrences, fate, and impacts of microplastics. *Ecotoxicology and Environmental Safety* [Internet]. 2019 Nov;184:109612. Available from: https://dx.doi.org/10.1016/j.ecoenv.2019.109612

Wu X, Zhao X, Chen R, Liu P, Liang W, Wang J, et al. Size-dependent long-term weathering converting floating polypropylene macro- and microplastics into nanoplastics in coastal seawater environments. *Water Research* [Internet]. 2023 Aug;242:120165. Available from: https://dx.doi.org/10.1016/j.watres.2023.120165

Yoon, MY, Kellis, J, Poulose, AJ. Enzymatic modification of polyester. *AATCC Review*. 2002;2(6):33–6.

Zhang J, Gao D, Li Q, Zhao Y, Li L, Lin H, et al. Biodegradation of polyethylene microplastic particles by the fungus *Aspergillus flavus* from the guts of wax moth *Galleria mellonella*. *Science of the Total Environment* [Internet]. 2020 Feb;704:135931. Available from: https://dx.doi.org/10.1016/j.scitotenv.2019.135931

Zhang K, Hamidian AH, Tubić A, Zhang Y, Fang JKH, Wu C, et al. Understanding plastic degradation and microplastic formation in the environment: A review. *Environmental Pollution* [Internet]. 2021 Apr;274:116554. Available from: https://dx.doi.org/10.1016/j.envpol.2021.116554

Zhang Z, Wu S, Stenoien DL, Paša-Tolić L. High-throughput proteomics. *Annual Review of Analytical Chemistry* [Internet]. 2014 Jun 12;7(1):427–54. Available from: https://dx.doi.org/10.1146/annurev-anchem-071213-020216

Zhou F, Hanson TE, Johnston MV. Intact protein profiling of *Chlorobium tepidum* by capillary isoelectric focusing, reversed-phase liquid chromatography, and mass spectrometry. *Analytical Chemistry* [Internet]. 2007 Aug 22;79(18):7145–53. Available from: https://dx.doi.org/10.1021/ac071147c

Zhu B, Wang D, Wei N. Enzyme discovery and engineering for sustainable plastic recycling. *Trends in Biotechnology* [Internet]. 2022 Jan;40(1):22–37. Available from: https://dx.doi.org/10.1016/j.tibtech.2021.02.008

18 Plastisphere as a Vector for Pathogenic Microbes and Antibiotic Resistance

Dhammika N. Magana-Arachchi, Tharindi Bandara, and Shashini Gunathilaka

18.1 INTRODUCTION

Scientists have identified microplastics (MPs) in all environments they have studied in the last decade, from the deepest ocean depths to the highest remote mountain peaks, tropical regions, and polar areas. These MPs are generously found in the air, soils, food, digestive tracts, and also in lungs (Ghosh et al., 2023). MPs are widespread and could be categorized as the most ubiquitous contaminator of the earth (Magana-Arachchi and Wanigatunge, 2023).

MPs' persistent nature creates a novel habitat for microbial colonization, which leads to biofilm formation. Biofilms are complex aggregates of microorganisms sheathed in a self-produced matrix that facilitates their adherence to surfaces and protects them from environmental stressors. This environment on MPs attracts a range of microorganisms, including pathogenic microbes and bacteria that carry antibiotic-resistance genes (ARGs). The biofilm matrix on MPs helps infective microbes to survive in water. This increases the chance that the microbes will interact with other living things. Second, the proximity of diverse microbes within the biofilm facilitates horizontal gene transfer (HGT) and allows bacteria to exchange genetic material, including ARGs. This interaction within the biofilm can accelerate the spread of antibiotic resistance. The plastisphere, an ecosystem of plastic debris suspended in water surfaces with living microorganisms, could be considered a vector for infectious bacteria and antibiotic-resistant gene transmission in several ways. Hence, this chapter explores the interrelation between microorganisms, biofilm formation, and MPs within the plastisphere, highlighting the plastisphere's potential to act as a vector for developing ARGs and transmitting antibiotic-resistant bacteria (ARB).

18.2 THE PLASTISPHERE CONCEPT

The environmental pollution caused by man-made polymers has been a significant concern for many years. Plastics are made out of synthetic or semi-synthetic organic material, and about 60% of this plastic is disposed of in landfills or recycled, while the rest is in the environment (Muñoz Meneses et al., 2022). Large plastic waste is the primary source of MPs, which are small particles that measure 5 mm or less. MPs are categorized into two groups based on their original sizes. Plastic microbeads, industrially produced particles, and powders (<5 mm diameter) are considered primary MPs, and in contrast, large plastic fragments break down through physical, biological, and chemical processes into smaller fragments, known as secondary MPs.

MPs exhibit various shapes, including foils, foams, fibres, pellets, fragments, and microbeads. Plastics are characterized by chemical diversity. Materials such as polyamide (PA), polyvinyl chloride (PVC), and polyethylene terephthalate (PET) have higher densities than water, so their settlement rates in sediments are increasing. Conversely, plastics like polystyrene (PS), high-density

DOI: 10.1201/9781032717814-18

polyethylene (HDPE), low-density polyethylene (LDPE), polypropylene (PP), and polyurethane (PUR) have lower densities and tend to float predominantly on water (Du et al., 2022).

In 2013, Zettler and his team defined the plastisphere as a microbial community found on plastic waste in aquatic ecosystems (Zettler et al., 2013). Since the 1970s, when scientists first observed bacteria and diatoms growing on plastic debris in the Sargasso Sea, it has become clear that plastic in the ocean quickly becomes home to various microorganisms and small animals (Didier et al., 2017). Certain microbes, such as members of the fungal phylum Chytridiomycota and bacterial phyla Bacteroidetes, Proteobacteria, and Cyanobacteria, are highly prevalent in the plastisphere. It still needs to be determined if certain types of microbes only live on specific kinds of plastics (Du et al., 2022). Further, the plastisphere community has exhibited significant metabolic pathways compared to the sea-water-free living microbes. It remains to be seen if different sorts of plastic attract and support diverse communities of microbes and small creatures. Plastispheres are significant in harbouring and disseminating pathogenic microorganisms and ARB. Hence, understanding and monitoring the plastisphere is essential for assessing its potential risks to aquatic ecosystems and human health.

18.3 THE MICROPLASTICS IN BIOTA

Microorganisms tend to inhabit the surfaces of MPs. The interaction between microorganisms and MPs is influenced by several surface properties of the MPs, including size, shape, roughness, and hydrophobicity. The interaction is primarily driven by hydrophobic and electrostatic forces, further modulated by environmental factors such as temperature, pH, and ionic strength.

For bacteria, initial attachment to MPs is driven by electrostatic forces. MPs can serve as conducive substrates for microbial colonization. It provides essential nutrients, such as metal ions like zinc, iron, and copper, by adsorption from the surroundings. Various microorganisms, including bacteria such as *Aeromonas*, *Rhodococcus*, *Pseudomonas*, *Enterobacter*, *Halomonas*, *Mycobacterium*, *Photobacterium*, and *Shigella*, as well as fungi, have been discovered on MP surfaces (Gkoutselis et al., 2021; Parsaeimehr et al., 2023). The composition and characteristics of MPs, biodegradability, and varying hydrophobicity levels influence microbial colonization. It indicates a degree of selectivity (Sooriyakumar et al., 2022). Proteobacteria, Bacteroidota, and Actinobacteria were the most abundant phyla in beach sand MPs (Vidal-Verdú et al., 2022).

Some evidence indicates that viruses can adhere to plastic surfaces. Studies have shown that SARS-CoV-2, the virus responsible for COVID-19, can adhere to PP surfaces and remain viable for over 72 hours (Ashokkumar et al., 2023). It is longer than on materials like copper or cardboard. Viruses can adhere to naked PP plastic surfaces through non-ionic forces (Ashokkumar et al., 2023). In certain instances, adhesion to plastic surfaces leads to viral inactivation, as observed with Poliovirus 1 when stored in a hydrophobic PP container in groundwater. Therefore, the fate of viruses interacting with MPs depends on the specific type of plastic and the nature of the virus (Gassilloud and Gantzer, 2005). While the coexistence of bacterial colonies and viruses on MPs seems plausible, experimental confirmation of this phenomenon has yet to be achieved.

Studies have shown that MPs could carry more potential pathogenic microbes and ARB than the natural substrate. In 2019, Miao and colleagues investigated biofilm growth on several substrates under laboratory conditions; cobblestone, wood, polyethylene (PE), and PVC were compared. Among those, the natural substrate had a lower species abundance. In contrast, the MP substrate had a higher species abundance. Table 18.1 lists a diverse array of microorganisms found in different substrates.

18.4 MICROPLASTIC-ASSOCIATED BIOFILM FORMATION

MPs are susceptible to colonization by microorganisms when they come into contact with water, leading to the formation of biofilms. Naturally, biofilm development involves a complex three-phase process: attachment, maturation, and detachment. Temperature, light/dark conditions, oxygen levels,

TABLE 18.1
Comparison of Microbial Communities on Different Substrates

Location	Natural/Plastic Substrate	Microbial Community		References
		Non-Pathogenic	Potential Pathogenic	
China	Wood –cultured under the laboratory condition	Deltaproteobacteria, Verrucomicrobiae, Anaerolineae, Cytophagia, Oscillatoriophycideae, Phycisphaerae, Planctomycelia, Synechococcophycideae, Chloracidobacteria	Alphaproteobacteria, Betaproteobacteria, Gammaproteobacteria, Flavobacteria	Miao et al. (2019)
China	Glass – cultured under the laboratory condition	Rhodobacterales, Planctomycetales	Frankiales, Micrococcales, Rhizobiales, Sphingomonadales, Burkholderiales, Cytophagales, Sphingobacteriales	Ogonowski et al. (2018)
United Kingdom	Coastal sand – collected from site	Verrucomicrobia, Deltaproteobacteria	Bacteroidetes, Proteobacteria, Actinobacteria, Gammaproteobacteria	Gobet et al. (2012)
China	Leaf –collected from site	*Pseudomonas oryzae*, *Pseudomonas benzenivorans*, *Pseudomonas graminis*, *Pseudomonas peli*	*Pantoea vagans*, *Pantoea agglomerans*	Wu et al. (2019)
China	Cobblestone – cultured under the laboratory condition	Cytophagia, Oscillatoriophycideae, Planctomycelia, Synechococcophycideae, Chloracidobacteria	Alphaproteobacteria, Betaproteobacteria, Gammaproteobacteria, Flavobacteria	Miao et al. (2019)
Kandy lake, Sri Lanka	Microplastic – collected from lake waters in an urban city	*Acinetobacter baumannii*, *Herbaspirillum huttiense*	*Serratia marcescens*, *Stenotrophomonas maltophilia*, *Bacillus anthracis*	Welagedara et al. (2023)
China	Microplastics – collected from water in urban area	Denococcus-Thermus Acidobacteria, Nitospirae, Actinobacteria	Proteobacteria	Huang et al. (2021)
China	Microplastics – collected from site	*Pseudomonas chlororaphis*, *Pseudomonas rhizosphaerae*, *Pseudomonas denitrificans*	*Pantoea agglomerans*, *Leclercia adecarboxylata*	Wu et al. (2019)
China	Microplastic – cultured under the laboratory condition	*Aquabacterium* sp. *Loktanella* sp.	*Hyphomonas* sp. *Hydrogenophaga* sp.	Ogonowski et al. (2018)

and nutrient availability can influence biofilm formation. The hydrophobicity and texture of attaching surfaces also contribute to biofilm formation. Microorganisms colonize particles with rough surfaces more readily than smooth surfaces (Rajcoomar et al., 2024). They are exposed to various physical and chemical factors that may influence biofilm development in different conditions.

Recent findings indicate that microorganisms attached to MPs can be transported and dispersed over significant distances. Biofilm-forming bacteria have physiological features that facilitate biofilm formation, including flagella, fimbriae, and pili. Gram-negative and Gram-positive bacteria can form biofilms. However, MP surfaces' most frequently reported biofilm-forming families include Pseudomonadaceae, Moraxellaceae, Enterobacteriaceae, and Comamonadaceae (Cholewińska et al., 2022).

18.4.1 Role of Extracellular Matrix

The Lifshitz–van der Waals forces and electrostatic interactions potentially occur between the cell surface and the substrate at the first bacterial adhesion phase (Carniello et al., 2018). They act to strengthen adhesion via adhesion receptors. In the subsequent irreversible adhesion stage, bacteria produce extracellular polymers, lipopolysaccharides (LPS), or proteins that lead to the proliferation of the initial adherent cells and the enlargement of microcolonies. Bacterial cells are connected through extracellular polymeric substances (EPSs), consisting of polysaccharides, proteins, nucleic acids, surfactants, lipids, and water, with the exact composition depending on the bacterial species. EPS mainly supports the formation and functioning of biofilms. Such as forming a protective barrier that enables tolerance to bactericidal factors, avoidance of host immune responses and predator attacks, bacterial cell aggregation, increased density, and mutual recognition with the transfer of genetic material (Rajcoomar et al., 2024). Nutrient supply to cells within the biofilm is facilitated by the formation of numerous tubules within the structure, allowing fluid flow around the microcolonies. This system delivers essential nutrients and oxygen while removing unnecessary metabolic products (Tu et al., 2020).

18.4.2 Quorum-Sensing

The expression of quorum-sensing genes by bacteria accumulated on surfaces, including MPs, facilitates further cell-to-cell adhesion, maturation, and dispersion of biofilms. This process increases bacterial resistance (Nath et al., 2023). Quorum sensing involves the production of different signalling molecules known as autoinducers. Bacteria produce these molecules in proportion to their population density. As the concentration of these molecules increases, they regulate gene expression, coordinating communal activities such as biofilm formation. On MPs, quorum sensing triggers the production of EPSs, which form the biofilm matrix. That enables bacteria to adhere firmly to the surface and each other (Zhao et al., 2020). This biofilm environment protects from environmental stressors, antibiotics, and the immune responses of host organisms, enhancing bacterial survival and persistence.

18.4.3 MP Surface Chemistry on Biofilm Formation

Several physical and chemical properties of the plastic substrates, including polymer type, surface characteristics, and particle sizes, influence the formation of biofilms on MPs. For instance, PVC and PE promote more bacterial adhesion compared to PP and PET due to their larger specific surface area and roughness (Cai et al., 2019). The surface properties of MPs, such as hydrophobicity, significantly affect bacterial colonization. Surface roughness has been identified as a significant factor influencing cell attachment. Furthermore, the number of attached cells increases with roughness (Joo et al., 2021). Differences in bacterial attachment and community composition on MPs depend on surface hydrophobicity (Zhao et al., 2024). Additionally, surface charges and sizes of MPs also influence microbial attachment and biofilm formation (Moyal et al., 2023). This biofilm formation can also lead to the partial degradation of MPs, enhancing the release of leachates and toxic substances into the environment (Nath et al., 2023).

Several studies have reported that biofilm formation can increase the density of MPs, thereby affecting their environmental transport (Kaiser et al., 2017; Semcesen and Wells, 2021). The growth of biofilms on MPs alters their density, affecting their mobility in the environment, whether in water or soil. The biofilm formation can elevate particle density, enhancing sinking behaviour in aquatic environments (Semcesen and Wells, 2021). The sinking velocity of PS MPs increases significantly after biofilm formation, with an 81% increase in marine water and a 16% increase in estuarine water (Kaiser et al., 2017).

18.4.4 MP Biophysical Forces on Biofilm Formation

Microorganisms are attracted to MP surfaces via hydrophilic interactions, electrostatic interactions, or Van der Waals forces facilitated by functional groups expressed in the ecocorona due to the sorption of biological materials into plastics (Chen et al., 2022; Galloway et al., 2017; Zhang et al., 2022). Microbial cells may adhere to surfaces through both specific and non-specific interactions. Hydrophobic interactions are typically the strongest among all long-range non-covalent interactions in biological systems (Fu et al., 2021).

Upon contacting surfaces, cell attachment occurs in two phases. The initial attachment is reversible and occurs rapidly. It involves hydrodynamic and electrostatic interactions. During this period, the adhesive force between bacteria and surfaces increases quickly. The second phase of attachment is irreversible, occurs over several hours, and involves van der Waals interactions between the hydrophobic regions of the outer cell wall and the surface (Tuson and Weibel, 2013). Most bacteria exhibit a negative surface charge, particularly in the stationary phase of their cell development stage. That influences their interactions with surfaces. This negative charge leads to better interactions with positively charged surfaces, and it has been shown that aliphatic plastics attract unique microorganisms compared to other plastics (Martínez-Campos Gutiérrez, 2022).

18.4.5 Temporal Dynamics of Plastisphere

The temporal dynamics of the plastisphere depend on several factors, including colonization patterns, microbial succession, and changes in community composition. These factors influence the abundance and diversity of ARBs and ARGs associated with MPs. Furthermore, factors such as nutrient availability, temperature fluctuations, and exposure to antimicrobial compounds also play a significant role.

18.4.5.1 Seasonal Variations

Long-term monitoring studies have revealed fluctuations in ARG abundance over time, with seasonal variations and anthropogenic activities, that impact microbial communities and their resistance profiles (Yang et al., 2021). In temperate regions, microbial colonization may peak during the warmer months when environmental conditions favour microbial growth and activity. Conversely, microbial biomass and diversity may decline during colder periods because it reduces metabolic rates and nutrient availability (Zhang et al., 2022). Certain conditions may promote the proliferation of ARB and ARGs, while others may lead to their decay or removal from the plastisphere environment (Lenaker et al., 2019).

18.4.5.2 Anthropogenic Activities

Anthropogenic activities, such as urbanization, agricultural runoff, and industrial pollution, can also influence the temporal dynamics of the plastisphere by introducing additional stressors and pollutants into aquatic ecosystems (Ng et al., 2018). High levels of organic matter and nutrients from wastewater effluent can stimulate microbial growth on MP surfaces, which may lead to shifts in community composition and the enrichment of ARBs and ARGs (Sooriyakumar et al., 2022).

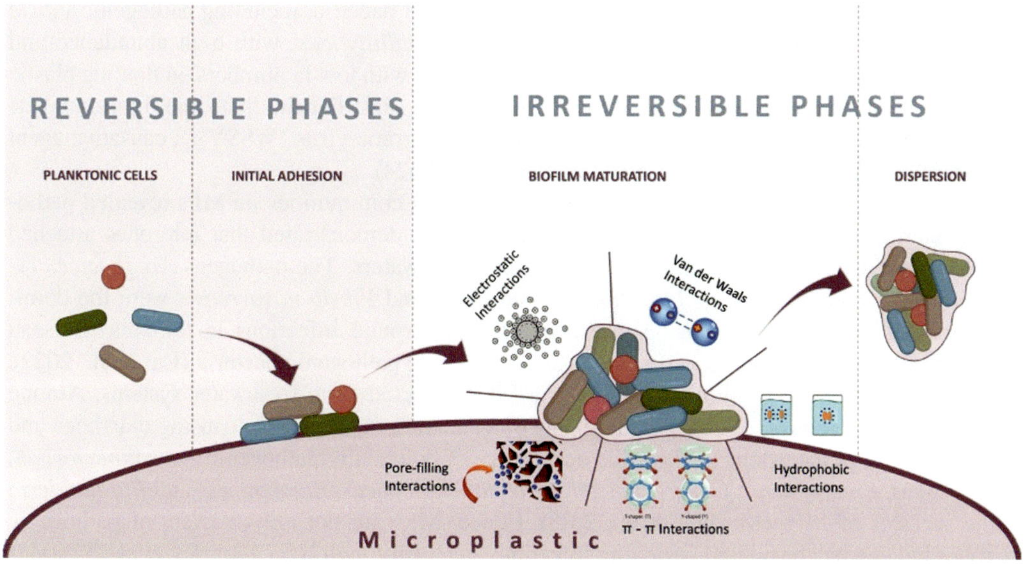

FIGURE 18.1 The formation of the plastisphere on microplastic surfaces through microbial colonization and biofilm formation.

18.4.5.3 Presence of Pharmaceuticals

The presence of pharmaceuticals on MPs can further modulate the temporal dynamics of the plastisphere by utilizing selective pressures on microbial communities. Antibiotics and other antimicrobial agents may promote the proliferation of ARBs, which leads to the accumulation of ARGs over time (D'agata and Mitchell, 2008). Fluctuations in pharmaceutical concentrations or exposure to environmental stressors may drive changes in microbial community structure and resistance phenotypes. Environmental stressors such as nutrient depletion, UV radiation, and fluctuations in salinity may inhibit microbial growth and reduce the abundance of ARB and ARGs in the plastisphere. These stressors can disrupt microbial communities, leading to composition shifts and ARG loss over time (Yang et al., 2021).

Microbial turnover rates, biofilm detachment, and sedimentation processes could influence the fate and persistence of ARGs in the plastisphere. Some ARGs may persist on MP surfaces for extended periods, while others undergo degradation or transformation through microbial metabolism and environmental factors (Nguyen et al., 2023). Understanding these dynamics is essential for elucidating the long-term implications of MP pollution on antimicrobial resistance (AMR) in aquatic environments (Figure 18.1).

18.5 ROLE OF MP AS A VECTOR FOR PATHOGENS

MPs are recognized as facilitators for the emergence of human diseases (Loiseau and Sorci, 2022). The biofilms formed on MPs have unique features compared to those on natural materials such as rocks and wood. Hence, the relative proportions of pathogenic bacteria such as *Chlamydiae and Pseudomonadota* were higher in biofilms formed on MPs than in biofilms formed on natural substrates (Wu et al., 2019). The global records on the presence of pathogenic microbes on plastics collected from oceans are increasing, and according to researchers, *Vibrios* are plentiful in the plastisphere, though only some are pathogenic (Bowley et al., 2021). Furthermore, according to Bowley et al. (2021), with rising plastic waste in oceans, *Vibrio* and other pathogens may get transported and transmitted into potential hosts, causing disease outbreaks.

A study by Siddiqui et al. (2024) revealed the presence of bacteria, including pathogenic *Vibrio* species, namely *V. campbelli*, *V. alginolyticus*, and *V. coralliilyticus*, with high abundance and viruses Herpesvirales, Caudovirales, and Poxviridae groups with low in numbers on floating plastic surfaces collected from the two sites of the Mediterranean seas. For the first time, in the marine plastisphere, the researchers recorded the White Spot Syndrome virus (WSSV), a causative agent for epidemics in the aquaculture industry (Siddiqui et al., 2024).

Metagenomic sequencing studies of residential bacterial communities on MPs revealed pathogenic bacteria in urban wastewater samples. Studies have demonstrated that microbes attached to MPs are the bacterial communities in the surrounding waters. The pathogens *Arcobacter cryaerophilus*, *Aeromonas salmonicida*, *Vibrio areninigrae*, and *Vibrio navarrensis* were the dominant bacterial communities, causing gastroenteritis and wound infections in humans (Loiseau and Sorci, 2022). In wastewater, MPs are a carrier for these pathogenic bacteria (Lai et al., 2022). Proteobacteria were the leading bacteria on the MPs collected from freshwater systems. Among numerous *Proteobacteria*, *Escherichia coli*, and *Shigella* are noteworthy for causing diarrhoea and fever. Moreover, observations confirm the occurrence of potentially pathogenic *Faecalibacterium*, *Enterococcus*, *Enterobacter*, *Campylobacteraceae*, *Rumunicoccus*, *Romboutsia*, and *Burkholderia* on the surfaces of MPs (Murphy et al., 2018). Hence, MPs are not only carriers of pathogenic microbes but can be considered enriching infectious strains through HGT (Bowley et al., 2021), as described elsewhere in the chapter. Furthermore, it is not only humans but other species that are also affected by the MPs with pathogenic bacteria. Many aquaculture and fishery reports demonstrate how these infectious organisms on MPs cause diseases.

18.6 THE MICROPLASTICS AND ANTIBIOTIC RESISTANCE CONNECTION

High density and close physical contact of MPs with biofilms increase the transfer of ARGs within bacteria and their ability to persist in the environment. The biofilms act as sinks for contaminants, including heavy metals and antibiotics. Recent studies have reported the role of different plastic types in biofilm formation and their impact on ARG composition. For example, PET plastic-associated biofilms have been found to significantly influence the composition of ARGs (Ventura et al., 2024). The specific characteristics of PET, including its surface properties and polymer structure, create a distinct environment that affects microbial colonization and the types of ARGs present. During the COVID-19 epidemic, personal protective equipment (PPE), including gloves, facemasks, and protective suits, has accumulated in the natural environment. Among those, as facemasks became mandatory, inappropriately dumped facemasks accumulated in waste and provided substrates for microorganisms, including potential pathogens.

18.6.1 Antibiotic-Resistance Genes Identified in Microplastics

The most common ARGs identified in MPs confer resistance to multiple critical classes of antibiotics. These include tetracycline, sulfonamide, aminoglycoside, β-lactam, MLSB, and vancomycin (Itzhari et al., 2024). Among MP samples, the most abundant ARGs from the drift line zone that confer resistance to tetracycline are tetW, tetG, and tetM for sulfonamide, sul1 and sul2 and aminoglycoside, aaC3-VI. The β-lactam responsible genes include blaKPC and blaTEM_1, and for macrolide, lincosamide, and streptogramin B (MLSB), it is mphA (Merline and Dhinagaran, 2023). These antibiotic classes are among the most widely administered globally and have been used to treat bacterial infections for over half a century. The presence of blaKPC, blaCTX-M, and tetM in MPs is particularly concerning because these genes are associated with multidrug resistance and may be linked to other antimicrobial-resistance genes (Merline and Dhinagaran, 2023).

Examples of commonly detected ARB are included in Table 18.2.

These genes are often found on bacterial plasmids, mobile genetic elements (MGEs) that can be easily transferred between bacteria, facilitating the rapid spread of resistance (Lerminiaux and

TABLE 18.2
Antibiotic-Resistance Genes Associated with the Plastisphere in Various Ecosystems

Source	Major MP Polymer Type	ARGs	Possible Microbes/ Microbial Communities	References
Recirculating aquaculture system	PET	*tetG*, *qnrS*, *sul1*, *sul2*, *ermF*	Proteobacteria *Mycobacterium* sp. *Marinobactor* sp. *Pseudophaeobactor* sp. *Alcanivorax* sp.	Lu et al. (2019)
Sewage	PE, PVC	*tetA-02*, *bla*, *tetW*	*Clostridum* sp. *Mycobacterium* sp. *Neisseria* sp. *Legionella* sp. *Arcobactor* sp.	Wang et al. (2021)
Sediments/soil of mangroves	PHA, PET, PE, PFP, PP, PS, PF	*bcrA*, *macB*, *rpoB2*, *vanR*	*Acinetobacter baumannii* *Enterococcus faecium* *Haemophilus influenzae* *Helicobacter pylori* *Klebsiella pneumoniae* *Neisseria gonorrhoeae* *Pseudomonas aeruginosa*	Sun et al. (2021)
Tap water/ground water	PE, PET, PP, PA	*sul1*, *tetA*	*Klebsiella pneumonia* *E. coli* *Enterobacter cloacae* *Pseudomonas aeruginosa*	Bergeron et al. (2017), Gambino et al. (2022)
Synthetic wastewater/ membrane-filtered seawater	PS	*tetA*, *tetB*, *tetC*, *tetX*, *tetO*, *tetQ*, *acrB*, *blaKPC*, *blaCTX-M*, *tetM*, *mdtE*, and *acrB_1* *mexB*, *mexD*, *intI1*, and *intI* *107 ARGs*	*Escherichia coli* *Enterococcus* sp.	Dasí et al. (2024), Zhao et al. (2020)
Wastewater treatment plant	PS	*Antibiotic-resistant plasmid RP4*	*Raoultella ornithinolytica* *Stenotrophomonas maltophilia* Heterotrophic *Novosphingobium* Filamentous *Flectobacillus*	Pham et al. (2021)
Lake water/river water	PE, PVC, PET, PP, PB	*Sul1*, *sul2*, *i-tetA*, *i-TetC*, *iTetO*, *i-sul1* *e-tetA*, *e-bla*	*Flavobacterium* sp. *Rhodoferax* sp. *Pseudomonas* sp. *Janthinobacterium* sp.	Perveen et al. (2022)
Sewage	PE, PVC	*tet,bla*	*Proteobacteria* sp. *Mycobacterium* sp. *Neisseria* sp. *Arcobacter* sp.	Wang et al. (2021)

Abbreviations: PET, polyethylene terephthalate; PE, polyethylene; PVC, polyvinyl chloride; PHA, polyhydroxyalkanoate; PFP, plant fibre plastic; PP, polypropylene; PS, polystyrene; PF, phenol formaldehyde; PA, polyamide; PB, polybutylene.

Cameron, 2019). Therefore, MPs could be described as co-selective agents for AMR (Stevenson et al., 2024). The genetic linkage of antibiotic resistance and virulence genes on multidrug-resistant plasmids can facilitate the co-selection of AMR genes. In biofilms, ARGs can spread to pathogenic bacteria via HGT through conjugation (plasmid-borne ARGs), transformation (free extracellular ARGs), or transduction (phage-transported ARGs) (Tao et al., 2022). The distribution and transfer mechanisms of ARGs within biofilms remain unclear.

Flavobacterium spp. and *Chryseobacterium* spp. frequently carry ARGs on MPs (Merline and Dhinagaran, 2023; Tuvo et al., 2023). There is a significant correlation between MP concentration and the integron integrase class-1 gene, an MGE that promotes ARG transfer (Pham et al., 2021). According to a study, the conjugation rates of ARGs were more significant in the plastisphere than in control wastewater communities (Stevenson et al., 2024) and also higher than in planktonic communities.

Considering transformation, the presence of PS nanoplastics significantly increased the transformation frequency of *E. coli* (Stevenson et al., 2024). Furthermore, transformation frequency was significantly greater in the plastisphere than in free-living and natural particle controls (Stevenson et al., 2024). The increased transformation rate observed in the plastisphere is associated with several factors. These factors include bacterial density, EPS content (such as extracellular DNA (eDNA)), and the expression of genes related to biofilm formation or DNA uptake, such as flagella motility and bacterial adhesion (Stevenson et al., 2024).

The uptake frequency of plasmids in bacterial communities was twice as high in MP biofilms compared to free-living bacteria. The reported HGT on MPs could lead to many MGEs with a diverse ARG composition (Stevenson et al., 2024). Arias-Andres et al. (2018) observed that mobile elements carrying resistance genes could develop faster in biofilms than in planktonic bacterial communities. It additionally provided evidence that MPs could accelerate gene exchange between the biofilms of attached communities and the surrounding water environments (Arias-Andres et al., 2018).

Non-biodegradable MPs like PET have a higher ARG binding capacity than biodegradable MPs such as polyhydroxyalkanoates (PHAs) (Sun et al., 2021). *Pseudomonas* spp. with multidrug or glycopeptide-resistance genes was found on PET and PHA. The *Desulfovibrio* spp. resistant to macrolide–lincosamide–streptogramin was mainly isolated on PET (Tuvo et al., 2023). The environmental ageing of MPs, caused by mechanical abrasion, solar radiation, and biodegradation, enhances their ability to bind ARG-carrying bacteria or free-floating genetic materials. UV-C radiation exposure for 5–20 days increases the pore size and total volume of MPs, and it promotes ARG binding, biofilm formation, and gene exchange (Tuvo et al., 2023). However, releasing chemical compounds from MPs can alter microorganism membrane permeability, facilitating ARG transfer. Additives or pollutants on MPs, such as copper and zinc, may also favour ARG binding (Trevisan et al., 2022). These pollutants and polycyclic aromatic hydrocarbons apply selective pressure on ARG transfer through co-selection or cross-selection (Tuvo et al., 2023).

MPs may transport pathogens and ARGs in the aquatic environment (Gaylarde et al., 2023). In marine aquacultures, ARB on MP surfaces is 100–5,000 times higher than in the surrounding water (Zhang et al., 2020). MPs may not solely serve as vehicles for microbial pathogens; they could enrich pathogenic strains that have acquired pathogenicity islands (PAIs) and other antimicrobial properties through HGT (Bowley et al., 2021).

18.6.2 Plastispheres as Hotspots for AMR in Soil

In parallel, recent research has highlighted that plastispheres, which form on MPs in soil environments, are critical hotspots for AMR. These plastispheres show a significantly higher abundance of ARGs compared to the surrounding soils. Long-term exposure to plastic mulch applications in Chinese soils has led to an increase in antibiotic resistance. The researchers discovered the presence of ARGs and thousands of MGEs, including integrons, insertions, and plasmids (Wu et al., 2019).

These findings illustrate the significant environmental and health risks posed by the interaction of MPs with pathogenic and ARBs. In the terrestrial environment, plastisphere communities comprised 12.4 times more potential pathogens than ambient soil (Stevenson et al., 2024). In the agricultural soil, environmental factors like manure addition, elevated soil temperatures, and increased moisture levels provide a nutrient-rich environment and a protective biofilm structure; within plastispheres, it creates an ideal setting for the proliferation and exchange of ARGs through HGT. All of which have been shown to significantly enhance ARG concentrations within plastispheres. These findings underscore the role of plastispheres as reservoirs of ARGs, which poses a substantial risk to soil health and ecosystem balance (Zhu et al., 2021). Doxycycline (DOX), when combined with manure, significantly enhances biofilm formation on MPs within the soil plastisphere, with low doses of DOX increasing biofilm concentration and the correlation between ARGs and biofilm development while also boosting the abundance of potential pathogens on MPs and promoting biofilm production of MP-associated strains (Liu et al., 2023).

18.6.3 The Co-occurrence of Heavy Metals and Antibiotics in Microplastics

Due to their hydrophobic nature, MPs provide extensive surfaces for the adsorption of hydrophobic pollutants, altering their bioavailability and chemical properties. Current research indicates that concentrations of heavy metals and organic contaminants on MP surfaces can be up to 10^6 times higher than in the surrounding environment (Campanale et al., 2020). Creating combined toxicity as chemicals adhered to MP biofilms could be released during natural weathering processes. The heavy metals, antibiotics, and other xenobiotics make MPs potential carriers of pollution and multidrug resistance in humans.

There is a higher prevalence of metal resistance and multidrug-resistance genes in bacteria isolated from MPs than in free-living strains (Piergiacomo et al., 2023) and these contaminated particles can enter humans indirectly. People may consume seafood such as crustaceans, bivalves, fish, or sea salt containing MPs that transmit antibiotic-resistant pathogens and metal-driven multi-resistances. Approximately 80% of fish have been found to contain MPs in their stomachs, which can cause tissue necrosis, inflammation, cell necrosis, cytotoxic complications, and oxidative stress in humans.

In one study, PVC MPs pre-adsorbed with copper and/or tetracycline were incubated in an artificially activated sludge system (Stevenson et al., 2024). The antimicrobials selectively enriched ARGs in the plastisphere and surrounding sludge. Additionally, exposure to PVC leachate significantly increases the relative abundance of ARGs and virulence genes in a marine bacterial community (Vlaanderen et al., 2023). The hydrocarbonoclastic bacteria, capable of producing plastic-degrading enzymes, have been previously documented in the plastisphere (Delacuvellerie et al., 2019). The adsorption of oxytetracycline to biodegradable MPs is enhanced in the presence of a biofilm, compared to virgin MPs (Sun et al., 2021).

MPs can be vectors for harmful microorganisms, often closely associated with other microbes and sorbed contaminants, including metals. This can lead to MPs functioning as a microcosm for more effective gene exchange between bacteria, potentially facilitating the spread of antibiotic resistance (Bowley et al., 2021). Differences in microbial community composition on plastic buried at various depths suggest that environmental conditions and dominant redox processes control plastic colonization. Organisms with metabolic flexibility are found throughout changing redox conditions of the plastisphere (Dodhia et al., 2023). The variability in microbial communities on plastics may be attributed to thicker biofilm development on plastics in the water column compared to those in the sediment and selective sorption of iron and manganese oxides and minor metals. These metals and minerals can serve as templates for microbial attachment and catalysts for further reactions, influencing the composition of the biofilm (Dong et al., 2022) (Figure 18.2).

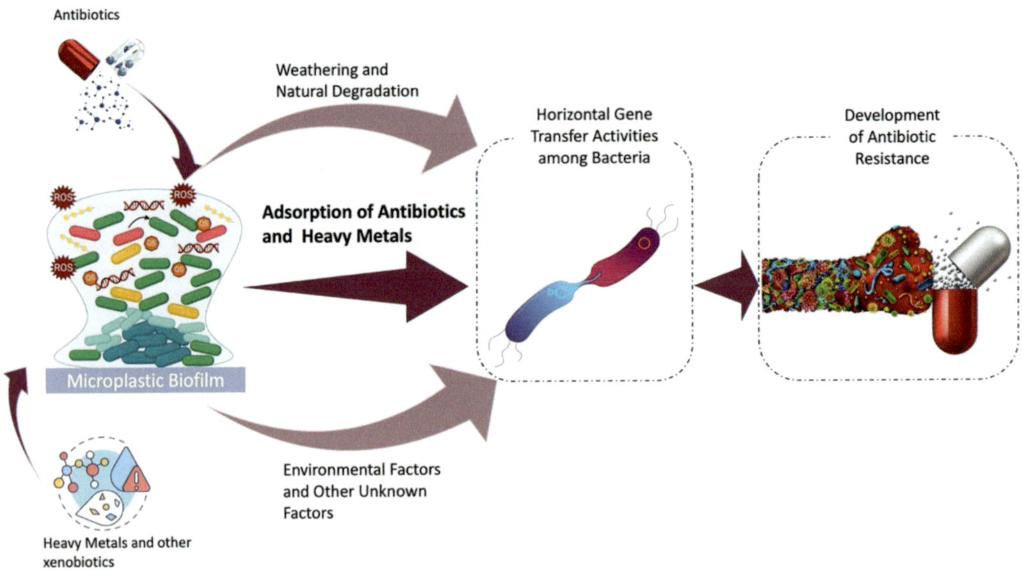

FIGURE 18.2 The interconnection between microplastic biofilm formation and the development and environmental transmission of antibiotic resistance.

18.7 MITIGATION OF ENVIRONMENTAL HEALTH RISKS

Mitigating the environmental health risks of pharmaceuticals adsorbed on MPs and the spread of ARB in natural environments require an interdisciplinary approach. The risks involve comprehensive strategies to manage pollution and control the transport of MPs over large areas. Effective mitigation strategies must address the complex interactions between pharmaceuticals, MPs, and microbial communities (Caban and Stepnowski, 2021).

18.7.1 Source Reduction and Proper Disposal of Pharmaceuticals

The release of pharmaceuticals at the source is the one major factor in concluding pharmaceutical pollution in the environment (Larsen et al., 2004). Minimizing unnecessary pharmaceutical use and promoting proper disposal of unused or expired medications could be practised as preventive measures to reduce pollution.

18.7.2 Implementation of Wastewater Treatment Technologies

Wastewater treatment plants (WWTPs) are supposed to be able to remove pharmaceuticals and MPs from effluent streams before they are discharged into aquatic ecosystems. Progressive treatment technologies include ozonation, activated carbon adsorption, membrane filtration, and UV disinfection (Khasawneh and Palaniandy, 2021).

However, conventional WWTPs may not be adequately equipped to effectively remove pharmaceuticals and MPs, as these compounds can persist through treatment and enter receiving waters at relevant concentrations. To address this challenge, upgrades and optimization of existing treatment facilities are needed, as well as the development of novel treatment technologies specifically designed to target pharmaceuticals and MPs.

18.7.3 Development of Environmentally Friendly Alternatives to Conventional Plastics

Reducing the accumulation of MP in aquatic habitats requires a shift toward using environmentally friendly alternatives to conventional plastics. Biodegradable polymers, plant-based plastics, and other sustainable materials offer potential solutions for reducing plastic pollution (Song et al., 2009). However, it is essential to ensure that alternative materials are environmentally sustainable and do not pose unintended consequences, such as increased resource consumption or habitat disruption.

18.7.4 Promotion of Responsible Antibiotic Use Practices

The proliferation of ARB and ARG in aquatic environments is mainly due to improper healthcare, agriculture, and aquaculture antibiotic practices. Antibiotic overuse contributes to the emergence and spread of antibiotic resistance, creating selective pressures that promote the survival of resistant bacterial strains (Chokshi et al., 2019).

Public health campaigns, antibiotic awareness programs, and regulatory measures can help promote appropriate antibiotic prescribing practices and minimize unnecessary antibiotic use. In agricultural and aquaculture settings, alternative disease management strategies, such as vaccination, biosecurity measures, and probiotic supplements, can reduce the reliance on antibiotics and mitigate the risk of AMR in food production systems (Adorka et al., 2013).

18.7.5 Continued Research and Monitoring

Continued research and monitoring are essential for improving our understanding of pharmaceuticals, ARGs, and ARBs about MPs, transport, and ecological impacts on the environment. Long-term monitoring programs with field data can track changes in pollutant concentrations, microbial community dynamics, and antibiotic resistance patterns over time, providing a valuable understanding of the effectiveness of mitigation measures. Interdisciplinary research collaborations are needed to address knowledge gaps and develop innovative solutions.

18.8 CONCLUSION

MPs contribute to ecosystems by providing substrates for colonizing microbes in natural environments. This chapter highlights that PS can act as a vector for ARB, thereby causing the emergence and spread of human diseases. The adsorption of antibiotics on MPs facilitates the persistence and spread of ARG, which enhances the frequency of plasmid transfer among PS-associated bacteria with the help of unique features of biofilm formation around them. MP shows substantial abundance and potential pathogen diversity compared to biofilm formed around natural materials such as rocks and wood. These unique features resulted from higher relative proportions of chemical and physical characteristics of MPs. These findings underscore that PS acts as living vectors, facilitating the movement and persistence of pathogenic bacteria within ecosystems and across communities, thus posing substantial risks to human health.

ACKNOWLEDGEMENT

We are expressing our sincere appreciation to Mr. Yasasa Mendis for the technical support in preparing the figures.

REFERENCES

Adorka, M., Honoré, M.K., Lubbe, M., Serfontein, J., Allen, K., (2013). The impact of appropriate antibiotic prescribing on treatment evaluation parameters. *Journal of public health in Africa*, 4(1), p. e2.

Arias-Andres, M., Klümper, U., Rojas-Jimenez, K., Grossart, H.P., (2018). Microplastic pollution increases gene exchange in aquatic ecosystems. *Environ Pollut.* (237), pp. 253–261. https://doi.org/10.1016/j.envpol.2018.02.058

Ashokkumar, S., Kaushik, N.K., Han, I., Uhm, H.S., Park, J.S., Cho, G.S., Oh, Y.J., Shin, Y.O., Choi, E.H., (2023). Persistence of coronavirus on surface materials and its control measures using nonthermal plasma and other agents. *Int J Mol Sci.* (24), p. 14106. https://doi.org/10.3390/ijms241814106

Bergeron, S., Raj, B., Nathaniel, R., Corbin, A., LaFleur, G., (2017). Presence of antibiotic resistance genes in raw source water of a drinking water treatment plant in a rural community of USA. *Int Biodeterior Biodegradation.* (124), pp. 3–9. https://doi.org/10.1016/j.ibiod.2017.05.024

Bowley, J., Baker-Austin, C., Porter, A., Hartnell, R., Lewis, C., (2021). Oceanic hitchhikers – Assessing pathogen risks from marine microplastic. *Trends Microbiol.* (29), pp. 107–116. https://doi.org/10.1016/j.tim.2020.06.011

Caban, M., Stepnowski, P., 2021. How can pharmaceuticals be decreased in the environment? A review. *Environ Chem Lett.* (19), pp. 3115–3138. https://doi.org/10.1007/s10311-021-01194-y

Cai, L., Wu, D., Xia, J., Shi, H., Kim, H., (2019). Influence of physicochemical surface properties on the adhesion of bacteria onto four types of plastics. *Sci Total Environ.* (671), pp. 1101–1107. https://doi.org/10.1016/j.scitotenv.2019.03.434

Campanale, C., Dierkes, G., Massarelli, C., Bagnuolo, G., Uricchio, V.F., (2020). A relevant screening of organic contaminants present on freshwater and pre-production microplastics. *Toxics.* (8), pp. 1–21. https://doi.org/10.3390/toxics8040100

Carniello, V., Peterson, B.W., van der Mei, H.C., Busscher, H.J., (2018). Physico-chemistry from initial bacterial adhesion to surface-programmed biofilm growth. *Adv Colloid Interface Sci.* (261), pp. 1–14. https://doi.org/10.1016/j.cis.2018.10.005

Chen, Y., Shen, Z., Li, G., Wang, K., Cai, X., Xiong, X., Wu, C., (2022). Factors affecting microplastic accumulation by wild fish: A case study in the Nandu River, South China. *Sci Total Environ.* (847), pp. 157486. https://doi.org/10.1016/J.SCITOTENV.2022.157486

Chokshi, A., Sifri, Z., Cennimo, D., Horng, H., (2019). Global contributors to antibiotic resistance. *J Glob Infect Dis.* (11), pp. 36–42. https://doi.org/10.4103/jgid.jgid_110_18

Cholewińska, P., Moniuszko, H., Wojnarowski, K., Pokorny, P., Szeligowska, N., Dobicki, W., Polechoński, R., Górniak, W., (2022). The occurrence of microplastics and the formation of biofilms by pathogenic and opportunistic bacteria as threats in aquaculture. *Int J Environ Res Public Health.* (19), pp. 8137. https://doi.org/10.3390/ijerph19138137

D'agata, E., Mitchell, S.L., (2008). Patterns of antimicrobial use among nursing home residents with advanced dementia. *Arch Intern Med.* 168(4), pp. 357. https://doi.org/10.1001/archinternmed.2007.104.

Dasí, D., Camaró-Sala, M.L., González, A., García-Ferrús, M., Jiménez-Belenguer, A.I., Castillo, M.Á., (2024). Antibiotic resistance in seawater samples from east coast of Spain. *Appl Sci.* (14), pp. 1965. https://doi.org/10.3390/app14051965

Delacuvellerie, A., Cyriaque, V., Gobert, S., Benali, S., Wattiez, R., (2019). The plastisphere in marine ecosystem hosts potential specific microbial degraders including *Alcanivorax borkumensis* as a key player for the low-density polyethylene degradation. *J Hazard Mater.* (380), pp. 120899. https://doi.org/10.1016/j.jhazmat.2019.120899

Didier, D., Anne, M., Alexandra, T.H., (2017). Plastics in the North Atlantic garbage patch: A boat-microbe for hitchhikers and plastic degraders. *Sci Total Environ.* (599–600), pp. 1222–1232. https://doi.org/10.1016/j.scitotenv.2017.05.059

Dodhia, M.S., Rogers, K.L., Fernández-Juárez, V., Carreres-Calabuig, J.A., Löscher, C.R., Tisserand, A.A., Keulen, N., Riemann, L., Shashoua, Y., Posth, N.R., (2023). Microbe-mineral interactions in the Plastisphere: Coastal biogeochemistry and consequences for degradation of plastics. *Front Mar Sci.* (10), pp. 1134815. https://doi.org/10.3389/fmars.2023.1134815

Dong, H., Huang, L., Zhao, L., Zeng, Q., Liu, X., Sheng, Y., Shi, L., Wu, G., Jiang, H., Li, F., Zhang, L., Guo, D., Li, G., Hou, W., Chen, H., (2022). A critical review of mineral-microbe interaction and co-evolution: mechanisms and applications. *Natl Sci Rev.* (9), pp. 10. https://doi.org/10.1093/nsr/nwac128

Du, Y., Liu, X., Dong, X., Yin, Z., (2022). A review on marine plastisphere: Biodiversity, formation, and role in degradation. *Comput Struct Biotechnol J.* (20), pp. 975–988. https://doi.org/10.1016/j.csbj.2022.02.008

Fu, L., Li, J., Wang, G., Luan, Y., Dai, W., (2021). Adsorption behavior of organic pollutants on microplastics. *Ecotoxicol Environ Saf.* (217), pp. 112207. https://doi.org/10.1016/j.ecoenv.2021.112207

Galloway, T.S., Cole, M., Lewis, C., (2017). Interactions of microplastic debris throughout the marine ecosystem. *Nat Ecol Evol.* (1), pp. 0116. https://doi.org/10.1038/s41559-017-0116

Gambino, I., Bagordo, F., Grassi, T., Panico, A., De Donno, A., (2022). Occurrence of microplastics in tap and bottled water: Current knowledge. *Int J Environ Res Public Health.* 19(9), pp. 5283. https://doi.org/10.3390/ijerph19095283

Gassilloud, B., Gantzer, C., (2005). Adhesion-aggregation and inactivation of Poliovirus 1 in groundwater stored in a hydrophobic container. *Appl Environ Microbiol.* (71), pp. 912–920. https://doi.org/10.1128/AEM.71.2.912-920.2005

Gaylarde, C.C., de Almeida, M.P., Neves, C.V., Neto, J.A.B., da Fonseca, E.M., (2023). The importance of biofilms on microplastic particles in their sinking behavior and the transfer of invasive organisms between ecosystems. *Micro* (3), pp. 320–337. https://doi.org/10.3390/micro3010022

Ghosh, S., Sinha, J.K., Ghosh, S., Vashisth, K., Han, S., Bhaskar, R., (2023). Microplastics as an emerging threat to the global environment and human health. *Sustainability (Switzerland).* (15), pp. 10821–10821. https://doi.org/10.3390/su151410821

Gkoutselis, G., Rohrbach, S., Harjes, J., Obst, M., Brachmann, A., Horn, M.A., Rambold, G., (2021). Microplastics accumulate fungal pathogens in terrestrial ecosystems. *Scientific Reports*, 11(1), p. 13214.

Gobet, A., Böer, S.I., Huse, S.M., Van Beusekom, J.E.E., Quince, C., Sogin, M.L., Boetius, A., Ramette, A., (2012). Diversity and dynamics of rare and of resident bacterial populations in coastal sands. *ISME J.* (6), pp. 542–553. https://doi.org/10.1038/ismej.2011.132

Huang, D., Li, X., Ouyang, Z., Zhao, X., Wu, R., Zhang, C., Lin, C., Li, Y., Guo, X., (2021). The occurrence and abundance of microplastics in surface water and sediment of the West River downstream, in the south of China. *Science of the Total Environment*, 756, p. 143857.

Itzhari, D., Shuai, W., Hartmann, E.M., Ronen, Z., (2024). Heterogeneous antibiotic resistance gene removal impedes evaluation of constructed wetlands for effective greywater treatment. *Antibiotics.* 13(4), pp. 315–315. https://doi.org/10.3390/antibiotics13040315

Joo, S.H., Liang, Y., Kim, M., Byun, J., Choi, H., (2021). Microplastics with adsorbed contaminants: Mechanisms and treatment. *Environ Chall.* (3), pp. 100042. https://doi.org/10.1016/j.envc.2021.100042

Kaiser, D., Kowalski, N., Waniek, J.J., (2017). Effects of biofouling on the sinking behavior of microplastics. *Environ Res Lett.* (15), pp. e0228209. https://doi.org/10.1088/1748-9326/aa8e8b

Khasawneh, O.F.S., Palaniandy, P., (2021). Occurrence and removal of pharmaceuticals in wastewater treatment plants. *Process Saf Environ Prot.* (150), pp. 532–556. https://doi.org/10.1016/j.psep.2021.04.045

Lai, K.P., Tsang, C.F., Li, L., Yu, R.M.K., Kong, R.Y.C., (2022). Microplastics act as a carrier for wastewater-borne pathogenic bacteria in sewage. *Chemosphere.* (301), pp. 134692. https://doi.org/10.1016/J.CHEMOSPHERE.2022.134692

Larsen, T.A., Lienert, J., Joss, A., Siegrist, H., (2004). How to avoid pharmaceuticals in the aquatic environment. *J Biotechnol.* 113(1–3), pp. 295–304. https://doi.org/10.1016/j.jbiotec.2004.03.033

Lenaker, P.L., Baldwin, A.K., Corsi, S.R., Mason, S.A., Reneau, P.C., Scott, J.W., (2019). Vertical distribution of microplastics in the water column and surficial sediment from the Milwaukee River Basin to Lake Michigan. *Environ Sci Technol.* (53), pp. 12227–12237. https://doi.org/10.1021/acs.est.9b03850

Lerminiaux, N.A., Cameron, A.D.S., (2019). Horizontal transfer of antibiotic resistance genes in clinical environments. *Can J Microbiol.* (65), pp 34–44. https://doi.org/10.1139/cjm-2018-0275

Liu, S., Deng, W.K., Niu, S.H., Mo, C.H., Liao, X.D., Xing, S.C., (2023). Doxycycline combined manure microbes to enhances biofilm formation of the soil plastisphere and increases the surface bio-risk of microplastics vehicle. *Chem Eng J.* 454, p. 140530. https://doi.org/10.1016/j.cej.2022.140530

Loiseau, C., Sorci, G., (2022). Can microplastics facilitate the emergence of infectious diseases? *Sci Total Environ.* (823), pp. 153694. https://doi.org/10.1016/j.scitotenv.2022.153694ï

Lu, J., Zhang, Y., Wu, J., Luo, Y., (2019). Effects of microplastics on distribution of antibiotic resistance genes in recirculating aquaculture system. *Ecotoxicol Environ Saf.* (184), pp. 109631. https://doi.org/10.1016/j.ecoenv.2019.109631

Magana-Arachchi, D.N., Wanigatunge, R.P. (2023). Microplastics in the atmosphere and their human and eco risks. In: *Microplastics in the Ecosphere: Air, Water, Soil, and Food* Edited by Meththika Vithanage, Majeti Narasimha,and Vara Prasad. John Wiley & Sons Ltd, pp. 97–112.

Martínez-Campos Gutiérrez, S., (2022). Plastics as vector of microorganisms in the aquatic environment. Ph. D. Thesis Universidad de Alcalá. Available at: https://doi.org/10.13140/RG.2.2.35164.87688 (Accessed: 20 May 2024).

Merline, S.A., Dhinagaran, G., (2023). Prevalence of microplastics, antibiotic resistant genes and microplastic associated biofilms in estuary-A review. *Environ Eng Res.* (8), pp. 220325. https://doi.org/10.4491/eer.2022.325

Miao, L., Wang, P., Hou, J., Yao, Y., Liu, Z., Liu, S., Li, T., (2019). Distinct community structure and microbial functions of biofilms colonizing microplastics. *Sci Total Environ.* (650), pp. 2395–2402. https://doi.org/10.1016/j.scitotenv.2018.09.378

Moyal, J., Dave, P.H., Wu, M., Karimpour, S., Brar, S.K., Zhong, H., Kwong, R.W.M., (2023). Impacts of biofilm formation on the physicochemical properties and toxicity of microplastics: A concise review. *Rev Environ Contam Toxicol.* (261), pp. 8. https://doi.org/10.1007/s44169-023-00035-z

MuñozMeneses, R.A., Cabrera-Papamija, G., Machuca-Martínez, F., Rodríguez, L.A., Diosa, J.E., Mosquera-Vargas, E., (2022). Plastic recycling and their use as raw material for the synthesis of carbonaceous materials. *Heliyon.* 8(3), pp. e09028. https://doi.org/10.1016/j.heliyon.2022.e09028

Murphy, A.C., Muldoon, S.F., Baker, D., Lastowka, A., Bennett, B., Yang, M., Bassett, D.S., (2018). Structure, function, and control of the human musculoskeletal network. *PLoS Biol.* 16(1), pp. e2002811. https://doi.org/10.1371/journal.pbio.2002811

Nath, J., De, J., Sur, S., Banerjee, P., (2023). Interaction of microbes with microplastics and nanoplastics in the agroecosystems—Impact on antimicrobial resistance. *Pathogens.* 12(7), pp. 888. https://doi.org/10.3390/pathogens12070888

Ng, E.L., Huerta Lwanga, E., Eldridge, S.M., Johnston, P., Hu, H.W., Geissen, V., Chen, D., (2018). An overview of microplastic and nanoplastic pollution in agroecosystems. *Sci Total Environ.* (627), pp. 1377–1388. https://doi.org/10.1016/j.scitotenv.2018.01.341.

Nguyen, M.K., Lin, C., Nguyen, H.L., Le, V.R., KL, P., Singh, J., Chang, S.W., Um, M.J., Nguyen, D.D., (2023). Emergence of microplastics in the aquatic ecosystem and their potential effects on health risks: The insights into Vietnam. *J Environ Manage.* (344), pp. 118499. https://doi.org/10.1016/J.JENVMAN.2023.118499

Ogonowski, M., Motiei, A., Ininbergs, K., Hell, E., Gerdes, Z., Udekwu, K.I., Bacsik, Z., Gorokhova, E., (2018). Evidence for selective bacterial community structuring on microplastics. *Environ Microbiol.* (20), pp. 2796–2808. https://doi.org/10.1111/1462-2920.14120

Parsaeimehr, A., Miller, C.M., Ozbay, G., (2023). Microplastics and their interactions with microbiota. *Heliyon.* 9(4), pp. e15104. https://doi.org/10.1016/j.heliyon.2023.e15104

Perveen, S., Pablos, C., Reynolds, K., Stanley, S., Marugán, J., (2022). Microplastics in fresh- and wastewater are potential contributors to antibiotic resistance – A minireview. *J Hazard Mater Adv.* (6), pp. 100071. https://doi.org/10.1016/j.hazadv.2022.100071

Pham, D.N., Clark, L., Li, M., (2021). Microplastics as hubs enriching antibiotic-resistant bacteria and pathogens in municipal activated sludge. *J Hazard Mater Lett.* (2), pp. 100014. https://doi.org/10.1016/j.hazl.2021.100014

Piergiacomo, F., Brusetti, L., Pagani, L., (2023). Understanding the interplay between antimicrobial resistance, microplastics and xenobiotic contaminants: A leap towards one health? *Int J Environ Res Public Health.* 20(1), pp. 42. https://doi.org/10.3390/ijerph20010042

Rajcoomar, S., Amoah, I.D., Abunama, T., Mohlomi, N., Bux, F., Kumari, S., (2024). Biofilm formation on microplastics in wastewater: Insights into factors, diversity and inactivation strategies. *Int J Environ Sci Technol.* (21), pp. 4429–4444. https://doi.org/10.1007/s13762-023-05266-0

Semcesen, P.O., Wells, M.G., (2021). Biofilm growth on buoyant microplastics leads to changes in settling rates: Implications for microplastic retention in the Great Lakes. *Mar Pollut* Bull. 170, pp. 112573. https://doi.org/10.1016/j.marpolbul.2021.112573

Siddiqui, R., Luzia Lacerda, A., Briand, J.-F., Lenoble, V., Quadro Oreste, E., Kessler, F., Luiza Pedrotti, M., (2024). Assessing the plastisphere from floating plastics in the Northwestern Mediterranean Sea, with emphasis on viruses. *Microorganisms.* (12), pp. 444. https://doi.org/10.3390/microorganisms

Song, J.H., Murphy, R.J., Narayan, R., Davies, G.B.H., (2009). Biodegradable and compostable alternatives to conventional plastics. *Philos Trans R Soc B Biol Sci.* 364(1526), pp. 2127–2139. https://doi.org/10.1098/rstb.2008.0289

Sooriyakumar, P., Bolan, N., Kumar, M., Singh, L., Yu, Y., Li, Y., Weralupitiya, C., Vithanage, M., Ramanayaka, S., Sarkar, B., Wang, F., Gleeson, D.B., Zhang, D., Kirkham, M.B., Rinklebe, J., M Siddique, K.H., (2022). Biofilm formation and its implications on the properties and fate of microplastics in aquatic environments: A review. *J Hazard Mater Adv.* (6), pp. 100077. https://doi.org/10.1016/j.hazadv.2022.100077

Stevenson, E.M., Buckling, A., Cole, M., Lindeque, P.K., Murray, A.K., (2024). Selection for antimicrobial resistance in the plastisphere. *Sci Total Environ.* (908), pp. 168234. https://doi.org/10.1016/j.scitotenv.2023.168234

Sun, Y., Cao, N., Duan, C., Wang, Q., Ding, C., Wang, J., (2021). Selection of antibiotic resistance genes on biodegradable and non-biodegradable microplastics. *J Hazard Mater.* (409), pp. 124979. https://doi.org/10.1016/j.jhazmat.2020.124979

Trevisan, R., Ranasinghe, P., Jayasundara, N., Di Giulio, R.T., (2022). Nanoplastics in aquatic environments: Impacts on aquatic species and interactions with environmental factors and pollutants. *Toxics.* (10), pp. 326. https://doi.org/10.3390/toxics10060326

Tao, S., Chen, H., Li, N., Wang, T., Liang, W., (2022). The spread of antibiotic resistance genes in vivo model. *Canadian Journal of Infectious Diseases and Medical Microbiology*, (1), p. 3348695.

Tu, C., Chen, T., Zhou, Q., Liu, Y., Wei, J., Waniek, J.J., Luo, Y., (2020). Biofilm formation and its influences on the properties of microplastics as affected by exposure time and depth in the seawater. *Sci Total Environ.* (734), p. 139237. https://doi.org/10.1016/j.scitotenv.2020.139237

Tuson, H.H., Weibel, D.B., (2013). Bacteria-surface interactions. *Soft Matter.* (9), pp. 4368. https://doi.org/10.1039/c3sm27705d

Tuvo, B., Scarpaci, M., Bracaloni, S., Esposito, E., Costa, A.L., Ioppolo, M., Casini, B., (2023). Microplastics and antibiotic resistance: The magnitude of the problem and the emerging role of hospital wastewater. *Int J Environ Res Public Health.* (20), pp. 5868–5868. https://doi.org/10.3390/ijerph20105868

Ventura, E., Marín, A., Gámez-Pérez, J., Cabedo, L., (2024). Recent advances in the relationships between biofilms and microplastics in natural environments. *World J Microbiol Biotechnol.* (40), 220. https://doi.org/10.1007/s11274-024-04021-y

Vidal-Verdú, À., Latorre-Pérez, A., Molina-Menor, E., Baixeras, J., Peretó, J., Porcar, M., 2022. Living in a bottle: Bacteria from sediment-associated Mediterranean waste and potential growth on polyethylene terephthalate. *Microbiologyopen.* (11), pp. 1259. https://doi.org/10.1002/mbo3.1259

Vlaanderen, E.J., Ghaly, T.M., Moore, L.R., Focardi, A., Paulsen, I.T., Tetu, S.G., (2023). Plastic leachate exposure drives antibiotic resistance and virulence in marine bacterial communities. *Environ Pollut.* (327), p. 121558. https://doi.org/10.1016/j.envpol.2023.121558

Wang, Z., Gao, J., Zhao, Y., Dai, H., Jia, J., Zhang, D., (2021). Plastisphere enrich antibiotic resistance genes and potential pathogenic bacteria in sewage with pharmaceuticals. *Sci Total Environ.* (768), p. 144663. https://doi.org/10.1016/j.scitotenv.2020.144663

Welagedara, W., Gunathilaka, H., Lakmali, M., Karunathilaka, R., Jayarathne, I.L., Magana-Arachchi, D., (2023). Determining total bacterial load on Microplastics sampled from Kandy Lake and inlets: A pilot study. In *Proceedings of National Aquatic Resources Research and Development Agency (NARA),* 13th December 2023. http://www.nara.ac.lk/?page_id=13207

Wu, P., Huang, J., Zheng, Y., Yang, Y., Zhang, Y., He, F., Chen, H., Quan, G., Yan, J., Li, T., Gao, B., (2019). Environmental occurrences, fate, and impacts of microplastics. *Ecotoxicol Environ Saf.* (184), p. 109612. https://doi.org/10.1016/j.ecoenv.2019.109612

Yang, H., Chen, G., Wang, J., (2021). Microplastics in the marine environment: Sources, fates, impacts and microbial degradation. *Toxics.* (9), pp. 41. https://doi.org/10.3390/toxics9020041

Zettler, E.R., Mincer, T.J., Amaral-Zettler, L.A., (2013). Life in the "plastisphere": Microbial communities on plastic marine debris. *Environ Sci Technol.* (47), pp. 7137–7146. https://doi.org/10.1021/es401288x

Zhang, P., Liu, Y., Zhang, L., Xu, M., Gao, L., Zhao, B., (2022). The interaction of micro/nano plastics and the environment: Effects of ecological corona on the toxicity to aquatic organisms. *Ecotoxicol Environ Saf.* (243), p. 113997. https://doi.org/10.1016/j.ecoenv.2022.113997

Zhang, Y., Lu, J., Wu, J., Wang, J., Luo, Y., (2020). Potential risks of microplastics combined with superbugs: Enrichment of antibiotic resistant bacteria on the surface of microplastics in mariculture system. *Ecotoxicol Environ Saf.* (187), p. 109852. https://doi.org/10.1016/j.ecoenv.2019.109852

Zhao, H., Hong, X., Chai, J., Wan, B., Zhao, K., Han, C., Zhang, W., Huan, H., (2024). Interaction between microplastics and pathogens in subsurface system: What we know so far. *Water (Switzerland).* (16), pp. 499–499. https://doi.org/10.3390/w16030499

Zhao, X., Yu, Z., Ding, T., (2020). Quorum-sensing regulation of antimicrobial resistance in bacteria. *Microorganisms.* (8), p. 425. https://doi.org/10.3390/microorganisms8030425

Zhu, D., Ma, J., Li, G., Rillig, M.C., Zhu, Y.G., (2021). Soil plastispheres as hotspots of antibiotic resistance genes and potential pathogens. *ISME J.* 16, pp. 521–532. https://doi.org/10.1038/s41396-021-01103-9

19 Plastisphere Nomads: An Overview of Organism Dispersal via Marine Plastic Litter

Peter J. Barry, Amelia Curd, Gabin Droual, and Morgan le Moigne

19.1 INTRODUCTION

19.1.1 The Risk

Millions of tonnes of plastic litter pollute our oceans every year, creating multiple issues for wildlife. These range from entanglement and ingestion through to the release of toxins (Bugoni et al., 2001; Miranda and De Carvalho-Souza, 2016; Thacharodi et al., 2024) but another underappreciated danger to biodiversity exists – a new pathway for introduced species. Species usually evolve in isolation or within broadly isolated communities, separated from other ecosystems by natural barriers like geographic distance or unsuitable environmental conditions in between (Luiz et al., 2011; Blackburn et al., 2011; Hagen et al., 2024). If a species is transported across these barriers and enters a new environment, as a result of human activities (either intentional or unintentional), it is recognised as an *introduced* species and since it now exists in a new environment where it did not evolve, it is considered a *non-native species* (NNS) (Soto et al., 2024). If this introduced NNS is able to settle and start to spread and that impacts the receiving ecosystem, it is recognised as an invasion and the NNS as an *invasive* species. Invasive species are defined as being both introduced beyond their natural range through human activities *and adversely impacting people and nature* (Roy et al., 2024). In the manner of other pollution (i.e. physicochemical pollution), some authors recognise these invasive species as a form of 'biological pollution', which can threaten ecosystem health in the recipient ecosystem (Olenin et al., 2024).

Humans moving plants and animals as a commodity, or accidental translocation through contamination are recognised invasion mechanisms and pathways (Turbelin et al., 2022). Control measures are put in place to mitigate such accidental organism transfers for many of the 44 pathway subcategories identified by the Convention on Biological Diversity (CBD, 2014). Marine plastic pollution is a mode of accidental transfer and falls under two pathway subcategories: as a contaminant of 'habitat material' and as an enabler of secondary dispersal. The first step in any invasion process is the movement or translocation of species (Blackburn et al., 2011), which requires a transport mechanism. To understand how plastic serves as a vector for introducing NNS, this chapter will cover the natural movement of aquatic species across barriers, then explore how plastic pollution has increased movement opportunities and created a new mobile habitat on an oceanic scale (Barnes et al., 2009; Audrézet et al., 2021).

 DOI: 10.1201/9781032717814-19

19.1.2 Rafting in the Marine Environment

In the marine environment, population spread can occur via natural forces such as oceanic currents that move organisms or their larvae. Species with indirect development, where offspring go through a larval stage, are often limited to dispersal by larval drift in the water column, which can only transport them for a few hundred kilometres at most (Watts et al., 1998; D'Aloia et al., 2015). Sessile (i.e. permanently attached to a substrate) benthic organisms can sometimes artificially expand their range by attaching to and hitchhiking on 'rafts' of buoyant materials (Jokiel, 1990; Póvoa et al., 2021). These rafts comprise natural or artificial materials and are usually large, solid or semi-solid objects that allow organisms to settle on them but are still free-floating and mobile (Thiel and Gutow, 2005). Natural rafts have traditionally been composed of seaweed clumps, fallen trees or even small buoyant materials such as cuttlebones or lumps of pumice or tar (Minchin, 1996; Whitehead et al., 2011; Bryan et al., 2012). Organisms that reach such rafts can survive much longer in open water than those that drift in the plankton alone (Scheltema, 1977; Póvoa et al., 2021). For some animals that don't normally rely on ocean currents and brood their young to a crawling stage, transport via these rafts is a novel dispersal opportunity, offering far greater reach than they could have achieved unaided (Highsmith, 1985; Jokiel, 1990; Soares et al., 2023).

Biotic rafts are normally ephemeral and rarely survive trans-oceanic drifts. Objects like logs risk becoming waterlogged, overgrown by heavy fauna, degraded by the action of the rafters (e.g. bored by shipworms) or broken up by wave action which ultimately causes them to sink (Thiel and Gutow, 2005). Once sunk, there is little chance that they will be resuspended. Recently, however, degradable biotic rafts are being supplanted by buoyant plastic waste, which forms a much more durable rafting platform (Figures 19.1 and 19.2). The common plastic polymers – polyethylene (PE) and polypropylene (PP) – are less dense than seawater and therefore float (Andrady, 2015). Conventional plastic products are designed to be non-biodegradable, so drifting plastic persists over a far greater timescale than biotic rafts. Attaching organisms will not seriously undermine the integrity of the plastic raft, and if sunk either by currents or by the weight of attached organisms (Ryan, 2015) it can still resurface (Ye and Andrady, 1991; Pinochet et al., 2024), even from great depths (van Sebille et al., 2020). With plastic rafts lasting longer, the net effect is a great expansion among the range and diversity of hitchhiking species in our oceans (Gregory, 2009). For example, coastal species that were typically restricted to areas close to land are now being found hundreds of miles offshore, living on artificial islands of anthropogenic waste (Haram et al., 2023) or some animals that never adapted to rafting have been found clinging to pieces of plastic, far beyond their natural range (e.g. Tutman et al., 2017).

19.1.3 Multiple Plastisphere Vectors

Vectors are transfer mechanisms for the introduction and spread of invasive species (Ruiz and Carlton, 2003; Sinclair et al., 2020). Plastic pollution washes up on beaches all around the world (e.g. Nelms et al., 2017; Gracia et al., 2018; Rodrigues et al., 2019) and while this type of plastic pollution is the most visible vector, it is not the only one. Plastic vectors might be separated into three broad categories (Figure 19.3): drifting including floating objects (all macro-litter visible from the surface); subsurface (all macro-litter invisible from the surface but with a neutral or negative buoyancy) including seafloor plastic; and beached plastic litter (flotsam which has been washed up on a beach, but may yet be washed back out to sea). A fourth category of micro-litter could be added but is outside the remit of this chapter and is not treated in detail here. Please see Chapters 1 and 2 of this book for more details on micro-litter.

Drifting and beached plastic is the most visible (and studied), but subsurface transport, though harder to monitor, also occurs. Subsurface transfers can happen in two ways: either organisms are directly carried on sunken litter that is remobilised into the water column (Figure 19.3) or they transfer their larvae between fixed items embedded in seafloor sediments. Some seafloor litter items can still be moved by currents (tumbleweeding along the seafloor) (Mantelatto et al., 2020) picking up other epifauna as they wash along (De Wolff, 2017). If resuspended fully, the plastic can re-enter

FIGURE 19.1 Fishing buoy washed up on a beach in Vendée, France in February 2024. Among the goose barnacles (*Lepas* spp. Linnaeus, 1758), there were 17 specimens of the truly pelagic Columbus crab *Planes minutus* (Linnaeus, 1758), (see inset) consisting of 8 females, 6 males and 3 juveniles. (Credit: Luc Chaillot.)

surface waters and get pushed onto the shore or washed back out to sea. Normally, only the beached category is visible when it arrives on the shoreline, but it could have had a long life cycle of transport between open sea and seafloor or beach to beach prior to this stage, picking up and transporting organisms throughout. Thus, the three main plastic macro-litter vectors are interlinked.

Organisms travelling with anthropogenic objects is not a novel concept, with a long history of fouling species attaching to vessel hulls (Seebens et al., 2013). The difference is that those artificial objects (i.e. boats) usually travel along established routes from port to port and there are numerous efforts and recommendations in place to control that vector (International Maritime Organization, 2023). In contrast, with tonnes of plastic drifting freely in the world's oceans (Eriksen et al., 2014),

FIGURE 19.2 Bait pot with encrusting animals and containing a community of invertebrates inside. Inset shows *Eoacmaea pustulata* (Helbling, 1779), a Caribbean intertidal limpet that was found attached to the inside of the pot. Recovered on Weymouth Beach, England in October 2023. (Credit: Peter Barry.)

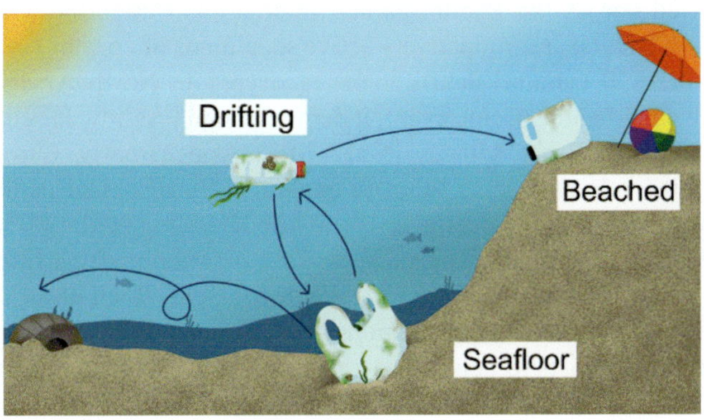

FIGURE 19.3 A schematic drawing explaining the three main plastic macro-litter vectors found in the marine environment.

the plastic vector is not confined to established shipping routes and goes largely unmonitored. At present, there is little understanding of the scale and scope of what could be transported via this vector but what is known will be reviewed in this chapter.

19.2 MARINE FAUNA AND THE PLASTISPHERE: A TIMELINE OF INVESTIGATIONS

To fully grasp the impact of plastic pollution on species' movement, it's essential to delve into how awareness of plastic as a vector has evolved over time. Ever since plastic was first produced at scale and waste plastic started entering the world's oceans – with current estimates of plastic particles load now in the trillions (van Sebille et al., 2015) – plants and animals that had evolved to settle on natural substrates began to colonise this novel surface area (Zhao et al., 2021). With a great proliferation of plastic entering the oceans from the 1950s onward (Audrézet et al., 2021), there was a lag period, with the first reports of marine organisms settling on plastic debris emerging in the following decades.

In the early 1970s, diatoms and hydroids were first reported as living on plastic waste floating in the Sargasso Sea (Carpenter and Smith, 1972). A few years later, plastic pellets with bryozoan colonies were detected washed up on the shores of New Zealand, including one with a *Membranipora tuberculata* colony. That bryozoan species was unknown from New Zealand waters at the time (Gregory, 1978). By the 1980s and 1990s, it was clear that more and more sessile organisms were quickly adapting to the plastic substrate (Harms, 1990; Ye and Andrady, 1991). Reports from around the world confirmed a range of fauna on plastic waste, with data provided mostly from debris on beaches (Winston et al., 1997). It wasn't until the early 2000s that the surprising discovery was made that healthy bryozoan colonies could be found in open water thousands of miles from their source and many degrees of latitude beyond their native range, attached to floating plastic (Barnes, 2002).

With a growing number of large-distance transport events being reported (Barnes and Fraser, 2003; Lewis et al., 2005), a marine plastic vector started to be suspected for some erratic 'jump' dispersal events (Mineur et al., 2010), when species appear on both sides of a natural barrier, such as the open ocean. However, with only sporadic reports over the following years, floating plastic was seldom listed as a viable pathway for species introductions. Dispersal by plastic flotsam was ranked far below other well-studied pathways such as vessel fouling and aquaculture in many invasive species lists (e.g. Haydar, 2010; Minchin et al., 2013). Then, a dramatic event changed perceptions of what could raft and just how far they could travel.

The 2011 Tohoku earthquake off Japan and subsequent tsunami washed tons of debris across the Pacific and onto the west coast of North America. The diversity of debris pieces and attached organisms they carried across the Pacific gained international attention (Carlton et al., 2017; Therriault et al., 2018). Diverse, reproductively active specimens of coastal Japanese species were transported among the 630 tsunami debris items recovered in North America (Carlton et al., 2017). Although an isolated and extreme event, it demonstrated just how far coastal species could actually travel (>7,000 km; Clarke Murray et al., 2018). Concurrently, European studies have started investigating the role of marine litter, to explain NNS dispersal patterns in the eastern Mediterranean (Katsanevakis and Crocetta, 2014). While Kiessling et al. (2015) published a summary of what could be found in marine plastics, the link between drifting litter and NNS wasn't fully realized until a crucial study appeared the following year. Rech et al. (2016) warned that lost aquaculture gear could be the vector responsible for dispersing pest species, and this alert prompted conservation biologists to begin investigating where biofouled plastic was washing up and what was on it.

Data from beach surveys then started to be compiled, detailing rich communities on plastic waste (see Mghili et al., 2023 for a review). Benthic communities on seafloor litter were also investigated, albeit by relatively few surveys (e.g. Masó et al., 2016; Crocetta et al., 2020; Mancini et al., 2021), possibly because of the cost and difficulties in collection. Since there is a high likelihood that focusing on beached litter alone could provide a distorted picture (see below), drifting litter in open water is now being targeted (Canning-Clode et al., 2021).

Finally, a more recent example of a large-scale pollution event resulted from the COVID-19 pandemic, which led to a dramatic increase in the use of plastic-based personal protective equipment (PPE) worldwide. In 2020, it was estimated that 1.5 billion protective masks ended up in the oceans (Mghili et al., 2023). The plastic structure of these masks makes them suitable candidates for the development of specific marine microbial communities and for the transport of macrofaunal NNS (e.g. Mohamadi et al., 2022), so many more examples of organisms rafting on PPE are expected.

19.3 WHAT IS ATTACHING TO PLASTIC? WHICH MACROFAUNA FORM THE PLASTISPHERE NOMADS?

Often the first life form to attach to marine plastic is microorganisms that form a biofilm. The development of biofilms is a critical factor in the settlement of some macrofauna larvae (Li et al., 2016). The advent of scanning electron microscopy and molecular analysis is driving further investigation into the primary microorganism community that attaches first and forms the biofilm (Audrézet et al., 2021). There is a field of ecology that studies the transport of pathogens which colonise microplastics (Goldstein et al., 2014; Ormsby et al., 2024; Magana-Arachchi, Chapter 24) some of which are even able to digest plastic (review made by Shaji et al., 2024; Jayasanka, Chapter 21), but these are beyond the scope of this chapter. There is no consensus yet on whether the biofilm that attaches first is driven by substrate texture, chemical signal or other factors (Oberbeckmann and Labrenz, 2020); however, once it develops, biofilm presence has been observed to facilitate secondary settlement by other larger organisms (Hadfield, 2011).

In a typical colonisation sequence (see Thiel and Gutow, 2005), larger macrofauna then settle and provide new micro-habitats for smaller animals, bringing more complexity to the plastic surface and thus making it available for nestling organisms that occupy gaps and crevices (Enrichetti et al., 2021). For example, cementing bivalves or bryozoans (referred to as 'pioneer' or 'foundation species' (Floerl et al., 2004) will form the basal layers of the community and many 'strata' can build up on top of these (Wood et al., 2012). Other epizoic species are attracted by the chemical signals the pioneers release and settlement progresses in successive layers (Leonard, 2015). On the upper layers, the immediate interaction between plastic surface and settlers is bypassed, so restrictions imposed by surface area or reaction to texture/polymers can be overcome as epifauna accrete and turn essentially a flat 2D plastic into a complex 3D habitat (Thiel and Gutow, 2005).

With the increasing number of datasets containing a repeated cohort of the same broad taxonomic groups, ecologists have begun to explore whether differences in litter composition and the ensuing exposure to plastic leachates, size, shape, flexibility and texture influence which species settle on what (De-la-Torre et al., 2021; Barry et al., 2024). A rough or rugose texture can create micro-habitats that will attract larval settlement of certain species (Brewer, 1984; Zheng et al., 2021; Hirose and Sensui, 2021), while other groups such as bryozoans appear to favour smooth surfaces such as glass (Gracia et al., 2018). It must be noted that weathering of plastic at sea can degrade the surface layers (Ryan, 2015) and therefore alter the surface chemical composition and rugosity, which could also drive settlement patterns (Chase et al., 2016). If a link could be discovered between high-risk NNS associated with a certain litter type then that combination could be prioritised in order to help focus mitigation efforts (De-la-Torre et al., 2021; Abel et al., 2023). However, the diverse nature of plastic waste, with a multitude of complex shapes, erosion stages and different communities presents a particularly complex problem (Fernandez et al., 2022).

Until there is an understanding of what drives these associations, data collection must continue by cataloguing species most often found on plastic macro-litter (see García-Gómez et al., 2021). Items with different characteristics attract different communities: neustonic, intertidal or subtidal benthic organisms with their own variability in their dependence on the plastic. For example, epifaunal communities on seafloor plastics are usually more diverse since it may have been fixed in place for longer, allowing time for a denser or mature community to develop (Katsanevakis et al., 2007; Crocetta et al., 2020). Conversely, beached litter should be treated with caution, since anything mobile could leave the plastic once it beaches, leaving just sessile organisms for investigators to collect (Winston et al., 1997).

19.4 PLASTISPHERE NOMAD LIFE TRAITS

In a recent review, De-la-Torre et al. (2023), gathered macrofaunal data from across 39 studies, covering mostly rafting communities washed up on beaches or collected in the open ocean, so the data was predominantly rafting species. The top four macrofaunal groups were Thecostraca (barnacles), Polychaeta, Bivalvia and Bryozoa, particularly the bryozoan Class Gymnolaemata. García-Gómez et al. (2021) highlighted that most examples of this *common cohort* were united in having cosmopolitan distributions, suspension feeding strategies and were sessile with direct larval development.

The nature of this cohort suggests that the difficulties imposed by rafting would select for hard-shelled encrusting filter-feeders or nestlers that can find shelter among the larger species (Mesaglio et al., 2021), but another well-represented group were the Cnidarians (De-la-Torre et al., 2023), including anemones and hydroids which repeatedly occur on plastic macro-litter. While not hard-shelled, they are often found in folds or corners of plastic packaging (Barry et al., 2024), which provide shelter (Wilson et al., 2021). Among the most frequently observed are intertidal species with physiological adaptations that help them withstand the harsh environment of the open ocean, such as a resistance to broad temperature fluctuations, intense solar radiation and physical damage caused by weather or currents (i.e. storms) (Bravo et al., 2011). The neopelagic community described by Haram et al. (2023) on the floating plastics of the 'Great Pacific garbage patch' in the North Pacific subtropical gyre revealed that 70.5% of the waste they collected was colonised by coastal taxa. Most of these could perform asexual reproduction, which may explain their long-term persistence on the rafts (over several years and very long distances). Therefore, further coastal species migrating from one coast to another is likely, increasing the likelihood of them encountering favourable conditions upon arrival. Far fewer studies have examined organisms colonising seafloor litter (but see Madricardo et al., 2020; Barry et al., 2024), so further work is needed on this type of plastic dispersal vector and their associated communities.

19.5 WHAT TYPE OF LITTER IS COLONISED?

Since larvae have been found on a diverse range of plastics (Pinochet et al., 2020), there is no appreciable limit to what this life history stage can exploit. All sizes of plastic are colonised along an increasing size scale (Figure 19.4) from microplastics (<1 mm) (Oberbeckmann and Labrenz, 2020;

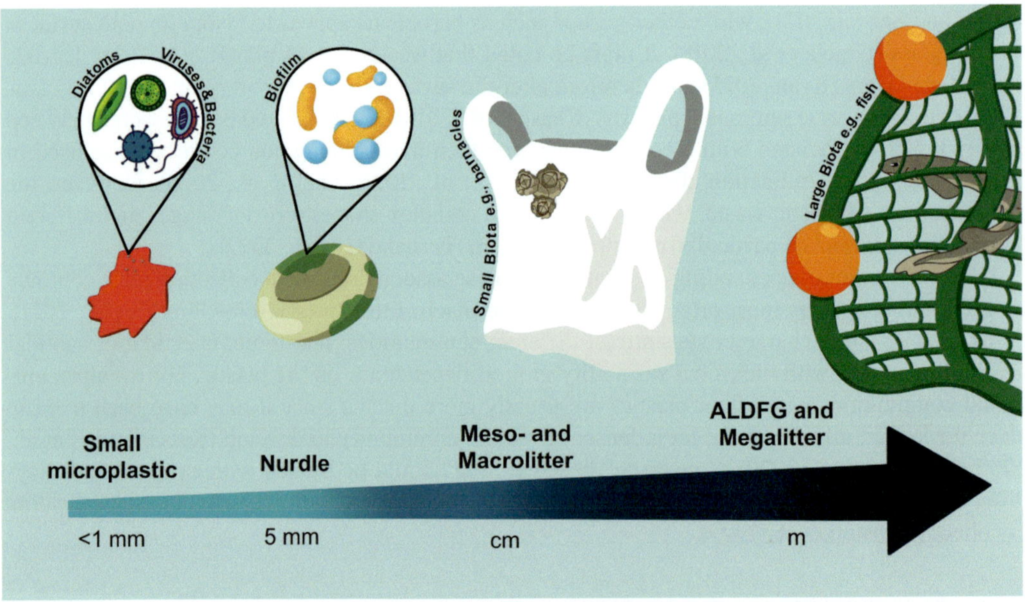

FIGURE 19.4 Infographic illustrating the broad span in nomadic organisms and plastic litter sizes in the growing realm of the plastisphere. From left to right: microplastic fragments colonised by viruses and bacteria, a nurdle colonised by bacterial/phytoplankton biofilm, a submerged plastic bag colonised by macrofauna (e.g. barnacles), a 'netball' – an ocean tumbleweed of lost fishing nets and lines.

Kreczak et al., 2021), agglomerations of small fragments (2 mm), fragments of fishing line (3 mm in length) (Sempere-Valverde et al., 2024) and nurdles (5 mm in diameter) (Rodrigues et al., 2019) to meso- and macro-litter (Turner et al., 2021), all the way up to small, abandoned fibreglass vessels (Turk, 1988).

Abandoned, Lost or otherwise Discarded Fishing Gear (ALDFG) can harbour a diverse biota as well as continuing to ghost fish (Kiessling et al., 2015). Several NNS have been detected on ALDFG and aquaculture material (Holmes et al., 2015; Campbell et al., 2017). Some studies (e.g. Goldstein et al., 2014; Shabani et al., 2019) have found that larger items host more diverse communities, but again, this is dependent on drift duration and community succession state (Tsikhon-Lukanina et al., 2001).

19.6 WHERE ARE THEY FOUND?

Plastic marine litter is now being found in a range of environments including beaches close to urban areas (Barnes et al., 2009), near riverine systems (Rech et al., 2014), coastal areas or even deep-sea trenches (Abel et al., 2023; Xie et al., 2024). There is an estimated 80:20 split between the amount of litter coming from terrestrial runoff and litter generated at sea through maritime activities (Jambeck et al., 2015), so the majority of plastic entering the sea will have passed through a coastal area. This increases the likelihood that coastal species (including NNS) will be picked up and transported to other coastlines (Miralles et al., 2018). Most of the litter is terrestrial in origin and introduced to the marine system through rivers (Vriend et al., 2020; Wilson et al., 2021), while some are discarded or lost items from aquaculture (Campbell et al., 2017). Further offshore, a large proportion of the litter will be ALDFG (Ivkić et al., 2019; Battaglia et al., 2019; Enrichetti et al., 2021).

While plastic pollution is a global crisis, limited investigations focused on specific regions hinder our ability to fully understand the global scope of the problem. Grouping 52 surveys conducted worldwide (after De-la-Torre et al., 2023) it is clear that the majority were concentrated in either Europe (including the Mediterranean) or South America, with relatively few in Asia (Figure 19.5), although the number has recently started to increase.

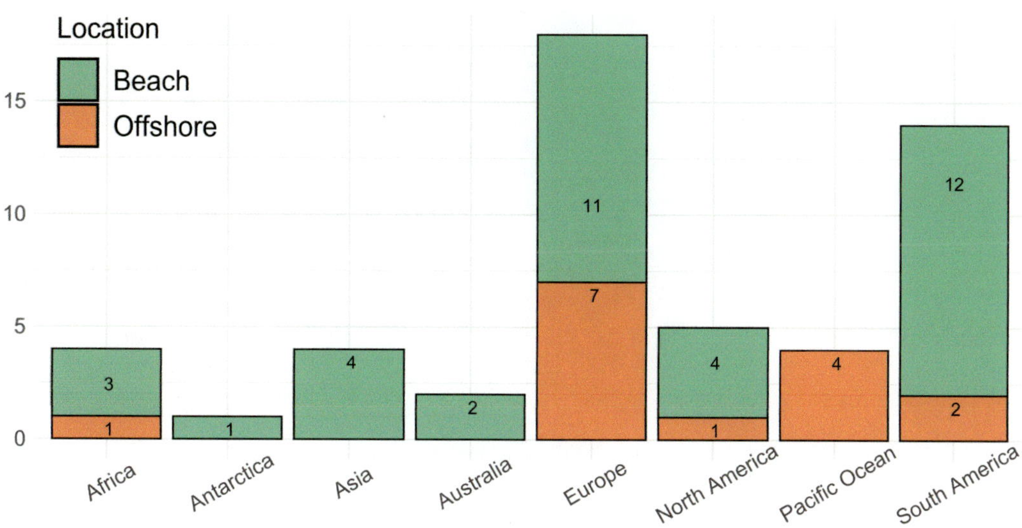

FIGURE 19.5 Regions where 52 studies of NNS on marine litter were focused, updated from De-la-Torre et al. (2023). Offshore denotes litter either trawled from the seabed or collected from the sea surface.

19.7 RECENT ADVANCES

Studying marine litter and NNS to determine where they come from or what could arrive next involves untangling patterns from a complex interplay of various factors (Figure 19.6).

However, recent advances in dispersal modelling, combined with novel genetic tools such as environmental DNA (eDNA)-based methods for community analysis or individual species detection, through to attenuated total reflection-Fourier transform infrared (ATR-FTIR) spectroscopy for substrate analysis, provide a panel of monitoring tools, which when used together give a more holistic understanding of NNS pathways and mechanisms.

19.7.1 Tracing Marine Litter Sources and Pathways

Determining sources is key to our understanding of what animals could be transported (Goldstein et al., 2014) and helps investigators model future scenarios. Analysis of the plastic piece's physical characteristics, including its dimensions, buoyancy and floating behaviour can provide clues about both the origin and how it was transported to where it was intercepted (Bravo et al., 2011). Studies on the drifting behaviour of the pieces themselves are scarce (van Sebille et al., 2020) but can be substituted with simulated particle tracking modelling to estimate the history of the drift (e.g. Rosas et al., 2021; Al-Qattan et al., 2023; Barry et al., 2023).

Early estimates of a drifting object's time in the water were very uncertain. In the 1940s, a log beached in Scotland with goose barnacles attached had an estimated drift duration put at between 3.5 and 7 months (a 1:2 ratio of the base estimate) (Anton, 1949) but nowadays numerical modelling has greatly improved these estimates (Madricardo et al., 2020; Guerrini et al., 2021). If the object is found mid-flow and the interception point is known, then drift trajectory can be calculated using preceding current patterns up to that point (Hoeksema et al., 2012).

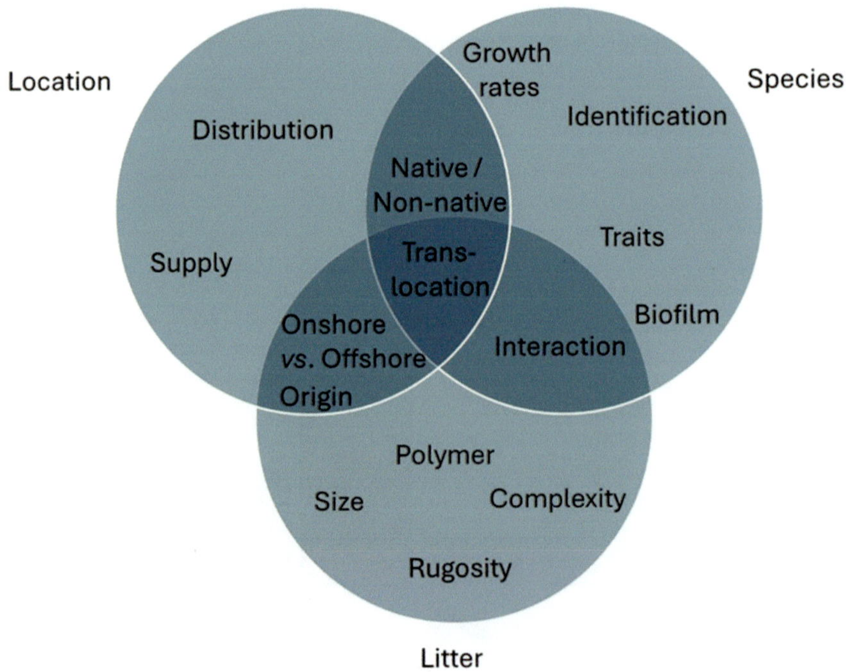

FIGURE 19.6 Venn diagram depicting the factors involved in the association of NNS with the litter, which must be considered when tracing origins or predicting what will be transported.

19.7.2 Substrate Analysis

Investigating an organism's chemical affinity with the substrate is another growing area of research with the rollout of ATR-FTIR spectroscopy. Polymers that the organisms are interacting with (i.e. the surface layer of plastic) can be determined and are increasingly included in studies of fouling communities (e.g. Gündoğdu et al., 2017; Int-Veen et al., 2021).

19.7.3 Community Analysis

For taxonomic studies of the fouling community, the rapid development of analytical techniques allows characterisation of the entire assemblage using DNA barcoding. The technique of swabbing all epifauna and using quantitative polymerase chain reaction (Ibabe et al., 2020) or targeting single organisms (see Audrézet et al., 2021 for a review of techniques) is now common practice. Alongside efforts to use the identification of species to determine their natural range, corroborating estimates of drift duration, trajectories etc. techniques using measurements of the maturity of the community or the size of attached organisms (Barry et al., 2023; Al-Qattan et al., 2023) are now being employed to gauge drift times.

These advances are timely as there is a growing awareness of the increased number of NNS moving around the world's oceans (Cassey et al., 2018). The potential impacts this can bring are outlined in the next section.

19.8 THE EFFECT OF SPECIES INTRODUCTIONS

Invasive NNS is one of the five direct drivers of biodiversity loss identified by the Intergovernmental Science-Policy Platform on Biodiversity and Ecosystem Services (IPBES), the other four being changing use of sea and land, direct exploitation of organisms, climate change and pollution (Roy et al., 2023). It is imperative to monitor the introduction of NNS since some have traits that allow them to establish and rapidly colonise a new environment. This puts pressure on the recipient ecosystem by introducing resource competition (Chesson, 2000; Pyšek et al., 2012). Invasions often result in negative impacts, leading to biodiversity loss, degradation of ecosystem services (including socio-economic impacts) and even impacts on human health (Galil, 2018). Invasive species competing for resources may be responsible for pathogen introductions (e.g. Tan et al., 2023) and some called 'ecosystem engineers' can even alter the physical habitat (Crooks, 2009). In extreme cases, such disruptions can lead to 'invasional meltdown' (Simberloff and von Holle, 1999), a cascading effect that ultimately threatens food chain sustainability and economic well-being (Gallardo et al., 2019). There are, however, rare cases of NNS positively affecting their invaded environment through habitat modification, trophic subsidy, pollination, competitive release and predatory release (Rodriguez, 2006). Predation (Noè et al., 2018) or native habitat complexity may also play a facilitating role even if some counterexamples exist (Cheebaan et al., 2023; Leclerc and Viard, 2018). The imbalance between positive and negative documented consequences of invasive species might be a consequence of funds dedicated to studies evaluating negative impacts (Anton et al., 2019); however, positive or negative, it is clear that invasive NNS leads to some level of ecosystem disruption.

19.9 THE ROLE OF PLASTIC POLLUTION IN ECOSYSTEM INVASIONS

Although there are compelling cases (see Lindstrom, 2018), a direct link between plastic transport all the way through to successful invasion is yet to be established; there is evidence that marine plastic can *facilitate* invasions at different stages of the process (Figure 19.7). Factors enabling a species' introduction (rapid, widespread dispersal) may not necessarily confer an advantage once they arrive. Different factors and natural barriers come into play when species are trying to establish

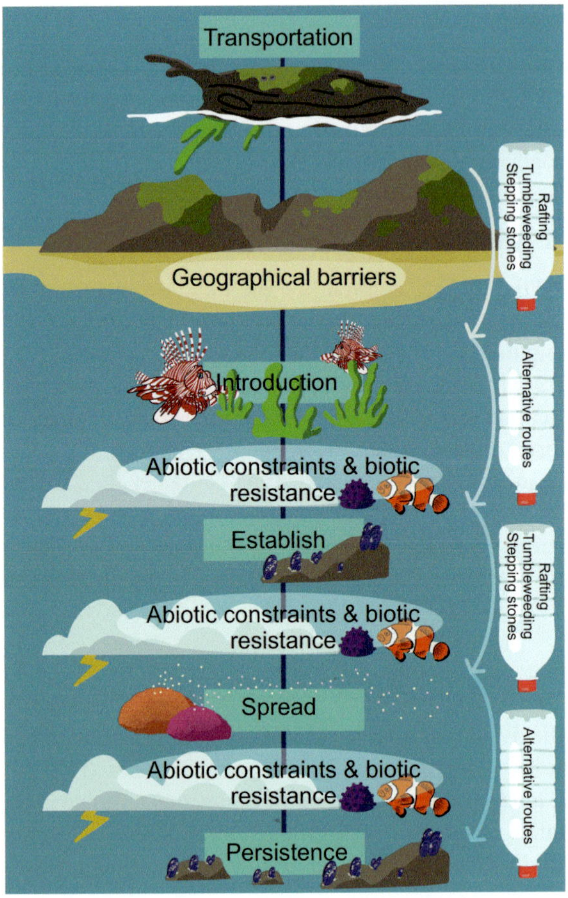

FIGURE 19.7 Stages in the invasion process, with vertical arrows joining each. Natural barriers are represented by graphics between each stage. To the right are the various mechanisms provided by plastic pollution which enable NNS to circumvent the barriers. (Adapted from Byun et al., 2018.)

in a new environment (Vermeij, 1996) but plastic pollution can supplement the invasion process by serving as both vector to introduce and then settlement platform once the NNS has arrived.

The accepted stages of a successful biological invasion include transport, introduction, establishment and spread (Table 19.1), each with its own challenges (Blackburn et al., 2011). An additional stage is the persistence of NNS in the new environment. The section below outlines how marine plastic pollution can facilitate the different stages of the invasion process.

19.9.1 Transport

Transport is a critical early stage as it moves the species out of their natural environment into one where they are absent or present at a low level. The transport mechanism will determine the number, condition and propagule pressure newly introduced species can exert upon the receiving environment (Lockwood et al., 2005). Environmental barriers remain unchanged, but the condition and complexity of the plastic will influence whether the species can overcome them and in what physical condition. Most of the transport in the marine environment is accounted for by buoyant plastic physically transposing populations but seafloor litter also plays a role (see Section 1.2).

TABLE 19.1
Stages in Invasion Sequence of a Newly Introduced Species, the Barriers They Have to Cross and How Plastic Pollution Can Support Them

Invasion Stage	Transport	Introduction	Establishment	Spread	Persistence
Barrier	Geography	Native species resistance	Survival/ reproduction	Dispersal/ environmental	Lack of space/ competition
Influence of plastic litter	Mobile platform, including shelter and support	Suitable platform, refuge	Large surface area, complex habitat, shelter	Propagule pressure, direct transport	Space, if it can be exploited first
Vector	Mobile plastic – floating or mobilised seafloor	Seafloor or drifting	Seafloor or drifting	Seafloor or drifting	Seafloor or drifting

Source: Adapted from Blackburn et al. (2011).

19.9.2 Introduction

'Propagule pressure' covers the number of individuals released into an environment, including not only the number of individuals but also the number of repeated introduction events (Lockwood et al., 2005). Biofouled plastic can introduce multiple individuals repeatedly into the same general area, increasing the likelihood that a successful invasion event will occur. A recent example is the influx of tropical Caribbean bivalves that continue to be collected in the British Isles inside plastic fishing pots (e.g. Holmes et al., 2015; Barry and Taylor, 2024). Bait pots create an ideal sheltered enclosure that can travel across the Atlantic and interceptions of live or recently dead Caribbean bivalves continue to be recorded over and over on the southern coasts of Britain and Ireland.

19.9.3 Establishment

Plastic pollution can help newly introduced NNS overcome resistance by abiotic (lack of settlement space) and biotic (native species' resistance) factors. Habitat provision is a key factor promoting invasions (Hufbauer et al., 2011) and anthropogenic activities are supplying this through the large surface area of plastic in coastal waters that could be colonised (Maes et al., 2018). Invasive species readily exploit new resources, such as artificial structures (Bulleri and Chapman, 2010; Simkanin et al., 2012), including plastic surfaces. Plastic provides a novel substrate for colonising organisms to settle on when they arrive (Landschoff et al., 2013) which serves the same basic function as the natural hard substrates they have evolved on (Beermann et al., 2020). It offers settlement opportunities in soft or shifting sediments where epifaunal/epifloral species wouldn't otherwise survive (Crocetta et al., 2020), even in the deep seas (Song et al., 2021) and provides shelter for many others (De-la-Torre et al., 2021).

If suitable niches are unavailable the invasion may fail (Shea and Chesson, 2002). However, species introduced through plastic vectors can thrive in a highly polluted environment if they find comparable habitats (Castro et al., 2022). The likelihood that species that have arrived on a plastic surface will spread to other plastic surfaces is high (Hufbauer et al., 2011) and has been reported in many seafloor studies (see Mancini et al., 2021). This is because successful NNS demonstrates flexible phenotypic responses to environmental conditions (Jofré-Madariaga et al., 2024). Trans-oceanic plastic debris is more likely to be colonised by generalists (Dolédec et al., 2000) than by species with more specialised requirements. Artificial structures are 'equally foreign to the evolutionary history of all species' (Tyrrell and Byers, 2007:55), but invaders often gain an advantage by rapidly exploiting this 'additional' space more efficiently than their native counterparts (Olyarnik et al., 2009; Ricciardi et al., 2021).

19.9.4 Spread

Those generalist species that make some of the most effective NNS can exploit multiple vectors to spread further (Cardeccia et al., 2018). The same mechanisms that first introduced NNS can return them into circulation and into other environments (Fernandez et al., 2022). We've seen that introduced species may leverage plastic objects as stepping stones for further spread (Glasby et al., 2007). The spread of the invasive ascidian *Didemnum vexillum* is thought to have been exacerbated by its ability to spread among artificial structures (including plastic) in harbours (Tyrrell and Byers, 2007). This *secondary spread* can be facilitated by plastics for shorter-distance travel following the initial invasion (Minchin et al., 2009; Lehtiniemi et al., 2015). Evidence suggests that invasive species brought into ports by shipping or aquaculture are being spread to surrounding areas via plastic waste washed out of harbours (Campbell et al., 2017; Miralles et al., 2018).

19.9.5 Persistence

Plastic pollution is a broad term that encompasses many different types of plastic items. Different polymer shapes and behaviour in the water all control or promote settlement (Gil and Pfaller, 2016). Colonisation is highly variable and driven by a range of factors (Li et al., 2016), but the ubiquity of available plastics, combined with the traits of some generalist species raises the likelihood that some area of overlap will be found and exploited (Garcia-Vazquez et al., 2018). A prime example is the case of the invasive coral *Tubastraea* spp. which was introduced to Brazil on oil and gas platforms but has recently been found on a wide range of plastic litter (Mantelatto et al., 2020). Secondary expansion along the Brazilian coast has been attributed to surface, subsurface and seafloor spread on buoyant and sunken plastics. Such a wide dispersal of different colonies further ensures persistence of the species in Brazil, as a greater spread of different populations will increase the chance of survival if one or more populations are wiped out (Drake et al., 2005).

19.9.6 Plastic as a Reservoir

While the majority of investigations have focused on large offshore or harbour infrastructure and their role in spreading NNS (Simkanin et al., 2012; McLean et al., 2022; Elliott and Birchenough, 2022) there might be a physically smaller, but more widespread agent providing the same level of settlement facilitation. A recent study has demonstrated that sessile assemblages found on floating habitats host a much greater proportion of NNS than on non-floating counterparts (Leclerc et al., 2020). Plastic pollution is a highly dispersed collection of micro-habitats, forming a vast network with shorter distances between 'nodes' (Zhao et al., 2021). This network allows for a higher probability of dispersal success and probably allows some taxa to go through several cycles of survival and reproduction in the open sea. Investigations of the great plastic garbage patches (Eriksen et al., 2014) revealed the formation of novel ecosystems on plastics (Haram et al., 2021). These 'neopelagic communities' can be concentrated or dispersed, but represent a great reservoir of organisms (and propagules) surviving on plastic.

The combined effect of artificial structures and the plastic vector is an increasing connectivity between marine ecosystems (Cabezas et al., 2013). With the exponential increase in the rate of detections of NNS in the last few decades (Seebens et al., 2017) combined with the evidence that NNS is using plastic as a vector, the need for strict pollution control and responsible management is urgent. The current regulations to tackle the problem are outlined below.

19.10 PREVENTION, REGULATION AND MANAGEMENT

Marine litter (including plastic pollution) and NNS are largely treated as separate issues. While national, regional and global legislation exist for both, scientific and policy groups end up working in silos. For example, in 2008, the term 'marine litter' was highlighted by the European Union's flagship *Marine Strategy Framework Directive* (MSFD-2008/56/EC), a policy introduced to maintain clean, healthy, productive and resilient marine ecosystems. It stipulates that 'the quantities of plastic and microplastic should not cause harm in the marine environment' (Descriptor 10). This same directive also requires that 'Non-indigenous species introduced by human activities are at levels that do not adversely alter the ecosystems' (Descriptor 2). It is, however, worth noting that both marine litter and NIS are addressed as separate impacts in European Seas, with very little interaction between the technical groups and experts tasked with monitoring and assessing both descriptors.

The Joint Research Center (JRC), as co-chair of the EU Technical Group on Marine Litter, released in December 2023 a new edition of the Guidance on Monitoring of Marine Litter in European Seas (https://mcc.jrc.ec.europa.eu/main/dev.py?N=41&O=466), first published in 2013 (Galgani et al., 2023). The publication provides recommendations on how to sample, analyse and record marine litter with the aim of making the data comparable and usable across countries. It gives advice on how to monitor litter objects and particles of different sizes on the coastline, seafloor, floating on the surface and ingested by animals. It also recommends how to assess the presence of organisms wrapped or trapped in marine litter such as ALDFG. The new guidance will help national authorities implement the MSFD and protect the marine environment against harm caused by plastic pollution.

In March 2022, nearly 200 nations endorsed a historic resolution to develop a new global and legally binding agreement or treaty by the end of 2024 (UNEP, 2022). An Intergovernmental Negotiating Committee has been tasked with developing an international legally binding instrument on plastic pollution, including in the marine environment, which could include both binding and voluntary approaches, based on a comprehensive approach that addresses the full life cycle of plastic. Although only a draft text at this point, there is however no mention of the role that plastic pollution plays as a pathway for introducing and dispersing NNS.

19.11 FUTURE WORK

The Convention on Biological Diversity (2014) urges governments to develop early warning systems and action plans to identify priority pathways. A possible course of action would be to evaluate each stage in Table 19.1 and identify which could be regulated or used as a starting point for designing management strategies (Geburzi and McCarthy, 2018). This will require focusing on the following four points:

1. **Furthering Strategic Data Collection:**

   To gain meaningful insight into the role of plastic, it is important to assess the true extent of these introductions (Carvalho et al., 2023). One example is the baseline made of litter types and abundance in the Madeira Archipelago (Álvarez et al., 2023), which was then interrogated for the presence of NNS on the litter (Canning-Clode et al., 2021). Physical and biological data from interceptions should be collated: the location, abundance and environmental conditions of the interception, the identity and biological traits of the NNS recovered. Online data collection tools such as DeNIS (https://www.denis-db.com) (Canning-Clode et al., 2023) will be particularly helpful in identifying patterns more effectively.

   In addition to fieldwork, post-hoc analysis of introductions and their circumstances will inform future scenarios. Documented cases should be studied for clues to the 'next' invasion. The number of review papers summarising cases of species moving on or with marine litter (e.g. Thiel and Gutow, 2005; Kiessling et al., 2015; De-la-Torre et al., 2023) is growing, so a broad overview will soon be possible.

2. **Filling Knowledge Gaps**:

   Approximately half of the plastics produced are heavier than seawater; consequently, the seafloor is an important long-term sink for marine macro- and micro-litter (Canals et al., 2021). A range of different approaches may be used to cost-effectively monitor seafloor areas, such as using simple litter quantification protocols based on existing fish stock surveys (Galgani et al., 2022). Going the extra step however, to photographing, cataloguing and identifying species colonising plastic seafloor litter has yet to be regularly implemented (Barry et al., 2024). If there is a certain trait or attribute that governs successful settlement and is driven by a characteristic of the litter (Li et al., 2016), then identifying these dynamics should be prioritised (Gündoğdu et al., 2017). The biggest challenge is the complexity of the interactions, and so far, most studies have only focused on the macrofauna, without capturing the biofilm which could be driving settlement patterns. Also, there is a geographic knowledge gap. Although marine plastic litter is an inherently global issue, regions that do not have adequate data or representation (see Figure 19.5) will need to be prioritised for further data collection. Major gaps exist in sampling effort and taxonomic expertise, with small-sized organisms in species-rich taxonomic groups (i.e. Crustacea and Mollusca) remaining poorly studied. Together, these impair society's ability to discover new species and identify and understand species of economic and ecological importance (Costello et al., 2010).
3. **Developing Critical Tools**:

   In response to the threat of the plastic vector, several recently developed tools are ready for wider implementation. A **standardised approach** to collection, classification and analysis combining morphological taxonomy and molecular methods has been promoted (De-la-Torre et al., 2023) but often these methods are employed in limited studies in isolated regions of the world. **Identifying source to sink** routes could benefit from numerical modelling to predict or even back-track the probable path of NNS back to their source (e.g. Barry et al., 2023). A **multi-vector approach** should identify a list of high-profile species, litter types and litter hotspots that can be identified, and flows into and out of them modelled (Di Febbraro et al., 2021). **Aggregated data** can be used in predicting what type of species may become invasive through 'ecological profiling' (Fournier et al., 2019) or Ecological Niche Modelling (Laeseke et al., 2020), but this requires a large amount of data.
4. **Strengthening Research Networks**:

   With plastic pollution a global problem, accessing adequate coverage is always going to be a challenge. Citizen science participation can help expand geographic sampling coverage (Larson et al., 2020), and local stakeholders should be consulted since they will have background knowledge of certain regions (Sagarminaga et al., 2023). Different specialists from various fields (i.e. oceanography, numerical modelling, waste management and marine policy) will need to be involved.

19.12 CONCLUSION

Without action, plastic production could triple by 2060 (Lau et al., 2020), thus providing a greater total surface area of artificial habitat. Even today, the total colonisable surface area of floating, drifting or sinking plastic substrate is simply staggering. For biota, settlement on the plastisphere is a novel interaction, with highly complex and as yet, imprecisely understood associations (Wilson et al., 2021). It is, however, clear that it poses a significant biosecurity issue (Ibabe et al., 2020).

Current legislation underestimates the threat marine plastic poses as an introduction vector and needs to be modified (Audrézet et al., 2021). Since there will be stochastic introduction events, the precautionary principle should be adopted and each interception treated as a potentially successful introduction (Hulme, 2011). Aggregating data from a range of interceptions (seafloor, drifting and beached) will inform effective management strategy formulation by exploring what NNS is on what type of plastic and where they are arriving, thereby suggesting what to prioritise. Not all plastics will host NNS, and not all invasions will be successful (or at least not at first (Spear et al., 2021)), but the scale and density of plastic in the oceans make this relationship a global experiment (Carlton and Fowler, 2018) with the outcome not yet clear.

Thus, for both NNS and plastics, the focus must be on prevention: by reducing propagule pressure for the former and ensuring any plastic items produced are essential, safe and sustainable for the latter. Only through these joint preventive actions will any reduction in the rate and numbers of plastisphere nomads hitchhiking around the oceans occur.

ACKNOWLEDGEMENTS

This study was supported by the Interreg Atlantic Area CleanAtlantic programme (EAPA-46/2016) and represents a contribution to the term of reference 'f' ('Investigate the role of human produced marine debris as a vector and facilitator for the introduction and spread of non-indigenous species (NIS)') of the ICES working group on Introductions and Transfers of Marine Organisms (WGITMO). We would like to thank Alex McGoran (Cefas) for providing the infographics.

REFERENCES

Abel, S. M., Wu, F., Primpke, S., et al. (2023). Journey to the deep: plastic pollution in the hadal of deep-sea trenches. *Environmental Pollution*, (333), 122078. https://doi.org/10.1016/j.envpol.2023.122078

Al-Qattan, N., Herbert, G. S., Spero, H. J., et al. (2023). A stable isotope sclerochronology-based forensic method for reconstructing debris drift paths with application to the MH370 crash. *AGU Advances*, (4), 1–12. https://doi.org/10.1029/2023AV000915

Álvarez, S., Gestoso, I., Ramalhosa, P., et al. (2023). Regional Studies in Marine Science Exploring visual methods for monitoring marine non-indigenous species colonizing artificial structures in the Madeira archipelago (NE Atlantic). *Regional Studies in Marine Science*, (69), 103293. https://doi.org/10.1016/j.rsma.2023.103293

Andrady, A. L. (2015). Persistence of plastic litter in the oceans, in: M. Bergmann, L. Gutow, and M. Klages (Eds.) *Marine Anthropogenic Litter.* Springer International Publishing, Cham, pp. 57–72. https://doi.org/10.1007/978-3-319-16510-3_3

Anton, J. F. (1949). Stalked barnacles on the shores of the British Isles. *The Scottish Naturalist*, (61), 65–67.

Anton, A., Geraldi, N.R., Lovelock, C. E., et al. (2019). Global ecological impacts of marine exotic species. *Nature Ecology & Evolution,* (3), 787–800. https://doi.org/10.1038/s41559-019-0851-0

Audrézet, F., Zaiko, A., Lear, G., et al. (2021). Biosecurity implications of drifting marine plastic debris: Current knowledge and future research. *Marine Pollution Bulletin*, (162) 111835. https://doi.org/10.1016/j.marpolbul.2020.111835

Barnes, D. K. A. (2002). Invasions by marine life on plastic debris. *Nature*, (416), 808–809. https://doi.org/10.1038/416808a

Barnes, D. K. A., & Fraser, K. P. P. (2003). Rafting by five phyla on man-made flotsam in the Southern Ocean. *Marine Ecology Progress Series*, (262), 289–291.

Barnes, D. K. A., Galgani, F., Thompson, R. C., et al. (2009). Accumulation and fragmentation of plastic debris in global environments. *Philosophical Transactions of the Royal Society B: Biological Sciences*, (364), 1985–1998. https://doi.org/10.1098/rstb.2008.0205

Barry, P. J., & Taylor, D. (2024). Discarded fishing gear aids trans-Atlantic transport of species from the Caribbean to UK shores. *Porcupine Marine Natural History Society Bulletin*, (21), 19–28.

Barry, P. J., Beraud, C., Wood, L. E., et al. (2023). Modelling of marine debris pathways into UK waters: Example of non-native crustaceans transported across the Atlantic Ocean on floating marine debris. *Marine Pollution Bulletin,* (186), 114388. https://doi.org/10.1016/j.marpolbul.2022.114388

Barry P. J., Silburn, B., Bakir, A., et al. (2024) Seafloor macrolitter as a settling platform for non-native species: A case study from UK waters. *Marine Pollution Bulletin*, (204), 116499. https://doi.org/10.1016/j.marpolbul.2024.116499

Battaglia, P., Consoli, P., Ammendolia, G., et al. (2019). Colonization of floats from submerged derelict fishing gears by four protected species of deep-sea corals and barnacles in the Strait of Messina (central Mediterranean Sea). *Marine Pollution Bulletin* (148), 61–65. https://doi.org/10.1016/j.marpolbul.2019.07.073

Beermann, J., Hall-Mullen, A. K., Havermans, C., et al. (2020). Ancient globetrotters—connectivity and putative native ranges of two cosmopolitan biofouling amphipods. *PeerJ* (8), e9613. https://doi.org/10.7717/peerj.9613

Blackburn, T. M., Pyšek, P., Bacher, et al. (2011). A proposed unified framework for biological invasions. *Trends in Ecology & Evolution,* (26), 333–339. https://doi.org/10.1016/j.tree.2011.03.023

Bravo, M., Astudillo, J. C., Lancellotti, D., et al. (2011). Rafting on abiotic substrata: Properties of floating items and their influence on community succession. *Marine Ecology Progress Series*, (439), 1–17. https://doi.org/10.3354/meps09344

Brewer, R. H. (1984). The influence of the orientation, roughness, and wettability of solid surfaces on the behaviour and attachment of planulae of *Cyanea* (Cnidaria: Scyphozoa). *Biological Bulletin*, (166), 11–21. https://doi.org/10.1086/277797

Bryan, S. E., Cook, A. G., Evans, J. P., et al. (2012). Rapid, long-distance dispersal by pumice rafting. *PLOS ONE*, (7), e40583. https://doi.org/10.1371/journal.pone.0040583

Bugoni, L., Krause, L., & Virgínia Petry, M. (2001). Marine debris and human impacts on sea turtles in Southern Brazil. *Marine Pollution Bulletin* (42), 1330–1334. https://doi.org/10.1016/S0025-326X(01)00147-3

Bulleri, F., & Chapman, M. G. (2010). The introduction of coastal infrastructure as a driver of change in marine environments. *Journal of Applied Ecology,* (47), 26–35. https://doi.org/10.1111/j.1365-2664.2009.01751.x

Byun, C., de Blois, S., & Brisson, J. (2018). Management of invasive plants through ecological resistance. *Biological Invasions*, (20), 13–27. https://doi.org/10.1007/s10530-017-1529-7

Cabezas, M. P., Navarro-Barranco, C., Ros, M., et al. (2013). Long-distance dispersal, low connectivity and molecular evidence of a new cryptic species in the obligate rafter *Caprella andreae* Mayer, 1890 (Crustacea: Amphipoda: Caprellidae). *Helgoland Marine Research*, (67), 483–497. https://doi.org/10.1007/s10152-012-0337-9

Campbell, M. L., King, S., Heppenstall, L. D., et al. (2017). Aquaculture and urban marine structures facilitate native and non-indigenous species transfer through generation and accumulation of marine debris. *Marine Pollution Bulletin*, (123), 304–312. https://doi.org/10.1016/j.marpolbul.2017.08.040

Canals, M., Pham, C. K., Bergmann, M., et al. (2021) The quest for seafloor macrolitter: a critical review of background knowledge, current methods and future prospects. *Environmental Research Letters*, 16, 023001

Canning-Clode, J., Freitas, R., Barry, P., et al. (2023). Introducing "DeNIS": a global database on anthropogenic marine Debris and Non-Indigenous Species. https://doi.org/10.22541/au.168381770.04204050/v1

Canning-Clode, J., Ramalhosa, P., Gestoso, I., et al. (2021). Clean Atlantic. Tackling Marine Litter in the Atlantic Area. WP5: Monitoring and Data Management. DELIVERABLE 5.5. - Evaluation of marine litter as transport facilitator for nuisance biota. https://www.cleanatlantic.eu/wp-content/uploads/2021/09/CleanAtlantic-ARDITI-D5.5_FINAL.pdf

Cardeccia, A., Marchini, A., Occhipinti-Ambrogi, A., et al. (2018). Assessing biological invasions in European Seas: Biological traits of the most widespread non-indigenous species. *Estuarine, Coastal and Shelf Science* (201), 17–28. https://doi.org/10.1016/j.ecss.2016.02.014

Carlton, J. T., & Fowler, A. E. (2018). Ocean rafting and marine debris: A broader vector menu requires a greater appetite for invasion biology research support. *Aquatic Invasions*, (13), 11–15.

Carlton, J. T., Chapman, J. W., Geller, J. B., et al. (2017). Tsunami-driven rafting: Transoceanic species dispersal and implications for marine biogeography. *Science* (357), 1402–1406. https://doi.org/10.1126/science.aao1498

Carpenter, E. J., & Smith Jnr., K. L. (1972). Plastics on the Sargasso Sea Surface. *Science*, (175), 1240–1241.

Carvalho, S., Shchepanik, H., Aylagas, E., et al. (2023). Hurdles and opportunities in implementing marine biosecurity systems in data-poor regions. *BioScience* (73), 494–512. https://doi.org/10.1093/biosci/biad056

Cassey, P., García-Díaz, P., Lockwood, J. L., et al. (2018). Invasion biology: Searching for predictions and prevention, and avoiding lost causes, pp. 3–13, in: J. Jeschke and T. Heger (Eds.) *Invasion Biology: Hypotheses and Evidence*. CABI Invasive Series, 9. CABI.

Castro, N., Gestoso, I., Marques, C. S., et al. (2022). Anthropogenic pressure leads to more introductions: Marine traffic and artificial structures in offshore islands increases non-indigenous species introduction. *Marine Pollution Bulletin*, (181), 113898. https://doi.org/10.1016/j.marpolbul.2022.113898

Chase, A. L., Dijkstra, J. A., & Harris, L. G. (2016). The influence of substrate material on ascidian larval settlement. *Marine Pollution Bulletin*, (106), 35–42. https://doi.org/10.1016/j.marpolbul.2016.03.049

Cheebaan, S., Freestone, A. L., Des Pérez, A., et al. (2023) Predation facilitates the abundance of biofouling non-indigenous species in estuarine marinas in NE Atlantic Portugal. *Marine Pollution Bulletin, (*188), 108541. https://doi.org/10.1016/j.ecss.2023.108541

Chesson, P. (2000). Mechanisms of maintenance of species diversity. *Annual Review of Ecology and Systematics*, (31), 343–366. https://doi.org/10.1146/annurev.ecolsys.31.1.343

Clarke Murray, C., Maximenko, N., & Lippiatt, S. (2018). The influx of marine debris from the Great Japan Tsunami of 2011 to North American shorelines. *Marine Pollution Bulletin*, (132), 26–32. https://doi.org/10.1016/j.marpolbul.2018.01.004

Convention on Biological Diversity (CBD) (2014). Pathways of introduction of invasive species, their prioritization and management. https://www.cbd.int/doc/meetings/sbstta/sbstta-18/official/sbstta-18-09-add1-en.pdf

Costello, M. J., Coll, M., Danovaro, R., et al. (2010). A census of marine biodiversity knowledge, resources, and future challenges. *PLoS One* (5), e12110. https://doi.org/10.1371/journal.pone.0012110

Crocetta, F., Riginella, E., Lezzi, M., et al. (2020). Bottom-trawl catch composition in a highly polluted coastal area reveals multifaceted native biodiversity and complex communities of fouling organisms on litter discharge. *Marine Environmental Research*, (155), 104875. https://doi.org/10.1016/j.marenvres.2020.104875

Crooks, J. A. (2009). The role of exotic marine ecosystem engineers, pp. 287–304, in: G. Rilov and J. A. Crooks (Eds.) *Biological Invasions in Marine Ecosystems. Ecological, Management, and Geographic Perspectives*. Springer-Verlag, Berlin Heidelberg. 641 pp.

D'Aloia, C. C., Bogdanowicz, S. M., Francis, R. K., et al. (2015). Patterns, causes, and consequences of marine larval dispersal. *Proceedings of the National Academy of Sciences*, 112 (45), 13940–13945. https://doi.org/10.1073/pnas.1513754112

De Wolff, K. (2017). Plastic nature cultures: Multispecies ethnography and the dangers of separating living from nonliving bodies. *Body and Society*, (23), 23–47. https://doi.org/10.1177/1357034X17715074

De-la-Torre, G. E., Bele, M., Alisson, V., et al. (2023). Marine litter colonization: Methodological challenges and recommendations. *Frontiers in Marine Science*, (10), 1070575. https://doi.org/10.3389/fmars.2023.1070575

De-la-Torre, G. E., Dioses-Salinas, D. C., Pérez-Baca, B. L., et al. (2021). Marine macroinvertebrates inhabiting plastic litter in Peru. *Marine Pollution Bulletin*, (167), 112296. https://doi.org/10.1016/j.marpolbul.2021.112296

Di Febbraro, M., Frate, L., de Francesco, M. C., et al. (2021). Modelling beach litter accumulation on Mediterranean coastal landscapes: An integrative framework using species distribution models. *Land*, (10), 1–17. https://doi.org/10.3390/land10010054

Dolédec, S., Chessel, D., & Gimaret-Carpentier, C. (2000). Niche separation in community analysis: A new method. *Ecology*, (81), 2914. https://doi.org/10.2307/177351

Drake, J. M., Baggenstos, P., & Lodge, D. M. (2005). Propagule pressure and persistence in experimental populations. *Biology Letters*, (1), 480–483. https://doi.org/10.1098/rsbl.2005.0375

Elliott, M., & Birchenough, S. N. R. (2022). Man-made marine structures – Agents of marine environmental change or just other bits of the hard stuff? *Marine Pollution Bulletin*, (176), 1–3. https://doi.org/10.1016/j.marpolbul.2022.113468

Enrichetti, F., Bavestrello, G., Betti, F., et al. (2021). Fate of lost fishing gears: Experimental evidence of biofouling colonization patterns from the northwestern Mediterranean Sea. *Environmental Pollution*, (268), 115746. https://doi.org/10.1016/j.envpol.2020.115746

Eriksen, M., Lebreton, L. C. M., Carson, H. S., et al. (2014). Plastic pollution in the world's oceans: More than 5 trillion plastic pieces weighing over 250,000 tons afloat at sea. *PLoS One* (9), e111913. https://doi.org/10.1371/journal.pone.0111913

Fernandez, S., Ibabe, A., Rayon-Viña, F., et al. (2022). Flotsam, an overlooked vector of alien dispersal from ports. *Estuarine, Coastal and Shelf Science*, (271), 107879. https://doi.org/10.1016/j.ecss.2022.107879

Floerl, O., Pool, T. K., & Inglis, G. J. (2004) Positive interactions between nonindigenous species facilitate transport by human vectors. *Ecological Applications,* (14), 1724–1736.

Fournier, A., Penone, C., Pennino, M. G., et al. (2019). Predicting future invaders and future invasions. *Proceedings of the National Academy of Sciences*, (116), 7905–7910. https://doi.org/10.1073/pnas.1803456116

Galgani, F., Michela, A., Gérigny, O., et al. (2022). Marine litter, plastic, and microplastics on the seafloor, pp.151–197, in: Anthony L. Andrady (Ed.) *Plastics and the Ocean: Origin, Characterization, Fate, and Impacts*. Wiley, 516 pp. https://doi.org/10.1002/9781119768432

Galgani, F., Ruiz Orejon Sanchez Pastor, L., Ronchi, F., et al. (2023). Guidance on the monitoring of marine litter in European seas. Publications Office of the European Union, Luxembourg, 2023, doi:10.2760/59137, JRC133594.

Galil, B. S. (2018). Poisonous and venomous: Marine alien species in the Mediterranean Sea and human health, pp. 1–15, in: G. Mazza and E. Tricarico (Eds.) *Invasive Species and Human Health. CABI Invasive Series,* 10. CABI.

Gallardo, B., Bacher, S., Bradley, B., et al. (2019) InvasiBES: Understanding and managing the impacts of invasive alien species on biodiversity and ecosystem services. *NeoBiota*, (50), 109–122.

García-Gómez, J. C., Garrigós, M., & Garrigós, J. (2021). Plastic as a vector of dispersion for marine species with invasive potential. A review. *Frontiers in Ecology and Evolution*, (9), 1–28. https://doi.org/10.3389/fevo.2021.629756

Garcia-Vazquez, E., Cani, A., Diem, A., et al. (2018). Leave no traces – Beached marine litter shelters both invasive and native species. *Marine Pollution Bulletin,* (131), 314–322. https://doi.org/10.1016/j.marpolbul.2018.04.037

Geburzi, J. C., & McCarthy, M. L. (2018). How do they do it ? – Understanding the success of marine invasive species, pp. 109–124, in: *YOUMARES 8-Oceans Across Boundaries: Learning from each other*. https://doi.org/10.1007/978-3-319-93284-2

Gil, M. A., & Pfaller, J. B. (2016). Oceanic barnacles act as foundation species on plastic debris: Implications for marine dispersal. *Scientific Reports*, (6), 6–12. https://doi.org/10.1038/srep19987

Glasby, T. M., Connell, S. D., Holloway, M. G., et al. (2007). Nonindigenous biota on artificial structures: Could habitat creation facilitate biological invasions? *Marine Biology*, (151), 887–895. https://doi.org/10.1007/s00227-006-0552-5

Goldstein, M. C., Carson, H. S., & Eriksen, M. (2014). Relationship of diversity and habitat area in North Pacific plastic-associated rafting communities. *Marine Biology*, (161), 1441–1453. https://doi.org/10.1007/s00227-014-2432-8

Gracia, A., Rangel-Buitrago, N., & Flórez, P. (2018). Beach litter and woody-debris colonizers on the Atlantico department Caribbean coastline, Colombia. *Marine Pollution Bulletin,* (128), 185–196. https://doi.org/10.1016/j.marpolbul.2018.01.017

Gregory, M. R. (1978). Accumulation and distribution of virgin plastic granules on New Zealand beaches. *New Zealand Journal of Marine and Freshwater Research*, (12), 399–414. https://doi.org/10.1080/00288330.1978.9515768

Gregory, M. R. (2009). Environmental implications of plastic debris in marine settings-entanglement, ingestion, smothering, hangers-on, hitch-hiking and alien invasions. *Philosophical Transactions of the Royal Society B: Biological Sciences*, (364), 2013–2025. https://doi.org/10.1098/rstb.2008.0265

Guerrini, F., Mari, L., & Casagrandi, R. (2021). The dynamics of microplastics and associated contaminants: Data-driven Lagrangian and Eulerian modelling approaches in the Mediterranean Sea. *Science of the Total Environment*, (777), 145944. https://doi.org/10.1016/j.scitotenv.2021.145944

Gündoğdu, S., Çevik, C., & Karaca, S. (2017). Fouling assemblage of benthic plastic debris collected from Mersin Bay, NE Levantine coast of Turkey. *Marine Pollution Bulletin*, (124), 147–154. https://doi.org/10.1016/j.marpolbul.2017.07.023

Hadfield, M. G. (2011) Biofilms and marine invertebrate larvae: What bacteria produce that larvae use to choose settlement sites. *Annual Review of Marine Science*, (3), 453–470.

Hagen, O., Viana, D. S., Wiegand, T., et al. (2024). The macro-eco-evolutionary interplay between dispersal, competition and landscape structure in generating biodiversity. *Philosophical Transactions of the Royal Society* B, (379), 20230140. https://doi.org/10.1098/rstb.2023.0140

Haram, L. E., Carlton, J. T., Centurioni, L., et al. (2021). Emergence of a neopelagic community through the establishment of coastal species on the high seas. *Nature Communications*, (12), 6885. https://doi.org/10.1038/s41467-021-27188-6

Haram, L. E., Carlton, J. T., Centurioni, L., et al. (2023). Extent and reproduction of coastal species on plastic debris in the North Pacific Subtropical Gyre. *Nature Ecology & Evolution,* (7), 687–697. https://doi.org/10.1038/s41559-023-01997-y

Harms, J. (1990). Marine plastic litter as an artificial hard bottom fouling ground. *Helgoländer Meeresuntersuchungen*, (44), 503–506. https://doi.org/10.1007/BF02365483

Haydar, D. (2010). What is natural? The scale and consequences of marine bioinvasions in the North Atlantic Ocean, (PhD Thesis) University of Groningen, 184 pp.

Highsmith, R. (1985). Floating and algal rafting as potential dispersal mechanisms in brooding invertebrates. *Marine Ecology Progress Series*, (25), 169–179. https://doi.org/10.3354/meps025169

Hirose, E., & Sensui, N. (2021). Substrate selection of ascidian larva: Wettability and nano-structures. *Journal of Marine Science and Engineering*, (9), 634. https://doi.org/10.3390/jmse9060634

Hoeksema, B. W., Roos, P. J., & Cadée, G. C. (2012). Trans-Atlantic rafting by the brooding reef coral *Favia fragum* on man-made flotsam. *Marine Ecology Progress Series*, (445), 209–218. https://doi.org/10.3354/meps09460

Holmes, A. M., Oliver, G. P., Trewhella, S., et al. (2015). Trans-Atlantic Rafting of inshore mollusca on macro-litter: American molluscs on British and Irish shores, new records. *Journal of Conchology*, (42), 1–9.

Hufbauer, R. A., Facon, B., Ravigné, V., et al. (2011). Anthropogenically induced adaptation to invade (AIAI): Contemporary adaptation to human-altered habitats within the native range can promote invasions. *Evolutionary Applications* (5), 89–101. https://doi.org/10.1111/j.1752-4571.2011.00211.x

Hulme, P. E. (2011) Biosecurity: The changing face of invasion biology, pp. 301–314, in: Richardson, D. M. (Ed.) *Fifty Years of Invasion Ecology.* The Legacy of Charles Elton. Wiley-Blackwell, 357 pp.

Ibabe, A., Rayón, F., Martinez, J. L., et al. (2020). Environmental DNA from plastic and textile marine litter detects exotic and nuisance species nearby ports. *PLoS One*, (15), 1–20. https://doi.org/10.1371/journal.pone.0228811

Int-Veen, I., Nogueira, P., Isigkeit, J., et al. (2021). Positively buoyant but sinking: Polymer identification and composition of marine litter at the seafloor of the North Sea and Baltic Sea. *Marine Pollution Bulletin*, (172), 112876. https://doi.org/10.1016/j.marpolbul.2021.112876

International Maritime Organization. (2023). Guidelines for the control and management of ships' biofouling to minimize the transfer of invasive aquatic species. Adopted by resolution MEPC 378 (80)/17/Add.1. (Vol. 378). https://wwwcdn.imo.org/localresources/en/KnowledgeCentre/IndexofIMOResolutions/MEPCDocuments/MEPC.378(80).pdf

Ivkić, A., Steger, J., Galil, B. S., et al. (2019). The potential of large rafting objects to spread Lessepsian invaders: the case of a detached buoy. *Biological Invasions*, (21), 1887–1893. https://doi.org/10.1007/s10530-019-01972-4

Jambeck, J. R., Geyer, R., Wilcox, C., et al. (2015). Plastic waste inputs from land into the ocean. *Science* (347), 768–771. https://doi.org/10.1126/science.1260352

Jofré-Madariaga, D., Aguilera Moya, M. A., Alves-de-Souza, C., et al. (2024). Non-indigenous species and their realized niche in tidepools along the South-East Pacific coast. *Marine Environmental Research,* 199, 106541. https://doi.org/10.1016/j.marenvres.2024.106541

Jokiel, P. L. (1990). Long-distance dispersal by rafting: reemergence of an old hypothesis. *Endeavour,* (14), 66–73. https://doi.org/10.1016/0160-9327(90)90074-2

Katsanevakis, S., & Crocetta, F. (2014). Pathways of introduction of marine alien species in European waters and the Mediterranean – A possible undermined role of marine litter. *CIESM Workshop Monographs*, (46), 61–68. Retrieved from https://publications.jrc.ec.europa.eu/repository/handle/JRC90466

Katsanevakis, S., Verriopoulos, G., Nicolaidou, A., et al. (2007). Effect of marine litter on the benthic megafauna of coastal soft bottoms: A manipulative field experiment. *Marine Pollution Bulletin*, (54), 771–778. https://doi.org/10.1016/j.marpolbul.2006.12.016

Kiessling, T., Gutow, L., & Thiel, M. (2015). Marine litter as habitat and dispersal vector, pp. 141–181, in: M. Bergmann, L. Gutow, and M. Klages (Eds.) *Marine Anthropogenic Litter.* Springer International Publishing, Cham. https://doi.org/10.1007/978-3-319-16510-3_1

Kreczak, H., Willmott, A. J., & Baggaley, A. W. (2021). Subsurface dynamics of buoyant microplastics subject to algal biofouling. *Limnology and Oceanography*, (66), 3287–3299. https://doi.org/10.1002/lno.11879

Laeseke, P., Schiller, J., Letschert, J., et al. (2020). Theories, vectors, and computer models: Marine invasion science in the anthropocene, pp. 195–209, in: *YOUMARES 9-* The Oceans*:* Our Research, Our Future. https://doi.org/10.1007/978-3-030-20389-4

Landschoff, J., Lackschewitz, D., Kesy, K., et al. (2013) Globalization pressure and habitat change: Pacific rocky shore crabs invade armored shorelines in the Atlantic Wadden Sea. *Aquatic Invasions*, (8), 77–87. https://doi.org/10.3391/ai.2013.8.1.09

Larson, E. R., Graham, B. M., Achury, R., et al. (2020). From eDNA to citizen science: emerging tools for the early detection of invasive species. *Frontiers in Ecology and the Environment*, (18), 194–202. https://doi.org/10.1002/fee.2162

Lau, W. W. Y., Shiran Y., Bailey, R. M., et al. (2020) Evaluating scenarios toward zero plastic pollution. *Science,* (369), 1455–146. https://doi.org/10.1126/science.aba9475

Leclerc, J.-C., & Viard, F. (2018) Habitat formation prevails over predation in influencing fouling communities. *Ecology and Evolution,* (8), 477–492. https://doi.org/10.1002/ece3.3654

Leclerc, J., Viard, F., González Sepúlveda, E., et al. (2020). Habitat type drives the distribution of non-indigenous species in fouling communities regardless of associated maritime traffic. *Diversity and Distributions*, (26), 62–75. https://doi.org/10.1111/ddi.12997

Lehtiniemi, M., Ojaveer, H., David, M., et al. (2015). Dose of truth—Monitoring marine non-indigenous species to serve legislative requirements. *Marine Policy*, (54), 26–35. https://doi.org/10.1016/j.marpol.2014.12.015

Leonard, K. (2015). Does epibiosis facilitate the invasion success of marine benthic invertebrates? (PhD Thesis) University of Tasmania, 215 pp.

Lewis, P. N., Riddle, M. J., & Smith, S. D. A. (2005). Assisted passage or passive drift: a comparison of alternative transport mechanisms for non-indigenous coastal species into the Southern Ocean. *Antarctic Science*, (17), 183–191. https://doi.org/10.1017/s0954102005002580

Li, H. X., Orihuela, B., Zhu, M., et al. (2016). Recyclable plastics as substrata for settlement and growth of bryozoans *Bugula neritina* and barnacles *Amphibalanus amphitrite*. *Environmental Pollution*, (218), 973–980. https://doi.org/10.1016/j.envpol.2016.08.047

Lindstrom, S. C. (2018). An undescribed species of putative Japanese *Pyropia* first appeared on the central coast of British Columbia, Canada, in 2015. *Marine Pollution Bulletin*, (132), 70–73. https://doi.org/10.1016/j.marpolbul.2017.07.016

Lockwood, J. L., Cassey, P., & Blackburn, T. (2005). The role of propagule pressure in explaining species invasions. *Trends in Ecology and Evolution*, (20), 223–228. https://doi.org/10.1016/j.tree.2005.02.004

Luiz, O. J., Robertson, D. R., Floeter, S. R., et al. (2011). Ecological traits influencing range expansion across large oceanic dispersal barriers: insights from tropical Atlantic reef fishes. *Proceedings of the Royal Society of London* B, (279), 1033–1040. https://doi.org/10.1098/rspb.2011.1525

Madricardo, F., Ghezzo, M., Nesto, N., et al. (2020). How to deal with seafloor marine litter: An overview of the state-of-the-art and future perspectives. *Frontiers in Marine Science*, (7), 1–16. https://doi.org/10.3389/fmars.2020.505134

Maes, T., Barry, J., Leslie, H. A., et al. (2018). Below the surface: Twenty-five years of seafloor litter monitoring in coastal seas of North West Europe (1992–2017). *Science of The Total Environment*, (630), 790–798. https://doi.org/10.1016/j.scitotenv.2018.02.245

Mancini, E., Miccoli, A., Piazzolla, D., et al. (2021). Macrozoobenthic fauna associated with benthic marine litter (Northern Tyrrhenian Sea, Italy) and first report of two bryozoan species in Italian waters. *Regional Studies in Marine Science*, (47), 101912. https://doi.org/10.1016/j.rsma.2021.101912

Mantelatto, M. C., Póvoa, A. A., Skinner, L. F., et al. (2020). Marine litter and wood debris as habitat and vector for the range expansion of invasive corals (*Tubastraea* spp.). *Marine Pollution Bulletin*, (160), 111659. https://doi.org/10.1016/j.marpolbul.2020.111659

Marine Strategy Framework Directive (MSFD). (2008). Directive 2008/56/EC of the European Parliament and of the Council of 17 June 2008 establishing a framework for community action in the field of marine environmental policy (Marine Strategy Framework Directive). https://eur-lex.europa.eu/legal-content/EN/ALL/?uri=CELEX:32008L0056

Masó, M., Fortuño, J. M., De Juan, S., et al. (2016). Microfouling communities from pelagic and benthic marine plastic debris sampled across Mediterranean coastal waters. *Scientia Marina*, (80), 117–127. https://doi.org/10.3989/scimar.04281.10A

McLean, D. L., Ferreira, L. C., Benthuysen, J. A., et al. (2022). Influence of offshore oil and gas structures on seascape ecological connectivity. *Global Change Biology*, (28), 3515–3536. https://doi.org/10.1111/gcb.16134

Mesaglio, T. P., Schilling, H. T., Adler, L., et al. (2021). The ecology of Lepas-based biofouling communities on moored and drifting objects, with applications for marine forensic science. *Marine Biology*, (168), 1–16. https://doi.org/10.1007/s00227-021-03822-1

Mghili, B., Keznine, M., Analla, M., et al. (2023). The impacts of abandoned, discarded and lost fishing gear on marine biodiversity in Morocco. *Ocean and Coastal Management*, (239), 106593. https://doi.org/10.1016/j.ocecoaman.2023.106593

Minchin, D. (1996). Tar pellets and plastics as attachment surfaces for lepadid cirripedes in the North Atlantic Ocean. *Marine Pollution Bulletin*, (32), 855–859. https://doi.org/10.1016/S0025-326X(96)00045-8

Minchin, D., Cook, E. J., & Clark, P. F. (2013). Alien species in British brackish and marine waters. *Aquatic Invasions*, (8), 3–19. https://doi.org/10.3391/ai.2013.8.1.02

Minchin, D., Gollasch, S., Cohen, A.N., et al. (2009) Characterizing Vectors of Marine Invasions, pp. 109–116, in: G. Rilov and J. A. Crooks (Eds.) *Biological Invasions in Marine Ecosystems. Ecological, Management, and Geographic Perspectives.* Springer-Verlag, Berlin Heidelberg. 641 pp.

Mineur, F., Davies, A. J., Maggs, C. A., et al. (2010). Fronts, jumps and secondary introductions suggested as different invasion patterns in marine species, with an increase in spread rates over time, *Proceedings of the Royal Society B: Biological Sciences*, (277), 2693–2701. https://doi.org/10.1098/rspb.2010.0494

Miralles, L., Gomez-Agenjo, M., Rayon-Viña, F., et al. (2018). Alert calling in port areas: Marine litter as possible secondary dispersal vector for hitchhiking invasive species. *Journal for Nature Conservation*, (42), 12–18. https://doi.org/10.1016/j.jnc.2018.01.005

Miranda, D. D. A., & De Carvalho-Souza, G. F., 2016. Are we eating plastic-ingesting fish? *Marine Pollution Bulletin,* (103), 109–114. https://doi.org/10.1016/j.marpolbul.2015.12.035

Mohamadi, S., Madadi, R., Jahan, R., et al. (2022). Abundance and characterization of personal protective equipment (PPE) polluting Kish Island, Persian Gulf. *Science of the Total Environment*, (854), 158678.

Nelms, S. E., Coombes, C., Foster, L. C., et al. (2017). Marine anthropogenic litter on British beaches: A 10-year nationwide assessment using citizen science data. *Science of the Total Environment*, (579), 1399–1409. https://doi.org/10.1016/j.scitotenv.2016.11.137

Noè, S., Gianguzza, P., Di Trapani F., et al. (2018) Native predators control the population of an invasive crab in no-take marine protected areas. *Aquatic Conservation: Marine and Freshwater Ecosystems,* (28), 1229–1237. https://doi.org/10.1002/aqc.2921

Oberbeckmann, S., & Labrenz, M. (2020). Marine microbial assemblages on microplastics : Diversity, adaptation, and role in degradation. *Annual Review of Marine Science*, (12), 209–232.

Olenin, S., Elliott, M., Minchin, D., et al. (2024). Marine ecosystem health and biological pollution: Reconsidering the paradigm. *Marine Pollution Bulletin*, (200), 116054. https://doi.org/10.1016/j.marpolbul.2024.116054

Olyarnik, S. V., Bracken, M. E. S., Byrnes, J. E., et al. (2009). Ecological factors affecting community invasibility, pp: 215–238, in: G. Rilov and J. A. Crooks (Eds.) *Biological Invasions in Marine Ecosystems. Ecological, Management, and Geographic Perspectives*. Springer-Verlag, Berlin Heidelberg. 641 pp.

Ormsby, M. J., Woodford, L., & Quilliam, R. S. (2024). Can plastic pollution drive the emergence and dissemination of novel zoonotic diseases? *Environmental Research,* (246), 118172. https://doi.org/10.1016/j.envres.2024.118172

Pinochet, J., Thiel, M., & Urbina, M., 2024. How plastic litter sunk by biofouling recovers buoyancy – The role of benthic predation. *Science of the Total Environment*, (952), 175910. https://doi.org/10.1016/j.scitotenv.2024.175910

Pinochet, J., Urbina, M. A., & Lagos, M. E. (2020). Marine invertebrate larvae love plastics: Habitat selection and settlement on artificial substrates. *Environmental Pollution*, (257), 113571. https://doi.org/10.1016/j.envpol.2019.113571.

Póvoa, A.A., Skinner, L.F., De Araújo, F.V., (2021). Fouling organisms in marine litter (rafting on abiogenic substrates): A global review of literature. *Marine Pollution Bulletin*, (166), 112189. https://doi.org/10.1016/j.marpolbul.2021.112189

Pyšek, P., Jarošík, V., Hulme, P. E., et al. (2012). A global assessment of invasive plant impacts on resident species, communities and ecosystems: The interaction of impact measures, invading species' traits and environment. *Global Change Biology*, (18), 1725–1737. https://doi.org/10.1111/j.1365-2486.2011.02636.x

Rech, S., Borrell, Y., & García-Vazquez, E. (2016). Marine litter as a vector for non-native species: What we need to know. *Marine Pollution Bulletin*, (113), 40–43. https://doi.org/10.1016/j.marpolbul.2016.08.032

Rech, S., Macaya-Caquilpán, V., Pantoja, J. F., et al. (2014). Rivers as a source of marine litter – A study from the SE Pacific. *Marine Pollution Bulletin*, (82), 66–75. https://doi.org/10.1016/j.marpolbul.2014.03.019

Ricciardi, A., Iacarella, J. C., Aldridge, D. C., et al. (2021). Four priority areas to advance invasion science in the face of rapid environmental change. *Environmental Reviews*, (29), 119–141. https://doi.org/10.1139/er-2020-0088

Rodrigues, A., Oliver, D. M., McCarron, A., et al. (2019). Colonisation of plastic pellets (nurdles) by *E. coli* at public bathing beaches. *Marine Pollution Bulletin*, (139), 376–380. https://doi.org/10.1016/j.marpolbul.2019.01.011

Rodriguez, L. F. (2006). Can invasive species facilitate native species? Evidence of how, when, and why these impacts occur. *Biological Invasions*, (8), 927–939. https://doi.org/10.1007/s10530-005-5103-3

Rosas, E., Martins, F., & Janeiro, J. (2021). Marine litter on the coast of the Algarve: Main sources and distribution using a modeling approach. *Journal of Marine Science and Engineering*, (9), 412. https://doi.org/10.3390/jmse9040412

Roy, H.E., Pauchard, A., Stoett, P., et al. (2023). *IPBES Invasive Alien Species Assessment: Summary for Policymakers.* Zenodo. https://doi.org/10.5281/ZENODO.7430692

Roy, H. E., Pauchard, A., Stoett, P. J., et al. (2024). Curbing the major and growing threats from invasive alien species is urgent and achievable. *Nature Ecology & Evolution*, (8), 1216–1223. https://doi.org/10.1038/s41559-024-02412-w

Ruiz, G. M., & Carlton, J. T. (2003). Invasion vectors: A conceptual framework for management, pp. 459–504, in: G. M. Ruiz and J. T. Carlton (Eds.) *Invasive Species: Vectors and Management Strategies*. Island Press, Washington DC. xiii+520 pp.

Ryan, P. G. (2015). A brief history of marine litter research, pp. 1–25, in: M. Bergmann, L. Gutow, and M. Klages (Eds.) *Marine Anthropogenic Litter.* Springer International Publishing, Cham.Sagarminaga, Y., Garcés, E., Francé, J., et al. (2023). New tools and recommendations for a better management of harmful algal blooms under the European Marine Strategy Framework Directive. *Frontiers in Ocean Sustainability*, (1), 1298800. https://doi.org/10.3389/focsu.2023.1298800

Scheltema, R. S. (1977). Dispersal of marine invertebrate organisms: Paleobiogeographic and biostratigraphic implications, pp 73–108, in: E. G. Kauffman and J. E. Hazel (Eds.) *Concepts and Methods of Biostratigraphy*. Dowden, Hutchinson and Ross, Stroudsburg, PA. xiii+658 pp.

Seebens, H., Blackburn, T. M., Dyer, E. E., et al. (2017). No saturation in the accumulation of alien species worldwide. *Nature Communications*, (8), 14435. https://doi.org/10.1038/ncomms14435

Seebens, H., Gastner, M. T., & Blasius, B. (2013). The risk of marine bioinvasion caused by global shipping. *Ecology Letters*, (16), 782–790. https://doi.org/10.1111/ele.12111

Sempere-Valverde, J., Saenz-Arias, P., Martynova, A., et al. (2024). Plasticlusters: A marine litter microhabitat in a marina of Tunisia, N Africa. *Marine Pollution Bulletin,* (202), 116389. https://doi.org/10.1016/j.marpolbul.2024.116389

Shabani, F., Nasrolahi, A., & Thiel, M. (2019). Assemblage of encrusting organisms on floating anthropogenic debris along the northern coast of the Persian Gulf. *Environmental Pollution,* (254), 112979. doi: 10.1016/j.envpol.2019.112979

Shaji, A., Kamalesh, R., Dinakarkumar, Y., et al. (2024). Microbial degradation of marine plastic debris: A comprehensive review on the environmental effects, disposal, and biodegradation. *Biochemical Engineering Journal,* (201), 109133. https://doi.org/10.1016/j.bej.2023.109133

Shea, K., & Chesson, P. (2002). Community ecology theory as a framework for biological invasions. *Trends in Ecology & Evolution*, (17), 170–176.

Simberloff, D., & von Holle, B. (1999). Positive interactions of nonindigenous species: Invasional meltdown? *Biological Invasions*, (1), 21–32. https://doi.org/10.1023/A:1010086329619

Simkanin, C., Davidson, I. C., Dower, J. F., et al. (2012). Anthropogenic structures and the infiltration of natural benthos by invasive ascidians. *Marine Ecology,* (33), 499–511. https://doi.org/10.1111/j.1439-0485.2012.00516.x

Sinclair, J. S., Lockwood, J. L., Hasnain, S., et al. (2020). A framework for predicting which non-native individuals and species will enter, survive, and exit human-mediated transport. *Biological Invasions*, (22), 217–231. https://doi.org/10.1007/s10530-019-02086-7

Soares, M. O., Garcia, T. M., Giarrizzo, T., et al. (2023). Marine debris provide long-distance pathways for spreading invasive corals. *Science of the Total Environmen*t, (900), 165637. https://doi.org/10.1016/j.scitotenv.2023.165637

Song, X., Lyu, M., Xiaodi, Z., et al. (2021). Large plastic debris dumps: New biodiversity hot spots emerging on the deep- Sea floor. *Environmental Science and Technology*, (8), 148–154. https://doi.org/10.1021/acs.estlett.0c00967

Soto, I., Balzani, P., Carneiro, L., et al. (2024). Taming the terminological tempest in invasion science. *Biological Reviews*, (99), 1357–1390. https://doi.org/10.1111/brv.13071

Spear, M. J., Walsh, J. R., Ricciardi, A., et al. (2021). The invasion ecology of sleeper populations: Prevalence, persistence, and abrupt shifts. *BioScience*, (71), 357–369. https://doi.org/10.1093/biosci/biaa168

Tan, K., Ya, P., Tan, K., et al. (2023). Ecological impact of invasive species and pathogens introduced through bivalve aquaculture. *Estuarine, Coastal and Shelf Science,* (294), 108541. https://doi.org/10.1016/j.ecss.2023.108541

Thacharodi, A., Meenatchi, R., Hassan, S., et al. (2024). Microplastics in the environment: A critical overview on its fate, toxicity, implications, management, and bioremediation strategies. *Journal of Environmental Management*, (349), 119433. https://doi.org/10.1016/j.jenvman.2023.119433

Therriault, T. W., Nelson, J. C., Carlton, J. T., et al. (2018). The invasion risk of species associated with Japanese Tsunami Marine Debris in Pacific North America and Hawaii. *Marine Pollution Bulletin*, (132), 82–89. https://doi.org/10.1016/j.marpolbul.2017.12.063

Thiel, M., & Gutow, L. (2005). The ecology of rafting in the marine environment. II. The rafting organisms and community. *Oceanography and Marine Biology: An Annual Review*, (43), 279–418.

Tsikhon-Lukanina, E. A., Reznichenko, O. G., & Nikolaeva, G. G. (2001). Ecology of invertebrates on oceanic floating substrata in the North-Western Pacific. *Russian Academy of Sciences. Oceanology*, (41), 550–555.

Turbelin, A. J., Diagne, C., Hudgins, E. J., et al. (2022). Introduction pathways of economically costly invasive alien species. *Biological Invasions*, (24), 2061–2079. https://doi.org/10.1007/s10530-022-02796-5

Turk, S. M. (1988) Christopher Columbus and *Pteria columbus*. *The Conchologist's Newsletter*, (105), 93–95.

Turner, A., Williams, T., & Pitchford, T. (2021). Transport, weathering and pollution of plastic from container losses at sea: Observations from a spillage of inkjet cartridges in the North Atlantic Ocean. *Environmental Pollution*, (284), 117131. https://doi.org/10.1016/j.envpol.2021.117131

Tutman, P., Kapiris, K., Kirin, M., et al. (2017). Floating marine litter as a raft for drifting voyages for *Planes minutus* (Crustacea: Decapoda: Grapsidae) and *Liocarcinus navigator* (Crustacea: Decapoda: Polybiidae). *Marine Pollution Bulletin*, (120), 217–221. https://doi.org/10.1016/j.marpolbul.2017.04.063

Tyrrell, M. C., & Byers, J. E. (2007). Do artificial substrates favor nonindigenous fouling species over native species? *Journal of Experimental Marine Biology and Ecology*, (342), 54–60. https://doi.org/10.1016/j.jembe.2006.10.014

UNEP. (2022). United Nations Environment Programme, 5th session, 2022. UNEA Resolution 55/14 entitled "End plastic pollution: Towards an international legally binding instrument", UN Doc UNEP/EA.5/Res.14 adopted on 2 March 2022.

van Sebille, E., Aliani, S., Law, K. L., et al. (2020). The physical oceanography of the transport of floating marine debris. *Environmental Research Letters,* (15), 023003. https://doi.org/10.1088/1748-9326/ab6d7d

van Sebille, E., Wilcox, C., Lebreton, L., et al. (2015). A global inventory of small floating plastic debris. *Environmental Research Letters,* (10), 124006. https://doi.org/10.1088/1748-9326/10/12/124006

Vermeij, G. J. (1996). An agenda for invasion biology. *Biological Conservation*, (78), 3–9. https://doi.org/10.1016/0006-3207(96)00013-4

Vriend, P., van Calcar, C., Kooi, M., et al. (2020). Rapid assessment of floating macroplastic transport in the rhine. *Frontiers in Marine Science*, (7), 1–8. https://doi.org/10.3389/fmars.2020.00010

Watts, P. C., Thorpe, J. P., & Taylor, P. D., (1998). Natural and anthropogenic dispersal mechanisms in the marine environment: A study using cheilostome Bryozoa. *Philosophical Transactions of the Royal Society B: Biological Sciences*, (353), 453–464. https://doi.org/10.1098/rstb.1998.0222

Whitehead, T. O., Biccard, A., & Griffiths, C. L. (2011). South African pelagic goose barnacles (Cirripedia, Thoracica): Substratum preferences and influence of plastic debris on abundance and distribution. *Crustaceana*, (84), 635–649. https://doi.org/10.1163/001121611X574290

Wilson, H. L., Johnson, M. F., Wood, P. J., et al. (2021). Anthropogenic litter is a novel habitat for aquatic macroinvertebrates in urban rivers. *Freshwater Biology*, (66), 524–534. https://doi.org/10.1111/fwb.13657

Winston, J. E., Gregory, M. R., & Stevens, L. M. (1997). Encrusters, epibionts, and other biota associated with pelagic plastics: A review of biogeographical, environmental, and conservation issues, pp. 81–97, in: J. M. Coe and D. B. Rogers, (Eds.) *Marine Debris.* Springer *Series on Environmental Management.* Springer, New York. Marine Debris: Sources, Impacts, and Solutions. https://doi.org/10.1007/978-1-4613-8486-1_9

Wood, A. C. L., Probert, P. K., Rowden, A. A., et al. (2012). Complex habitat generated by marine bryozoans: A review of its distribution, structure, diversity, threats and conservation. *Aquatic Conservation: Marine and Freshwater Ecosystems*, (22), 547–563. https://doi.org/10.1002/aqc.2236

Xie, J., Chen, C., Luo, M., et al. (2024). Hidden dangers: High levels of organic pollutants in hadal trenches. *Water Research*, (251), 121126. https://doi.org/10.1016/j.watres.2024.121126

Ye, S., & Andrady, A. L. (1991). Fouling of floating plastic debris under Biscayne Bay exposure conditions. *Marine Pollution Bulletin*, (22), 608–613. https://doi.org/10.1016/0025-326X(91)90249-R

Zhao, S., Zettler, E. R., Amaral-Zettler, L. A., et al. (2021). Microbial carrying capacity and carbon biomass of plastic marine debris. *The ISME Journal*, (15), 67–77. https://doi.org/10.1038/s41396-020-00756-2

Zheng, S., Bawazir, M., Dhall, A., et al. (2021). Implication of surface properties, bacterial motility, and hydrodynamic conditions on bacterial surface sensing and their initial adhesion. *Frontiers in Bioengineering and Biotechnology*, (9), 1–22. https://doi.org/10.3389/fbioe.2021.643722

20 Data Gaps in Assessing the Risks of Microplastics into the Food Chain

Craig Baker-Austin and David I. Walker

Microplastic pollution is a significant and growing environmental problem, especially in marine and aquatic systems. There is a clear need for a better understanding of the impact of microplastic pollution on both the environment and public health, particularly concerning its potential pathways into human food sources. In 2022, we published a large critical review outlining the known risks of microplastics in the human food chain. Here, and utilising the outputs from this critical review, we summarise the known risks of microplastics into the human food chain, highlighting and focussing on key data gaps recently identified in this area. Limited research suggests an association between microplastics and pathogenic organisms, raising concerns about potential entry into the food chain. However, the consequence of this and comparison to non-plastic vectors have been insufficiently studied to create a clear picture of the risk posed to public health. Data and carefully controlled studies comparing microplastics to non-plastic particles and quantifying dietary intake are in general scarce in the published literature. The risk posed to public health remains unclear due to significant data gaps and insufficient research. However, by addressing these deficiencies, we are able to better understand these issues.

Plastics have become an important component of everyday life since the start of their mass production in the 1950s. However, it is estimated that several million tonnes of plastic waste have been released into the world's oceans between 2010 and 2025 alone (Figure 20.1). Nano- and microplastics (NMPs) are therefore a global concern as environmental contaminants and have been found in aquatic and terrestrial environments and they are ingested by both food and non-food-producing animals and potentially can contaminate crops. However, a current evidence gap exists in the actual risk to consumers (Walker et al., 2022). Consuming microplastics in foods poses numerous potential health risks that have garnered increasing attention in recent years. Microplastics are defined as plastic particles between 5 mm and 1 μm in size, and nanoplastics are between 1 μm and 1 nm in size. Both have become ubiquitous in the environment, including in the food chain, raising concerns about their impact on human health (Lee et al., 2019; De-la-Torre, 2019). Various estimates of both the abundance and geographical extent of plastic contamination have been produced in the last two decades. These studies demonstrate that contamination is geographically widespread, longstanding and likely to increase in the future (Franzellitti et al., 2019; Bowley et al., 2021).

Various estimates suggest that several trillions of microplastic particles are present in our global oceanic system and that these are readily ingested by a wide range of marine and freshwater species across feeding modes and ecological niches (Bowley et al., 2022).

Microplastics may pose a threat to human health through various pathways, primarily through the ingestion of contaminated food. For instance, when consuming seafood, individuals may inadvertently ingest microplastics, which can lead to health risks due to the combined effects of microplastics and absorbed contaminants. While the ingestion of microplastics through contaminated food sources, particularly seafood, has emerged as a significant issue due to the potential adverse health consequences associated with these particles (Kristanti et al., 2022), understanding the

DOI: 10.1201/9781032717814-20

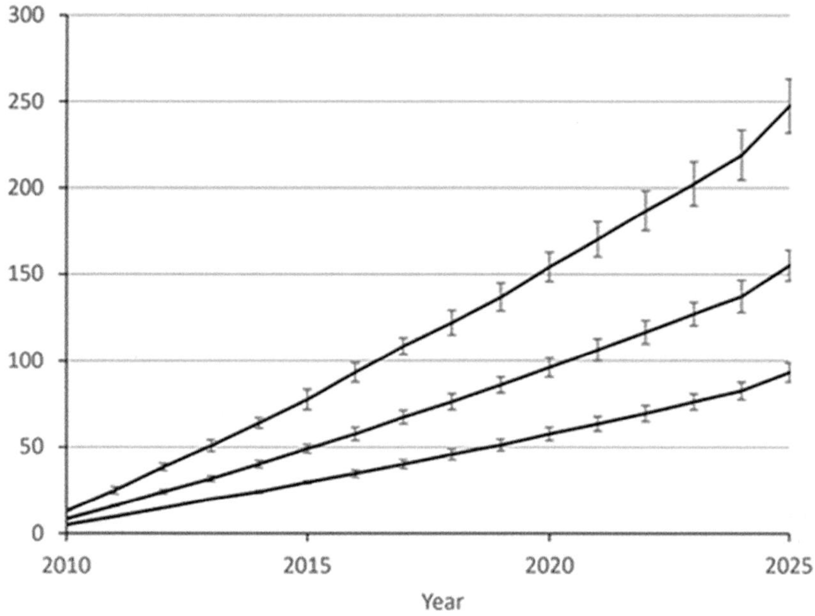

FIGURE 20.1 Estimated mass of mismanaged plastic waste (millions of metric tonnes) input to the ocean by populations living within 50 km of a coast in 192 countries, (2015) plotted as a cumulative sum from 2010 to 2025. (Figure adapted from Jambeck et al. 2015). Estimates reflect assumed conversion rates of mismanaged plastic waste to marine debris (high, 40%; mid, 25%; low, 15%).

possible health risks from consuming microplastics in foods requires a comprehensive examination of the available evidence from various studies.

To address these key evidence gaps, in 2019, the UK Foods Standards Agency (FSA) commissioned Cefas to produce a critical literature review of the scientific evidence by considering both peer-reviewed and grey literature in this area. The focus of this critical review was to assess the scientific evidence concerning the diversity of microorganisms that colonise NMPs, the key pathways that microbiologically contaminated NMP could enter the food chain from environmental sources and the subsequent risk these pathogens could pose to the consumer. The documents we considered for this critical review covered published research from 1980 to 2020 searchable via Web of Science and specified grey literature databases. This chapter outlines many of the important and key data gaps that were elucidated from this critical review.

20.1 POTENTIAL RISKS FROM FOOD

One of the primary concerns regarding the consumption of microplastics in foods is the potential for trophic transfer of contaminants. NMP, along with absorbed contaminants, can accumulate in lower-trophic organisms such as zooplankton, leading to the transfer of these contaminants up the food chain (Cole et al., 2013). This trophic transfer can result in bioaccumulation of harmful substances in higher trophic organisms, posing risks to human health through the consumption of contaminated seafood (Cole et al., 2013). The presence of NMP in aquatic organisms, including shrimp, without proper removal before consumption, highlights the potential health risks to humans due to the combined effects of NMP and absorbed contaminants (Kristanti et al., 2022).

Studies have also addressed the possible human health effects of consuming NMP through different food sources. The ingestion of NMP, particularly through seafood, has been associated with physical damage and chemical toxicity, leading to health issues such as inflammation, DNA damage, and cancer (De-la-Torre, 2019). Additionally, the presence of NMP in drinking water and seafood raises concerns about the ingestion of these particles and their potential impact on human health (Yu, 2023). The accumulation of NMP in the gastrointestinal tract can lead to inflammation, disruption of gut microbiota (e.g. dysbiosis), and the release of hazardous chemicals, underscoring the health risks associated with NMP exposure (De-la-Torre, 2019). Furthermore, the interaction between NMP and marine organisms can have implications for human health. Marine organisms exposed to NMP can accumulate these particles, potentially transferring them to humans through the consumption of seafood (Ziccardi et al., 2016). The bioaccumulation of hydrophobic organic chemicals (HOCs) in NMP and their subsequent uptake by aquatic organisms raise concerns about the adverse effects on ecological and human health (Ziccardi et al., 2016). The ingestion of NMP by aquatic animals can lead to contamination of the food web (Lim, 2023). The potential health risks from consuming NMP in foods extend beyond physical effects to include immunotoxicity and intestinal impacts. NMPs have been shown to interact with cells and tissues, potentially leading to immunotoxic effects and dysbiosis in the gastrointestinal microbiome (Hirt & Body-Malapel, 2020). Exposure to NMP can disrupt the immune system and intestinal health, highlighting the need for further research to understand the full extent of these impacts on human health (Hirt & Body-Malapel, 2020). Additionally, the leaching of chemical additives or adsorbed persistent organic pollutants (POPs) from NMP can have consequences for human health, emphasising the importance of assessing the ecosystem function in relation to NMP exposure (Prinz & Korez, 2019).

The increasing presence of NMP in the environment has emerged as a significant and growing concern. Although research has demonstrated a connection between NMP and pathogenic organisms, the available data is limited, particularly concerning dietary intake across most food and drink categories. Existing research predominantly focuses on fish and shellfish, often examining inedible parts or early stages of food preparation. While there is evidence suggesting that pathogenic organisms that colonise NMP as biofilms could enter the human food chain through NMP, the implications and comparisons with non-plastic vectors have not been thoroughly studied, leaving the risk assessment unclear. Additionally, while bacterial colonisation of microplastics has been demonstrated, the possibility of viral absorption to nanoplastics remains understudied. Gaps and scarcity of data are apparent in many aspects of this field, particularly in understanding the toxicological impacts of NMP, their effects on gut microflora, and their presence across a wide variety of foods and beverages. Additional key areas needing exploration include the significance of NMP size, the likelihood of certain pathogens being vectored more frequently than others, and whether these factors are related to the polymer type or size.

20.2 2022 CRITICAL REVIEW

Our 2022 published critical review examined the significance of microplastics and microbial contamination of the human food chain and highlighted that plastics are pervasive in our lives, leading to significant amounts of NMP entering the environment (Walker et al., 2022). Since the 1950s, millions of tonnes have accumulated, breaking down into smaller particles, where they can enter the food chain (Rochman, 2020). These NMPs interact with the environment, creating new substrates for biofilms, which can harbour human pathogens and facilitate horizontal gene transfer, including antimicrobial resistance (AMR) genes. The work determined that NMP can enter the human food chain, potentially transferring pathogens to the gut, but data on these effects are limited. Indeed, research on NMP is lacking, and existing studies vary in quality, making detailed comparisons difficult. The review was split into four separate evidence chapters as summarised in Table 20.1. A summary of the data gaps identified across these four chapters is subsequently outlined and discussed.

TABLE 20.1
A Summary of the Four Chapters Included in a 2022 Critical Review by Walker et al. (2022)

Chapter	Description
An overview of NMP in the environment	Examines the presence and behaviour of NMP in various environments, highlighting the need for standardised research methods and definitions.
Interactions between NMP and microorganisms	Explores how NMP interacts with microorganisms, forming biofilms that may select for or transport pathogenic bacteria.
Pathways of **colonised** NMP into the human food chain	Investigates how NMP colonised by bacteria can enter the food chain and the potential health risks they pose.
NMP-specific microbial risks to consumers	Discusses the risks associated with human pathogens found on NMP and their implications for human health

20.3 MICROPLASTICS IN THE ENVIRONMENT: KEY DATA GAPS

The first evidence chapter of the critical reviewed current available knowledge about NMP in general into context before looking into the potential pathways of pathogens into the human food chain. Several critical issues surrounding the study of non-microplastic particles in the environment were noted. First, there is a general lack of standardisation in sampling protocols, analytical methods, and units for measuring NMP abundance, which hampers consistent data collection and comparison across studies (Peng et al., 2020). Moreover, there is no universally accepted definition of NMP (Frias & Nash, 2019), with some studies including cellulose-based polymers while others do not, underscoring the need for a clear and internationally agreed-upon definition, or set of definitions for these particles. Although numerous studies have explored the fate of NMPs in marine environments, research on their presence and behaviour in other environments is somewhat limited. Indeed, the vast majority of studies are based on marine settings compared to other environmental components (Figure 20.2).

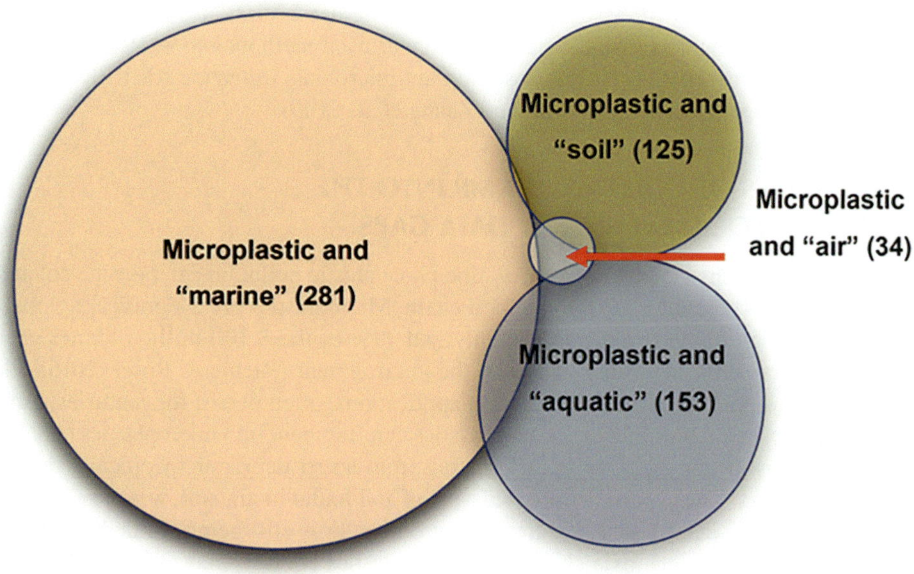

FIGURE 20.2 Published scientific studies with the term 'microplastic' in the title alongside other environmental search terms (until February 2021) available on Microplastic and 'marine' (281) Microplastic and 'aquatic' (153) Microplastic and 'soil' (125) Microplastic and 'air' (34) published on PubMed. (Figure courtesy Walker et al. 2022.)

It should be noted, however, that this gap is gradually being filled, with a noticeable increase in studies on NMPs in less-studied environments like the Arctic, Antarctic, and other "pristine" areas over the past few years (Bergman et al., 2019). Another significant issue is the limited understanding of plastic transformations in seawater, including the degradation timescales and the ultimate sinks of these materials. Additionally, there is a lack of standardised, laboratory-based data on the long-term fate of NMPs, which is crucial for understanding their environmental impact (Abed et al., 2017; Syranidou et al., 2019). The review also notes that while NMPs have been widely detected in certain food commodities but there is a scarcity of detailed studies examining their effects at ecosystem levels. Furthermore, data on the cycling of NMP contaminants between different environmental compartments is lacking (Galloway et al., 2017). Like the nitrogen or carbon cycles, the NMP cycle is complex and only beginning to be understood, and there are significant data gaps, particularly regarding how NMPs move between different environmental compartments and over what timescales.

20.4 MICROPLASTICS AND THEIR INTERACTIONS WITH MICROBES: KEY DATA GAPS

Chapter 4 of the report looked at the number of papers that deal with interaction between NMP and microorganisms. Once microplastics enter the environment, they quickly begin interacting with microorganisms. The extent of this interaction varies, ranging from brief contact to the formation of complex biofilms. These biofilms are structures composed of microorganisms and their associated extracellular matrices, which attach to the surface of the microplastics (Yin et al., 2019). Several key data gaps were identified from this chapter. First, and as with published studies, there is a lack of standardised methods for studying the interactions between plastics and microorganisms (Galloway et al., 2017). In addition, the absence of surface property characterisation makes it challenging to identify key factors for colonisation (Cai et al., 2017). Non-plastic substrates are often excluded from controlled exposure experiments, limiting our understanding of the factors influencing colonisation and whether plastic substrates drive selective processes (Walker et al., 2022). More research is needed to understand how NMP-associated biofilms may select for or transport pathogenic bacteria (Bowley et al., 2022), emphasising the need for standardised methodologies. Research should be expanded to investigate whether NMPs or sorbed antimicrobials influence AMR through species sorting, mutations, or horizontal gene transfer (Zhang et al., 2019).

20.5 PATHWAYS OF COLONISED NMP INTO THE HUMAN FOOD CHAIN: KEY DATA GAPS

Research Chapter 5 of the critical review examined how plastic pollution can become colonised by bacteria and how this can enter the human food chain. Microplastics are a significant global issue, with over 350 million tonnes produced annually and 79% of the 8,300 million tonnes generated in the past 60 years being landfilled or lost to the environment (Richie & Roser, 2016). Plastics, known for their low cost, durability, and diverse applications, often persist for centuries, especially in low-light, low-oxygen areas like the ocean. Plastics can fragment into microplastics (smaller than 5 mm) and nanoplastics (under 100 nm), originating from larger debris or specifically designed as primary microplastics. These particles are widespread and found in air, soil, water, and even inside organisms. The review determined that the impact on health is still being studied, with concerns about their ability to transport toxic chemicals. Microplastics are now part of daily life, appearing in the air we breathe and the food we consume.

Currently, there are limited published studies on the presence of NMP (nanoparticle materials) in ready-to-consume foods and beverages, although this field is evolving rapidly, and most existing research is still in the preliminary stages. More focused studies on foods at the point of consumption are needed to assess potential risks to human health. There is a scarcity of research addressing the transfer of NMPs and pathogens to humans. Many traditional food preparation methods, such as cooking, smoking, and pickling, effectively neutralise pathogens that NMPs might carry, and there is evidence (also scarce) of differential loads of NMPs between processed and unprocessed foods (Davidson & Dudas, 2017; Li et al., 2016). To identify potential risks, research should prioritise foods and drinks that do not undergo these processes before consumption. Linking pathogens on NMPs directly to those on human food and beverages is challenging due to the shared environment and processes involved. Pathogens may originate from different sources and attach to both food and NMPs simultaneously. Therefore, research should examine both the food or drink and the NMPs in conjunction. Further investigation is needed into the behaviour of co-occurring NMP-sized particles to distinguish their effects from those of other small particles. Many studies currently overlook this aspect, leading to gaps in understanding (Walker et al., 2022). As with other research chapters (see above), there is a lack of standardisation in analytical methods and reporting of particle sizes, quantities, detection limits, and other quality control parameters. This inconsistency complicates comparisons between studies and obscures whether observed differences are significant. Ongoing discussions aim to address these issues, and improvements in standardisation are expected in the near future.

20.6 NMP-SPECIFIC MICROBIAL RISKS TO CONSUMERS: KEY DATA GAPS

Pathogens of human health relevance have been found on NMP and have been the subject of numerous peer-review publications (Kirstein et al., 2016; Bowley et al., 2021). Recent peer-reviewed studies suggest that NMP particles could create environments for microorganisms to thrive, potentially affecting the food chain and human health. Indeed, pathogens of human health relevance have been found on NMP in a variety of published scientific studies. Marine NMP research has found pathogens like *Vibrio* spp. in NMP biofilms, raising concerns that NMP might facilitate long-distance pathogen transport in marine environments (Kirstein et al., 2016). If human pathogens also colonise NMP, they might enter the human food chain more easily. Our review explored several research areas: dysbiosis due to stress-induced changes in microbial flora, antibiotic-resistant bacteria on NMP, the presence of key human pathogens and novel ones on NMP, and the associated human disease risks. Dysbiosis can be influenced by the presence of NMP, which may alter the microbial landscape in both marine and terrestrial environments (Roman et al., 2018; Maharramov et al., 2019). Few studies were published with regard to dysbiosis and human health, with a gamut of studies focussed on model organisms instead (e.g. Jin et al., 2019; Qiao, R., et al., 2018). With regard to AMR, many studies suggest that NMP can be colonised by antibiotic-resistant bacteria and that densities are frequently higher than in the surrounding natural environment (Zhang et al., 2020). Again, a lack of comparative studies using appropriate controls reduces the robustness of these comparisons. Because AMR studies tend to lack appropriate controls, risk analysis based on these approaches is challenging.

20.7 SUMMARY

This critical review highlights several key data gaps, which may be very useful for future scientific endeavours focussed on more fully establishing the specific risks associated with plastics, food, and human health. First, to fully understand the scale of issues related to NMP, studies must be conducted across a broader range of environments and matrices. This need applies to all areas of NMP-related science discussed in the critical report. Second, developing standardised methodologies and reporting parameters are essential for comparing studies across different institutions and matrices. Global agreements or normalised practices are needed in the following areas: (a) defining which polymers qualify as plastics, (b) determining the size threshold for classifying plastics as nanoplastics, and (c) establishing suitable mesh sizes for sampling. Third, studies must include appropriate controls, such as similarly sized non-plastic particles, to ensure observed impacts are due to NMP and not just small particles and fibres, which can occur naturally. Lastly, there is a lack of research on the potential links between NMP and pathogens, including their respective fates. Critical data on viruses, interrelationships between organisms and sorbed chemicals, plastic additives, AMR acquisition, and dysbiosis effects in compromised subjects are either missing or anecdotal, making it difficult to establish direct links to health outcomes.

REFERENCES

Abed, R., T. Muthukrishnan, M. A. Khaburi, F. Al-Senafi, A. Munam, and H. Mahmoud, 'Degradability and biofouling of oxo-biodegradable polyethylene in the planktonic and benthic zones of the Arabian Gulf', *Marine Pollution Bulletin*, vol. 150, p. 110639, 2020, https://doi.org/10.1016/j.marpolbul.2019.110639.

Bergmann, et al., 'Microplastic in snow from European and Arctic sites', 2019. [Online]. Available: https://doi.pangaea.de/10.1594/PANGAEA.901447

Bowley, J., C. Baker-Austin, A. Porter, R. Hartnell, and C. Lewis, 'Oceanic hitchhikers – Assessing pathogen risks from marine microplastic', *Trends in Microbiology*, vol. 29, no. 2, pp. 107–116, 2021, https://doi.org/10.1016/j.tim.2020.06.011

Cai, L., D. Wu, J. Xia, H. Shi, and H. Kim, 'Influence of physicochemical surface properties on the adhesion of bacteria onto four types of plastics', *Science of the Total Environment*, vol. 671, pp. 1101–1107, 2019, https://doi.org/10.1016/j.scitotenv.2019.03.434

Cole, M., P. Lindeque, E. Fileman, C. Halsband, R. Goodhead, J. Moger, J., … and T. Galloway, 'Microplastic ingestion by zooplankton', *Environmental Science & Technology*, vol. 47, no. 12, pp. 6646–6655, 2013, https://doi.org/10.1021/es400663f

Davidson K., and S. E. Dudas, 'Microplastic ingestion by wild and cultured Manila clams (*Venerupis philippinarum*) from Baynes Sound, British Columbia', *Archives of Environmental Contamination and Toxicology*, vol. 71, no. 2, pp. 147–156, 2016, https://doi.org/10.1007/s00244-016-0286-4

De-la-Torre, G., 'Microplastics: An emerging threat to food security and human health', *Journal of Food Science and Technology*, vol. 57, no. 5, pp. 1601–1608, 2019, https://doi.org/10.1007/s13197-019-04138-1

Franzellitti, S., L. Canesi, M. Auguste, R. H. G. R. Wathsala, and E. Fabbri, 'Microplastic exposure and effects in aquatic organisms: A physiological perspective', *Environmental Toxicology and Pharmacology*, vol. 68, pp. 37–51, 2019, https://doi.org/10.1016/j.etap.2019.03.009

Frias and R. Nash, 'Microplastics: Finding a consensus on the definition', *Marine Pollution Bulletin*, vol. 138, pp. 145–147, 2019, https://doi.org/10.1016/j.marpolbul.2018.11.022

Galloway, T.S., M. Cole, and C. Lewis, 'Interactions of microplastic debris throughout the marine ecosystem', *Nature Ecology & Evolution*, vol. 1, no. 5, 2017, https://doi.org/10.1038/s41559-017-0116

Hirt, N., and M. Body-Malapel, 'Immunotoxicity and intestinal effects of nano- and microplastics: A review of the literature, *Particle and Fibre Toxicology*, vol. 17, no. 1, 2020, https://doi.org/10.1186/s12989-020-00387-7

Jambeck, J., et al., 'Plastic waste inputs from land into the ocean', *Science*, vol. 347, no. 6223, pp. 768–771, 2015, https://doi.org/10.1126/science.1260352

Jin, Y., L. Lu, W. Tu, T. Luo, and Z. Fu, 'Impacts of polystyrene microplastic on the gut barrier, microbiota and metabolism of mice', *Science of the Total Environment*, vol. 649, pp. 308–317, 2019, https://doi.org/10.1016/j.scitotenv.2018.08.353

Kirstein I. V., et al., 'Dangerous hitchhikers? Evidence for potentially pathogenic *Vibrio* spp. on microplastic particles', *Marine Environmental Research*, vol. 120, pp. 1–8, 2016, https://doi.org/10.1016/j.marenvres.2016.07.004

Kristanti, R., W. Wong, Y. Darmayati, A. Hatmanti, N. Wulandari, M. Sibero, … and F. Lopez-Martinez, 'Characteristics of microplastic in commercial aquatic organisms', *Tropical Aquatic and Soil Pollution*, vol. 2, no. 2, pp. 134–158, 2022, https://doi.org/10.53623/tasp.v2i2.134

Lee, H., A. Kunz, W. Shim, and B. Walther, 'Microplastic contamination of table salts from Taiwan, including a global review', *Scientific Reports*, vol. 9, no. 1, https://doi.org/10.1038/s41598-019-46417-z

Li, H.-X., et al., 'Microplastics in oysters *Saccostrea cucullata* along the Pearl River Estuary, China', *Environmental Pollution*, vol. 236, pp. 619–625, 2018.

Lim, A., 'Microplastic ingestion in aquatic animals in south east Asia', *Tropical Environment, Biology, and Technology*, vol. 1, no. 1, pp. 25–35, 2023, https://doi.org/10.53623/tebt.v1i1.223

Maharramov, A., U. Hasanova, I. Suleymanova, G. Osmanova, and N. Hajiyeva, 'The engineered nanoparticles in food chain: Potential toxicity and effects', *Sn Applied Sciences*, vol. 1, no. 11, 2019, https://doi.org/10.1007/s42452-019-1412-5

Peng, L., D. Fu, H. Qi, C. Q. Lan, H. Yu, and C. Ge, 'Micro- and nano-plastics in marine environment: Source, distribution and threats – A review', *Science of the Total Environment*, vol. 698, p. 134254, 2020, https://doi.org/10.1016/j.scitotenv.2019.134254

Prinz, N., and Korez, Š, 'Understanding how microplastics affect marine biota on the cellular level is important for assessing ecosystem function: A review', In: Jungblut, S., Liebich, V., Bode-Dalby, M. (eds) *YOUMARES 9 - The Oceans: Our Research, Our Future*. Springer, pp. 101–120, 2019, https://doi.org/10.1007/978-3-030-20389-4_6

Qiao, R., et al., 'Accumulation of different shapes of microplastics initiates intestinal injury and gut microbiota dysbiosis in the gut of zebrafish', *Chemosphere*, vol. 236, 2019, DOI: 10.1016/j.chemosphere.2019.07.065

Ritchie, H., and M. Roser, 'Plastic Pollution', Our World in Data, 2018 152. Halle, A. T., et al., 'Understanding the fragmentation pattern of marine plastic debris', *Environmental Science and Technology*, 2016, https://doi.org/10.1021/acs.est.6b00594

Rochman, C.m., 'The story of plastic pollution', *Oceanography*, vol. 33, no. 3, pp. 60–70, 2020, https://doi.org/10.5670/oceanog.2020.308

Roman, M., C. Contado, C. Micheletti, I. Olivato, T. Erik, and B. Federico, 'Advanced approaches for efficacy evaluation and risk assessment of nanomaterials in food', In: Roman, M., Contado, C., Micheletti, C., Olivato, I., Tedesco, E., Benetti, F. (eds) *Nanotechnology Applications in the Food Industry*. CRC Press, pp. 483–501, 2018, DOI: 10.1201/9780429488870-22.

Syranidou, et al., 'Biodegradation of weathered polystyrene films in seawater microcosms', *Scientific Reports*, vol. 7, 2017, https://doi.org/10.1038/s41598-017-18366-y.

Walker, D. I., et al., A critical review of microbiological colonisation of nano- and microplastics (NMPs) and their significance to the food chain | Food Standards Agency, 2022. https://doi.org/10.46756/sci.fsa.xdx112

Yin, W., Y. Wang, L. Liu, and J. He, 'Biofilms: The microbial "protective clothing" in extreme environments', *International Journal of Molecular Sciences*, vol. 20, no. 14, 2019, https://doi.org/10.3390/ijms20143423

Yu, R., 'Microplastic pollution: Threats and impacts on global marine ecosystems', *Sustainability*, vol. 15, no. 17, pp. 13252, 2023, https://doi.org/10.3390/su151713252

Zhang, Y., J. Lu, J. Wu, J. Wang, and Y. Luo, 'Potential risks of microplastics combined with superbugs: Enrichment of antibiotic resistant bacteria on the surface of microplastics in mariculture system', *Ecotoxicology and Environmental Safety*, vol. 187, 2020, https://doi.org/10.1016/j.ecoenv.2019.10985

Ziccardi, L., A. Edgington, K. Hentz, K. Kulacki, and S. Driscoll, 'Microplastics as vectors for bioaccumulation of hydrophobic organic chemicals in the marine environment: A state-of-the-science review, *Environmental Toxicology and Chemistry*, vol. 35, no. 7, pp. 1667–1676, 2016, https://doi.org/10.1002/etc.3461

Index

Note: **Bold** page numbers refer to tables and *italic* page numbers refer to figures.

For Product Safety Concerns and Information please contact our
EU representative GPSR@taylorandfrancis.com Taylor & Francis
Verlag GmbH, Kaufingerstraße 24, 80331 München, Germany